DESIGNING AND DEPLOYING RFID APPLICATIONS

Edited by **Cristina Turcu**

Designing and Deploying RFID Applications
Edited by Cristina Turcu

Published by InTech

Janeza Trdine 9, 51000 Rijeka, Croatia

Publishing Process Manager Davor Vidic

Technical Editor Teodora Smiljanic

Cover Designer InTech Design Team

Additional hard copies can be obtained from orders@intechopen.com

Designing and Deploying RFID Applications, Edited by Cristina Turcu
p. cm.
ISBN 978-953-307-265-4

For EU product safety concerns:
IN TECH d.o.o., Prolaz Marije Krucifikse Kozulić 3, 51000 Rijeka, Croatia,
info@intechopen.com or visit our website at intechopen.com.

Meet the editor

Cristina Turcu is an Associate Professor of Software Engineering and Artificial Intelligence at Stefan cel Mare University of Suceava, Romania. She received her Diploma (M.Sc.) in Automatics and Computers and her Doctorate (Ph.D.) in Automatics, in 1991 and 2000, respectively, both from the Gheorghe Asachi Technical University of Iasi, Romania where she has been Head of Computer Department since 2004. Her research interests include software engineering, RFID applications for the end-consumer, and intelligent systems. Dr. Turcu is an Editor of four books and has served on various program committees of conferences in computing and RFID systems. She also has served as a reviewer for numerous referred journals and conferences. She is the Editor in Chief of the International Journal of Radio Frequency Identification & Wireless Sensor Networks. Dr. Turcu has published over 80 publications in books or book chapters, refereed journals, technical reports, and refereed conference/workshop/seminar proceedings.

Contents

Preface

Radio Frequency Identification (RFID), a method of remotely storing and receiving data using devices called RFID tags, brings many real business benefits to today world's organizations. It exerts a major influence in different life areas like inventory tracking, modern supply chain management, automated manufacturing, healthcare, etc. The benefits are multiple and include expedited data capture/lead retrieval, accurate and trusted data, reduced cost, time and work processes, increased speed, productivity and business efficiency, improved security, etc.

Today, there are many companies offering RFID hardware, RFID tags, or dedicated solutions for both. It is a joint effort of researchers and scientists to enable customers from different areas to deploy high performance solutions by understanding their demands. These demands are converted into innovative products which are delivered as viable and reliable solutions, at competitive prices.

Over the years, RFID research has resulted in many concrete achievements and also contributed to the creation of communities that bring scientists and engineers together with users. This book includes valuable research studies of the experienced scientists in the field of RFID, including most recent developments.

Chapter 1 aims at illustrating and clarifying the impact of RFID on business models, an essential fact for companies to create a business value by using RFID technology in order to gain competitive advantage.

Chapter 2 examines the perceptions of RFID among Australian RFID suppliers/integrators, and the role and importance that perceptions play in the actual adoption process. The authors' interview-based research study shows that integrators' perceptions can affect the adoption process. Thus, integrator perceptions can act upon present expectations of RFID technology.

Chapter 3 deals with the role of RFID technology in supply chain risk management. This research shows that RFID technology holds great promise for managing supply disruptions and for containing their harmful ripple effects. Also, a review of the literature is conducted to identify specific risks associated with RFID's capability to provide supply chain visibility. The research goes on to examine the existing mitigation approaches for dealing with RFID's visibility-related risks. Finally, the management implications of RFID use in supply chain risk management are outlined, based on both advantages and risks.

Chapter 4 describes the main features and challenges of humanitarian logistics, along with the role of RFID technology in disaster supply chains and the implementation and deployment of secure RFID.

Chapter 5 introduces two typical application cases of RFID technology in the executive process of complex product assembly. The first one solves the asynchrony problem between the logistics stream and the information stream in the executive process of complex product assembly. The second one addresses the guidance of the on-spot assembly operation, and achieves dynamic matching mechanism between 3D models and real-size counter parts. Both these cases are discussed from methodology and implementation. They illustrate the potential of applying RFID technology in enhancing the controlling and monitoring methods of the executive process of complex product assembly.

The broad objective of chapter 6 is to show how RFID technology can be used to simplify the retail processes. The major aim of the considered study is to create a simple RFID-based process model for retailers. Also, the authors present a case study of Turkish retail industry.

In chapter 7 the authors analyze and discuss the technology and issues on RFID applications in supply chains for track & trace services, such as item identification, event capture and management, information storage and sharing among all participants in a supply chain. Also, they introduce an RFID-based track & trace solution with security concerns in supply chains based on EPCglobal standards.

In chapter 8, the authors examine closely the discovery services in the EPC network. EPCglobal provides an infrastructure to increase visibility and efficiency throughout the supply chain as well as to guarantee higher quality information flow between companies and their trading partners. The present chapter shows real world use cases that require a discovery service. Furthermore, the authors derive functional as well as non-functional requirements from these use cases. They also discuss the implications of these requirements regarding possible discovery service designs.

In chapter 9 the authors analyze the advantages of implementing the Discovery Services (DS) component of the EPCglobal Network architecture using a Distributed Hash Table (DHT) application. Moreover, they demonstrate the possibility of developing new applications over the DHT-based discovery services.

The current plasterboard disposal situation was introduced in chapter 10 and also, the logistic problem, a barrier to an increased recycling rate, was addressed in the same context. A prototype system for waste management was outlined.

Chapter 11 deals with an RFID-based monitoring system of laboratory equipment to effectively monitor the in-out equipment from a laboratory. The main aim of the research is to identify a generic approach of monitoring items in a several room location.

Chapter 12 presents a Mobile RFID-based Maintenance Management system that integrates RFID technology and mobile devices to improve the effectiveness and convenience of information flow during maintenance in a construction lab. The case study the authors examine is applying their system in order to improve the process of work inspection and maintenance of a construction lab in Taiwan.

In chapter 13 the authors present their research results regarding the expected investments for RFID-enablement and operating models for an independent service provider dealing with anti-counterfeiting. They considered RFID technology as the key-enabler for an entire pharmaceutical supply chain. Their research work is motivated by the increasing number of detected pharmaceutical counterfeits in the world-wide pharmaceutical industry.

Chapter 14 deals with privacy preservation in RFID-enabled supply chain management. The authors review the current literature on RFID security and privacy issues in the supply chain. Also, a complete methodology is presented regarding both the best and easiest technique to use, and the approach or guideline in dealing with counterfeiting. Finally, counterfeiting in an RFID based-wine supply chain is used as a case study. The authors demonstrate how privacy preservation and security protection through prevention and detection can be provided in an open-loop RFID supply chain, such as the wine industry.

Chapter 15 proposes an RFID-based anti-counterfeiting, track and trace solution. The presented system helps small, medium companies and enterprise organizations to improve productivity and provide better service to their customers.

In chapter 16 the authors propose a methodology in which misuse detection employs a knowledge base built upon a "track & trace" model. This model relies on the notion of "tag location" to gather all information required to identify an attack tag cloning. Also, the presented research aims to investigate whether it is feasible to integrate the principles of ontology modeling and reasoning in the intrusion detection paradigm.

Chapter 17 outlines a study on implementation and service of digital watermark technology architecture for distribution management. The author proposes a solution that uses digital watermarking technology to identify certain copyrighted content and the related rights during the distribution or after distribution process.

Chapter 18 introduces RFID middleware and its design issues and presents some existing middleware solutions. Also, the authors detail a middleware framework that they developed as a highly scalable and easily deployable middleware. The smart library application developed to show the usefulness of the designed middleware solution is also presented. Furthermore, the scenarios of integrating this middleware in an inventory management application have been set.

Chapter 19 addresses some important issues related to the influence of RFID tagging on circulation services and collection management. Also, the authors present a case study of the Taipei Public Library.

In chapter 20 the authors share the experience on tag selections based on the research and studies that have been performed in the past few years at the Run Run Shaw Library at the City University of Hong Kong *(CityU HK Library)*.

In chapter 21 the authors present some considerations about RFID application in info-documentary systems. Also, they propose a low-cost model using open sources software, that offers a useful tool for the university researchers.

The book offers new insights, solutions and ideas for the design of efficient RFID architectures and applications. While not aiming to be comprehensive, its wide coverage may be appropriate not only for RFID novices, but also for engineers, researchers, industry personnel, and all interested in producing new and valuable results in RFID domain.

Cristina Turcu
Stefan cel Mare University of Suceava
Romania

1

Impacts of RFID on Business Models

Ela Sibel Bayrak Meydanoğlu
Marmara University
Turkey

1. Introduction

Business model describes the business logic a company or network of companies use to generate revenue and create customer and network value. It enables to identify how inputs of a company or network of companies are transformed to value-adding outputs (Kamoun, 2008). RFID affects business models through guiding new business models (e.g. RFID-enabled pay-per-use business model) or reshaping existing ones and thereby enables value creation. The number of studies that deals with the mentioned impact of RFID on business models is limited. Understanding the impact of RFID on business models is essential for companies to create a business value by using RFID in order to gain competitive advantage. This study aims to illustrate and clarify the mentioned impact. Thereby it provides an important contribution to the limited studies in the relevant literature.

In this study initially the term "business model" is defined. Subsequently the business model framework adopted in the study is presented and the major components of RFID systems are reviewed briefly. This is followed by brief explanations about RFID business models. Furthermore it is discussed how RFID systems influence the components of the adopted business model framework and how they contribute to reshape existing business models or create new ones. The study concludes with the main findings and implications.

2. Methodology

This conceptual study is a basic research, which is executed based on the previous studies about business models, business model frameworks, benefits of RFID systems as well as the limited studies about the impact of RFID systems on the components of business model frameworks. Its aim is to reorganize the existent ideas in order to give an insight how companies can create business value from RFID technology.

3. Business models

3.1 Definition of business models

Different definitions for the term business model and its building blocks exit in the relevant literature. Table 1 below includes the definitions of some authors.

Based on the definitions in Table 1 business model can be defined as a model that describes how a company or network of companies creates value from new products, innovations, activities for business partners and customers.

Author(s)	Definition
Timmers (1998), (as cited in Schweizer, 2005)	Timmers defines business model as *"an architecture for the product, service and information flows including a description of the various business actors and their roles, a description of the potential benefits for the various business actors and a description of the sources of revenues."*
Hamel (2000), (as cited in Schweizer, 2005)	Hamel defines four elements that form a business model: customer interface, core strategy, strategic resources and value network. The customer interface and the value network represent the relation between buyer and supplier side. The core strategy represents the mission of the company and the scope of production. Strategic resources explain competitive advantage gained through competencies and assets deployed. Competencies and assets support the underlying strategy through customer benefits resulting from the core strategy and through company boundaries intermediating between the strategic resources and the value network where a company is positioned.
Hoppe and Kollmer (2001), (as cited in Schweizer, 2005)	Hoppe and Kollmer define business model as an integrated and consistent picture of a company that illustrates the way it aims to generate revenues.
Magretta (2002), (as cited in Schweizer, 2005)	Magretta defines business model as a story that explains how companies work and that contains motivation and a plan that describes how value is delivered.
Betz (2002)	*"A business model is an abstraction of a business identifying how that business profitably makes money. Business models are abstracts about how inputs to an organization are transformed to value-adding outputs."*
Osterwalder, Pigneur and Tucci (2005)	*"A business model is a conceptual tool that contains a set of elements and their relationships and allows expressing the business logic of a specific firm. It is a description of the value a company offers to one or several segments of customers and of the architecture of the firm and its network of partners for creating, marketing and delivering this value and relationship capital to generate profitable and sustainable revenue streams."*
Kamoun (2008)	Kamoun defines business model as *"the logic a company or network of companies use to generate revenue and create customer and network value."* According to Kamoun business model can be defined as a blueprint that defines the way a business creates and captures value from new services, products or innovations.
Shi and Manning (2009)	Shi and Manning define business model as *"the outcome of management actions – planned, emergent or realized – in defining a firms's offerings and activities."*

Table 1. Business Model Definitions

3.2 Business model framework

Basic building blocks of a business model and the external forces that have an affect on these blocks are described in a business model framework. In the relevant literature various business model frameworks are proposed (e.g. framework of Kamoun (Kamoun, 2008), framework of Shi and Manning (Shi & Manning, 2009), framework of Osterwalder, Pigneur and Tucci (Osterwalder et al., 2005)). Among these frameworks the value-driven framework of Kamoun (see Figure 1), which is constructed considering the inherent use of business model in conceptualizing the value creation and money earning logic of a company or network of companies, is adopted in this study.

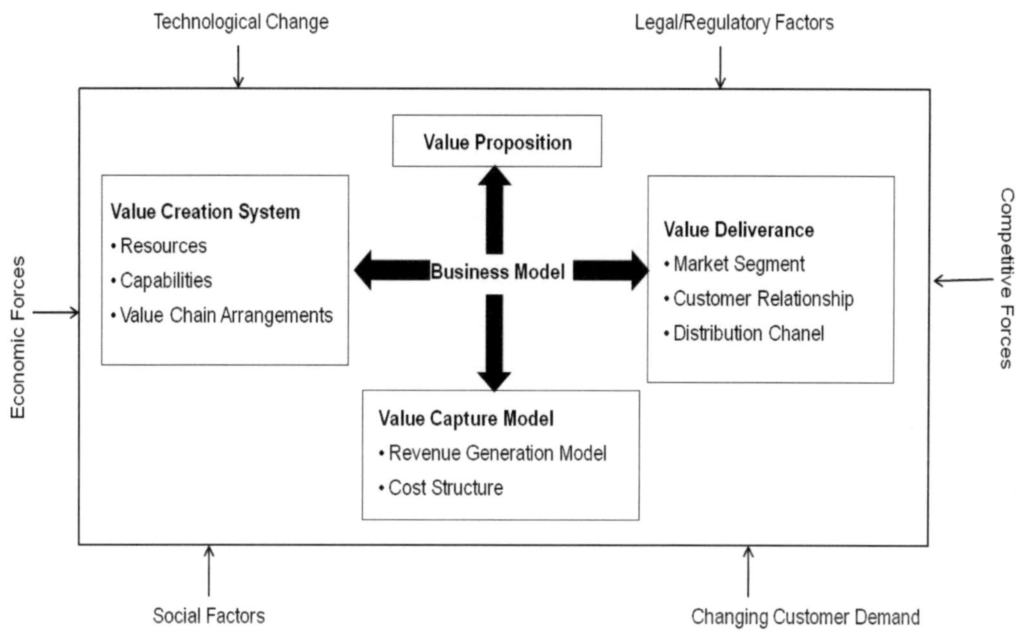

Fig. 1. Business Model Framework of Kamoun (Kamoun, 2008)

The business model framework of Kamoun encompasses four main components: value proposition, value creation system, value deliverance and value capture model. Each of these four components is further divided into subcomponents (see Table 2).

As shown in Figure 1 the components of the defined framework are influenced by external environmental forces "economy", "technological change", "legal/regulatory factors", "social factors", "competitive forces" and "customer demand".

Component	Description
Value Proposition	This component defines the added value that is offered by a company or network of companies to its customers via a bundle of new products, services or innovations.
Value Creation System	This component consists of subcomponents that are necessary to create the above mentioned added value.
Resources	Financial, physical, human, technological and organizational resources that are necessary to execute the business model and deliver the proposed added value.
Capabilities	Skills that are necessary to coordinate the resources.
Value Chain Arrangement	With this component the structure of the value chain is intended. Under the structure the inter-linked activities and alliances of the company with suppliers, partners and distributors have to be understood. These activities and alliances are necessary to execute the business model and deliver the proposed value.
Value Deliverance	This component consists of subcomponents that are relevant to deliver the proposed value.
Market Segment	This component represents the group of customers and geographic markets a company wants to deliver the proposed value. Different segments might have different needs. As a result of this, different products, services and value proposition might be required.
Customer Relationship	This component represents the link the company establishes with its customer to deliver the proposed value.
Distribution Channel	This component represents the way a company transmits its customers its products, services that have an added value for customers.
Value Capture Model	This component consists of subcomponents that are relevant to revenue and costs that arise from the proposed value.
Revenue Generation Model	This component defines how the income is generated. It defines the impact of the proposed value on revenue.
Cost Structure	This component defines the costs that arise to execute the business model and deliver the proposed value.

Table 2. Components of Business Model Framework of Kamoun (Kamoun, 2008)

4. RFID-systems

RFID is an Auto-ID technology that enables to identify tagged items by means of radio waves. Main components of a RFID system are:

- *Tag (Transponder)*: It consists of an antenna and a microchip. Microchip stores data about the tagged item. Antenna transmits the data about the tagged item to the reader by means of radio waves (Kavas, 2007).
- *Reader (Transceiver)*: It is a device that communicates with tags through radio waves and reads data on them (Karygiannis et al., 2007).
- *RFID Middleware*: It is a software that is used to consolidate, aggregate, process and filter raw RFID data received from multiple readers to generate useful information for end-users. It transmits also the processed data to backend enterprise applications (Kamoun, 2008).

- *RFID System Software*: It is necessary for the communication between tags and readers. This software enables to read tags, write on tags, detect and fix erroneous data as well as to realize authentication for security (Üstündağ, 2008).
- *Backend Enterprise Service*: This service enables to receive filtered RFID data from the middleware and integrate these with existing applications such as ERP, SCM or CRM systems through Application Programming Interfaces (APIs) (Kamoun, 2008).

5. Reshaping existing business models or creating new business models with RFID

RFID can reshape existing business models or create new ones. Instead of manual scanning of the products bought at paying counter, automatic scanning through RFID is an example for reshaping of a business model. Through reshaping, efficiency of business model can be increased. RFID is also used to reduce failure rate, shrinkage, operating stock and to enhance on-time delivery, shipment quality and so forth.

RFID technology can also be used to create a new business model that enables to generate new value creation opportunities and therewith to gain competitive advantage as well as to develop new ways to make money. Table 3 includes some examples for this type of RFID business models.

RFID Business Model	Description
RFID Infrastructure & Management Services Provider	Provider of RFID infrastructure-related products and solutions (e.g. tags, readers, data integration services, middleware)
RFID-enabled pay-per-use Business Model	According to this model a firm that has a huge number of assets in a given industry lets its trading partners to use these assets and to pay for the assets per use. The firm tracks its assets through RFID devices located throughout the supply chain.
RFID-based Security Provider	A firm that provides RFID-based tracking solutions for authentication, brand protection as well as to combat tampering, theft, counterfeiting.
Information and Business Intelligence Agent	A firm that uses RFID to offer new information-based services. It analyzes, for example with the use of data mining techniques, the extensive data collected from RFID systems and provides business intelligence solutions.

Table 3. Examples for RFID Business Models (Kamoun, 2008)

6. Impact of RFID on the components of the adopted business model framework

6.1 RFID and value proposition

RFID can provide a new or an added value for customers of companies that use this technology. As this value is a motive for customers to prefer the mentioned companies, it has a positive impact on the revenues of the companies. A new or an added value can be proposed by RFID:

- *Through creating a new way of conducting economic transactions among trading partners:* RFID enables to form new strategic networks such as buyer-supplier partnerships, pay-per-use business models. As RFID used in these networks provides various advantages (e.g. cost reduction, superior customer service level) for companies, they want to be a partner in such a network. For example, the tracking capability of RFID enables a company to realize pay-per-use business model that means invoicing the trading partner for the hired asset each time he uses the asset. Pay-per-use business models help trading partners, who participate in these models, to reduce purchasing, storage and maintenance costs as well as to eliminate losses due to stock thefts (Kamoun, 2008). For example, a furniture manufacturer can hire RFID-tagged sofas produced by him to a hotel according to pay-per-use model. He can monitor the usage of sofas with the help of RFID (e.g. a sofa can count the number of persons that sit on it, the person's weight and seating time) and create a monthly itemized billing statement to the hotel (Bohn et al., 2004). This model enables the hotel to exempt from purchasing and maintenance costs.
- *Through providing a superior customer service level:* Superior customer service level produced by using RFID increases customer satisfaction, which has a positive impact on the revenue stream of a company. For example, Metro aims to enrich the value of its product offerings through RFID-enabled smart shelves, smart dressing rooms that allow its customers to find the correct size, color and additional information about a displayed garment by touching a screen (Weber, 2003). Such a service can be a good motive for a consumer to go Metro for shopping.
- *Through reducing transaction costs by achieving transaction efficiencies:* RFID can increase the efficiency of executed transactions (Lin et al., 2006)). This causes reduction at transaction costs. Transaction efficiency and cost reduction can lead to lower prices that are an important motive for customers to prefer a company (Kamoun, 2008). For example, instead of scanning each product bought manually at the paying counter, a retailer can use RFID and scan automatically as well as instantaneously all bought products at the counter as the customer passes through a reader and exits the store. Thereby scanning and paying processes can be executed more quickly. This means time saving for customers and labor saving for companies that use RFID (Erickson & Kelly, 2007).

6.2 RFID and value deliverance

Among subcomponents of value deliverance RFID has an impact on customer relationship component. RFID can increase customer satisfaction that has a positive impact on revenue stream. For example in a store, which uses RFID, a customer can get information about the existence of a garment matching his style, size and color requirements and if the desired garment exist, the store's clerk can precisely locate the garment for him. If the garment is out of stock, customer can get information with the help of RFID-enabled system about the nearest store where the garment is available. Contactless checkouts are other examples that increase customer satisfaction by using RFID. This type of checkouts enables automatic scanning of RFID-tagged items in shopping carts by RFID readers at checkout counters. Readers enable also the automatic billing of customers using their RFID-tagged credit cards (Kamoun, 2008).
Gaining insights about consumer behaviours is essential to execute marketing activities successfully. Marketing activities aim to increase customer satisfaction and thereby to affect revenue stream positively. Data mining technologies enable to gain the necessary insights. It is possible to combine RFID with data mining technologies. Based on RFID-captured consumer behaviour valuable insights about consumer behaviour can be created with the help of data

mining technologies. These insights are used to enhance responsiveness of companies to their customer preferences (Kamoun, 2008; Hoffmann et al., 2005). Clothing retailer, for example, can make promotional offers based on collected and analyzed information of buying habit (size, favorite colours etc.) of a consumer who has a loyalty card (Erickson & Kelly, 2007).

6.3 RFID and value capture

6.3.1 RFID and revenue model

As illustrated above RFID can increase customer satisfaction, strengthen customer loyalty and thereby has a positive impact on revenue (Kamoun, 2008).

RFID can also be used to protect manufacturers against counterfeits, which are a real threat to the revenues of manufacturers (Kamoun, 2008). Product authentication plays an important role to combat counterfeiting and to detect counterfeit products. It enables to determine whether a given product is genuine or counterfeit. RFID-based product authentication is an important technological measure for checking the originality of a product that moves in a network of companies. Up to consumer each actor in a network can be the entry point of the counterfeit product. To realize a secure network each actor in a network has to verify the authenticity of the products on hand. To understand whether a given product is genuine or counterfeit, the insertion of a security feature into the product and the authentication of this feature are essential. RFID tags can be used for the authentication of security features (Filimon, 2008).

RFID increases network visibility, which affects revenues positively (Kamoun, 2008; Erickson & Kelly, 2007; Lin et al., 2006). Out-of-stock situations, which means loss of revenues, can be prevented through a better visibility. Better network visibility prevents also to retain great amount of stock that can be sold at discounted prices, if it is not sold by the end of the season (Kamoun, 2008).

6.3.2 RFID and cost model

Cost structure is also an important subcomponent of the value capture component. Utilization of RFID in companies or company networks can contribute much to reduce costs. Below this contribution is illustrated based on some examples:

- RFID enables perpetual inventory that is important to get information about current inventory level. Through perpetual inventory time and costs for physical inventory are saved. Stocking more or less items as a result of false inventory information can also be prevented (LakeWest Group & MeadWestvaco Intelligent Systems, 2003).
- As every item from warehouses to distribution centers and from these centers to retail shelves can be tracked through RFID, a reduction in stockouts can be ascertained by using RFID. In companies or company networks items leaving shelves or facilities are automatically recorded via RFID, computers are updated in terms of existing stocks and purchasing is executed if inventory levels drop too low. With less worries about stockouts companies or networks hold less safety stocks. Reduction at stocks means less inventory costs (Erickson and Kelly, 2007; LakeWest Group & MeadWestvaco Intelligent Systems, 2003).
- As demand for stocking decreases it will be possible to use store floor, which is used before for stocking, for new products that enrich the product range of a company or network of companies. In other words additional merchandise will be available for sale without costly requirements of store design and remodel (LakeWest Group & MeadWestvaco Intelligent Systems, 2003).

- As mentioned above RFID enhances transaction efficiency that gives rise also to cost reductions. For example, automation of some processes (e.g. checkouts, incoming goods control) via RFID causes a reduction both at labor costs and costs that incur to rectify errors caused by manual execution (Erickson and Kelly, 2007).

6.4 RFID and value creation

In order to create value with RFID, resources such as IT personnel with proper training, RFID infrastructure (tags, readers, printers, antennas, computers, networking equipment, middleware, application software, integration software etc.) must exist. A budget must also be allocated (Kamoun, 2008).

To operate RFID systems and to create a value some capabilities are also necessary. Following some examples for capabilities are listed (Kamoun, 2008):

- Ability to deal with erroneous tag reads, data redundancy, damaged tags
- Ability to manage reader and tag collision, signal inference and noise
- Ability to overcome great amount data generated by readers to enhance existing knowledge base and enable intelligent decision making.

RFID affects the structure of value chains in three ways (Kamoun, 2008):

- It eliminates inefficiencies in existing value chains. For example, RFID enables to track the movements of goods from the store's back door to the point of purchase in real time. This leads to better inventory visibility.
- It can enhance the collaboration between chain partners. For example, the collaboration between retailers and suppliers can be enhanced through RFID that enables suppliers to optimize production and replenishment scheduling based on real-time demand.
- It can give rise to new strategic networks. RFID enabled pay-per-use business model is an example for such a strategic network.

6.5 RFID and external environmental factors

Among external factors shown in Figure 1 technological change, competitive forces and legal/regulatory factors are relevant for RFID business models.

As at the present day RFID infrastructure is costly the use of this technology is limited. However through technological developments it is expected that infrastructure costs will decrease. For example, it is expected that with the help of nano technology silicone chip demand will disappear and through ink based RFID circuits costs of tags will decrease (Kış, 2006).

Adversaries of RFID declare that RFID threatens data privacy. According to them through RFID consumers can be tracked and thereby not only the consumption behaviours of consumers but also their private lives and other habits can be tracked (LakeWest Group & MeadWestvaco Intelligent Systems, 2003). However through some methods data privacy can be protected. Using lock command, kill command, press-to-activate switch, blocker tags, clipped tags and electromagnetic shielding, active jamming, frequency hopping, encryption of data in transit, encryption of data stored on tags, authentication are some examples for these methods. Furthermore regulative countermeasures can play an important role to eliminate fears about data privacy. Garfinkel's manifesto - named RFID Bill of Rights - is an example for regulative countermeasures. According to this manifesto consumers that consume tagged products have the following rights (Korkmaz et al., 2006):

- To know whether a product is tagged

- To have a choice to accept, discard, disable or remove the tag
- To know which information is saved on tags
- To know when, where and why tags are read
- Not to lose their rights (e.g. right of product return) even if they prefer to buy products without tags or deactivate tags with kill command.

The firm IDTechEx projects a very rapid growth in RFID. A market research report, which is executed by this firm, contains the predictions in Table 4 about the deployment of RFID tags from 2005 to 2015 (Raafat et al., 2007). In an environment, in which the use of RFID increases rapidly, companies have to use RFID in order to outmatch and not to lose market share.

Year Kategories in which tags are used	2005	2010	2015
Item	0.5	27.0	1,000.0
Pallet/Case	0.4	30.0	35.0
Other	0.4	5.7	12.5
All Kategories Total	1.3	62.7	1,047.5

Table 4. Forecast for global sales of RFID tags in billions between 2005-2015 (Raafat et al., 2007)

7. Conclusion

In order to gain advantages over competitors, companies or network of companies can use technological innovations to reshape their business models or to create new ones. RFID is one of these innovations. This technology reshapes business models through automating transactions (e.g. contactless checkouts). Thereby it contributes to increase transactional efficiency. RFID can also be used to create a new business model through developing new ways to make money. For example, a firm that offers business intelligence solutions with the use of data mining techniques can integrate its business intelligence software with RFID and start to analyze data collected from RFID systems. Thereby the firm can offer its customers a new information-based service that brings it money. RFID users can propose an added value or a new value for their customers. Added or new value increases customer satisfaction that has a positive impact on the revenue stream of RFID users. Naturally to create the mentioned value certain resources as well as capabilities and sometimes changes at organizational structures are required. As these requirements cause costs, it will be logical to invest in RFID if the impact of created value on revenue is greater than its impact on costs that arise to create the value. Only those organizations that consider this relation can benefit from RFID business models.

Technological developments contribute much to decrease especially the infrastructure costs of RFID technology. It is expected that this will increase the utilization of RFID in the near future. As a result of this companies will apply RFID technology to compete with their rivals. RFID adversaries declare that RFID threats data privacy. Despite this declaration it seems that the rapid deployment of RFID cannot be prevented, because both technical and regulative countermeasures are available to protect data privacy.

8. References

Betz, F. (2002). Strategic Business Models, *Engineering Management Journal*, Vol. 14, No. 1, (March, 2002), pp. 21-27.

Bohn, J., Coroamă, V., Langheinrich, M., Mattern, F. & Rohs, M. (2004). Living in a World of Smart Everyday Objects – Social, Economic, and Ethical Implications, *Human and Ecological Risk Assessment* , Vol. 10, Issue 5, (October, 2004), pp. 763-785, ISSN : 1080-7039 print / 1549-7680 online.

Erickson, G. S. & Kelly, E. P. (2007). Building Competitive Advantage With Radio Frequency Identification Tags, *Competitiveness Review : An International Business Journal incorporating Journal of Global Competitiveness,* Vol. 17, Issue 1/2, pp. 37-46, ISSN: 1059-5422.

Filimon, E. (2008). Anti-counterfeiting- prevention of counterfeit products with RFID, In: *Business Aspects of the Internet of Things,* Michahelles, F. (ed.), pp. 19-24, ETH (Eidgenössische Technische Hochschule) Seminarreport, Zürich.

Hoffmann, M., Jerzynek, D. & Weinand, R. (2005). Fiktives Gutachten zum RFID-Einsatz im Einkaufszentrum Mitte im Auftrag des Berliner Senat, 17.01.2011, Available from http://ig.cs.tu-berlin.de/lehre/w2004/ir1/uebref/gutachten/HoffWeinJerz-Rfid-einsatzImEinkaufszentrumMitte-2005-02-09.pdf.

Kamoun, F. (2008). Rethinking the Business Model with RFID, *Communications of the Association for Information Systems (CAIS),* Vol. 22, Article 35, (June, 2008), pp. 636-658.

Karygiannis, T.; Eydt, B.; Barber, G.; Bunn, L. & Phillips, T. (2007): *Guidelines for Securing Radio Frequency Identification (RFID) Systems - Recommendations of the National Institute of Standards and Technology (NIST),* NIST Special Publication 800-98, Computer Security Division Information Technology Laboratory National Institute of Standards and Technology, Gaithersburg.

Kavas, A. (2007): Radyo Frekans Tanımlama Sistemleri, *Elektrik Mühendisliği Dergisi,* Sayı 430, (Nisan, 2007), pp. 74-80.

Kış, M. (2006). RFID'nin Geleceği, 21.01.2011, Available from http://kodveus.blogspot.com/2006/05/rfidnin-gelecei.html.

Korkmaz, E., Üstündağ, A., Tanyaş, M. (2006). Standards, Security & Privacy Issues about Radio Frequency Identification (RFID), *Proceedings of the 4th International Logistics and Supply Chain Congress,* ISBN: 975-8789-08-2, İzmir, 29-30 November and 1 December 2006.

LakeWest Group, LLC & MeadWestvaco Intelligent Systems (2003). RFID In Retail - The Future Is Now, 11.06.2008, Available from http://www.lakewest.com/PDFdocs/RFID%20In%20Retail%20The%20Future%20Is%20Now_June%202003.pdf.

Lin, H.-T., L., W.-S. & Chiang, C.-L. (2006). Using RFID in Supply Chain Management for Customer Service, *Proceedings of 2006 IEEE International Conference on Systems, Man and Cybernetics,* pp. 1377-1381, ISBN: 1-4244-0099-6, Tapei, Taiwan, October 8-11, 2006.

Osterwalder, A.; Pigneur, Y. & Tucci, C. L. (2005). Clarifying Business Models: Origins, Present, and Future of the Concept, *Communications of the Association for Information Systems (CAIS),* Vol. 15, Tutorial, (May 2005).

Raafat, F. F., Sherrard, W. R., Meslis, L., Windt, J. (2007). Case Study: Applications of RFID in Retail Business, 25.01.2011, Available from http://www.swdsi.org/swdsi07/2007_proceedings/papers/695.pdf.

Shi, Y. & Manning, T. (2009). Understanding Business Models and Business Model Risks, *The Journal of Private Equity,* Vol. 12, No. 2, (Spring 2009), pp. 49-59.

Schweizer, L. (2005). Concept and Evolution of Business Models, *Journal of General Management,* Vol. 31, No. 2, (Winter 2005), pp. 37-56.

Üstündağ, A. (2008). *RFID ve Tedarik Zinciri,* Sistem Yayıncılık, ISBN 978-975-322-515-1, İstanbul

Weber, T. (2003). The Future of Shopping, BBC News Article, 04.01.2011, Available from http://news.bbc.co.uk/2/hi/business/3712261.stm.

2

Commercial and Implementation Issues Relating to the Widespread Acceptance and Adoption of Radio Frequency Identification Technology

Mark J. Rodrigues[1] and Kieran James[2]
[1]*Master of E-Business Graduate, Murdoch University School of Business, Murdoch University,*
[2]*Senior Lecturer in Accounting, School of Accounting, Economics & Finance, Faculty of Business, University of Southern Queensland, Australia*

1. Introduction

The adoption of Radio Frequency Identification Technology (RFID) is giving rise to major improvements for consumer goods manufacturers. RFID technology offers a huge spectrum of applications, through increased flexibility, transparency, and performance in supply chain management and warehouse execution systems. As a result of the expansion, marketing research companies have invaded the consumer market by predicting million-dollar investment, and unrealistic applications for today. They have caused attention to focus upon consumer privacy concerns that have reduced the adoption of the technology. Managers of large companies are encouraged to block out the hype and exploit the technology for its ability to increase return on investment in the supply chain.

RFID can be thought of as *Smart Labels* or *Silent Commerce*. RFID are the new-generation computer tags attached to an item and containing full product information which, when activated, transmit information to an RFID reader as the customer leaves the store with the product (Turban et al., 2006). This technology is most likely to replace the standard barcode in supermarkets and department stores (Turban et al., 2006, p.294) if the adoption process follows the standardized model suggested in Rogers (1995) and the adoption rate reaches 100% or close to 100%. The demand for RFID has been increasing over the past few years. The hype in the market-place and in some consumer circles suggesting everything will be tracked' is rapidly becoming a reality.

This chapter examines the perceptions of RFID among Australian RFID suppliers/integrators, and the role and importance that perceptions play in the actual adoption process. As the Chicago School of Sociology often proclaimed, perceptions are 'real in their consequences' (Thomas and Znaniecki, 1927, p.8, cited in Rogers, 1995, p.209). This project also examines how integrators' perceptions can act upon present expectations of RFID technology. An understanding of what leading integrators think at this moment may benefit vendors and others to create applications that will eventually secure more widespread acceptance. The research study discussed in this chapter was undertaken using qualitative data collection methods, i.e. personal interviews with a sample of leading RFID suppliers/integrators located in Australia.

2. Theoretical framework

This theoretical framework examines issues pertaining to adoption and barriers to adoption of RFID technology and related perception issues. Questions posed to the respondents in the research study were derived from Thomas Ehrmann's Business Model theory and are now used to analyze the perceptions of industry managers regarding RFID adoption.
Two other theories are used. Firstly, we use Efraim Turban's theory (Turban et al., 2002; 2006) about how companies can adopt a systematic approach to discover their Electronic Commerce (EC) opportunities in the market-place. Secondly, we refer to Everett Rogers' (1995) adoption theory which deals with the process through which an individual or other decision maker unit passes from first knowledge of an innovation, to a decision to adopt or reject, and then finally (if the innovation is not rejected) to implementation of the new idea. According to Ehrmann (in Jones, 2003), the following outlines the 'Appraisal of a Business Model' theory: The way to appraise a business model or proposition is to evaluate each of the following: value proposition, innovation, content, structure, and governance.

2.1 Value proposition

According to Ehrmann (in Jones, 2003, p.720), 'value proposition' asks customers what value does the product have on the entire supply chain? This, in turn, focuses upon the business idea, economic role, and the value that the product yields to the customer. Integration is another component of value, i.e. how successfully can the product integrate with new systems?
Turban et al. (2006, p.596) suggest that companies may be 'Market-Driven', waiting to observe what the competitors in their industry are doing. 'When one or more competitors starts [*sic*] to use EC, and it seems that they are doing well, it is time to follow suit' (Turban et al., 2002, p.691). This can be linked to Rogers' theory of *Diffusion of Innovations* (see Rogers, 1995) which suggests that, at an awareness stage, 'the individual is exposed to the innovation but lacks complete information about it' (CIA Advertising, 1998). In terms of Rogers' (1995) diffusion model with RFID in Australia we already have 'innovators' and 'early adopters' using, or at least trialing, the technology in primarily niche applications. However, we have not yet reached 'critical mass' (Rogers, 1995, pp.313-330). The 'early majority', 'late majority' and 'laggards' are all yet to come on board. Critical mass is especially vital in what Rogers (1995, p.313) terms 'interactive innovations', i.e. innovations, such as the Internet and Facebook, where each new user increases the benefits of adoption for all past users (by giving them access to more people) and also for all future users. RFID is essentially interactive since the system will clearly work best when all suppliers and all customer companies (other than end-consumers) own both readers and tags. In order to evaluate the RFID industry in this chapter, the questions posed to interview respondents in the research study are interpreted within the context of the Ehrmann Business model, as well as Rogers' (1995) theory.

2.2 Innovation

According to Ehrmann (Jones, 2003, p.720), the process of 'innovation' is defined as reducing costs of producing or offering existing goods or services through a business channel. Innovation deals with cost savings, and the consequent advantages these savings bestow upon the innovating firm relative to its competitors. As Rogers (1995, pp.413-414) makes clear, innovators and early adopters of a technology, such as the Iowa hybrid wheat farmers in the classic early diffusion study by Ryan and Gross (1943), typically reap windfall profits that are denied to more risk-averse later adopters. Innovation in the business model also considers production costs, and the market structures that are developed to support the product. This can be linked to Rogers' (1995) theory which suggests that, at the interest or information stage, 'the individual becomes interested in the new idea and seeks additional information about it' (CIA Advertising, 1998). Turban et al. (2006, p.596) suggest that innovation may be 'Problem-driven', i.e. 'Organizations have a problem such as inventory delays and deliveries. EC applications may be attempted in order to solve the problem' (Turban et al., 2002, p.691). Turban et al. (2002) also argue that much innovation occurs simply because organizations are fear and/or greed driven: 'Companies are either so scared that they are afraid that if they do not practice EC they will be big losers, or they think that they can make lots of money going EC' (Turban et al., 2002, p.691). Fear and greed are not specifically recognized in the Rogers' model but are consistent with Rogers' (1995) key argument that adoption of innovations is largely a social and psychological process. For her part, ethicist of new technology Cynthia K. West (2001, pp.124, 128) notes that fear of loss has been the major selling point used by retailers of biometric face, finger, and retina surveillance and identification technologies.

Rogers' (1995, Figure 7.2, p.262) theory views 'early adopters' as a group which comprises 13.5 percent of the total population. This group is comprised of highly educated and wealthy innovators who are highly visible and respected among their peers (Rogers, 1995, p.269). Early adopters play a key role in the adoption process for new technology, influencing very strongly the times when an innovation will be adopted by others. This category contains most of the 'opinion-leaders' who, largely through word-of-mouth among peer networks, extol the benefits of new technology to their less well-connected and influential peers (Rogers, 1995, p.264). Although Walmart, Gillette, and the United States of America (USA) Department of Defense are clearly innovators or early adopters of RFID they may be less able to function as opinion-leaders, hence slowing the rate of growth of the technology. Small businesses, foreign businesses, and those dealing in niche and/or luxury products may not necessarily have strong contacts at Walmart or be influenced significantly by what Walmart does. The same comment applies to Woolworths and Coles in Australia.

Early adopters in the manufacturing and technology sector have developed new methods to add RFID as a cost effective method. For example, a Motorola sponsored White Paper by IDC revealed that there has been a steady adoption of RFID technology. The highest rates of adoption are by: educational organizations (36%), followed by transport and logistics companies (33%) and then utilities (26%) (Motorola, 2010).

2.3 Content

According to Ehrmann (Jones, 2003, p.720), 'content' in the appraisal of business models refers to the goods and information that are being exchanged. This business model looks upon the individual capabilities required to enable exchanges in the supply chain. Content evaluates the information that is being exchanged in the supply chain, and examines new products. We can also reference Turban et al. (2002, p.691) where they state that '[t]echnology exists and the company is trying to use it. In doing so, the company may find problems that no one knew existed'. When this occurs people may modify or stop using the technology. Rogers (1995, p.320) explains that, just as critical mass is added quickly, it can also rapidly fall away as people abandon an innovation in droves.

2.4 Structure

According to Ehrmann (Jones, 2003, p.720), 'structure' refers to the actors that are linked in the value chain. The structure model analyzes customers at both ends of the business. Structure refers to the underlying partners, and focuses on a specific network rather than dealing with the entire value chain.

Turban et al. (2006, p.596) state that companies are frequently 'Problem-driven'. If the problem is to reduce inventory errors, then the advantage of RFID is in the accurate tracking of information. Rogers' (1995) theory suggests that adoption goes through a trial stage, as 'the individual makes full use of the innovation' (CIA Advertising, 1998). Many companies are presently adopting, trialing, or considering adopting or trialing, RFID technology in the supply chain.

2.5 Governance

Ehrmann (Jones, 2003, p.720) mentions 'Governance' which deals with the way in which exchanges are executed. The model looks at property rights that are allocated between parties to the transaction. Also, governance deals with the set-up of market roles, operations, and strategic tasks. The commercial RFID literature has repeatedly viewed RFID in the context of consumer privacy issues. Rogers (1995, chap.11) agrees that not all innovations are socially desirable. He cites the examples of missionaries introducing the steel ax into an Australian aboriginal tribe (which undermined the entire traditional social structure of the tribe and lessened respect for elders) and the introduction of snowmobiles amongst Finnish Lap communities. Rogers (1995, chap.3) warns against the 'pro-innovation' bias of many diffusion scholars and the many profit-fixated corporations who sponsor diffusion studies. The *consequences* of adopting an innovation remain an under-researched area in diffusion studies. Furthermore, Rogers (1995, p.412) explains that consequences can be desirable or undesirable, direct or indirect, and anticipated or unanticipated. In similar vein, West (2001, p.140, emphasis added) puts forward the view that '[a] discussion of *values* and *ethics* is needed both within the information technology industry as well as in the communities in which they [the technologies] are deployed'.

3. Research study method

The interview dialogues for our research study are presented below. Questions were posed to each research study participant based upon Ehrmann's Appraisal of Business Model theory. Respondents were asked to express their perceptions of the RFID industry as it presently stands.

For this research study, the first-mentioned author contacted leading suppliers/integrators working for organizations which were (at the time of the initial contact) leading integrators of RFID technology within Australia. Respondents were selected by purposeful sampling rather than by random sampling. A semi-structured interview approach was used as the data collection method. A pre-prepared list of 14 interview questions was first e-mailed to all respondents. The actual interviews were all conducted by telephone because the interviewer was based in Perth whereas the respondents were based in Sydney, Brisbane, and Melbourne. Length of interviews ranged from 30 minutes to 80 minutes. No tape-recorder was used. However, the interviewer took detailed shorthand notes. On the evening of each interview, the interviewer summarized responses and noted key themes that were emerging from the data. The first-mentioned author has 15 years of experience with RFID systems, including many years selling RFID systems for Paxar Canada to leading early adopters such as Walmart. Because of his industry experience, he was in a position to both carefully select the interviewees for the study and ask questions that were relevant and prescient from the practical industry perspective.

The 14 questions, sent to the interviewees in advance, were designed to meet the following objectives:

- Document the selected RFID integrator's perception on each issue; and
- Evaluate responses within the context of the RFID academic and commercial literature.

The five (5) interviewees were guaranteed strict confidentiality. Although interviews were conducted between July and December of 2004, informal conversations in late 2006 between the first-mentioned author and each of the interviewees confirmed that their views on the state of the RFID industry in Australia had not changed significantly between 2004 and 2006. None of the interviewees elected to modify, delete, or add to their original set of responses. Interviewees are denoted (A), (B), (C), (D), and (E) in the following Results section. We specifically incorporate new comments and references to bring our analysis up-to-date whilst still relying on our original results where they remain applicable.

4. Results for interviews

The respondents were provided with an opening vignette which appeared in the initial e-mail sent to the respondents above the list of interview questions. The vignette is as follows: 'RFID (Radio frequency identification) can be thought of as *Smart Labels* or *Silent Commerce*. The demand for RFID has increased over the past few years. The hype in the industry (that "everything will be tracked") is fast becoming a reality. At one end of the spectrum, RFID is viewed as a tracking and security device for enterprise application. At the other extreme, RFID is viewed as a true technological wonder that is going to transform the way that businesses will operate.'

In response to the 1st question 'What economic value will RFID Tags have on the business chain?' respondent (A) answers that economic value in the supply chain 'will amount to US$10 to US$100 million within 4 years [i.e. in Australia]'. (E) provides the longest response but he does not attempt to put a value on supply chain savings. He notes that '[i]t will have a big value. It will stop fraud and authenticate drugs, perfume and electronic goods. Read and write tags will make it database independent. Therefore, the cost of goods should come down in the supply chain'. (B) notes that a major positive feature of the technology, as compared to barcodes, is that '[a]ssuming enough read ranges, goods can be moved within the logistics without line of sight'. Barcodes are limited in that they cannot be read at a faster rate than 1 every 2 or 2.5 seconds. By contrast, an RFID reader reading 100 RFID tags per second is certainly possible. We conclude that the respondents have a realistic view of the impact of RFID on the supply chain, which can be compared to the more optimistic view (or 'hype') which has been frequently expressed within the commercial literature. For example, AIM Industry analyst firms predicted in 2003 that RFID would become a US$3 billion market globally by 2008 (AIM -RFID Connections, 2003). Recent research from IDTechEx reports that 'the next 10 years will see a rapid gain in market share of mainstream printed and chipless RFID tags, and that the numbers sold globally will rise from 40 million in 2009 to 624 billion in 2019 (Das and Harrop, 2010). ABI Research sees continuing strong growth potential in RFID markets worldwide and forecasts a total market size of about US$4.6 billion by the end of this year for RFID systems (hardware, software, and services) (ABI, 2010).

In response to the 2nd question 'What economic value will Smart Labels have on the consumer?' (A) notes that '[i]n the supply chain 30% of the savings will pass on to the consumers. The rest will ... [flow through as] stock and dividend profits which will be shared with consumers'. (C) comments that RFID will create '[m]ore choices for consumers. Easier to shop and locate products in store setting. Provide authentication of genuine goods.' In similar vein, (D) indicates that consumers will benefit from 'high security infrastructure, tracking the history of products'. (E) agrees, also emphasizing that RFID will ensure that '[p]roducts are not copies', i.e. there is 'an authentication guarantee'. Early implementations were lacking in several respects, and the technology influenced privacy advocates. Large-scale implementations stalled as a result of user apprehension and a declining economy. The benefits and cost savings of RFID seem real and significant but the technology must reach critical mass of adopters or the benefits may not fully materialize. For this to happen, costs of both readers and tags must come down (to be discussed shortly). In addition to this, we would expect supermarket customers would need to actively push for the technology to be introduced to replace barcodes which has not happened in Australia. Supermarket shoppers in Australia evidently do not yet feel the 'need' for faster check-out experiences. Rogers (1995, p.164) defines a 'need' as 'a state of dissatisfaction or frustration that occurs when one's desires outweigh one's actualities'. He notes that often a perceived need must precede adoption of an innovation, although, at times, an innovation is adopted without the prior perceived need. The need for RFID in the healthcare industry has been noted but not adopted on a large scale as at 2004.

Today, RFID in healthcare sector could significantly impact patient safety by decreasing medication errors and increasing efficiencies in locating hospital assets. Citing medication errors as a US$3.5 billion annual problem in the U.S.A., Young (2008) suggests that there is a need for uniform global healthcare standards.

In response to the 3rd question 'What is the RFID network size?' there is marked disagreement across respondents. Some respondents, i.e. (A), (C), (D) and (E), claim that the RFID network size is 'big'. For example, (E) states that '[i]t will have a big value. It will stop fraud and authenticate drugs, perfume and electronic goods'. (A) mentions the widespread use of RFID in animal tracking and the transport industry, while (E) refers to 'animal tracking and security applications' creating demand for RFID. Confirming his view that the industry will be 'big', (D) notes that '[i]t will require updating systems, and purchases of reader and writers. Microsoft involvement in new software will bring changes across the industry'. However, by contrast, (B) summarizes the industry as still being 'small, in its infancy, mainly propriety installations and pilot tests'. We conclude that, whilst the network size is *potentially* huge globally, in terms of actual realization the industry in Australia remains in its infancy at least in mainstream non-niche applications that involve the end-consumer. This is a reasonable conclusion given that neither Woolworths nor Coles (Australia's groceries duopoly) at the date of writing have actually implemented RFID systems (Mills, 2005; Walters, 2005).

A Woolworths' spokesperson has said that RFID adoption is not an immediate priority and that other project with 'more certain' patterns of perceived benefits will be pursued more vigorously than RFID adoption (Mills, 2005; Walters, 2005). A study of Woolworths' and Coles 2006, 2007, and 2008 annual reports by the second-mentioned author (Coles is part of the Wesfarmers Limited group of companies for the 2008 and following financial years) reveals no further mention of RFID or RFID trials in these reports. Woolworths' Managing Director, Michael Luscombe, claims that, even without RFID adoption, '[b]y lowering our costs of doing business, we have created a world-class model of efficiency and logistical expertise' (Managing Director's Report, Woolworths Limited, 2008 Annual Report, p.6). Woolworths, the 25th largest retailer in the world according to the corporation, does seem profitable and efficient enough without RFID. The second-mentioned author has computed Return on Equity (ROE), Return on Assets (ROA), and Inventory Turnover Days of 28.11%, 10.98%, and 29.76 days, respectively, for Woolworths based on publicly available consolidated accounting data taken from the independently audited Woolworths Limited 2008 Annual Report. For its part, Coles has undertaken RFID pilot tests but has generally viewed the technology as too expensive when compared to barcodes (Walters, 2005).

However, according to Swedberg (2010), this year (2010) seems to indicate a psychological change in the market. Five years ago (2005), people came to integrators and vendors skeptical that RFID could solve their problems. Now, there is confidence in the technology, so end users want to discuss things like software integration (Swedberg, 2010).

In response to the 4th question 'Do you think there is a demand for RFID Technology?' there is again marked diversity in the responses. Each commentator focuses on different perceived benefits and user groups. (A) provides the most detailed and quantified response as follows: 'By 2012 bar codes and RFID tags will equal each other in usage. The conversion from legacy systems on a grand scale will happen. By 2020, 20% of the supply chain will be used by bar codes, which becomes a niche market'. In terms of willingness to offer detailed projections of future developments, (A)'s response takes on Marxian proportions. His key dates for Australia are: 2012 (equal usage barcodes and RFID tags) and 2020 (barcodes a niche market; 20% barcodes; 80% RFID tags). While the projections are expressed in precise terms they indicate that market dominance for RFID tags (over 50% adoption rate) is still some years away. His predictions are simply that, predictions. Even the great philosopher Karl Marx, correct about so many things, was hopelessly wrong in his prediction of the worldwide triumph of communism. Future rates of adoption of RFID technology may surpass or underperform predictions. Rogers (1995) points out how new technologies often differ significantly in terms of the time it takes for the adoption rate to reach 100%. For example, in the education sphere in the USA, it took 50 years for kindergartens to be fully adopted, as opposed to 18 years for driver training, and only 5 years for modern math (Rogers, 1995, p.64). (D) is also optimistic, noting that '[t]he demand will depend upon the government, added security, fraud, and line of sight for identifying products. Also consumers are pushing the demand for cheaper and time saving retail experiences'. (C) is more circumspect, noting that whilst there is 'demand for information', '[t]he process is not in place with RFID tags'. He attempts to temper excessive enthusiasm by drawing upon history to note (correctly) that '[i]t took 20 years for bar codes to be accepted'. (B) also urges restraint and a wait-and-see approach: 'Only in niche industries at the moment [is there demand]; demand in retail will be led by large organizations such as Walmart, CML [Coles-Myer Limited, now Coles] here in Australia'. Critical mass has most definitely not been reached in Australia outside niche applications. These are usually in settings not involving direct dealings with end-consumers.

There still is a demand for RFID technology, as compared to six years ago, from the early adopters like Walmart and the U.S. Department of Defense which made their first RFID announcements in 2003. Growth in demand for RFID tags has been driven in part by Walmart's apparel tagging initiative. This has driven expected RFID tag growth rate for the industry. RFID tag demand growth exceeded manufacturer expectations in other sectors including: transport, storage, logistics, electronic payment, tracking medical devices, food safety systems, and asset management. Around the world there are several important examples of the growth in demand. For example, India's demand for RFID is apparent with expected 600 million unique ID cards, 50 million e-passports, 100 million health cards, 50 million transport and ticketing cards and 50 million banking cards likely to be issued over the next seven years (Reinhardt, 2010).

In response to the 5th question 'Is the market structure established for RFID?' there is also a diversity of responses. Both (D) and (E) refer to structure established with respect to specific applications. (D) notes 'a structure [exists] for example [in] the government control of animal tracking', whilst (E) refers to the auto-parts industry where '40 million RFID [tags] are used in the [Australian] auto industry each year'. (A)'s measured response notes the privacy concerns that consumer groups have expressed regarding RFID: 'The market has been established but the privacy issue has given RFID a bad start. There seems to be some confusion in consumer perceptions'. End-consumers do not seem unduly concerned about privacy issues regarding RFID usage in auto-parts most likely because the end product is not a standard retail shopping-mall item and people rarely feel any psychological or emotional closeness to purchased auto-parts.

In response to the 6th question 'Have other users in the industry caused interest in RFID?' most respondents refer to Walmart mandating RFID use for their Top 100 suppliers since January 2005 (Business Week Online, 2004a, 2004b; Kaiser, 2004; Lundquist, 2003; Turban et al., 2006, p.77; Walters, 2005). Other major users globally are Gillette and the U.S.A. Department of Defense (Turban et al., 2006, p.410) although Gillette is yet to mandate its use for suppliers. (C) refers to the 'Brazilian government use of RFID tags to track animals'. (A) notes that '…since 1995 I have been influenced by when Australia Post became interested in tracking mail'. More generally (B) comments that '[c]ertainly Walmart's drive has created interest in the retail sector', whilst (D) is cynical and wary: 'Initially [users] got fired up but [before long they] did not care'. In Rogers' terminology, Walmart, Gillette, and the U.S.A. Department of Defense can be classified as innovators or as early adopters. Mr Con Colovos, CIO of the Australian early adopter Moraitas Fresh (a supplier of tomatoes to the major supermarkets), has stated that the Walmart mandate means that widespread adoption of RFID in Australia is now 'inevitable' (Walters, 2005). Innovators and early adopters do tend to be much more upbeat than others about the prospects of rapid diffusion of an innovation. We should note that Walmart giving its suppliers no choice in the adoption decision means that adoption by its suppliers is an 'authority innovation-decision' (Rogers, 1995, p.29). Therefore, it is different from the classic innovation problems such as hybrid wheat adoption by Iowa farmers as studied by Ryan and Gross (1943). RFID adoption in Australia is unlikely to follow the 'mandate model'.

Of key significance is the demand for RFID tags in retail, which demands 300 million RFID labels in 2010. Tickets used for transit demands 380 million tags in 2010 and tagging of animals (such as pigs, sheep and pets) amounts to178 million tags being used for this sector in 2010. This is happening in regions such as China and Australasia. In total, 2.31 billion tags will be sold in 2010 versus 1.98 billion in 2009 (IDTechEx Ltd, 2010).

In response to the 7th question 'How and who will manage the information of RFID Technology?' and the 8th question 'What goods and information will be exchanged in the RFID tag?' the respondents note that ownership of information should not be exclusive to any one industry or organization. All managers of Information Technology will own the content for each good. The commercial literature explains that an Object Name Service (ONS), such as UPC (companies will need to maintain ONS servers locally), will store information for quick retrieval. The ONS will keep track of data for every EPC-labeled object (Shankland, 2002). As (C) explains: 'IT managers within the company will manage the information for goods entering the company; same as barcode item numbering systems. Proprietorship of information on the tag will be allowed by the manufacturer, e.g. authentication of a refrigerator for the disposal of product'. (D) points out that '[t]he retail industry will not be able to write tags'. (E) stresses that databases do exist for some niche application areas such as 'NLIS [the government-mandated National Livestock Identification Scheme for Australian cattle] and the Automotive Industry database'. (E) goes on to add that: 'RFID will provide for the maintenance history of machinery to be recorded on the tag for the [benefit of the] services industry'. Barcodes do not and cannot include such detailed information.

The respondents note that the information on the tag will specify the manufacturer, factory program, maintenance for service, and personal information of the product. This view is similar to viewpoints expressed in the commercial literature which state that the RFID tags will let you trace a particular unit of product through its life-cycle. However, it is not true that an item can be traced to a particular person. Current applications in the U.S.A. allow consumers to choose to 'kill' (de-activate) the tag after they exit the check-out. The data will have business intelligence, such as inventory reduction and total asset visibility (Rossi, Sommerville, and Brown, 2003). This raises the related issues of data integrity and privacy (to be discussed shortly), two potentially important 'consequences of innovation'.

Another important issue is that the speeds of the networks for retrieving tag identifiers have not been tested for large volumes. Interestingly, none of our research study respondents discussed this concern in their responses. Overall, the commercial literature has emphasized this concern, and has 'hyped' both the privacy issue and the large volume of retail tag usage issue.

Proper RFID governance is necessary if RFID is to become like the new wave of development of the Internet. Eventually, billions of smart devices will be interconnected into a global network communication infrastructure and managing this information has not been evaluated.

In response to the 9th question, 'What price do you expect RFID tags to cost in the coming years?' all five respondents note that the tag price will go down from dollars to cents in the next few years. For example, (D) notes that the retail tag price now (i.e. second half 2004) is A$1 (US$0.82 at 10 April 2007 exchange rate) landed, and could go down to A$0.40 (US$0.33). As (A) explains, the '[p]rice of tags will go down due to economies of scale. The more users that implement RFID the less the tag/label cost per unit. Tag prices will definitely go down to a few cents US when RFID equals bar codes share'. All respondents note that packing will be the costly item. The commercial literature states that tag costs in volume now (2004) 'could be in the range of (US) 18 to 35 cents each'. However, these costs depend on the type of product the tag is applied to and the kind of adhesive used to secure it to a package (Brewin, 2004).

Market research firm IDTechEx predicts that in 2019, the average price of an item level tag will be 1 cent, but chipless versions will cost less than that and especially when printed directly onto packaging (IDTechEx 2011). Despite the push from large retailers, analysts have predicted the demand for tags growing at double digit rates and 5¢ tags to come in the near future. Frost and Sullivan (2011) found that the total RFID market earned revenues of US$600-$800 million in 2009 and estimates this to be over US$2.0 billion by 2016, growing at a compound annual growth rate (CAGR) of 17.7 percent (Frost and Sullivan, 2011).

We conclude that the respondents perceive the tag pricing similarly to the commercial literature. Tag prices must come down for their usage to be more widespread which creates something of the 'chicken and egg' scenario that diffusion scholars are well aware of. Critical mass must be reached but this is by no means assured. Many people will adopt if costs come down but costs only come down as more people adopt.

In regards the crucial 11th question (we skip responses to Questions 10 and 12-14 for space reasons), 'Are you concerned with the privacy issues posed by RFID technology?' all integrators unanimously respond that they are 'not concerned' [(D) and (E)] and that there is 'no problem' (C). (A) offers the most detailed reply. As he explains: 'There has been bad publicity of RFID when it comes to privacy. As business integrators its does not matter, as all technologies have some negatives. Privacy will not pose an issue because consumers will be educated on the plan and usage of the product'. (C) is more specific in directly attempting to address consumers' *known concerns* as follows: 'Items do not get attached to the person so the retailer does not know who purchased the item'. In other words, the tags allow a product to be traced through its life cycle. However, the tag is not 'connected' to the buyer in any way that does not already occur under the barcode system.

Commercial articles (see, for example, Ferguson, 2002; Wired, 2004) have emphasized that there is a perception among privacy groups that RFID is a real threat to consumer privacy. For example, the mid-2000s announcement by Benetton of its planned adoption of RFID led to an immediate call by the U.S.A.-based Consumers against Super-market Privacy Invasion and Numbering (CASPIN) organization for a worldwide boycott of Benetton stores. The impact of this boycott caused the implementation of low-cost RFID systems in the retail market to be re-considered by some within the sector. We feel that this outlook is based upon two misconceptions: (a) that the tags contain personal information about the consumer (they do not), and (b) that tags can be read by a nearby reader *after* the consumer has taken the product back to home or office.

Recent articles suggest that privacy concerns were not high on the list for 2011. A few years ago retailers moved away from any mention of RFID because they feared adverse reactions from customers (Pleshek, 2011).

5. Summary and conclusions

We conclude that, despite the great potential of RFID, it is not as widely implemented as many would have predicted based upon the commercial literature around the year 2004. RFID has experienced many various roadblocks that have stunted the growth of the industry. Our interview-based research study, results for which have been discussed in this chapter, shows that integrators' perceptions can affect the adoption process. Integrator perceptions can act upon present expectations of RFID technology. Importantly, the interviewed industry integrators in 2004-2006 were generally more circumspect and realistic than the commercial literature of 2004 about the future prospects of RFID. In 2004-2006 they did not perceive that the consumer privacy concerns were insurmountable as oftentimes concerns have been based upon two misconceptions: (a) that the tags contain personal information about the consumer (they do not), and (b) that tags can be read by a nearby reader after the consumer has taken the product back to home or office (they cannot be). Also, to take further note, as at March 2011, the widespread adoption of RFID has been slow and one important reason for this delay has been the lack of uniform standards for network and data management. Cost and quality concerns have fractured the enthusiasm for RFID and reported high failure rates also exerted a dampening effect. In 2004 the suppliers had to absorb the cost of becoming RFID-compliant so the cost of doing business was risky. Despite this, the RFID hype in the commercial literature of 2004 has today become more realistic as the convergence of three technologies - Wireless Networks, RFID and Global Positioning Systems (GPS) – has occurred. The reality today, seven years on, is beginning to approach the wildly optimistic RFID growth forecasts in the 2004 commercial literature. Although practical problems still abound in this industry, the immediate future for consumer goods remains fit for speculation. There are benefits associated with global traceability to manufacturers.

6. References

ABI (2011). 'ABI Research: Item level retail tagging will drive double-digit growth for RFID in 2011'. Obtained through the internet:http://rfid24-7.com/rfidtalk/?cat=51. [accessed 03/11/ 2011].

AIM-RFID Connections (2003).Obtained through the internet:http://www.aimglobal.org/technologies/fid/resources/articles/nov03/industry.htl, [accessed 7/7/2004].

Brewin, B. (2004). 'No QuickROI from RFID, say Manufacturers',*Computerworld*. Obtained through the internet:http://www.computerweekly.com/Articl129677.htm, [accessed 12/8/2004].

Business Week Online. (2004a) 'Like it or not, RFID is coming', *Business Week Online*, 18 March, Obtained through the internet:http://www.businessweek.com/technology/content/mar2004/tc20040318_7698_tc121.htm, [accessed 17/2/2005].

Business Week Online. (2004b) 'Talking RFID with Wal-Mart's CIO', *Business Week Online*, 4 February.Obtained through the internet: http://www.businessweek.com/technoloy/content/feb2004/tc2004024_3168_tc15.htm, [accessed 12/2/2005].

CIA Advertising. (1998)'Diffusion of Innovation'. Obtainedthrough the internet: http://www.ciadvertising.org/studies/stdent/98_fall/theory/hornor/paper1.html,[accessed 21/2/2004].

Das R.and Dr. Harrop P. (2010) 'Printed and Chipless RFID Forecasts, Technologies & Players 2009-2019'.Obtained through the internet: http://www.idtechex.com/research/reports/printed_and_chipless_rfid_forecasts_technologies_and_players_2009_2019_000225.asp.[accessed 11March 2011]. http://www.idtechex.com/research/reports/printed_and_chipless_rfid_forecasts_technologies_and_players_2009_2019_000225.asp

Ferguson, G. (2002) 'Have yourObjects call my Objects', *HarvardBusiness Review*, June, pp.138-144.

Fishman, C. (2006) The Wal-MartEffect: How an Out-of-town SuperstoreBecame a Superpower, New York: AllenLane.
IDTechEx Ltd, (2010) 'RFID Forecasts, Players and Opportunities 2011-2021' Obtained through the internet:http://www.reportlinker.com/p0149567/RFID-Forecasts-Players-and-Opportunities.html, [accessed 11 March 2011].
Jones D. (Ed.) (2003) *The New Economy Handbook*, San Diego: Elsevier Science.
Kaiser, E. (2004) 'Wal-MartStarts RFID test', *Forbes.com*, 30 April.Obtained through the internet: http://www.forbes.com/home/newswire004/04/30/rtr1355059.html, [accessed23/2/2005].
Kinsella, B. (2003) 'Wal-Mart Factor',*Industrial Engineer*, November.
Lundquist, E. (2003) 'Wal-Mart gets itRight', *E-Week*, 14 July.
Mills, K. (2005) 'Radio Daze', *The Australian IT Business*, 19 July, p.1.
Mishra, D. (2004) 'Get Readyfor RFID'. Obtained through the internet: http://medialinenews.com/articles/publich/printer_518.shtml, [accessed12 July 2004].
Motorola (2010) 'Trends 2011: RFID takes off'. Obtained through the internet: http://www.gadget.co.za/pebble.asp?relid=2475, [accessed 11 March 2011].
Pleshek J.(2011) 'With privacy issues waning, RFID begins ramp up' Obtained through the internet: http://wistechnology.com/articles/8261/, [accessed 11 March 2011].
Reinhardt, S. (2010) Huge Demand in India for RFID Products'. Obtainedthrough the internet: http://rfidtechnews.wordpress.com/2010/03/07/huge-demand-in-india-for-rfid-products/, [accessed 03/12/ 2011].
Rogers, E. (1995) *Diffusion ofInnovations* (4th ed.), New York: TheFree Press.
Rossi, A., Sommerville, C. and Brown,O. (2003) 'RFID, The Growing Technology'. Obtained through the internet: http://www.personal.psu.edu/users/o/f/ob101/conclusion.htm, [accessed14/8/2004].
Ryan, B. and Gross, N. (1943) 'The Diffusion of Hybrid Seed Corn in TwoIowa Communities', *Rural* Sociology, Vol. 8, pp.15-24.
Shankland, S. (2002) 'Digital Dog Tags: Would you wear one?' Obtained through the internet: http://news.com.com/2100-1001833379.html?tag=nl, [accessed 17 June2004].
Spivey-Overby, C. (2004) 'RFID at what Cost? What Wal-Mart Compliancereally means', Forr Tel (webcast plustelephone), Forrester Research, 25 May.
Thomas, W. and Znaniecki, F. (1927)*The Polish Peasant in Europe* andAmerica, New York: Knopf.
Turban, E, King, D., Lee, J., Warkentin, M., Chung, H. and Chung, M. (2002) *Electronic Commerce: A Managerial Perspective*, Upper Saddle River: Prentice-Hall.
Turban, E., King, D., Viehland, D. andLee, J. (2006) *Electronic Commerce: AManagerial Perspective*, (Revisededition), Upper Saddle River: Prentice Hall.
Walters, K. (2005) 'Beyond the Barcode', *Business Review Weekly (Australia)*, 14-20 April, p.53.
West, C. (2001) *Techno-Human Mesh: The Growing Power of Information Technologies*, Westport:Quorum West.
Wired (2004) 'American Passports to get Chipped', 21 October.Obtained through the internet: http://www.wired.com/news/0,1294.6512.00.html, [accessed 22/10/2004].
Young, L. (2008) 'RFID: The Dialogue Continues', 01 October.Obtained through the internet: http://www.aimglobal.org/members/news/templates/template.aspx?articleid=3339&zoneid=24. [Accessed 11 March 2011].

3

The Role of RFID Technology in Supply Chain Risk Management

May Tajima
The University of Western Ontario
Canada

1. Introduction

Supply chain risks come in a variety of forms: disruptions to material flows, product quality problems, information systems breakdowns, and economic instability (Chopra & Sodhi, 2004; Zsidisin et al., 2000). The recent literature in supply chain management recognizes the importance of managing such risks in the age of global supply chains. Various researchers have discussed firms' increasing exposure to risks and the resulting, potentially severe negative impact on the firms' financial performances (e.g., Hendricks & Singhal, 2005).

One such risk to the supply chain, disruption of supply flows, can occur suddenly due to a number of unpredictable events. Even more unpredictable, however, is the ripple effect caused by the disruption. For example, the September 11th terrorist attacks of 2001 in New York and Washington, D.C., originally disrupted many supply chains on the United States (U.S.) East Coast, one of which was the Ford Motor Company's parts supply chain. The disruption eventually forced not one but five of Ford's assembly plants to cease production within a week of the incident (Zakaria, 2001). While Ford was experiencing parts shortages, Quanta Computer, a Taiwanese contract manufacturer for Dell and others, faced a pile-up of finished products when the U.S. airspace closed due to the attacks (Einhorn, 2001). One logistics service company in Europe estimated that the attacks cost the company £5 million (Parker, 2002). In this example, the ripple effects were extensive, affecting businesses in North America, Asia, and Europe. This high degree of impact clearly illustrates the importance of managing ripple effects as a part of supply chain risk management. In the first of two parts, this research shows that Radio Frequency Identification (RFID) technology, a relatively new development in supply chain management, holds great promise for managing supply disruptions and for containing their harmful ripple effects.

RFID — a wireless technology that uses transmitted radio signals to tag an item in order to track and trace its movement without human intervention — has superior capabilities over bar codes and promises many supply chain benefits, such as reductions in shrinkage, efficient handling of materials, increased product availability, and improved asset management (Angeles, 2005; Li & Visich, 2006; Taghaboni-Dutta & Velthouse, 2006). RFID has many applications in retail, healthcare, logistics, records management, and more, but so far its use in risk management has not been explored in the literature. To fill that gap, this research first addresses the following question:

Is RFID applicable in supply chain risk management; in particular, how is it useful for managing supply disruptions?

Based on RFID's technological capabilities, this research identifies three areas in which this technology could be utilized in the management of supply disruption risk: (i) monitoring for a disruption, (ii) responsiveness to the disruption, and (iii) the quality of decision-making involved in choosing corrective actions. Each of these three areas is discussed with a particular focus on how RFID could help to reduce the harmful ripple effects that are generated from supply disruptions. In order to provide support for these uses of RFID in risk management, this research presents case studies that originated from newspaper, magazine, and journal articles.

The discussion on RFID's risk management capabilities considers RFID as a source of advantages for firms that adopt the technology. However, the unprecedented level of supply chain visibility that is possible by the use of RFID can also be a source of risk. The literature has identified a number of concerns about this high degree of RFID-enabled visibility into supply chain activities. The concerns include consumer privacy invasion, corporate system security concerns, and industrial espionage (e.g., Juels, 2006; Shih et al., 2005). The second question in this research draws its motivation from the need to look at the other side of the same coin in order to gain a full understanding of RFID technology within the context of supply chain risk management:

What are the specific risks associated with RFID-enabled supply chain visibility, and how can these risks be mitigated?

The concerns associated with RFID's capability to provide supply chain visibility represent a timely and important research topic because similar concerns have been raised for other technologies that are capable of collecting, storing, and accessing huge amounts of data on individual items or people. For example, the Quit Facebook Day event in 2010 demonstrated Facebook members' concern for the privacy breach by the world's largest social networking website (CNN, 2010), which is capable of generating an unprecedented level of visibility into personal relationships. In the second part of this research, a review of the literature is conducted to identify specific risks associated with RFID's capability to provide supply chain visibility, and the research goes on to examine the existing mitigation approaches for dealing with RFID's visibility-related risks. Finally, the management implications are provided for the use of RFID in supply chain risk management based on both advantages and risks of its use.

The remainder of this chapter is organized as follows. Section 2 provides a review of the background literature. Section 3 presents the first part of this research, which focuses on RFID as a source of advantages in supply chain risk management, and Section 4 presents the second part, which focuses on RFID as a source of risks. Section 5 concludes the chapter with a summary of research contributions, limitations, and directions for future research.

2. Background

This section provides background information for this research. Two areas of the literature are particularly relevant: Section 2.1 reviews the capabilities and applications of RFID technology, and Section 2.2 reviews those risk management elements that are associated with supply disruptions.

2.1 RFID capabilities

RFID is an automatic identification technology that identifies specific items and gathers data on them without human intervention or data entry (Wyld, 2006). Item identification occurs when a reader scans an RFID tag that is tuned to the same frequency as that of the reader. Fundamentally, RFID technology can be summarized by the following characteristics: (a) RFID is wireless, (b) it provides unique identification to an object, and (c) it traces and tracks objects (Kärkkäinen & Holmström, 2002). Each of these fundamental characteristics leads to an advantage over the existing bar code technology and allows RFID to possess three distinct capabilities: (i) advanced process automation, (ii) closed-loop tracking, and (iii) supply chain visibility (Tajima, 2007). These capabilities and their related applications are discussed in turn, below.

First, RFID's wireless characteristic eliminates the need for product positioning that is associated with bar-code scanning. This allows for the contents of mixed pallets to be identified simultaneously without undoing the packaging. Hence, compared to bar codes, RFID can support a higher degree of automated material inspection and handling (McFarlane & Sheffi, 2003). This process-automation capability provides many benefits in the management of warehouses and logistics by reducing material handling time and human errors in operations, such as receiving, inventory counting, data entry, put-away, routing for cross-docking, and custom clearance for cross-border shipments (Rutner et al., 2004; Zebra Technologies, 2004).

Second, RFID's ability to provide a unique identifier to an object comes from the fact that an RFID tag has a higher data capacity than does a bar code. This higher data capacity provides RFID with advanced record keeping and retrieval capability, through which RFID enables closed-loop tracking of individual items and assets, an action that is not possible with bar codes, which refer only to a class of products (Wyld, 2006). Recently, a wide range of applications has been identified for RFID's closed-loop tracking, including the tracking of medical devices within a hospital; paper documents within a law firm; gaming chips in casinos; media players for rental cars; and flower-growing operations from seeds to blooms (RFID Update, 2006c, 2007a, 2007b, 2007d, 2008b).

Third, RFID's ability to track and trace objects provides supply chain-wide, real-time visibility of individual items. When combined with other real-time locating technologies, such as Global Positioning Systems (GPS), RFID can be used to capture product information such as a detailed description of the product, its manufacture and expiration dates, the time of its departure and arrival at various facilities, and the address and telephone number of its manufacturer (EPCglobal, 2004). RFID-generated product information can provide an unprecedented level of visibility in the supply chain when shared among supply chain partners, a level of visibility that is simply not obtainable from bar codes. In the retail industry, where inaccuracy of inventory data is a major problem (Raman et al., 2001), one of the major applications of RFID is to improve inventory visibility. RFID can also increase the visibility into shipment data, which can in turn improve demand visibility (Lapide, 2004; McCrea, 2005). Automatic replenishment using "smart shelves" is another application in the retail industry and is considered valuable by, for example, German retailer METRO and Finnish apparel manufacturer NP Collection (RFID Update, 2007c, 2007e). For the pharmaceutical industry, the degree of supply chain visibility provided by RFID is considered critical for anti-counterfeiting measures and product recall management (Wicks et al., 2006; Wyld & Jones, 2007).

As shown above, RFID technology has applications in a wide range of industries and settings, but it has not yet found a place in the area of risk management.

2.2 Supply chain risk management

As mentioned in the Introduction, supply chain risks come in a variety of forms. To limit the scope of discussion, however, this research focuses solely on supply disruptions. In this research, supply disruptions are, as defined in Craighead et al. (2007), the disruptions of the normal flows of goods and materials within a supply chain that are caused by unplanned and unanticipated events. These disrupting events come in the various forms, such as natural disasters, labor disputes, wars, power failures, supplier contract breaches, and infectious diseases (Chopra & Sodhi, 2004; Haksöz & Kadam, 2009; Tang, 2006). For the purpose of this research, a ripple effect of a disruption is defined as any other supply disruptions that occur at different locations and/or at later dates due to the original disruption.

Typical risk management consists of four elements: (i) risk source/driver identification, (ii) risk consequence and likelihood assessment, (iii) risk mitigation and treatment, and (iv) risk monitoring. For risk source identification, Helferich (2002) indicated that supply disruptions could occur from interruptions in production facilities, supplier networks, transportation networks, communication infrastructure, and electricity and water services. Global sourcing is particularly vulnerable to supply disruptions because it generally involves greater distance, longer transit time, limited transportation mode, and complex security protocols for border crossings (Prater et al., 2001; Zsidisin, 2003). The just-in-time system is also susceptible to supply disruptions because it operates under fast-cycle procurement and lean inventory (Aichlmayr, 2001).

For risk assessment, Haksöz and Kadam (2009) studied ways to assess the supply disruption risk that results from supplier contract breaches. In their study, a tool to assess the financial impact of contract breaches was developed.

Risk mitigation focuses on ways to avoid, reduce, eliminate, buffer, or hedge against risk. A variety of operational strategies for mitigating supply disruptions have been examined in the literature. Chopra and Sodhi (2004) discussed having redundant suppliers, adding capacity, and increasing responsiveness as possible mitigation strategies. Sheffi (2001) proposed a multiple sourcing strategy that allocates the bulk of the procurement volume to inexpensive offshore suppliers but also gives a fraction of the business to local suppliers as insurance against supply disruption. Prater et al. (2001) identified a number of advantages in using local logistics operators, such as their knowledge of regional transportation routes and their familiarity with the border-crossing procedures. Babich et al. (2007) studied a hedging strategy based on the pricing and ordering policies of multiple suppliers.

Some authors studied inventory-related strategies for mitigating supply disruptions. For example, Sheffi (2001) discussed the emergency designation for safety stock in order to discourage its use for day-to-day fluctuations. Martha and Subbakrishna (2001) suggested increasing safety stock for critical items only, such as those coming from a single international source or those whose shortage quickly leads to plant shutdowns. When transfer or production of goods is not possible within a reasonable time frame, a marketing strategy may be used to steer customers toward substitutes. This strategy was exercised by Dell in response to the September 11th terrorist attacks (Rocks, 2001): Dell salespeople searched online to see which configurations of computers were available and then steered customers accordingly. Finally, Craighead et al. (2007) identified two key capabilities for mitigating supply disruptions: the capability to detect and disseminate information pertaining to the disruptive event, and the capability to respond quickly and effectively to the disruption.

The next section, which presents the first of two parts in this research, highlights RFID's usefulness in supply chain risk management by demonstrating that RFID can improve some of the risk-mitigation strategies mentioned above.

3. RFID as a source of advantages

Is RFID applicable in supply chain risk management, and in particular, how is it useful for managing supply disruptions? This section addresses this research question by showing that an understanding of RFID's technological capabilities can lead to the discovery of RFID's risk management capabilities.

3.1 Methodology

First, it is shown that RFID's technological capabilities of closed-loop tracking, process automation, and supply chain visibility yield three specific risk management capabilities: increased monitoring capacity, increased response speed, and higher decision-making quality. Then, case studies are presented for all three risk management capabilities in order to provide support for their validity. As shown in Section 2.1, RFID application that is specific to the area of risk management has not yet been explored. Hence, any related case studies that could highlight the potential use of RFID in risk management were searched from newspaper, magazine, and journal articles. Below, RFID's three risk management capabilities are discussed in turn.

3.2 Monitoring capacity

Risk monitoring, as discussed in Section 2.2, is one of the typical elements in risk management, and it plays an important role in the management of unexpected supply disruptions. With an ability to monitor for and detect a disruption as it happens, corrective actions can begin sooner, the escalation of the disruption can be avoided, and the impact of the disruption, direct or indirect, can be reduced. Craighead et al. (2007) identified risk monitoring as one of the key capabilities needed for mitigating supply disruptions. It is shown below that closed-loop tracking, one of RFID's technological capabilities, can increase a firm's risk monitoring capacity.

As discussed in Section 2.1, the data capacity of RFID tags is higher than that of bar codes, and this higher data capacity allows for the closed-loop tracking of individual items and assets. RFID can be used to monitor not only cases and pallets but also individual raw materials, work-in-process inventories, and finished products. It can also monitor the use and condition of equipment and reusable assets. Therefore, with its closed-loop tracking capability, RFID can increase a firm's monitoring capacity by increasing the level of details that can be monitored.

The following case studies provide support for RFID's ability to increase a firm's monitoring capacity. At Nestlé, a large global food company, RFID was used to track the cleanliness of product trays (Bear, Stearns & Co. Inc., 2003). Such RFID-enabled tracking of reusable assets would extend Nestlé's capacity for detecting poor product quality to include the work-in-process items in addition to finished products. At the Wynn Hotel and Casino in Las Vegas, poker chips imprinted with RFID were used to monitor game play for possible cheating or gambling addiction (Wyld, 2008). In this case, RFID would increase the casino's capacity to detect problematic gaming behavior from the table/station level to the individual player.

RFID's closed-loop tracking capability can also increase a firm's risk monitoring capacity by providing the firm with an ability to monitor huge volumes of assets. RFID has already successfully managed a variety of assets with huge volumes. For example, a casino tracked 80,000 uniforms through the laundry process, and a beer company tracked three million beer kegs using RFID (Bear, Stearns & Co. Inc., 2003; Byrne, 2004). Byblos Amoreiras, a Portuguese book retailer, used RFID to track 150,000 books, periodicals, CDs, and other merchandise in its store (RFID Update, 2008a).

By increasing a firm's risk monitoring capacity, RFID can assist the firm with the identification of critical items. In Section 2.2, some risk mitigation strategies, such as an increase of safety stock and multiple sourcing, were discussed for critical items (Martha & Subbakrishna, 2001). With the use of RFID, firms could quickly identify critical items, such as those that run out first or those whose shortage causes a plant shutdown. Also, RFID's capability to monitor huge volumes of items can assist firms with the collection of historical data at the individual item level. These data would be useful in improving a firm's risk assessment in terms of estimating and updating the severity and likelihood of various supply disruptions.

3.3 Response speed

The previous section focused on RFID's ability to detect a disruption at the level of individual items and assets. Once detected, a firm's ability to respond quickly to the disruption becomes important in the containment of the ripple effects. Responsiveness has been identified as one of the key capabilities for managing supply disruptions in the literature (Chopra & Sodhi, 2004; Craighead et al., 2007). Another of RFID's technological capabilities, process automation, can increase a firm's response speed.

As discussed in Section 2.1, when compared to bar code technology, RFID's wireless characteristic allows for a higher degree of automation in the processes, such as material inspection and handling (McFarlane & Sheffi, 2003). For LCWaikiki, one of the largest apparel retailers in Turkey, the replacement of bar codes with RFID technology has resulted in the merchandise transfer from the back room to the shop floor being performed 70% faster and the merchandise receiving being performed 60% faster (RFID Update, 2008d). For Bloomingdale's, a large U.S. department store, an RFID pilot study resulted in a 96% reduction of cycle counting time for the store's inventories (RFID Update, 2009). For American Apparel, a large U.S. clothing manufacturer and retailer, with the use of RFID, the time required for store-level inventory count dropped from 120 work-hours to 15 work-hours (Avery Dennison Corporation, 2010). These case studies support RFID's ability to speed up some of the common responses to a supply disruption, such as recounting inventories, adjusting shipment data, and sending invoice reconciliations.

In reality, the response to a supply disruption cannot begin until key personnel within a firm are notified of the disruption. Once notified, these individuals can then authorize the start of corrective actions. In this leg of the process, RFID's process automation capability can increase a firm's response speed by facilitating the real-time alert for notifying key personnel in the event of a supply disruption. Throttleman, a Portuguese fashion retailer, has set up a real-time alert system using RFID in its distribution center (RFID Journal, 2007). Upon arrival at the distribution center, the contents of a box are automatically identified using RFID without opening the box. The captured contents are then compared to the items listed in an advance shipping notice that has been electronically sent by the garment manufacturer. If the received contents do not match with the advance shipping notice, then an alarm goes off for the center's personnel to physically deal with the discrepancy. In another instance, a real-time alert system has been implemented at several U.S. hospitals to notify staff immediately when a piece of equipment becomes misplaced (Emrich, 2008). Also, at Lincoln University, a Pennsylvania liberal arts college, valuable audio-visual equipment was tracked using RFID, and an alert notified the IT department as soon as a piece of equipment left its predetermined zone (RFID News, 2008). These case studies support RFID's ability to increase a firm's response speed by setting up a real-time personnel alert system.

3.4 Decision-making quality

Upon notification of a supply disruption, key personnel need to assess the extent of the disruption and decide on the appropriate risk mitigation strategies before corrective actions can actually begin. The quality of these strategic decisions can have a significant impact on the outcome of the corrective actions. For example, Hurricane Mitch in 1998 caused a supply interruption for two major banana producers in Central America, Dole and Chiquita (Martha & Subbakrishna, 2001). Dole's business suffered from this supply disruption, with the subsequent ripple effect lasting longer than a year. Chiquita, on the other hand, had a significantly different outcome: it was able to arrange alternative supply sources, and its revenue actually grew during the last quarter of 1998.

In general, management decisions are often made based on incomplete or old data (Lin et al., 2006). Therefore, an overall increase in information accessibility has significant potential to improve the quality of management decisions, including the ones that must be made in response to supply disruptions. As discussed in Section 2.1, compared to bar codes, RFID promises an unprecedented visibility into supply chain operations. Supply chain-wide visibility provides information such as inventory levels, shipment data, locations of stockpiles, and alternative suppliers throughout the extended enterprise. Such information is critical in providing a firm with the ability to redirect its inventories within its supply chain, or to steer customers toward substitute products based on informed decision-making, rather than based on incomplete or untimely data. Hence, the third of RFID's technological capabilities, supply chain-wide visibility, can improve the quality of a firm's decision-making in the selection of risk mitigation strategies by increasing the completeness and timeliness of information available for the decision-makers.

Several case studies provide support for RFID's ability to improve the quality of a firm's risk mitigation decisions. In the retail industry, electronic article surveillance (EAS) devices provide retailers with the knowledge of timing when something is stolen from a store, but it cannot reveal which item has been taken (RFID Journal, 2009). RFID, on the other hand, can provide the retailers with more complete information. At Sony Europe, a combination of RFID, EAS, and a video surveillance system was implemented in its largest European distribution warehouse, located in the Netherlands (RFID Journal, 2009). The system was designed to deter employee or professional theft by giving Sony as much information as possible on each theft: which item is stolen, when it is stolen, and who may be doing the stealing. In another case, the additional data obtained through RFID allowed one retailer to successfully link multiple thefts over a period of time to a single person (Arnstein, 2010). Moreover, RFID-generated data can support a targeted and cost-effective security strategy that provides different security levels for different products within the same store, such as a silent alarm for expensive items and an audible alarm for inexpensive ones (Arnstein, 2010). These case studies support RFID's ability to improve the quality of a firm's risk mitigation decisions by increasing the completeness of the information available for the decision-makers.

As discussed in Section 2.1, when combined with other real-time locating technologies, such as GPS, RFID is capable of capturing product information within a supply chain on a real-time basis. Hence, in addition to the increase in completeness of information, RFID-enabled supply chain visibility can increase the timeliness of information available for the decision-makers in a firm. For example, through the use of RFID, Dole Food Company, the world's largest producer and marketer of fresh fruits and vegetables, was able to initiate a voluntary, pre-emptive recall of packaged salads that were suspected of E. coli bacteria contamination *before* any consumers were reported ill (Uldrich, 2007). When the recall announcement was made, Dole also knew that a total of 5,058 bags of salad were most likely to have been exposed to the bacteria, of which 528 bags were distributed in Canada and 4,530 bags were distributed within eight U.S. states. The value of RFID in providing timely information was also discussed in a simulation study conducted by Kim et al. (2010). Their study showed that an RFID-based, vehicle-tracking system could significantly decrease the overall transfer time of finished vehicles from an automobile assembly plant to its shipment yard by providing the real-time availability of parking spots. The yard operators were then able to use real-time information to make their decisions more efficiently and effectively. Without RFID, the status of parking availability could be updated only periodically through a manual reporting process, and therefore, the yard operators had to make their decisions based on untimely data.

All the related case studies presented above support RFID's ability to improve a firm's risk management capabilities in terms of its monitoring capacity, response speed, and decision-making quality. As a result, this section clearly demonstrates RFID's applicability in supply chain risk management and its usefulness in managing supply disruptions.

4. RFID as a source of risks

The previous section focused on RFID as a source of advantages for firms that adopt the technology and use it for supply chain risk management. However, the use of RFID can also be a source of various risks in and of itself. Within the literature, numerous articles have discussed RFID-enabled supply chain visibility as a source of security and privacy risks. The concerns surrounding these risks are a timely and important research topic, from both industry and society perspectives.

From a society perspective, a general feeling of anxiety exists toward technology-enabled information visibility. As mentioned in the Introduction, the Quit Facebook Day event in 2010 stands out as a high-profile example of people's concerns for the breach of users' personal information. Another example concerns Google's Street View, which provides panoramic views of streets all over the world, as captured by a fleet of vehicles that are equipped with high-tech cameras and scanners. The main concern for this technology stems from the fact that these panoramic images are publicly accessible from the Google website and may contain personally identifiable details, such as people's faces, belongings, and cars on the driveways with visible license plate numbers, all matched to easily identifiable street addresses (Bradley, 2010). Digital medical records represent another high-profile example of the public's concerns for technology-enabled information visibility. On one hand, electronically accessible medical records offer many benefits, including the reduction of duplicate diagnostic testings and medical errors. On the other hand, medical records consist of highly personal information, from prescription records to family-health histories, and may even include DNA information in the future (Fox News, 2010). A concern exists among physicians in terms of how to protect such private, sensitive, and massive data from potential hacking. As can be seen from these examples, the concerns for technology-enabled information visibility are not unique to RFID, and at present, no absolute solutions or countermeasures exist to deal with these concerns.

From an industry perspective, the risks associated with RFID-enabled supply chain visibility constitute a timely and important research topic for RFID vendors, potential users, and corporate and public policy-makers mainly because RFID is still a developing technology (RFID Update, 2008c), whose industry adoption may easily be hindered by any risk related to its use. With other established supply chain technologies, such as bar codes, electronic data interchange (EDI), and enterprise resource planning (ERP), information sharing and the resulting visibility have not posed any serious issues since the scope of visibility has rarely been extended to involve individual items or consumers. Consequently, within the context of supply chain technologies, RFID has no obvious precedence to follow regarding how to deal with the risks related to information visibility, and this makes an understanding of these risks critical for RFID technology's future growth.

What, then, are the specific risks associated with RFID-enabled supply chain visibility, and how can these risks be mitigated? The remainder of this chapter focuses on addressing this research question.

4.1 Methodology

A review of published literature is provided on possible risks associated with RFID-enabled supply chain visibility. Two databases, ProQuest and Scholars Portal, were used to search relevant articles. The chosen search terms utilized various combinations of: RFID, risk, security, and privacy. The search dates were restricted to the years between 2003 and 2010. Due to insufficient resources for translation, the review was also restricted to English-language articles only. The search produced over 100 articles, covering more than 50 different journals from a variety of disciplines, such as business, engineering, information systems, economics, law, electronic commerce, marketing, production, and healthcare. Therefore, although the search was not exhaustive, the search range was considered sufficiently comprehensive in terms of the variety of articles, and further searches from other databases were deemed unnecessary.

Based on the articles found in the search described above, Section 4.2 provides the first result: an overview of two main categories of RFID's supply chain visibility risks, which are security risks and privacy risks. Section 4.3 then provides the second result: a classification of the existing mitigation approaches for dealing with RFID's supply chain visibility risks. Finally, Section 4.4 discusses the management implications of the use of RFID in supply chain risk management based on its advantages as well as risks.

4.2 Supply chain visibility risks

In the RFID literature, a variety of risks associated with the use of RFID have been discussed. For example, a risk to patients' health that might result from the altering of the chemical composition of a medication was discussed in the context of using RFID for pharmaceutical products (Symbol Technologies, 2006). A risk to the environment was discussed in relation to the disposal of non-biodegradable RFID tags (Li & Visich, 2006). A risk to the corporate information system was also discussed in terms of the vulnerability of RFID to computer viruses (RFID Update, 2006b). This section, however, focuses on providing an overview of the risks that are specifically associated with RFID-enabled supply chain visibility: security risks and privacy risks.

Security risks. Security risks in the RFID literature are often discussed as attacks against organizations by their competitors, opponents, or criminals. Various types of attacks are possible with the use of RFID. One type is referred to as "data eavesdropping," which is the interception of communications between RFID tags and readers. Through data eavesdropping, a military security breach may occur if enemy forces detect troop locations and monitor their movements by tracking RFID tags within a military supply chain (Juels, 2006; Zuo, 2010). Corporate espionage is also possible through data eavesdropping. For example, by tracking RFID tags through the retail supply chain, competitors may spy on one retailer's sensitive business data, such as sales trends, pricing trends, stock selections, and stock turnover rates (Juels, 2006; Li & Visich, 2006; Shih et al., 2005). A seller organization may also attempt to gain visibility into the downstream of the supply chain by monitoring RFID tags on the sold items after the seller no longer has physical access to the items (Kapoor et al., 2009).

Another type of attack against organizations is referred to as "data corruption," which erases or modifies RFID tag contents. If the tag contents include price information, then, through data corruption, hackers could lower the price of expensive retail items, and then use an RFID-enabled self-checkout counter to avoid detection by store employees (Li & Visich, 2006). Spoofing, another type of attack, involves the retrieval of confidential information by impersonating authentic readers (Shih et al., 2005). Spoofing can lead to, for example, counterfeiting of retail products by falsely authenticating fake products using stolen authentication information. Finally, denial of service is a type of attack that renders RFID tags temporarily or permanently incapacitated (Zuo, 2010). Denial of service can cause a loss of business data and operational disruptions to an organization.

Privacy risks. While security risks typically affect organizations and result in financial losses, privacy risks affect individuals and result in ethical issues. The literature discusses three main issues that are specifically related to RFID's ability to provide supply chain visibility that includes end-consumer information.

The first issue relates to the collection of personal data without an individual's knowledge or consent. This concern stems from the fact that the size of RFID tags can be as small as grains of sand, making it possible to inconspicuously attach the tags on products. Also, the scanning of RFID tags is a wireless process that cannot be detected by human eyes or ears. Hence, a retailer is technically able to conduct market research, for example, by tracking RFID tags on pre-sale items inside the store without the knowledge or consent of the consumers (Jones et al., 2004).

The second ethical issue relates to the infringement on individual anonymity. In the context of supply chain management, RFID tags are traditionally associated with product information but not with consumer information. However, since an RFID tag is capable of providing a unique identifier to a product, any association between the product and an individual can in turn become the unique identifier of the individual. For example, a female customer with a previously purchased item carried in a purse can be identified as a returning customer if the tag on the item is read upon her return to the store. Even if the retailer does not possess full information on her identity, the anonymity of this customer can still be infringed upon since it is possible to build a personal profile based on information such as the frequency of the store visit, the time and day of the visit, and the history of other purchases made by this customer (Wasieleski & Gal-Or, 2008). Such consumer profiles could then be exploited for price differentiation strategies or could be sold to third parties (Jones et al., 2004; Peslak, 2005). Moreover, if the item in question was, for example, a prescription drug bottle being carried by an individual, then the product information itself could represent a piece of sensitive personal data.

The third ethical issue, the surveillance of individuals, also stems from the association made between a product and an individual by the use of RFID. By tracking RFID tags on products that are owned by individuals, people may be tracked in the stores, on the streets, and even in people's homes (Jones et al., 2004; Rutner et al., 2004). This issue is often discussed in relation to the idea of "Big Brother," where the authority monitors civilians' every move.

All the ethical issues discussed above have been fueling an opposition to RFID from various consumer advocacy groups, such as Consumers Against Supermarket Privacy Invasion and Numbering (CASPIAN), the American Civil Liberties Union, the Privacy Rights Clearinghouse (PRC), all of which are generally against the use of RFID (Barut et al., 2006; Jones et al., 2004).

4.3 Risk mitigation approaches

The actual occurrence of security and privacy risks is still not common due to the limited and fragmented use of RFID, and the literature discusses these issues as potential risks of RFID. However, as mentioned previously, any risk related to RFID's use may negatively affect the growth of this technology. How, then, can these risks be mitigated? This section presents a classification of the existing mitigation approaches for dealing with RFID's supply chain visibility risks. An examination of the RFID literature revealed that four general treatments currently exist: technology-based countermeasures, business policies, consumer education, and legal measures. Each of these approaches is discussed below.

Technology-based countermeasures. In order to reduce the likelihood of the occurrence of security and privacy risks, some technology-based countermeasures have been proposed in the literature. One group of countermeasures is designed to protect RFID tags from unauthorized scanning, including: tag killing, to make tags permanently inoperative when the tags receive a "kill" command from a reader; tag sleeping, to make tags temporary inactive unless "woken" by authorized users; tag relabeling, to give tags different identifiers periodically; tag encryption, to use cryptography to encrypt tag data or identifiers; and hash locks, to make tags respond to data queries with only limited information when "locked" (e.g., Juels, 2006; Shih et al., 2005; Zuo, 2010). By protecting the tags, these countermeasures are intended to prevent security attacks, such as data corruption, spoofing, and denial of service; they can also prevent the surveillance of individuals.

Another group of technology-based countermeasures is designed to protect communications between RFID tags and readers. Most of these countermeasures are based on developing protocols for the search and authentication procedures that occur between the tags and readers (Zuo, 2010). Another approach to protect the tag-reader communication is to limit the tag-reading area, which can be accomplished by tag clipping, a process that shortens the antenna in a tag to reduce its read range (Kapoor et al., 2009), or by shielding the tag-reading area with metal screens to prevent the unauthorized scanning from outside (Swartz, 2007). The protection of communications between tags and readers would certainly be useful for the prevention of attacks that involve data eavesdropping and spoofing.

Another group of technology-based countermeasures focuses on informing individuals about unauthorized scanning. For example, a watchdog tag is supposed to be carried by an individual to monitor for any unsolicited scannings against the individual (Juels, 2006). A read-write tag, also carried by an individual, keeps a log of unauthorized scannings (Li & Visich, 2006), and a blocking tag on an individual is supposed to block any unsolicited scannings (Juels, 2006). By alerting individuals about the practice of unauthorized scanning, these countermeasures give consumers an opportunity to, for example, look for a suspicious reader or walk away from the area in question, therefore, helping to alleviate the risks, such as data eavesdropping, collection of personal data without the individuals' knowledge or consent, and the surveillance of individuals.

Business policies. The use of business policies offers another general approach for dealing with RFID's supply chain visibility risks. While technology-based countermeasures address both security and privacy risks, the business policies discussed in the literature focus on dealing with privacy risks. Common policies on the use of RFID include making RFID tags clearly visible to consumers, making tags easily removable by placing them on the product packaging or on price tags, and disabling tags at the point of product purchase (Jones et al., 2004; Li & Visich, 2006; Taghaboni-Dutta & Velthouse, 2006). As an example, H.D. Smith, a major pharmaceutical product distributor, officially requested that its customers (i.e., pharmacies and hospitals) remove RFID tags upon receipt of shipments from H.D. Smith (Downey, 2006). Business policies may also include statements about a firm's data-collection practices. For example, Pfizer, a large pharmaceutical manufacturer, made a public statement that it would not collect any patient information using RFID (RFID Update, 2006a). Such policy statements are used by firms mainly to reassure consumers that the subject firms intend to use RFID responsibly, thereby minimizing the privacy risks associated with its use.

Consumer education. Another general approach for dealing with RFID's supply chain visibility risks comes in the form of consumer education. As with the business policy approach, consumer education focuses on dealing with privacy risks. In 2004, a survey of 1,000 North American consumers showed that only one in four knew what RFID was (Jones et al., 2004). A study on the consumer attitude toward RFID revealed that, the less consumers were educated about RFID, the more hesitant they were about the use of RFID in businesses (Razzouk et al., 2008). In light of a general lack of consumer understanding regarding RFID, an industry group, the Association for Automatic Identification and Mobility (AIM), has recognized the need to convey accurate information about the technology to the community (Peslak, 2005). Information on RFID's technical capabilities, as well as its limitations and comparisons of RFID to other wireless technologies, could educate consumers on the likelihood of privacy risks. Moreover, consumers could be further educated on the likelihood of privacy risks through access to information on whether an RFID tag is embedded in a product, when the tag is read, and whether the tag is removed or deactivated upon purchase (Pottie, 2004).

Legal measures. The final approach that is discussed in the literature as a way to deal with RFID's supply chain visibility risks is the legal approach. This approach mainly focuses on the protection of information privacy, which is the right of an individual to retain control over the collection and use of personally identifiable information (Kelly & Erickson, 2005). In 2004, California passed a bill prohibiting the use of RFID to collect, store, use, or share personal information unless certain legal conditions were met (Taghaboni-Dutta & Velthouse, 2006). A proposal to extend the Fair Information Practices, originally promoted by the Federal Trade Commission in the U.S. to protect online privacy, has also been put forth for the use of RFID (Peslak, 2005). Fair Information Practices include business practices such as notifying consumers of the collection of personal information; giving consumers options concerning how information is used; giving consumers access to the collected information; providing security over the collected data; and providing penalties for non-compliance. The European Union does not have RFID-specific regulations. However, its existing regulations – the Data Protection Directive of 1995, the Electronic Commerce Directive of 2000, and the Privacy and Electronic Communications Directive of 2002 – do apply to the personal data collected by the use of RFID (Slettemeås, 2009). By establishing and enforcing the laws on RFID-generated data, the legal approach is intended to deter the occurrence of security and privacy risks and to provide individuals with a means of recourse in the case of a privacy breach.

4.4 Management implications

The first part of this two-part research demonstrated that RFID could be a source of tremendous advantage for firms that adopt the technology and use it for managing supply disruptions. The second part of this research, however, showed that RFID could also be a source of security and privacy risks, and attacks on the RFID system, such as data corruption and denial of service, may actually cause supply disruptions through a loss of business data or the disruptions to internal operations. Hence, from a management perspective, the use of RFID in supply chain risk management requires careful consideration of the risks in and of RFID itself. The security attacks in general may be alleviated by the use of technology-based countermeasures. However, the management must be aware that various shortcomings have been documented for these countermeasures. For example, tag killing can eliminate the security risks, but it also eliminates many post-sale benefits of having RFID tags on products, such as efficient warranty processing, easy handling of returns, and goods authentication (Juels, 2006). The use of cryptography may be computationally infeasible when RFID tags are employed on a mass scale (Zuo, 2010), and a security protocol is secure only until its loopholes are discovered (Kapoor et al., 2009).

Privacy risks associated with the use of RFID may not cause supply disruptions, but they may negatively influence the consumer attitude towards RFID (e.g., Slettemeås, 2009), thereby hindering the growth of RFID adoption. Hence, the management needs to address the privacy risks in general, but it must be aware that the mitigation of the privacy risks is a complex subject that requires a multi-faceted solution, for none of the existing mitigation approaches provides an all-encompassing solution on its own. Technology-based countermeasures, as mentioned above, come with various shortcomings. Business policies and consumer education do not actually prevent the incidence of unauthorized scanning. With the legal approach, the burden is placed on the plaintiff to prove that a privacy breach has taken place and it resulted in a high degree of shame, humiliation, mental illness, and so on (Willey, 2007).

A further examination of the privacy risks reveals that, when a firm considers the use of RFID in supply chain risk management, the collection of personal data without an individual's knowledge or consent and the surveillance of individuals may not be the first risks that the firm needs to address since the data utilized in the context of risk management are mostly inventory, shipment, equipment, asset, and supplier data, but not consumer data. On the other hand, since a unique product identifier given by an RFID tag can turn into a unique personal identifier as discussed previously, the infringement on individual anonymity should be addressed whenever a firm uses item-level information. In terms of risk mitigation, all of the current approaches focus on when or how to stop the collection of

personal data, but none of them effectively address the infringement on individual anonymity since they do not focus on what to do with the data that are already collected.
Based on this research, two suggestions are made for mitigating the infringement on individual anonymity when firms consider the use of RFID for supply chain risk management. First, the firms can utilize the consumer education approach to clearly communicate specific benefits for consumers resulting from the better management of supply disruptions. Based on the three risk management capabilities of RFID discussed previously, the consumers can expect benefits such as fewer and shorter business disruptions experienced by the consumers and increased public safety in certain cases (e.g., food recalls). Second, in addition to the business policies on whether or not certain data will be collected, the firms should consider adding policies on how they intend to utilize the collected data in order to come across as the responsible users of RFID in the eyes of consumers. For example, a firm may state that it will collect item-level product data via RFID for the purpose of detecting and mitigating supply disruptions.

5. Conclusion

The first part of this research demonstrated that RFID's three risk management capabilities – monitoring capacity, response speed, and decision-making quality – were applicable and useful in the management of supply disruptions. The second part of this research showed that the security and privacy risks were associated with RFID-enabled supply chain visibility, and that four general mitigation approaches exist at present: technology-based countermeasures, business policies, consumer education, and legal measures. Together, the two parts of this research provided a comprehensive understanding of the use of RFID in the context of supply chain risk management.
The main limitation of this research is that some practical issues related to the use of RFID in a real-life setting were not included in the discussion. One such issue is that a firm's corporate information system may not be capable of supporting the increased monitoring capacity that is promised by the use of RFID. A general increase of data processing needs has been discussed as one of the challenges associated with RFID implementation (Angeles, 2005). Another issue arises with the design of a real-time alerting system. The alerting of top personnel should ideally be reserved for severe supply disruptions only. This implies that the alerting system must recognize different levels of the disruptions in order to alert different levels of personnel. However, a firm may not have sufficient data on the actual supply disruptions with varying degrees. Also, as RFID improves the completeness and timeliness of information available for the key decision-makers, information overload may become an issue in general. Having abundant information may lead to the generation of many options for the decision-makers to assess, and consequently, it may slow the response to a supply disruption. Future research must address such practical issues in order to make RFID-based risk management a reality.
In conclusion, this research provided valuable insight into a novel application of RFID technology in the area of supply chain risk management. This insight was built from the balanced understanding of RFID as a source of advantages as well as a source of risks.

6. Acknowledgment

The author wishes to acknowledge Lelanya Perryman for her assistance in this research as a research assistant, whose work was funded by the Start-Up Research Grant at The University of Western Ontario.

7. References

Aichlmayr, M. (2001). The future of JIT – time will tell. *Transportation and Distribution*, Vol. 42, No. 12, pp. 18-22

Angeles, R. (2005). RFID technologies: supply-chain applications and implementation issues. *Information Systems Management*, Vol. 22, No. 1, pp. 51-65

Arnstein, L. (2010). How to serve the multi-channel consumer. *Material Handling Management*, December 1

Avery Dennison Corporation. (2010). American Apparel RFID case study: the challenges. Date of access, March 21, 2011, Available from: www.ibmd.averydennison.com/solutions/american-apparel-rfid-challenges.asp

Babich, V., Burnetas, A., & Ritchken, P.H. (2007). Competition and diversification effects in supply chains with supplier default risk. *Manufacturing and Service Operations Management*, Vol. 9, No. 2, pp. 123-146

Barut, M., Brown, R., Freund, N., May, J., & Reinhart, E., (2006). RFID and corporate responsibility: hidden costs in RFID implementation. *Business and Society Review*, Vol. 111, No. 3, pp. 287-303

Bear, Stearns & Co. Inc. (2003). Supply-chain technology: track(ing) to the future. Equity research report, Date of access, September 10, 2006, Available from: www.bearstearns.com/bscportal/pdfs/research/supplychain/technology_rfid.pdf

Bradley, T. (2010). Google Street View raises privacy concerns again. In: *PCWorld*, Date of access, February 18, 2011, Available from: www.pcworld.com/printable/article/id,190279/printable.html

Byrne, P.M. (2004). RFID: not just for Wal-Mart anymore. In: *Logistics Management*, September 1, Date of access, September 10, 2006, Available from: www.logisticsmgmt.com/archive

Chopra, S., & Sodhi, M. (2004). Managing risk to avoid supply-chain breakdown. *MIT Sloan Management Review*, Vol. 46, No. 1, pp. 53-61

CNN. (2010). Some quitting Facebook as privacy concerns escalate, In: *CNN Tech News*, Date of access, February 18, 2011, Available from: http://articles.cnn.com/2010-05-13

Craighead, C.W., Blackhurst, J., Rungtusanatham, M.J., & Handfield, R.B. (2007). The severity of supply chain disruptions: design characteristics and mitigation capabilities. *Decision Sciences*, Vol. 38, No. 1, pp. 131-156

Downey, L. (2006). An Interview with RFID Trailblazer H.D. Smith. *RFID Update*, August 21, Date of access, March 21, 2011, Available from: www.rfidupdate.com/articles/home.php

Einhorn, B. (2001). Laptop king; in a year that's decimated high tech, Taiwan's unstoppable Quanta is posting double-digit sales growth. *Business Week*, No. 3756, p. 48

Emrich, A.B. (2008). Wireless RFID product tracks hospital assets. *Grand Rapids Business Journal*, Vol. 26, No. 50, p. 14

EPCglobal. (2004). The EPCglobal Network and the Global Data Synchronization Network (GDSN). Date of access, September 10, 2006, Available from: www.epcglobalinc.org/news/position_papers.html

Fox News. (2010). WikiLeaks breach raises concern about privacy of electronic medical records. In: *Fox News*, Date of access, February 18, 2011, Available from: www.foxnews.com/politics/2010/12/07

Haksöz, Ç., & Kadam, A. (2009). Supply portfolio risk. *The Journal of Operational Risk*, Vol. 4, No. 1, pp. 59-77

Helferich, O.K. (2002). Securing the supply chain against disaster. *Distribution Business Management Journal*, Vol. 2, No. 3, pp. 10-14

Hendricks, K.B., & Singhal, V.R. (2005). An empirical analysis of the effect of supply chain disruptions on long-run stock price performance an equity risk of the firm. *Production and Operations Management*, Vol. 14, No. 1, pp. 35-52

Jones, P., Clarke-Hill, C., Hillier, D., Shears, P., & Comfort, D. (2004). Radio frequency identification in retailing and privacy and public policy issues. *Management Research News*, Vol. 27, No. 8/9, pp. 46-56

Juels, A. (2006). RFID security and privacy: a research survey. *IEEE Journal of Selected Areas in Communications*, Vol. 24, No. 2, pp. 381-394

Kapoor, G., Zhou, W., & Piramuthu, S. (2009). Challenges associated with RFID tag implementations in supply chains. *European Journal of Information Systems*, Vol. 18, No. 6, pp. 526-533

Kärkkäinen, M., & Holmström, J. (2002). Wireless product identification: enabler for handling efficiency, customization and information sharing. *Supply Chain Management: An International Journal*, Vol. 7, No. 4, pp. 242-252

Kelly, E.P., & Erickson, G.S. (2005). RFID tags: commercial applications v. privacy rights. *Industrial Management and Data Systems*, Vol. 105, No. 6, pp. 703-713

Kim, J., Ok, C.S., Kumara, S., & Yee, S.T. (2010). A market-based approach for dynamic vehicle deployment planning using radio frequency identification (RFID) information. *International Journal of Production Economics*, Vol. 128, No. 1, pp. 235-247

Lapide, L. (2004). RFID: what's in it for the forecaster? *Journal of Business Forecasting Methods and Systems*, Vol. 23, No. 2, pp. 16-19

Li, S., & Visich, J.K. (2006). Radio frequency identification: supply chain impact and implementation challenges. *International Journal of Integrated Supply Management*, Vol. 2, No. 4, pp. 407-424

Lin, D., Barton, R., Bi, H., & Freimer, M. (2006). Challenges in RFID enabled supply chain management. *Quality Progress*, Vol. 39, No. 11, pp. 23-28

Martha, J., & Subbakrishna, S. (2001). When just-in-time becomes just-in-case. *Wall Street Journal*, October 22, p. A18

McCrea, B. (2005). Reliable Foods lives up to its name. In: *Logistics Management*, July 1, Date of access, September 10, 2006, Available from: www.logisticsmgmt.com/archive

McFarlane, D., & Sheffi, Y. (2003). The impact of automatic identification on supply chain operations. *International Journal of Logistics Management*, Vol. 14, No. 1, pp. 1-17

Parker, R. (2002). Just-in-time freight plans survive 11 September crisis. *Supply Management*, Vol. 7, No. 5, p. 8

Peslak, A.R. (2005). An ethical exploration of privacy and radio frequency identification. *Journal of Business Ethics*, Vol. 59, No. 4, pp. 327-345

Pottie, G.J. (2004). Privacy in the global e-village. *Communications of the ACM*, Vol. 47, No. 2, pp. 21-23

Prater, E., Biehl, M., & Smith, M.A. (2001). International supply chain agility: tradeoffs between flexibility and uncertainty. *International Journal of Operations and Production Management*, Vol. 21, No. 5/6, pp. 823-839

Raman, A., DeHoratius, N., & Zeynep, T. (2001). Execution: the missing link in retail operations. *California Management Review*, Vol. 43, No. 3, pp. 136-152

Razzouk, N., Seitz, V., & Nicolaou, M. (2008). Consumer concerns regarding RFID privacy: an empirical study. *Journal of Global Business Technology*, Vol. 4, No. 1, pp. 69-78

RFID Journal. (2007). Throttleman adopts item-level tagging. In: *RFID Journal*, Date of access, January 19, 2011, Available from: www.rfidjournal.com/article/print/3580

RFID Journal. (2009). Retail theft is up - can RFID help? In: *RFID Journal*, Date of access, January 19, 2011, Available from: www.rfidjournal.com/blog/entry/5372

RFID News. (2008). Schools using RFID for security enhancement. In: *RFID News*, Date of access, January 19, 2011, Available from: www.rfidnews.org/2008/09/09/schools-using-rfid-for-security-enhancement

RFID Update. (2006a). Pfizer Shipping RFID-Tagged Viagra. In: *RFID Update*, January 9, Date of access, March 21, 2011, Available from:www.rfidupdate.com /articles/home.php.

RFID Update. (2006b). Study: RFID vulnerable to viruses and worms. In: *RFID Update*, March 15, Date of access, March 21, 2011, Available from: www.rfidupdate.com/articles/home.php

RFID Update. (2006c). Dutch casino operator bets on RFID chips. In: *RFID Update*, December 8, Date of access, March 21, 2011, Available from: www.rfidupdate.com/articles/home.php

RFID Update. (2007a). RFID blooms for Dutch flower tracking. In: *RFID Update*, July 18, Date of access, March 21, 2011, Available from: www.rfidupdate.com/articles/home.php

RFID Update. (2007b). Hospital manages thousands of patient files with RFID. In: *RFID Update*, August 1, Date of access, March 21, 2011, Available from: www.rfidupdate.com/articles/home.php

RFID Update. (2007c). Why METRO's item-level RFID deployment matters. In: *RFID Update*, September 24, Date of access, September 24, 2007, Available from: www.rfidupdate.com/articles/home.php

RFID Update. (2007d). RFID file tracking is heating up. In: *RFID Update*, November 29, Date of access, March 21, 2011, Available from: www.rfidupdate.com/articles/home.php

RFID Update. (2007e). Finnish retailer gets quick ROI on item-level RFID. In: *RFID Update*, November 16, Date of access, March 21, 2011, Available from: www.rfidupdate.com/articles/home.php

RFID Update. (2008a). World's largest item-level RFID application launches. In: *RFID Update*, Date of access, January 19, 2011, Available from: www.rfidupdate.com/articles/home.php

RFID Update. (2008b). RFID helps kids take Dora the Explorer on road trips. In: *RFID Update*, March 10, Date of access, March 21, 2011, Available from: www.rfidupdate.com/articles/home.php

RFID Update. (2008c). RFID market on target to reach $5.3B in 2008. In: *RFID Update*, June 12, Date of access, June 12, 2008, Available from: www.rfidupdate.com/articles/home.php

RFID Update. (2008d). Turkish retailer plans big item-level RFID expansion. In: *RFID Update*, Date of access, August 11, 2008, Available from: www.rfidupdate.com/articles/home.php

RFID Update. (2009). Bloomingdale's tracks strong results from RFID pilot. In: *RFID Update*, Date of access, January 19, 2011, Available from: www.rfidupdate.com/articles/home.php

Rocks, D. (2001). The Net as a lifeline; a tough economic environment makes the Web even more important for companies attempting to cut costs, generate new revenues, and better serve customers. *Business Week*, No. 3755, p. EB16

Rutner, S., Waller, M.A., & Mentzer, J.T. (2004). A practical look at RFID. In: *Supply Chain Management Review*, January/February, Date of access, September 10, 2006, Available from: www.manufacturing.net/scm/

Sheffi, Y. (2001). Supply chain management under the threat of international terrorism. *The International Journal of Logistics Management*, Vol. 12, No. 2, pp. 1-11

Shih, D.H., Lin, C.Y., & Lin, B. (2005). RFID tags: privacy and security aspects. *International Journal of Mobile Communications*, Vol. 3, No. 3, pp. 214-230

Slettemeås, D. (2009). RFID–the "next step" in consumer-product relations or Orwellian nightmare?: challenges for research and policy. *Journal of Consumer Policy*, Vol. 32, No. 3, pp. 219-244

Swartz, N. (2007). NIST issues RFID guidelines. *Information Management Journal*, Vol. 41, No. 4, p. 8

Symbol Technologies (2006). A prescription for RFID success in the pharmaceutical industry. Date of access, March 21, 2011, Available from: www.symbol.com/assets/files/rfid_uhf.pdf

Taghaboni-Dutta, F., & Velthouse, B. (2006). RFID technology is revolutionary: who should be involved in this game of tag? *Academy of Management Perspectives, Vol. 20, No. 4, pp. 65-78*

Tajima, M. (2007). Strategic value of RFID in supply chain management. *Journal of Purchasing and Supply Management*, Vol. 13, No. 4, pp. 261-273

Tang, C.S. (2006). Perspectives in supply chain risk management. *International Journal of Production Economics*, Vol. 103, No. 2, pp. 451-488

Uldrich, J. (2007). Dole: let us thank RFID technology. In: *The Motley Fool*, Date of access, February 17, 2011, Available from: www.fool.com/investing/value/2007/09/rfid-saves-the-dole.aspx

Wasieleski, D.M., & Gal-Or, M. (2008). An enquiry into the ethical efficacy of the use of radio frequency identification technology. *Ethics and Information Technology*, Vol. 10, No. 1, pp. 27-40

Wicks, A.M., Visich, J.K., & Li, S. (2006). Radio frequency identification applications in healthcare. *International Journal of Healthcare Technology and Management*, Vol. 7, No. 6, pp. 522-540

Willey, L. (2007). RFID and consumer privacy: let the buyer beware. *Journal of Legal, Ethical and Regulatory Issues*, Vol. 10, No. 2, pp. 25-37

Wyld, D.C. (2006). RFID 101: the next big thing for management. *Management Research News*, Vol. 29, No. 4, pp. 154-173

Wyld, D.C., & Jones, M.A. (2007). RFID is no fake: the adoption of radio frequency identification technology in the pharmaceutical supply chain. *International Journal of Integrated Supply Chain Management*, Vol. 3, No. 2, pp. 156-171

Wyld, D.C. (2008). Radio frequency identification: advanced intelligence for table games in casinos. *Cornell Hospitality Quarterly*, Vol. 49, No. 2, pp. 134-144.

Zakaria, F. (2001). Time to save "just in time". *Newsweek*, Vol. 138, No. 20, p. 38

Zebra Technologies. (2004). RFID: the next generation of AIDC. Date of access, September 10, 2006, Available from: www.integratedlabeling.com/rfid/white_papers/11315Lr2RFIDTechnology.pdf

Zsidisin, G.A. (2003). Managerial perceptions of supply risk. *The Journal of Supply Chain Management*, Vol. 39, No. 1, pp. 14-25

Zsidisin, G.A., Panelli, A., & Upton, R. (2000). Purchasing organization involvement in risk assessments, contingency plans, and risk management: an exploratory study. *Supply Chain Management: An International Journal*, Vol. 5, No. 4, pp. 187-197

Zuo, Y. (2010). Secure and private search protocols for RFID systems. *Information Systems Frontier*, Vol. 12, No. 5, pp. 507-519

4

Secure RFID for Humanitarian Logistics

Gianmarco Baldini[1], Franco Oliveri[1], Hermann Seuschek[2],
Erwin Hess[2] and Michael Braun[3]
[1]*Joint Research Centre - European Commission*
[2]*Siemens AG*
[3]*University of Applied Sciences, Darmstadt*
[1]*Italy*
[2,3]*Germany*

1. Introduction

Extreme events like hurricanes, flooding and earthquakes cause massive disruption to society, including large death tolls and property damage. In recent years, many events like the Katrina disaster Katrina (2004) have shown the importance of efficient disaster management to alleviate the resulting pain and suffering and to mitigate the consequences of the disaster. Disaster management includes a large set of activities including the care of the survivors needs, protection of assets from any further damage and provision of shelter, water, food, and medicines to dislocated people. The creation of an effective disaster supply chain to deliver necessary goods to disaster relief organizations is an essential function of disaster management. This function is also called humanitarian logistics. Humanitarian logistics is a wide term that covers the operations concerning supply chain strategies, processes, and technologies that will maintain the flow of goods and material needed for the humanitarian.
The management of the supply chain in disaster relief operations is considered an essential element in the resolution of a crisis since the Tsunami in South East Asia (December, 26th 2004) and the Katrina Hurricane (August, 2005). The scale of these disasters is huge both in geographical size and in severity. The Katrina Hurricane affected 92,000 square miles of land Gardner (2006) and hundreds of thousands of people were displaced from their homes.
In a recent report Fritz (2005), it was highlighted that most of the organizations involved in the 2004 tsunami disaster were lacking in supply chain expertise and technology. Humanitarian logistics is indeed a very challenging task for many organizations for a number or reasons, which will be described in this chapter. For example, natural disasters are usually characterized by a chaotic environment and by a general lack of transportation infrastructures, which are usually degraded or destroyed. Many different organizations may be involved with no a-priori coordination plan defined. All these challenges make the task of humanitarian relief organizations very difficult. Traditional mechanisms and processes implemented in commercial supply chains may not be directly adapted to humanitarian logistics because of these challenges and because of the different operational requirements. Timing constraints are much more severe in disaster supply chain than commercial supply chains because of the potential loss in human lives and assets if essential equipment is not distributed in time.
In other cases, specific processes and technologies can be tailored to humanitarian logistics. Radio-Frequency IDentification (RFID) technology has already been identified as a powerful

enabler to improve tracking and tracing in supply chain management. RFID is a device applied to a person or goods for identification and tracking purposes through radio waves. RFID could be used to create a "virtual infrastructure", which can be used to track cargo and goods and their delivery.
Security is a very important requirement in humanitarian logistics. In the aftermath of a disaster, many goods (e.g., medicine, foods), which are usually available in normal conditions, became extremely valuable and potential target of thieves.
In the commercial domain, theft reduction is considered the main expected benefit as it translates to cost savings, while it will be even more important in crisis situation where replacements for stolen goods, may not be readily available. As a consequence, the distributed goods must be protected and all the components of the supply chain should be made secure: RFID tags must not be tampered with and they should be resistant to security attacks (e.g., spoofing, eavesdropping and cloning) to ensure that the supply chain is not disrupted by criminals and that cargo and goods are not stolen. As a consequence, security of the RFID tags is an important element in humanitarian logistics.
This chapter will describe the main features and challenges of Humanitarian logistics, the role of RFID technology in disaster supply chains and the implementation and deployment of secure RFID.
The chapter has the following structure: section 2 describes the features of natural disasters and emergency crises, the main phases (mitigation, preparedness, response, recovery), the role of Humanitarian Logistics and the related challenges. Section 3 describes the role of RFID in Humanitarian Logistics. Section 4 describes security aspects in RFID. Section 5 describes the proposed system for secure RFID and the related authentication mechanism. Section 6 describes the system architecture and the deployment of RFID in humanitarian logistics. Section 7 describes the role of telecommunications in the disaster supply chains based on RFID. Finally section 8 provides suggestions for future developments in this area.

2. Disaster management and humanitarian logistics

2.1 Type of disasters and features

In this chapter, we will use the definition of natural disaster from Bankoff (2005):

> "a natural disaster is the effect of a natural hazard (e.g., flood, tornado, volcano eruption, earthquake or landslide) that affects the environment, and leads to financial, environmental and/or human losses. The resulting loss depends on the capacity of the population to support or resist the disaster, and their resilience"

it appears clear that what is relevant is the effect of the disaster on the human lives and activities in the affected area.
Natural disasters and emergency crises can have different types of classification based on their features. One main classification is natural and man-made disasters. Natural disasters are the consequences of natural hazards like earthquakes, flooding, avalanche or tsunami while man-made disasters are caused by human actions (e.g., terrorist attack) or human oversight.
Other taxonomies are based on the predictability of the event or the impact on the region. Table 1 provides an overview of the most typical disasters or emergency crises and their features from a qualitative point of view.
Each type of disaster have specific features, which require different responses and recovery actions, but they all produce devastating loss of lives and assets. Disaster management tries to minimize the impact of natural disasters through various activities, which include damage

Disaster Type	Predictability	Severity	Geographical impact
Earthquake	Low	High	National
Tsunami	Low	High	International
Storm/Hurricane	Medium	Medium/High	National
Vulcanic Eruption	Medium	High	National and International (i.e., dust clouds)
Pandemic Disease	Medium	Medium/High	Potentially Global
Terrorist Attack	Medium	Medium	Local/City
Transportation incident	Low	Medium	Local
Armed Conflict	Medium	High	International
Landslide	Medium	Low	Local
Avalanche	Medium	Low	Local
Chemical plant incident	Man-made	Low	Medium
Nuclear incident	Low	High	National and International (i.e., radioactive dust)

Table 1. Features of natural disasters and emergency crisis

assessment, reconstruction, emergency health services and the creation of supply chains to bring needed goods.

2.2 Phases of disaster management

Natural disaster management is usually split in four phases: prevention, preparedness, response and recovery. Such phases may be named in different way by different authors or different organizations, but usually the first two phases are related to activities to avoid the disaster (prevention) or to be prepared for the disaster (preparedness). The third phase (response) includes all the activities to be performed in the aftermath of a disaster, while the last phase (recovery) deals with the endeavor of bringing back a normal life to the people affected by the disaster.
From Altay and Green (2006), a number of activities are defined for each phase of disaster management, which are described in figure 1.

2.3 Supply chains in disaster management

The availability of materials such as medicines, food, shelter for the immediate needs after the disaster is usually not a problem and it is also quite viable to deliver them in the country affected by the disaster; what is a real challenge is to keep track of the shipped crates and ensure that medicines are distributed to the hospitals and the responders working in the disaster sites. Operators deploying the Supply Chain in disaster areas have to face, on top of the heavy disruption of the usual delivery systems as well as of the communication systems, the presence of every sort of looters; in the aftermath of a disaster, the usual physical protection of a sensitive area, such as a warehouse, may be limited. As a consequence, it is of paramount importance that the tools used to keep track of the delivered crates are well designed against tampering. The management of the supply chain is one of the most important activities during a natural disaster or an emergency crisis and an adequate preparation is the key for success in such phase, but it is obviously very difficult to invest large amount of money in the preparation of a response to events that may or may not

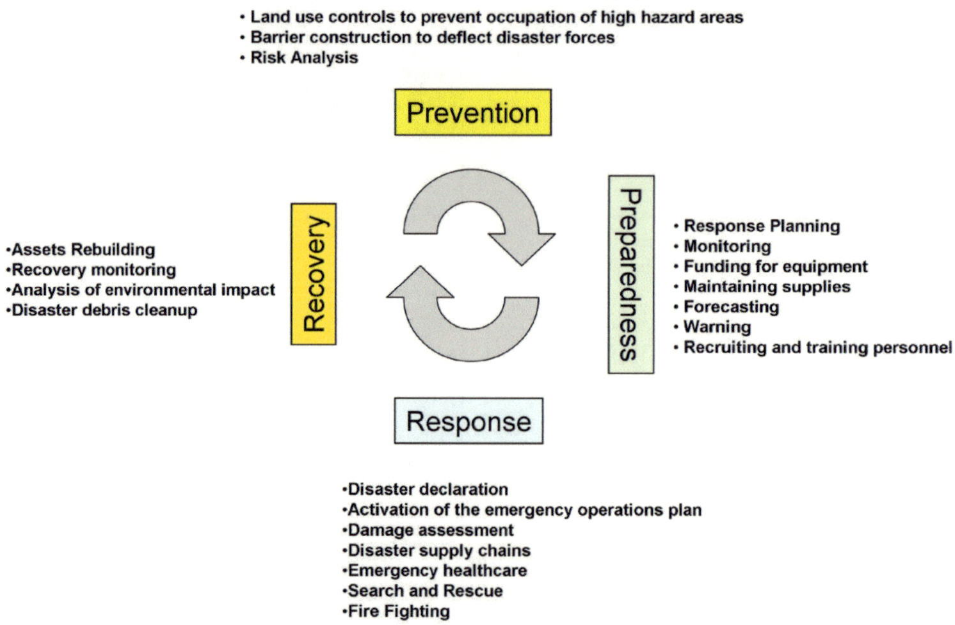

Fig. 1. Phases of Disaster Management

occur, therefore it is crucial to design a solution that is tamper proven, low cost and user friendly. The role of the logistics and the supply chain has also been highlighted by recent events. In the immediate aftermath of the 2004 Asian Tsunami, relief goods flooded airports and warehouses in the affected regions, aid agencies struggled to sort through, store and distribute the piles of supplies while disposing of those that were inappropriate. In Sri Lanka, the airports were overloaded by the large number of humanitarian cargo flights. At the distribution level, relief agencies struggled to identify warehouses to store excess inventory. In India, the transportation infrastructure was overloaded and bottlenecks were present at main conjunction points. In Indonesia, the damaged infrastructure combined with the flood of goods from many different agencies created huge logistics problems. Many participants to the relief efforts to the Tsunami disaster, claimed that logistics and efficient supply chain was more important than a large quantity of goods. Figure 1 describes the phases in disaster management and the related functions. Supply chains used in disaster management are usually called disaster supply chains.

Supply chain management for business applications had a long evolution and many companies have well established supply chains around the world but the strategic goal of commercial supply chains and disaster supply chains is different. Commercial supply chains are focused on quality and profitability, humanitarian supply chains must be focused on minimizing loss of life and suffering. Supply chain management may be present in the phases preparedness, response and recovery with different roles. In the preparedness phase, supply chains are used to stockpile and maintain disaster supplies and equipment, which may be used in disaster management. In preparedness phase, the management of the supply chain is relatively easy as the location of the stockpiling facilities and inventories is well known and the transfer of the materials is planned in advance. In the response phase, supply chains are an essential element in the resolution of the crisis. Depending on the features of the crisis as described in Table 1, supply chain management can become very complex with the presence of different stakeholders and large quantity of materials to be distributed. Because transportation infrastructures are degraded or destroyed, the distribution of the materials could be quite difficult. Furthermore, there are severe time constraints as people may die if goods are not distributed in time. In the recovery phase, disaster supply chains are needed to rebuild destroyed property and repair of essential critical infrastructures (e.g., energy, transportation and others). Supply chain management can still be hampered by degraded infrastructure in the area, but the timing constraints are usually more relaxed in comparison to the response phase.

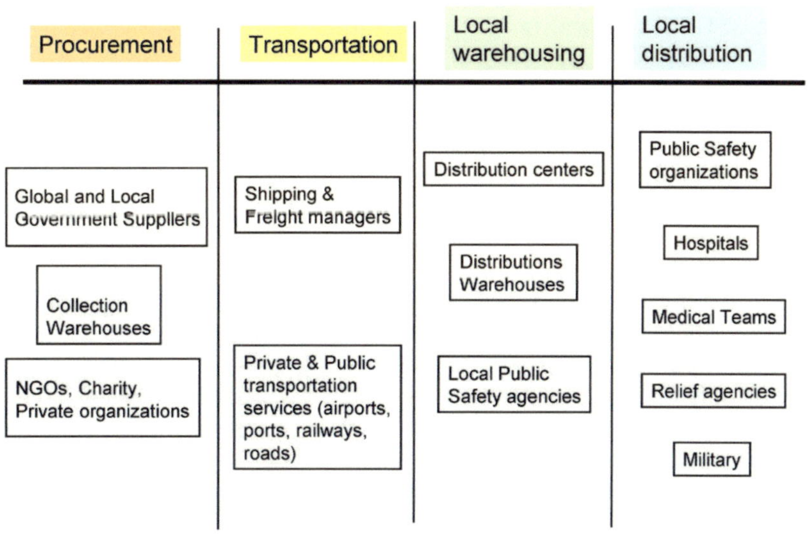

Fig. 2. Participants to Humanitarian Logistics

Another significant challenge for humanitarian logistics is the presence of many different types of organizations: from firefighters, military, relief organizations, non-government organizations (NGO) and others. The coordination among these organizations is essential to support disaster management operations and disaster supply chains in particular. Figure 2 describes the main participants, which are usually involved in the resolution of a large natural disaster or emergency crisis.

Technology can be essential in improving disaster supply chain management and in providing more capabilities to the partners involved in the resolution of the crisis.

One of the basic ingredients of supply chain management is information. Supply chain managers need to know what is the demand of the goods, where they are located at any time, when and where they will be shipped and so on. These tasks are already complex in generic commercial supply chains, but in humanitarian logistics they become even more difficult because many different partners are involved, which must share information among themselves in the severe time constraints imposed by the operational context. Interoperability issues may especially appear in unplanned situations.

An essential element is the proper identification of the goods and the distribution of this information to all the involved partners. In natural disasters, goods may come from any type of sources, because aid-agencies are sometimes not equipped to tag the material in the proper

way. Autier (1990) discusses the case of drug supplies, after the 1988 Armenian earthquake, when at least 5000 tons of drugs were sent by international relief operations but only one third was usable because it is was properly identified, relevant for the emergency situation and distributed in time. One fifth of the supplies had to be destroyed at the end of 1989.
One important aspect is security. As described in the challenges above, criminal entities may take advantage of the chaotic conditions to steal or redirect goods to the wrong destination. The information present in the supply chain management systems must be secured and protected so that it cannot be used for criminal purposes. Technology can improve the access, distribution and security of the information in a number of ways.

3. Application of RFID to disaster supply chains

3.1 RFID technology

Radio Frequency Identification (RFID) is said to be the future technology to improve and optimize supply chain management systems. In comparison to traditional bar codes, RFID technology provides better data security, does not need line of sight because it is based on radio propagation, provide computational capability, improve automation and thus enhances the operational efficiency of supply chains Lin (2009). Figure 3 shows the differences with barcode technology: RFID allows to write information on the tag without line of sight as they are based on radio propagation and they provide computational capability, which can be used to implement security features. Typical components of RFID systems are RFID tags attached to the assets, reading/writing devices, and back-end systems.
RFID tags are usually very small sized low cost devices which can be easily fixed on physical objects like asset boxes or even implanted in animals or persons. The interface to control RFID tags is a short range wireless link. The simplest tags just send a unique identification number (UID) on request. More sophisticated tags offer several bits to severals bytes of read and write memory. High end RFID tags are equipped with a computing device to execute tasks like cryptographic algorithms for authentication or encryption. Beside the classification concerning the feature set of RFID tags there are basically two possibilities for the power supply of RFID tags: passive and active. Passive tags harvest the energy received from the reader through the antenna and using that energy to power the device. Passive tags are less expensive than active tags, which include their own power supply like a battery. The battery powered tags offer the possibility to execute tasks while not connected to a reader. For example perishable goods may be monitored by active tags that integrate with thermometers to ensure the goods are kept at an acceptable temperature.
Devices for accessing RFID tags are called RFID readers. RFID readers can either be fixed devices with antennas mounted at points where assets pass by (e.g., reader gates) or readers can be handheld devices available for several operational environments. Some handheld devices are even equipped with a wireless communication link to access the back-end system. In the back-end system all the RFID data is aggregated, stored, and provided to the supply chain management. The main component of the back-end system is the RFID middleware which connects the reader infrastructure to databases and supply chain management tools.

3.2 Track and trace based on RFID

Systems based on RFID technology provide the track and trace capability to monitor the movement of tagged items from the suppliers to the emergency crisis through distribution channels. figure 4 shows the typical supply chain based on RFID technology. The crates carrying the necessary assets are equipped with RFID tags, which are read at every point

Fig. 3. Barcode and RFID

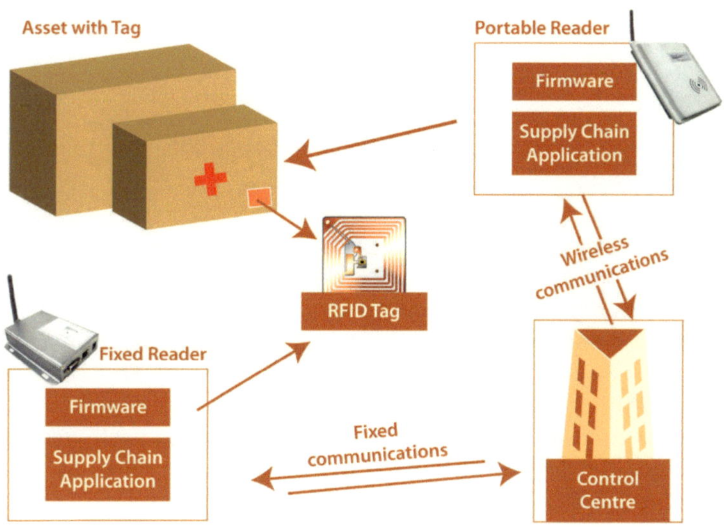

Fig. 4. Supply chains based on RFID technology

within the supply chain through automated systems equipped with RFID readers. The identification number provided by the RFID tag has to be unique for each item. The reading device aggregates the tag ID with its own ID and sent the data set to a central tracking server in a control center. Both fixed readers and mobile readers can be used to track the assets with RFID tags. Fixed readers are usually installed at main goverment and transportation centers (e.g., ports, airports). Mobile readers can be used by the government and relief agencies in the field or if the transportation centers themselves are destroyed by the disasters. Mobile readers may also provide their location through Global Navigation Satellite Systems (GNSS) like GPS. Control centers can use the position provided by the mobile readers to organize the distributions of goods in a more efficient way. There is the need to have a central tracking server, which stores the complete history of the RFID tags across all the disaster supply chain.

Various relief organizations and their own ICT systems can connect with the central tracking server to retrieve the information on the distributed goods as shown in figure 5. Currently the most promising approach for a track and trace solution is the Electronic Product Code (EPC) infrastructure. Designed and standardized by EPCglobal EPCGlobal (2003) it enables the exchange of RFID data using Internet protocols.

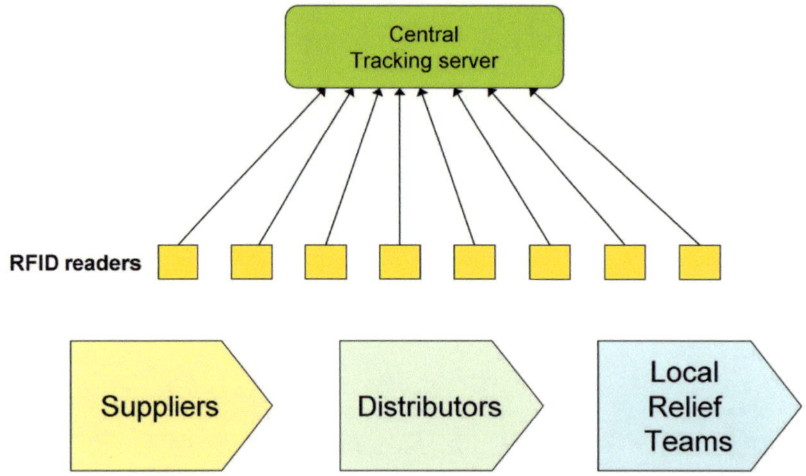

Fig. 5. Tracking system

At a first glance such a track and trace system seems to be a good approach, but there are some drawbacks. A precondition for track and trace techniques to work reliably is that each party involved in the distribution process must take part to the track and trace system. On the one hand all participants of the supply chain must be compliant with the chosen track and trace standard and they must also provide a consistent tracking data. This requires cooperation among all partners within the multi party supply chain. On the other hand in emergency crises the communication infrastructure can be degraded or even destroyed as consequence of the crisis itself. Hence, the item cannot be tracked along the complete supply chain in order to securely identify the object.

As written in the previous sections, security is an essential requirement. Ordinary RFID tags, with no security features, which are commonly used in commercial supply chains are simple tags, which only store an identification number in plain text. As a consequence the tags themselves can be susceptible to faking attacks. In addition all necessary information on the functionality of RFID is also available on the Internet or in the literature, e.g., the RFID handbook Finkenzeller (2003), as well as development tools. More information on the need for secure RFID in disaster supply chains in provided in section 5.1.

4. RFID security

Like other wireless technologies, RFID is vulnerable to a wide range of security threats, which have been identified in literature.

In Tanenbaum et al (2006), the authors identify the following threats to RFID technology:

1. *Sniffing or eavesdropping*, where RFID tags are read without the knowledge of the tag bearer. Even if RFID is a short-range wireless technology, RFID tag reading my happen

also at large distances using RFID readers equipped with directional antennas and power amplifiers.

2. *Spoofing*. Spoofing attacks supply false information that looks valid and that the system accepts. Attackers can create *authentic* RFID tags by writing properly formatted tag data on blank or rewritable RFID transponders.
3. *Tracking*. RFID readers in strategic locations can record sightings of unique tag identifiers.
4. *Denial of service*. Denial of Service (DoS) is when RFID systems are prevented from functioning properly. Tag reading can be hindered by Faraday cages or *Signal jamming*, both of which prevent radio waves from reaching RFID-tagged objects
5. *Replay attack* where a valid RFID signal is intercepted and its data is recorded; this data is later transmitted to a reader where it is played back. Because the data appears valid, the system accepts it.
6. *Cloning* where a RFID tag is duplicated with the same information.

Some of these RFID security threats are relevant to disaster supply chains. For example sniffing can be used to extract the information on the contents of the crates to understand if they contain valuables goods. By using long distance sniffers, malicious parties can collect the information on the distributed goods, without being detected by authorities, and plan a subsequent phyisical attack to steal valuable material. By using RFID replay attacks, thieves can make the theft more efficient. In a first phase, thieves intercept a valid RFID signal. Then they replace the crates and they use the replayed signal to mislead the RFID reader owned by the authorities. In another example, malicious parties can track the flow of goods of specific types to improve the planning for a subsequent theft.

While sniffing is relatively easy to implement, other RFID threats are more complex to implement and malicious parties may use them only for very valuable goods. For example Tanenbaum et al (2006) introduces a new type of RFID threat called RFID malware, where malicious software carried by an infected RFID tag can "'infect"' the backend of a RFID IT infrastructure during the reading phase. This type of attack is more complex to implement and may be limited to the commercial domain.

Security issues in the context of supply chain management has been investigated in Li and Ding (2007), which identifies the specific security requirements in supply chains and propose a practical design of RFID communication protocols that satisfy the security requirements.

5. Secure RFID in humanitarian logistics

5.1 Need for secure RFID

As described in the previous sections, a major issue in natural disasters and emergency crises is security.

Criminals like thieves and looters may take advantage of the chaotic environment to steal goods or to disrupt the supply chain to their advantage Cassidy (2003). In a natural disaster, the goods (medicines, food) brought by aid agencies and relief organizations are even more valuable because of their scarcity. In all disaster situations, there is the potential for loss through theft at all levels of the supply chain, and control systems must be established and supervised at all storage, hand-over and distribution points to minimize this risk. Even more dangerous of simple thieving is tampering: the use of unreliable medicines or rotten food can further endanger the life of the survivors, therefore it is crucial to be able to keep track of the origin of the goods along each step of their delivery. Security of the relief chains is

an important requirement in humanitarian logistics. Consequently, all the components of the supply chain should be made secure: RFID devices must not be tampered with and they should be resistant to security attacks (e.g., spoofing, eavesdropping and cloning) to ensure that the supply chain is not disrupted by criminals and that cargo and goods are not stolen.
Since ordinary RFID tags used for track and trace solutions are simple tags which only store an identification number in plain text the tags themselves are susceptible to faking attacks. It is a misbelief that tags which carry a unique identifier written during the manufacturing process can be used as security feature for unique identification. Usually RFID systems use standardized radio frequency communication protocols which are public domain. In addition all necessary information on the functionality of RFID is also available on the internet or in the literature, e.g. the RFID handbook (see Finkenzeller (2003)), as well as development tools. Cloning an original tag is not difficult with the proper tools.
Is the RFID is not secure, the following scenario is possible: A criminal party, duplicates tags as described and attaches them to goods. The shipping unit carrying the original RFID may be removed from the supply chain and sold using an illegal distribution channel. The goods carrying the cloned tags move within the supply chain without producing any inconsistence in the tracking history. In the worst case terrorists could replace drugs or food by worthless or even harmful units to sabotage disaster relief.
This chapter will analyze practical utilization of this type of device in the resolution of emergency crises to guarantee the reliability of sealing of the goods and their identification. The establishment of a logistics tracking framework based on secure RFID has the potential to greatly increase the effectiveness of future emergency crises response operations.
Track and trace systems using RFID allow to track the movement of tagged items from the suppliers to the emergency crisis through distribution. Each item is equipped with an RFID tag that can be read out automatically without any line-of-sight at every point within the supply chain. The read data provides detailed information on the corresponding item and it will then be sent via the internet to the central tracking server which stores the complete history of the RFID tag and checks its plausibility. Providing this electronic pedigree of each transport unit the barrier to disrupt the supply chain can be increased. Figure 5 shows the tracking system. For instance, the Electronic Product Code (=EPC) infrastructure by EPCglobal (see EPCGlobal (2003)) enables the exchange of RFID data via the internet and it is currently the most promising approach for a track and trace solution.

5.2 Cryptographic authentication

A track and trace only solution may not be sufficient for a secure identification of items. To obtain an appropriate security level that ensures authentication on item level, the RFID tags themselves must implement authentication mechanisms (see also Staake (2005)). This authentication mechanism must withstand the cloning attack as described in the previous sections. The approach is the commonly used *challenge response protocol*. The RFID tag contains its identification number, a secret key and a cryptographic unit. The reader transmits a randomly selected number, the so-called *challenge* and the tag calculates the corresponding *response* with the cryptographic algorithm using the secret key and the challenge. Then the tag sends this response back to the reader. Finally the reader, respectively the back end system, checks whether the response is correct or not. Note that the secret key itself is not transmitted over the radio channel and the correct response can only be generated with the aid of the secret key.

5.3 Public key authentication

A weakness of symmetric cryptography used in most of RFID system is that the tag and the reader share a common key to run the authentication protocol: the tag uses this secret key for response generation and the reader for the verification. This approach requires that the readers must store the secret keys of the RFID tags belonging to the application domain or an on-line connection from the reader to a server must be established to store the secret keys of the RFID tags in a secure and reliable back end system.

In public-key cryptography, the response generation is performed using a secret key, the so-called *private* key $priv_{id}$, but the response verification on the reader side can be performed *without* any secret key only with a public key pub_{id}, which needn't be protected against misuse. In order to avoid that each reader has to store the individual public keys pub_{id} of all tags belonging to the application, a Certification Authority (CA) issues a certificate $cert_{id}$ for every public key pub_{id} and only the CA knows the secret signature key (=PrivSigKey) necessary for the generation of the certificate. The corresponding public signature key (=PubSigKey) for verifying the certificates must be downloaded exactly one time to each reader within the system.

The authentication flow is following:

- the tag transmits its certificate $cert_{id}$ containing its public key pub_{id}.
- the reader verifies the authenticity of the sent public key pub_{id} with the public signature key.
- a challenge-response-protocol will be initialized. The reader generates a challenge C, transmits C to the tag upon which the tag computes the corresponding response R with its private key $priv_{id}$ using the public key operation.
- The tag sends R back to the reader and finally the reader checks the response with the tag's public key pub_{id} using the verification algorithm.

.

The major benefits of this approach are that:

- no secret key is needed for the authentication on the reader side, neither in the back end nor in the reader itself.
- the authentication process can be performed without any online connection which simplifies the system.

The disadvantage of the public key approach is the higher complexity in comparison to the symmetric key approach, which means a higher implementation effort in chip size and finally a lower performance and higher power consumption. Low-cost RFID tag based on elliptic curve cryptography (=ECC) are proposed in Wolkerstorfer (2005). Batina (2006) gave a further area optimization using a protocol based on zero knowledge.

5.4 Authentication protocol

An efficient authentication protocol for RFID tags is based on elliptic curves over binary finite fields $GF(2^n)$. An elliptic curve E is a set of points $P = (x_P, y_P)$ satisfying the Weierstraß equation $y^2 + xy = x^3 + ax^2 + b$ where $a, b \in GF(2^n)$. On an elliptic curve E one can define an addition $R = (x_R, y_R) = P + Q$ of elliptic curve points $P = (x_P, y_P)$ and $Q = (x_Q, y_Q)$ by the following formulae:

$P \neq Q$	$P = Q$
$x_R = \lambda^2 + \lambda + x_P + x_Q + a$	$x_R = \lambda^2 + \lambda + a$
$y_R = \lambda(x_P + x_R) + x_R + y_P$	$y_R = x_P^2 + (\lambda + 1)x_R$
$\lambda = \frac{y_P + y_Q}{x_P + x_Q}$	$\lambda = x_P + \frac{y_P}{x_P}$

The structure determined by the set of points and this addition operation allows public key operation which is the scalar multiplication $s * P$ of a scalar value s in binary representation $s = (s_\ell, \ldots, s_1)_2$ with a point $P = (x_P, y_P)$ on the curve E. An in deep introduction to this field of cryptography may be found in Hankerson (2004). The so-called elliptic curve point multiplication is the basis for our protocol. We implemented Montgomery's method for scalar multiplication Bock (2008); Hankerson (2004). This method has special characteristics preventing so-called side channel attacks and it is well suited for hardware efficient implementations since expensive inversions of finite fields elements can be avoided as projective coordinates of the x-coordinates are used Hankerson (2004).
The applied authentication protocol is based on a challenge-response-protocol, where the security is based on the Elliptic-Curve-Diffie-Hellman problem.
Now let P denote the base point on the elliptic curve E with order q. For each RFID tag an individual private key $priv_{id}$ is given, which is a random number d with $0 < d < q$. The corresponding public key pub_{id} is then the point Q given by the scalar multiplication of d and the base point P:

$$Q := d * P$$

As already pointed out in the previous section the RFID reader generates a challenge C. This will be done by choosing a random scalar k and multiplying it with P:

$$C := k * P$$

The corresponding response R is then calculated by the tag using its private key d:

$$R := d * C$$

The reader itself calculates $V := k * Q$ and checks if $R = V$. The verification works since the following chain of equations holds:

$$R = d * C = d * (k * P) = (dk) * P = k * (d * P) = k * Q = V$$

The complete authentication protocol is depicted in Figure 6.

6. System architecture

The application of secure RFID to Humanitarian logistics is depicted in figure 7.
The deployment of this system is based on the following steps:

1. In the first step of the disaster supply chain, the Certification Authority (CA) generates the key pars and store them in the RFID tags. This step has to be executed in a trustworthy environment; for example a logistic center of an humanitarian organization or a government agency. The CA is a server system which stores the private signature key

RFID Reader		RFID Tag
stores public signature key to verify the tag's certificate		stores private key d and certificate $cert_{id}$ containing the public key Q
	← $cert_{id}$	
verify certificate $cert_{id}$ pick random k compute $C := k * P$	C →	
		compute $R := d * C$
	← R	
compute $V := k * Q$ if $V = R$ accept tag else reject		

Fig. 6. The RFID Authentication Protocol based on Elliptic Curve Cryptography

PrivSigKey which has to be kept secret by the CA because this key is the cryptographic security anchor of the whole system. The associated public signature key PubSigKey may be publicly known and part of the CA certificate.

2. Certificates must be distributed to the main stakeholders as described in figure 8 to be installed on RFID readers (both fixed and mobile). Certificates can also be distributed in the mitigation phase using secure links over Internet or through secure communication links (e.g., VPN).
3. Then the RFID tags are applied to the relief goods, which are then transported to the disaster areas.
4. Relief agencies and other organizations can use the fixed and mobile RFID readers to track and trace the relief goods through all the nodes of the disaster supply chain. It is important that only trusted certificates are allowed to be installed on the readers.
5. At the disaster area the emergency responders may use handheld devices equipped with RFID readers to read the attached RFID tags, verify their authenticity and finally distribute the goods.

The proposed solution can be used to augment existing supply chains and it has a minimal impact on the organization structure and procedures of the relief organizations.
Figure 8 describes the deployment workflow of the proposed solution among the participants of the disaster supply chain.

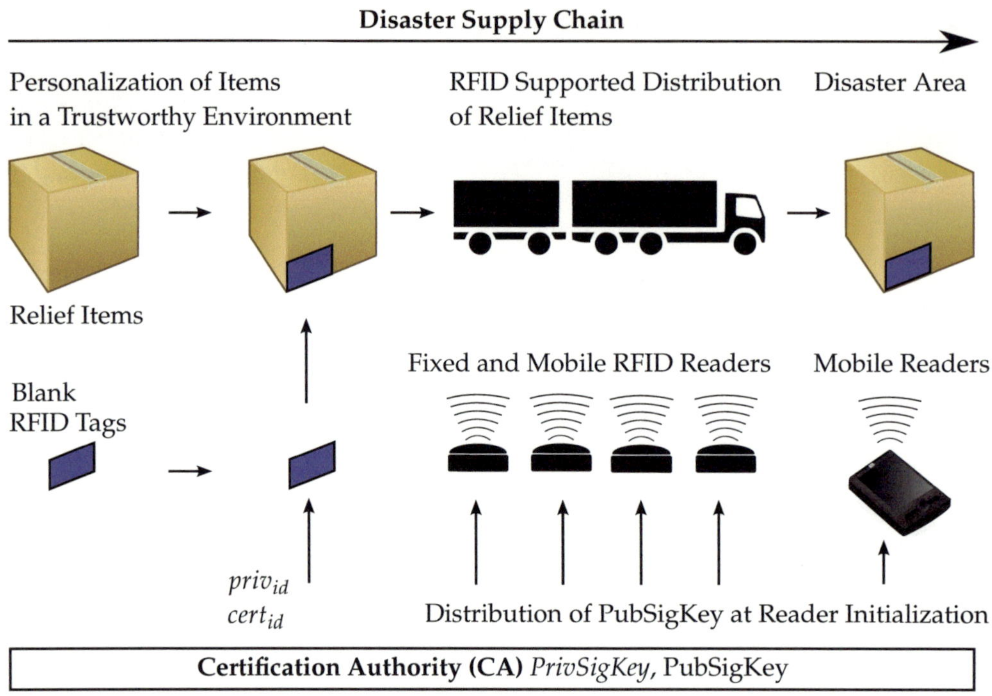

Fig. 7. Proposed system architecture for RFID secured relief item distribution

7. Communication infrastructure for humanitarian logistics

In order to fully exploit the capabilities of the RFID based Supply Chain Management, such system must be supported by an efficient and secure communication system as well as by a distributed data base management system. In a disaster area, most communications will be wireless because first responders need high mobility and because the fixed line infrastructure can be unavailable, e.g. destroyed, damaged or overloaded. The security of the communication link between the RFID tag and the reader/writer is described in another part of this paper, but it is very important to consider that any system is as secure as its weakest component; therefore the communication link between the reader/writer and its local or remote controller has to be considered and made secure. In order to make the system usable, it is very important to consider that the remote stations should be allowed to work without an *always-on connection* because it is unthinkable to have such connection available all the time. In the following we will provide a broad description of communication systems that could be implemented in a disaster situation to support the Supply Chain exploiting the security features described in the previous chapters. From the logical point of view, the logistic of the disaster supply chains is very similar to the Logistic of any Commercial Supply Chain, therefore we can assume that the basic concepts and the basic infrastructure remain the same, but few key features must be redesigned in order to cope with the peculiar operational environment of the Disaster Relief Operations. The first aspect to address is the lack of standard communication infrastructures (GSM/UMTS/PSTN) where crates of goods and people have to be dispatched, therefore the ideal situation is that any RFID reader used to acquire the information on the crates present in any intermediate station (e.g, warehouse)

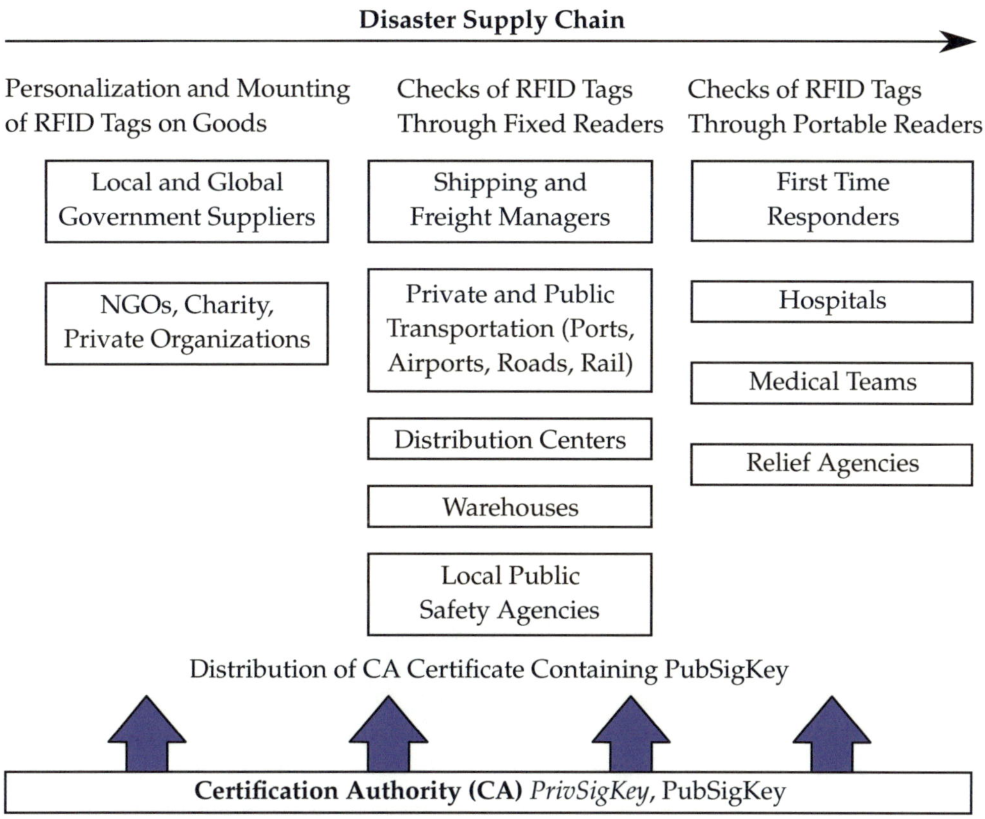

Fig. 8. Deployment workflow

in the disaster area is provided with a satellite link to transmit the data to the Logistic Control Centre as described in Figure 9.

An alternative solution (depicted in Figure 10) could be the establishment of a Wireless Local Area Network, to collect and manage data locally, connecting with the Central Logistic Control Center only when required. This solution presents some important pros, namely the possibility to operate without a permanent link with the logistic centre and a significant reduction in term of the cost of the communication equipment. The cons are the need to set up a local logistic control centre and the implementation of a secure client-server mechanism, between the control centres capable of surviving an unstable connection: usual commercial software, designed for a reliable "always on" environment may run into troubles facing frequent loss of connection. Furthermore the WiFi connection must be implemented with a reasonable level of security to avoid jeopardizing the secure RFIDs. An example of RFID sensor network for humanitarian logistics based on Zigbee communication technology is presented in Yang (2010).

In summary, the communication structure needed for such system should take into account some key issues:

- Distributed databases connected through potentially unreliable communication links

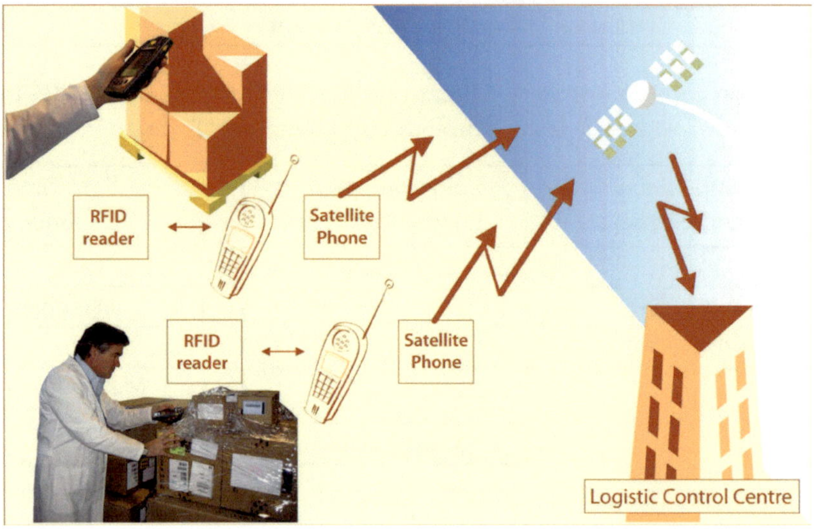

Fig. 9. RFID readers directly connected to the Logistic control center through Satellite Communications

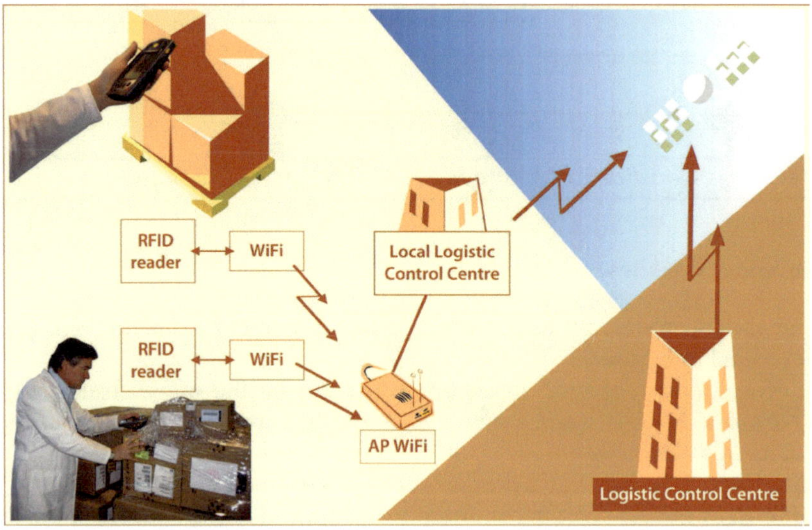

Fig. 10. RFID readers connected to a wireless Local Area Networks for local management

- Integrated and redundant communication systems using: a) Direct satellite links; b) Local wireless coverage (GSM and/or WiMAX and/or WiFi) plus satellite link
- Secure wireless links
- Store and forward protocols

8. Conclusions

The chapter has presented the application of secure RFID technology to the specific domain of humanitarian logistics. Because security is a important requirements in disaster management, we believe that relief organizations can benefit from this technology to ensure that goods are not stolen or tampered. A potential system architecture has been presented and described. Because, infrastructures are usually degraded or destroyed in a natural disaster or emergency crisis, mobile readers and fast deployable communication systems is an important component in the overall system architecture.

Future developments in this research area would be the integration of these technologies in the organizational and procedural frameworks of relief and government agencies. As described in this chapter, a central Certification Authority (CA) must be defined to provide the certificates, which must be installed in the fixed and mobile portable readers. Furthermore, an efficient disaster supply chain requires the set-up of a coordinated track and trace system in the prevention phase of disaster management.

9. References

Tom Gardner, Former FEMA director shoulders greater share of blame for Katrina failures, *ASSOC. PRESS*, Jan. 19, 2006

Melanie R. Rieback, Patrick N.D. Simpson, Bruno Crispo, Andrew S. Tanenbaum, RFID malware: Design principles and examples, *Pervasive and Mobile Computing*, Volume 2, Issue 4, Special Issue on PerCom 2006, November 2006, Pages 405-426, ISSN 1574-1192

Li, Y. and X. Ding, Protecting RFID communications in supply chains, in: *Proceedings of the 2nd ACM symposium on Information, computer and communications security, ASIACCS '07*, 2007, pp. 234Ű241

"An Entrepreneur Tackles the Logistics of Disaster". Available at URL: *http://www.globalenvision.org/library/*

Fritz Institute. "Logistics and the effective delivery of humanitarian relief". May 2005. Available at URL: *http://www.fritzinstitute.org/.*

A Failure of initiative - Final Report of the Select Bipartisan Committee to Investigate the Preparation for and Response to Hurricane Katrina

Autier P, Ferir MC, et al. "Drug supply in the aftermath of the 1988 Armenian earthquake", Lancet 1990;

Cassidy W. A logistics lifeline. *Traffic World*, October 2003.1. 335(8702):1388-1390.

L.C. Lin, "An integrated framework for the development of radio frequency identification technology in the logistics and supply chain management". *Computers and Industrial Engineering* (2009).

EPCglobal. available at URL: *http://www.epcglobalinc.org/home/.*

Infineon. available at URL: *http://www.infineon.com/.*

An asymmetric cryptosystem. available at URL: *http://www.ntru.com/.*

NXP. available at URL: *http://www.nxp.com/*

L. Batina, J. Guajardo, T. Kerins, N. Mentens, P. Tuyls, and I. Verbauwhede. Public key cryptography for RFID tags. In *RFIDSec 2006, Proceedings of the 2th Workshop on RFID Security*, July 2006.

H. Bock, M. Braun, M. Dichtl, J. Heyszl, E. Hess, W. Kargl, H. Koroschetz, B. Meyer, and H. Seuschek. A milestone towards RFID products offering asymmetric authentication based on elliptic curve cryptography. In *RFIDSec 2008, Proceedings of the 4th Workshop on RFID Security*, July 9-11 2008, Budapest, Hungary, July 2008.

Altay N., W. G. Green III W. G., OR/MS research in disaster operations management, European Journal of Operational Research, Volume 175, Issue 1, 16 November 2006, Pages 475-493, ISSN 0377-2217,

K. Finkenzeller. RFID-Handbook. Wiley & Son LTD, third edition,

Yang H,et al. Hybrid Zigbee RFID sensor network for humanitarian logistics centre management. *Journal of Network Computer Applications* (2010)

D. Hankerson, A. Menezes, and S. Vanstone. *Guide to elliptic curve Cryptography*. Springer, 2004.

T. Staake, F. Thiesse, and E. Fleisch. Extending the EPC Network — the potential of RFID in anti-counterfeiting. In *20th ACM symposium on Applied computing*, pages 1607–1612. ACM, March 2005.

J. Wolkerstorfer. Is elliptic curve cryptography suitable to secure RFID tags *Handout of the Ecrypt Workshop on RFID and Lightweight Crypto*, July 2005.

G. Bankoff, G. Frerks, D. Hilhorst (eds.) (2003). Mapping Vulnerability: Disasters, Development and People. ISBN ISBN 1-85383-964-7

5

Applications of RFID Technology in the Complex Product Assembly Executive Process

Huibin Sun
Key Laboratory of Contemporary Design and integrated Manufacturing Technology (Northwestern Polytechnical University), Ministry of Education, China

1. Introduction

As we know, a complex product assembly process involves numerous parts, complex processes and high precision demands. But most assembly operation is carried out manually, and on-site assembly data is recorded on paper. Due to lack of advanced technology methods, some shortness exists as follows:

a. Because of the asynchrony problem lies between the logistics stream and the information stream, assembly tasks are always assigned prior to materials' preparation and transportation. Then the need of adjusting assembly task assignments according to materials' state dynamically can't be met.
b. The associated relationship between materials and assembly tasks can't be established automatically. Then operators are in charge of determining relationship between them. And parts are often misassembled, especially in the mixed flow production mode.
c. The assembly executive process is a collaborative process among different operators and assembly workstations. And the operation order is established on operators' work habit or spontaneity, which throws impediments in the way of assembly executive process monitoring, controlling and tracing.

Although adoption of barcodes can solve these problems to some certain extend, but they are difficult to be read, easy to be destroyed and unable to be rewritten. Therefore, barcode technology can't meet the need of automatic, fast and smart material identifying. Due to the advantage of non-contact far distance reading/writing, RFID (Radio Frequency Identification) technology can not only make material identifying more convenient, but also turn real-time on-site monitoring into reality. On the other hand, because of the outstanding autonomy and collaboration characters, multi agent technology is widely used in manufacturing resource encapsulation and integration in the agile manufacturing system, manufacturing task scheduling in the collaborative design system, etc.

Under this circumstance, if we use the mobile agent technology to dispatch and recall assembly task, and use the RFID technology to identify material automatically, assembly executive process monitoring and controlling dynamically in real time will be enhanced, and misassembling phenomenon will be eliminated furthest. Therefore, based on multi agent technology and RFID technology, this paper aims to propose a complex product assembly executive monitoring and controlling method to achieve synchrony between the logistics stream and the information stream, and to match materials with assembly tasks dynamically. Then the automatic level and intelligent level of complex the product assembly executive process can both be improved.

2. Literature review

In recent years, more and more attention has been paid into manufacturing system monitoring and controlling related fields. Some of strong correlation researches are cited as follows.

The multi agent technology's usage in advance manufacturing system was a hot topic. Many researchers focused on agent-based task scheduling, resource integration, workshop management, cell controlling, etc. Among them, Kyung-Hyun Choi et al. (2007) proposed a multi-agent-based task assignment system for virtual enterprises, which attempted to address the selection of partners and the process of assigning tasks to them. Jose Barata et al. (2008) discussed the design and implementation of a multi agent-based control architecture to support modular reconfigurable production systems. Moreover, the mobile agent technology could enhance the flexibility and adaptability of multi agent-based system. In this field, Guanghui Zhou and Pingyu Jiang (2005) put forward a mobile agent-based framework for the manufacturing resource encapsulation and integration. They implemented the re-configuration and encapsulation for legacy manufacturing resources, and realized information interaction and acquirement. Hossein Tehrani Nik Nejad et al. (2008) put forward an agent-based architecture for process planning and scheduling in the flexible manufacturing systems. Coordination agents were adopted to generate a suitable job assignment to the machine tool agents at each step of the negotiation. In summary, most researches listed above used agent technology in task scheduling, resources integration and encapsulation. These agents executed there logic individually. Without a whole process management model, the dynamic triggering mechanism among agents was unable to come into being. This weak point prevented multi agent technology, especially mobile agent technology, from using in more practical fields.

Petri net is suitable to describe and analyze systems' asynchrony, concurrency, competition and randomicity characters. It is widely used in modeling, simulating and scheduling of discrete event dynamic systems (DEDSs). For example, some researchers adopted it to model and schedule the assembly and disassembly process. Among them, Fu-Shiung Hsieh (2006) studied the robustness of a class of controlled Petri nets, called controlled assembly/disassembly Petri net (CADPN), for assembly/disassembly processes with unreliable resources. He characterized different types of tolerable resource failures allowed for a nominal marking of a live CADPN. Weijun Zhang, et al. (2005) proposed a scheduling model for optimal production sequencing in a flexible assembly system. The assembly process was modeled using timed Petri nets and task scheduling was solved with a dynamic programming algorithm. Tang Xinmin et al. (2006) and Zhong Shisheng et al. (2006) put forward a timed colored Petri net to model the aero-engine assembly procedure. Based on the notion that assembly Petri net was reversed disassembly Petri net, they also proposed a Petri net reduction method for the disassembly Petri net. In summary, researches cited above adopted Petri net in assembly/disassembly process modeling and simulating. Only assembly/disassembly nodes were involved in the model, but logistics modes were excluded. Relationship between assembly/disassembly nodes and logistics nodes were also ignored. Such a model was unable to support the whole assembly/disassembly process. Moreover, besides modeling and simulating, how to use these models to monitor and control the assembly/disassembly process in practice was still a pendent problem.

The RFID technology is changing our life and production remarkably. Its usage in manufacturing system will benefit building of real time factory. In this field, George Q. Huang (2007a, 2007b) presented an approach to shop-floor performance improvement by using RFID technology for the collection and synchronization of the real-time field data from manufacturing workshops. His emphasis was placed upon how to deploy RFID technology for managing work-in progress (WIP) inventories in manufacturing job shops with typical functional layouts. He also studied how to deploy RFID technology in a walking-worker fixed-position flexible assembly islands where products were placed at fixed position work centers in the shop-floor, the workers moved from one work centre to another, and tools and components were brought to the work centre for assembly according to the process and production plan. To bridge the gap between shop floor automation and factory information systems, Robin G (2007) proposed an RFID-based framework to enable the instant delivery of pertinent data and information on a uniquely identifiable job/product at point-of-need across factories. Lu B. H. et al. (2007) reviewed the fundamental issues, methodologies, applications and potential of RFID enabled manufacturing, outlined a simulated RFID machining process application case study, and discussed a proposed methodology, framework and five-step deployment process aimed at developing a holistic approach to implementing RFID enabled manufacturing in manufacturing enterprises in detail. In summary, these researches used RFID technology to implement real time manufacturing workshops. RFID tags' wireless, long distance properties were fully exerted. But RFID tags can also be a carrier of manufacturing executive state. They can be a bridge between the information stream and the logistics stream. According to information taken back by them, triggering and controlling of manufacturing executive process can be implemented in a more automatic mode.

As discussed above, many researchers have studied assembly executive process modeling, monitoring and controlling methods. Petri net, multi agent technology and RFID technology have been adopted to solve the problem to a certain extent respectively. But each of them can't solve all problems individually. As a result, compound of these technologies may break a new path for implementation of a timely and intelligent complex product assembly digitalization system.

3. Assembly executive process control (huibin sun, 2009a)

3.1 Assembly executive process petri nets model

In the complex product assembly executive process, materials' states belong to the discrete set {assembly state, transport state}. Conversion between these two states is determined and triggered by events as "material drawn", "transport finished", "assembly finished", and so on. From this perspective, the complex product assembly executive process is a discrete event dynamic system. And it is suitable to be modeled, analyzed, and controlled via Petri net theories and methods. Therefore, an assembly executive process Petri net (AEPPN) will be proposed and discussed in detail here.

Definition: AEPPN is a 7-element set as AEPPN={P, T, C, I, O, m_0, D}. Among them, T={t_1, t_2, ···, t_m} is the transition set, which is composed by assembly transitions and logistics transitions. An assembly transition refers to a group's activity of executing and finishing an assembly task. A logistics transition refers to the logistics operators' activity of executing and finishing a transport task. $P=\{p_1, p_2, \cdots, p_n\}$ is the set of places, which describes events in the assembly executive process. C is the color set of transitions and places. It is used to distinguish products. Assuming s stands for the amount of products, then,

$$\forall p_i \in P : C(p_i) = \{a_1, a_2, \cdots, a_s\}, i = 1,2,\cdots,n$$

$$\forall t_j \in T : C(t_j) = \{a_1, a_2, \cdots, a_s\}, j = 1,2,\cdots,m$$

while a_1, a_2, ..., a_s are color types. In practice, each of them can be replaced by a product's unique ID code. The mapping relationship between the product set and the color set is 1:1.
I (p, t) is the input function from place p to transition t: $C(p)\times C(t)\rightarrow N$ (non-negative integer). It corresponds to the colored directional line from p to t. I (p, t) is an s-by-s matrix here. O (p, t) is the output function from transition t to place p: $C(t)\times C(p)\rightarrow N$ (non-negative integer). It corresponds to the colored directional line from t to p. O (p, t) is an s-by-s matrix here too. M_0 is the initial mark, which stands for the amount of token with certain color in the place p. $D=\{d_1, d_2, \cdots, d_m\}$ is the time delay set of all transitions'. For example, d_j stands for time delay of transition t_j. If t_j is an assembly transition, d_j equals to the correspondent assembly task's time consumption. If t_j is a logistics transition, d_j equals to the correspondent logistics task's time consumption. Every transition has a constant time delay, and there is no correlativity relationship lies between a transition's color and its time delay, as

$$\forall t_j \in T : D(t_j) = d_j, j = 1,2,\cdots,m$$

The input line from the place p_i with the color a_h to the transition t_j with the color a_k can be expressed by the scalar quantity $I(a_{i,h}, a_{j,k})$. Similarly, the scalar quantity $O(a_{i,h}, a_{j,k})$ expresses the correspondent output line.
In each place, the amount of token with certain color is no more than 1, as

$$\forall p_i \in P, a_j \in C(p_i) : m(a_{i,j}) \leq 1, i = 1,2,\cdots,n$$

Under the mark M, the transition t_j is enabled by the color a_k, if and only if

$$\forall p_i \in \bullet t_j : M(a_{i,h}) \geq I(a_{i,h}, a_{j,k})$$

When the transition t_j is just triggered, comes out a new mark M′ as

$$\forall p_i \in \bullet t_j : M'(a_{i,h}) = M(a_{i,h}) - I(a_{i,h}, a_{j,k})$$

After the transition t_j has been triggered for time delay d_i, comes out a new mark M″ as

$$\forall p_i \in t_j \bullet : M''(a_{i,h}) = M(a_{i,h}) + O(a_{i,h}, a_{j,k})$$

3.2 Multiple agent-based Implementation model

Assembly transitions and logistics transitions in the AEPPN model are distributed, dynamic and autonomy. And they are suitable to be implemented and controlled through agent technology. Therefore, these two types of transition are regarded as self-governed entities that are entitled to certain privileges and can intercommunicate with each other. Each of them has its own structure and mode, and can finish its task driven by local data. On the other hand, RFID tags can not only be used to identify materials. When a batch of material is drawn from the inventory, or an assembly task is finished, a new RFID tag is created to identify the material or the new assembly. When the assembly task or the material's transport task is finished, the RFID tag is updated. RFID tags' state changes are in line with the assembly executive process events, and RFID tags' states can be used to describe the assembly executive process states. Therefore, the AEPPN model can be implemented by an RFID-based multi agent system, in which assembly agents and logistics agents are included. Function model of these two types of agent is shown in figure 1. The main functions of assembly agents are listed as follows:

a. Get information from RFID tags, and promote task information;
b. Clean information saved in RFID tags that identify materials;
c. Get task information and 3D assembly process from database;
d. Guide the assembly operation process and control the quality check process;
e. Update task information in database, and create new RFID tag to identify the new assembly.

Main functions of logistics agent are listed as follows:

a. Get information from RFID tags, and promote task information;
b. Get task information and material information from database;
c. Guide the transportation process;
d. Update information saved in database and RFID tags.

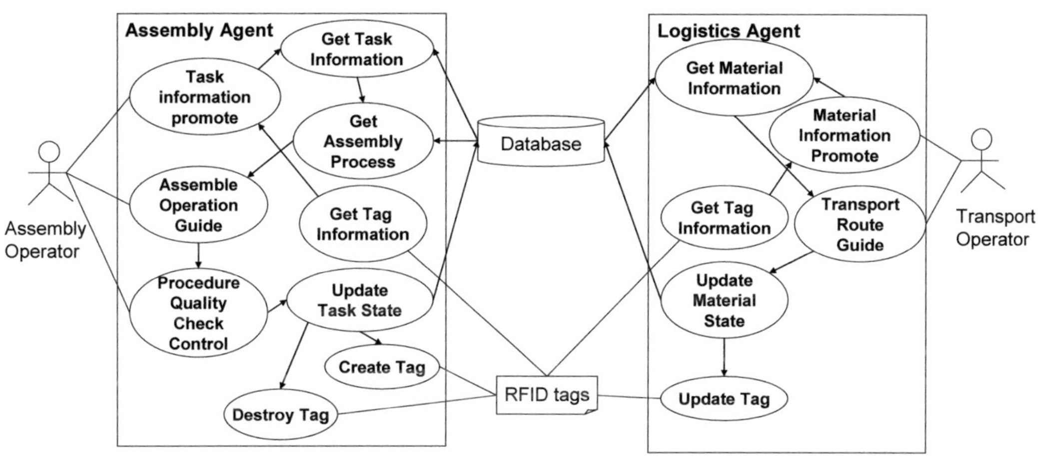

Fig. 1. Function model of assembly agent and logistics agent

As shown in figure 1, there is no direct communication channel lies between assembly agents and logistics agents. And RFID tags and database play the role of sharing blackboard between them. Each RFID tag has it unique Electronic Product Code (EPC), and saves encoded information, such as material's current state, the next operation instruction, in its storage space. For example, when an assembly task is finished, a new RFID tag is used to identify the new assembly. An ASCII string is saved in the tag's storage space, and what it means is decomposed as table 1 shows.

Information Type	Content
Current state	Assembly finished
Current station	Accessory assembly workstation
Next operation	Transport
Next operation method	Manually
Next station	Final assembly workstation
Related process	None
Deadline	2008-04-09 10:22:00

Table 1. A data structure example

All data related to the assembly process and associated relationship between RFID tags and material ID are stored in the database. Communication types among agents, RFID tags and database include "Get", "Create", "Update" and "Delete". Along a typical assembly executive process, what these communication types do is listed in figure 2. In each communication type, every arrow indicates the direction of information transmission.

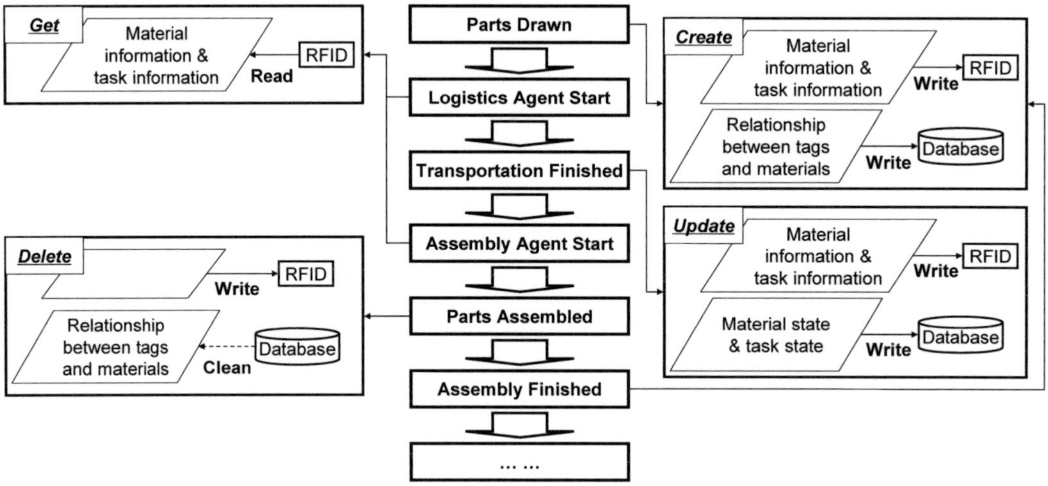

Fig. 2. Communication types

3.3 Mobile agent-based assembly digitalization framework

Agents can be decomposed into agent templates and agent instances. An agent template is product type-related. It defines every agent's default process, check rule, assembly group, sequence, and so on. An agent instance results from an agent template's instantiation. It describes practical process, check rule, assembly group, material RFID tag, triggering relationship among agent instances, and so on. If an agent instance's all parameters are satisfied, it will be dispatched to the correspondent assembly workstation to guide the assembly operation. When the assembly task is finished, the agent returns to the server side with dynamic data included.

Based on above analysis, the mobile agent-based assembly digitalization framework is shown in figure 3. The framework can be divided into the server layer and the assembly workstation layer. They are interconnected via computer network.

The server layer includes the ADS (Assembly Digitalization System) server and the mobile agent server. The ADS server's main functions include maintaining ADS's logic, providing agent information about task, process, users, manufacturing resources, etc, receiving and saving dynamic data from agents. The mobile agent server's main functions include providing running environment for agent instances, maintaining logic sequence among agents, triggering, dispatching, retracting and destroying agents.

The assembly workstation layer is composed by the mobile agent server and the RFID R/W equipments. The mobile agent server provides running time environment for agent instances, and the RFID R/W equipment is in charge of identifying materials and editing information saved in RFID tags' storage space.

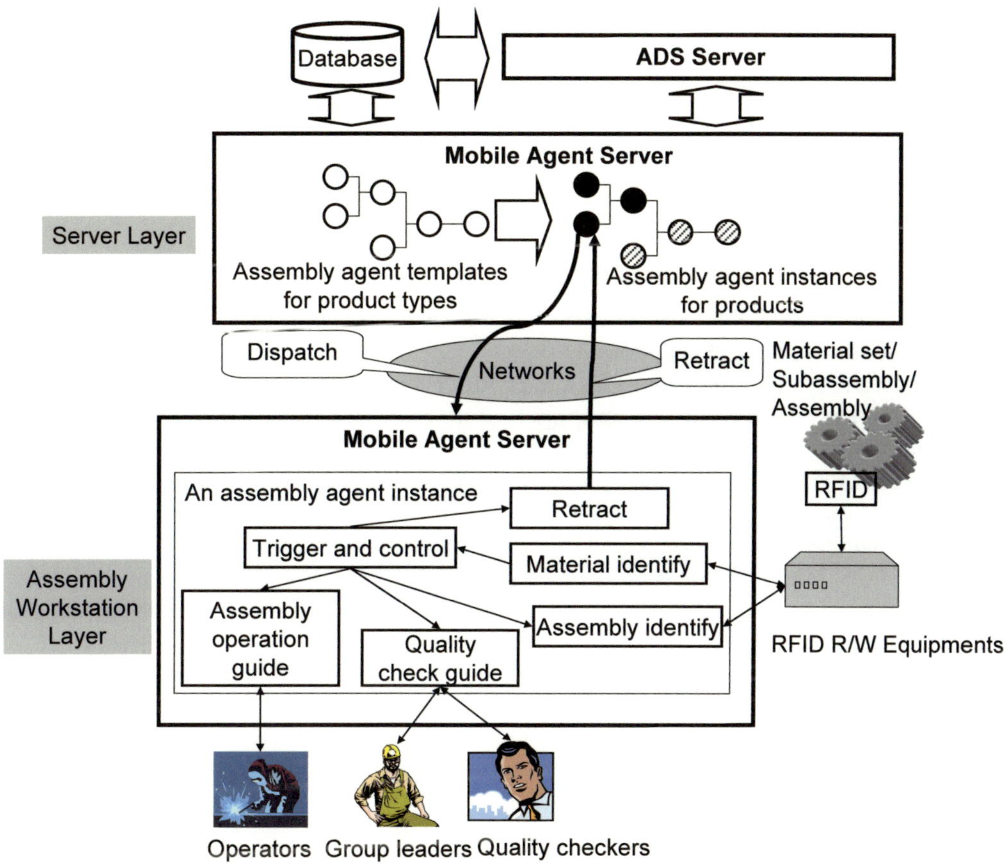

Fig. 3. The mobile agent-based assembly digitalization framework

Here, RFID tags' main functions include:

a. Identify part set, subassembly, assembly and product. Quantity relationship between RFID tag and subassembly, assembly or product is 1:1. And quantity relationship between RFID tag and part is 1:n, which means a tag can be used to identify a set of parts.
b. Mark the assembly executive state. For example, when parts for an assembly task are obtained from the inventory, the RFID tag's state changes to "material drawn". When an RFID tag is created to identify the new assembly, its state is set as "assembly finished".

c. Trigger the assembly agent instance. An agent can obtain the assembly executive state by reading the RFID tags' state. When material for an assembly task is all ready, the correspondent assembly agent instance is triggered and dispatched to the assembly workstation. When the material is transported to the assembly workstation, the agent instance is triggered to guide the assembly operation process and quality check process. When the assembly task is finished, the agent's retraction event is triggered.
d. Guide and trace the work-in-progress logistics. Exact position of material can be traced through RFID tags. Comparison between practical route and expected route is helpful for transportation guide.
e. Save and exchange information. A communication channel can be established among assembly workstations through RFID tags' storage space. It is very important for offline information exchanging.

In summary, RFID tags can be adopted to not only identify materials, but also describe assembly executive states. They can be used to guide and trace WIP logistics, or save and exchange information too. Compared with barcode technology, RFID tags have prominent technological advantages.

As discussed above, assembly tasks' logic and data can be encapsulated by assembly agents. Assembly tasks' execution and control, assembly operation's guide and trace, on-site data's collection and exchange, can also be implemented by assembly agents. As to a practical complex product assembly task, assembly agents' flow includes.

a. Create assembly agent templates for the product type,
b. Instantiate the assembly agent templates, and create assembly agent instances for the product.
c. Dispatch the agent to assembly workstation, if the necessary RFID tags and other parameters are satisfied.
d. At the assembly workstation, check material state through identifying RFID tags. If the answer is OK, trigger the assembly operation guide process.
e. The operators execute the assembly operation guided by the assembly agent's 3D assembly process.
f. Operators, assembly group leaders and checkers execute quality check process guided by the assembly agent.
g. Create a new RFID tag to identify the new assembly when the assembly task is finished.
h. Retract the assembly agent, and save assembly process data, quality check data and new RFID tag's EPC in the database.
i. Write new RFID tag's EPC into the next agent according to the logic sequence among agents.
j. Dispatch the agent to the correspondent assembly workstation if necessary condition is all ready.
k. Repeat above steps, until the complex product assembly task is finished.

To implement above flow, assembly agents must encapsulate some basic data and extending data involved in the assembly executive process. Among them, extending data is used to describe user defined information, and basic data is composed of basic parameters, input parameters and output parameters. Basic parameters describe assembly agents' basic attributions; input parameters define input information of certain assembly task. Output parameters encapsulate dynamic information produced in the assembly executive process. For example, an assembly agent's input/output parameters are listed in figure 4.

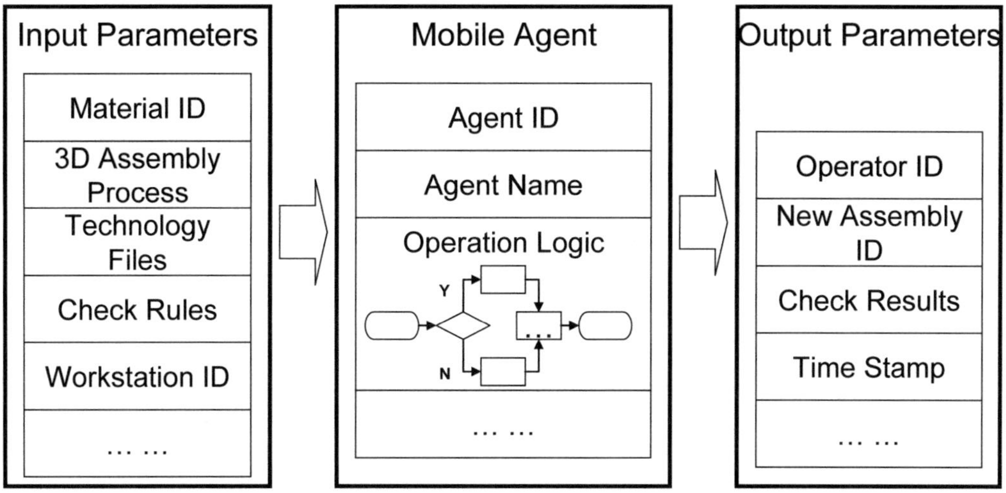

Fig. 4. A mobile agent's input/output parameters

3.4 An example

The aero-engine is a typical complex product. Commonly, its final assembly task is carried out at the final assembly workstation, which assembles the splitter lip, the lube pump and other assemblies together. Among them, the splitter lip is composed of three subassemblies as the upper gearing, middle gearing and lower gearing. Its assembly task and its subassemblies' assembly tasks are carried out at the splitter lip assembly workstation. And the lube pump assembly task is carried out at the accessory assembly workstation. All these assembly tasks are executed in a mixed flow production factory. An AEPPN model is established as figure 5 shows. To explain the issue without loss of generality, other assembly tasks have been simplified. Meanings of places and transition in the model are listed in table 2 and 3.

Now, two aero-engines are being assembled. They are numbered 0295 and 0318 respectively. Therefore, transitions and places have and only have two color types: 0295 and 0318, as

$$\forall p_i \in P : C(p_i) = \{0295, 0318\}, i = 1, 2, \cdots, 17$$

$$\forall t_j \in T : C(t_j) = \{0295, 0318\}, j = 1, 2, \cdots, 12$$

And current state is marked by M. Because of

$$\forall p_i \in \bullet t_7 : M(0318) = 1 \geq I(0318, 0318) = 1,$$

Transition t_7 is enabled by color 0318. It aims to assemble the splitter lip assembly, and can be carried out assisted by the splitter lip assembly agent. The agent obtains information saved in every tag's storage space by the "get" method at first. Then it cleans information saved in these tags by the "delete" method. And it also sets the splitter lip assembly task's state as "assembling" in the database. Here, the mark M′ comes out. When the splitter lip assembly task is finished, the transition t_7 creates a new tag to identify the splitter lip assembly by the "create" method. At the same time, it sets the splitter lip assembly task's state as "assembled". Here, the mark M″ comes out.

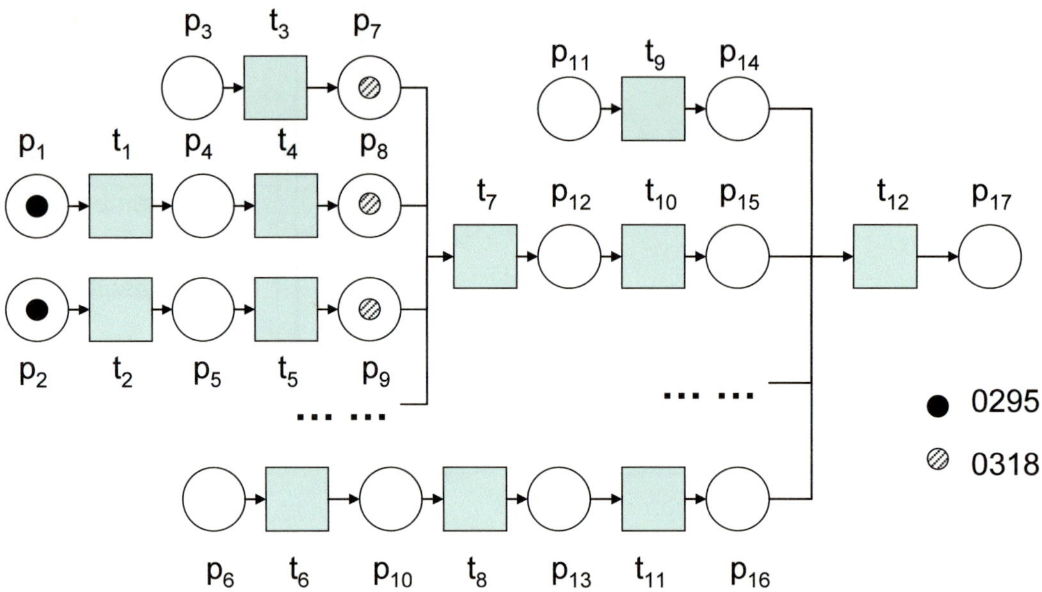

Fig. 5. The AEPPN model of an aero-engine assembly task

$$M=\begin{matrix} \\ p_7 \\ p_8 \\ p_9 \\ \\ p_{12} \\ \\ \end{matrix}\begin{bmatrix} \cdots \\ 0318 \\ 0318 \\ 0318 \\ \cdots \\ 0 \\ \cdots \end{bmatrix} \quad M'=\begin{matrix} \\ p_7 \\ p_8 \\ p_9 \\ \\ p_{12} \\ \\ \end{matrix}\begin{bmatrix} \cdots \\ 0 \\ 0 \\ 0 \\ \cdots \\ 0 \\ \cdots \end{bmatrix} \quad M''=\begin{matrix} \\ p_7 \\ p_8 \\ p_9 \\ \\ p_{12} \\ \\ \end{matrix}\begin{bmatrix} \cdots \\ 0 \\ 0 \\ 0 \\ \cdots \\ 0318 \\ \cdots \end{bmatrix}$$

Based on the aglets toolkit of IBM Japan, an aglet-based aero-engine assembly digitalization prototype is developed. Its user interfaces and flow are shown in figure 6. In step 1, the user sets the basic information and triggering condition for the product type's aglet templates. Among them, the relationship among assembly tasks, assembly processes and assembly groups are defined in the basic information section. Sequences among assembly aglets are also defined. And the triggering condition section defines associated relationship between assembly tasks and materials, especially necessary materials to trigger the assembly tasks' assignment and execution event. In step 2, the user sets the basic information and triggering condition for the product's aglet instances. Among them, the basic information section confirms information in the aglet templates, and the triggering condition section records the RFID tags' EPC (96 bits). When materials for the assembly task is drawn, the correspondent aglet is dispatched to the assembly workstation. The material's transportation state is monitored by the RFID reader. Here, the RFID tags' frequency is 13.56MHz, and their storage space is 8KB. When the material is arrived, the aglet starts the assembly operation process. And guided by the 3D process and check flow provided by the aglet, operators can finish assembling, checking and data recording. When the assembly task is finished, the aglet creates a new RFID tag to identify the new assembly. Then the aglet is retracted, and on-site data will be carried back to update the database. When necessary condition for another aglet is satisfied, the flow from step 3 to step 10 will run again.

Place	Meanings	State of RFID tag
p_1	Material for the upper gearing assembly task is drawn	A tag is created to identify materials for the upper gearing assembly task.
p_2	Material for the middle gearing assembly task is drawn	A tag is created to identify materials for the middle gearing assembly task
p_3	Material for the splitter lip assembly task is drawn.	A tag is created to identify materials for the splitter lip assembly task.
p_4	Material transportation for the upper gearing assembly task is finished.	The tag of materials for the upper gearing assembly task has been updated
p_5	Material transportation for the middle gearing assembly task is finished.	The tag of materials for the middle gearing assembly task has been updated.
p_6	Material for the lube pump assembly task is drawn	A tag has been created to identify materials for the lube pump assembly task.
p_7	Material transportation for the splitter lip assembly task is finished	The tag of materials for the splitter lip assembly task has been updated
p_8	The upper gearing assembly task is finished	A tag has been created to identify the upper gearing subassembly
p_9	The middle gearing assembly task is finished	A tag has been created to identify the middle gearing subassembly
p_{10}	Material transportation for the lube pump assembly task is finished	The tag of materials for the lube pump assembly task has been updated
p_{11}	Material for the aero-engine assembly task is drawn	A tag has been created to identify material of the aero-engine assembly task.
p_{12}	The splitter lip assembly task is finished	A tag has been created to identify the splitter lip assembly task.
p_{13}	The lube pump assembly task is finished	A tag has been created to identify the lube pump assembly task.
p_{14}	Material transportation for the aero-engine assembly task is finished	The tag of materials for the aero-engine assembly task has been updated
p_{15}	The splitter lip arrived at the final assembly workstation.	The tag of the splitter lip has been updated
p_{16}	The lube pump arrived at the final assembly workstation.	The tag of the lube pump has been updated
p_{17}	The aero-engine assembly task is finished	A tag has been created to identify the new aero-engine

Table 2. Place list

Transition	Meanings	Agent Type
t_1	Move materials for the upper gearing assembly task from inventory to the splitter lip assembly workstation	Logistics Agent
t_2	Move materials for the middle gearing assembly from inventory to the splitter lip assembly workstation	Logistics Agent
t_3	Move materials for the splitter lip assembly task from inventory to the splitter lip assembly workstation	Logistics Agent
t_4	Assemble the upper gearing subassembly	Assemble Agent
t_5	Assemble the middle gearing subassembly	Assemble Agent
t_6	Move materials for the lube pump assembly task from inventory to the accessory assembly workstation	Logistics Agent
t_7	Assemble the splitter lip assembly.	Assemble Agent
t_8	Assemble the lube pump assembly.	Assemble Agent
t_9	Move materials for the aero-engine assembly task from inventory to the final assemble workstation.	Logistics Agent
t_{10}	Move the splitter lip from the splitter lip assembly workstation to the finial assemble workstation	Logistics Agent
t_{11}	Move the lube pump from the accessory assembly workstation to the finial assemble workstation	Logistics Agent
t_{12}	Assemble the aero-engine.	Assemble Agent

Table 3. Transition list

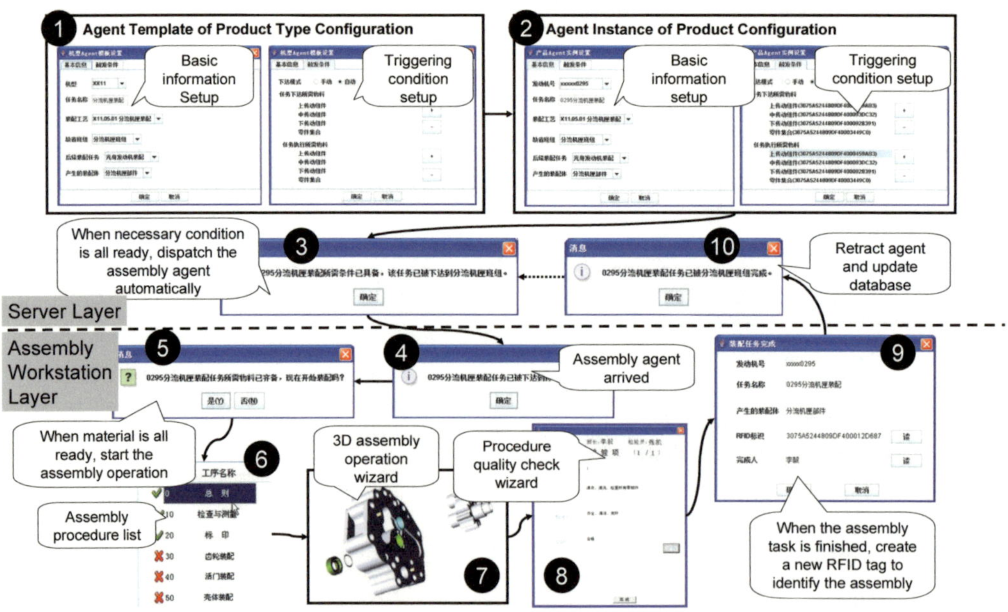

Fig. 6. The prototype's user interfaces and flow

4. Interactive 3D assembly operation guide (huibin sun, 2009b)

In the virtual assembly environment, the assembly sequence is designed, simulated and validated in 3D mode. As a result, the 3D assembly process can be wizard for operators at the assembly workstation. But mistakes couldn't be eliminated, because there is no relationship lies between models in the 3D assembly process and real manufacturing resources. Whether a part is assembled correctly or not can't be recognized automatically, and key parts' assembly history can't be traced under the repeatable assembly condition. To overcome above section, this paper aims to enhance the 3D assembly process's guide ability via establishing interactive mechanism between virtual models and real manufacturing resources. Then each manufacturing resource can be validated and checked automatically. And misassembly phenomenon can be avoided furthest.

4.1 The extended assembly step model

In traditional 3D assembly process model, detailed assembly operation order is encapsulated by steps. Manufacturing resources as operator, part, equipment and clamp are modeled. But traditional 3D assembly process is product type related. The same assembly process is referred by all products' assembly executive process with the same product type. But in fact, two manufacturing resources with the same type may differ from each other in different product assembly executive processes. Then the mapping relationship between a manufacturing resource and its model in traditional 3D assembly process is not 1:1. Several manufacturing resources with the same type may share the same model in the 3D assembly process. This fact prevents the traditional 3D assembly process from guiding each product executive process individually and interactively. Complex and important parts' assembly history can't be recorded and traced. To overcome the problem, an extended assembly step model is proposed here. Its components and structure are shown in figure 7.

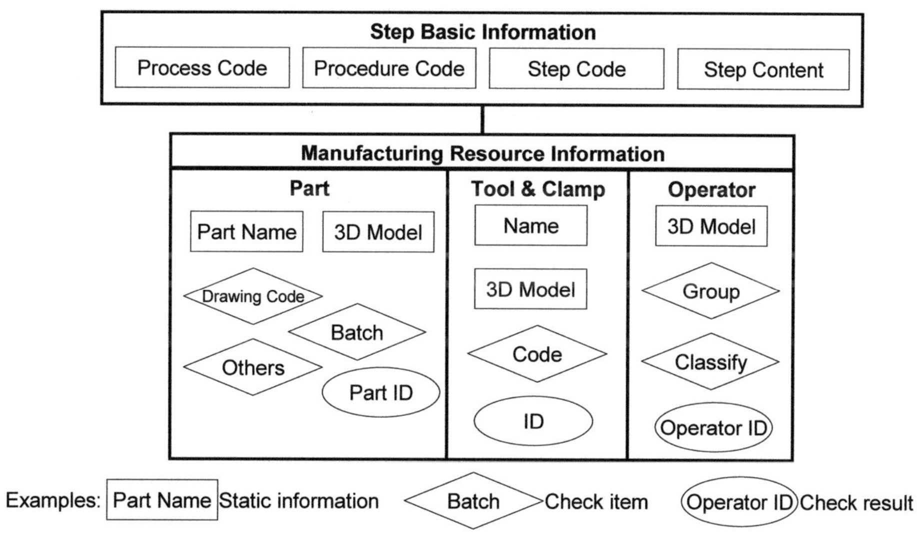

Fig. 7. The extended assembly step model

As shown in figure 7, the extended assembly step model is composed of step basic information section and manufacturing resource information section. The step basic information section describes the step's process code, procedure code and step content. The manufacturing resource information section describes manufacturing resources, such as part, equipment, clamp, tool, operator, and so on. Each manufacturing resource involves three kinds of information, static information, check item and check result. Static information describes a manufacturing resource type's general information, such as name and 3D model. It is unchangeable for all manufacturing resources with the same type whenever. A check item involves some information that can be checked, which means a check operation can be acted to validate individual manufacturing resource's validity. Whether an item should be checked is configurable. Some check items are changeable, such as drawing code. Others may vary in different assembly executive process, such as batch code. The check result encapsulates each manufacturing resource's identity information and other dynamic. It is used to feed back each manufacturing resource's assembly operation history.

4.2 Automatic matching mechanism between virtual models and real materials

In the extended assembly step model, the 3D model is a necessary attribute in the manufacturing resource section. This means that each manufacturing resource is mapped to a virtual solid model in the 3D assembly process. The manufacturing section also defines some check items. They are used to find out a manufacturing resource that fulfills the check conditions. If necessary check items are satisfied, a manufacturing resource is chosen and its ID will be record in the check result section. Then the virtual manufacturing resource model in the 3D assembly process is instanted by a real manufacturing resource. In practice, an automatic match mechanism can be implemented by a flow as figure 8 shows. In the flow, the assembly spot wizard is triggered by the assembly executive system, a software system in charge of assembly executive process monitoring and controlling. The assembly operation is guided by the 3D assembly process. When a 3D model of a manufacturing resource appears, the extended step model is introduced to decide whether a check operation is needed. If the answer is true, the 3D model is highlighted to wait for the check result. The automatic identification system (RFID technology) is started by the assembly executive system, and information about current manufacturing resource is obtained. Whether the check operation is passed is determined by the extended assembly step model via judging each check item is satisfied by current manufacturing resource information. For a passed check operation, the assembly executive system will fill the check result into the extended assembly step model, and the 3D model of the manufacturing resource will appear normally to guide sequent assembly operations. Otherwise, the flow is pending until the check operation is passed or the flow is stopped. As to a non check operation needed manufacturing resource model, it will appear normally to guide sequent assembly operations. The above steps will be executed again and again until the 3D assembly process finishes.

As shown in figure 9, the framework is composed of the server layer and the assembly workstation layer. At the server side, an assembly executive system server is in charge of assembly executive scheduling and monitoring, and on-spot data management. It gets the 3D assembly process from the MPM (Manufacturing Process Management) system and manufacturing resource information from ERP (Enterprise Resources Planning) system. At the workstation side, an assembly executive system client is in charge of assembly operation guide, manufacturing resource validation and on-spot data collection. To obtain the manufacturing resource information automatically, the RFID (Radio Frequency Identification) technology is adopted. Each RFID tag is used to identify a manufacturing resource, such as operator, part, equipment and clamp. Additional information for check operation is saved in the RFID tag's storage space. Therefore, a manufacturing resource's check operation can be acted offline without communicating with database. Although barcode technology can also be used to identify manufacturing resource automatically, an online check operation is necessary because more information must be saved in the database other than the barcode tag.

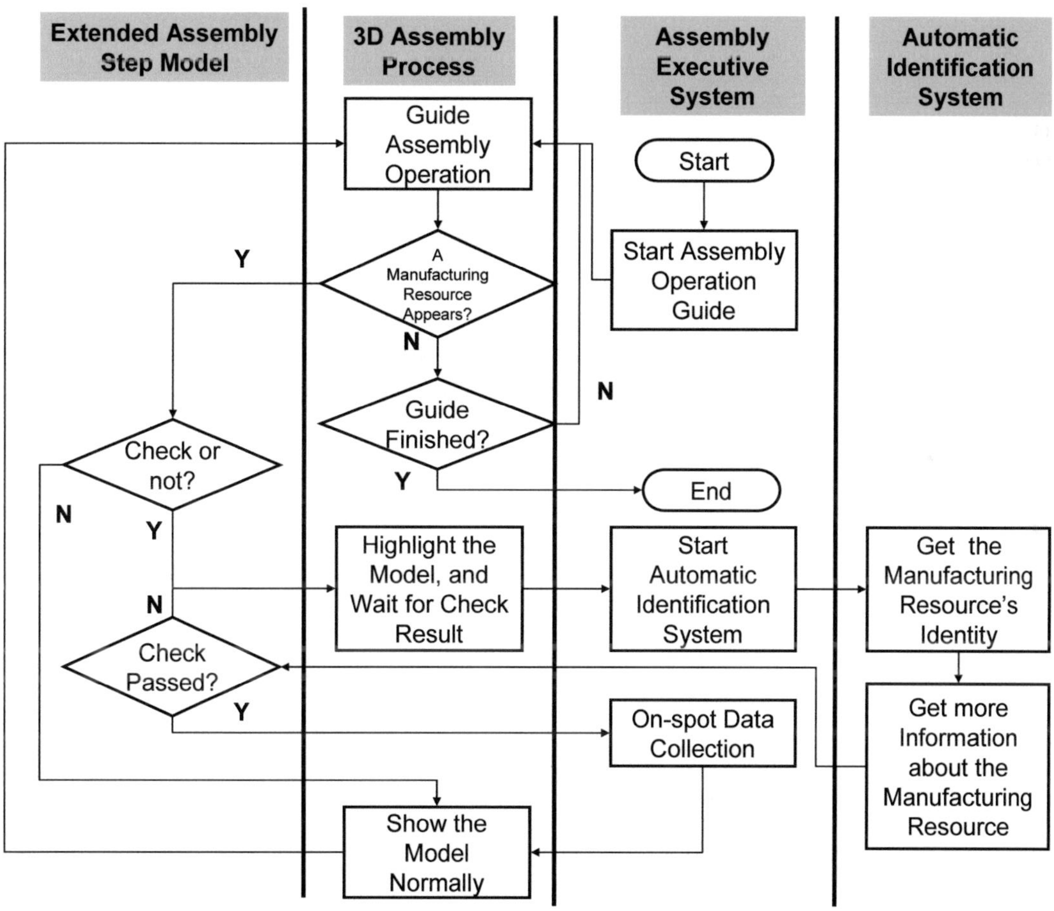

Fig. 8. The automatic matching flow

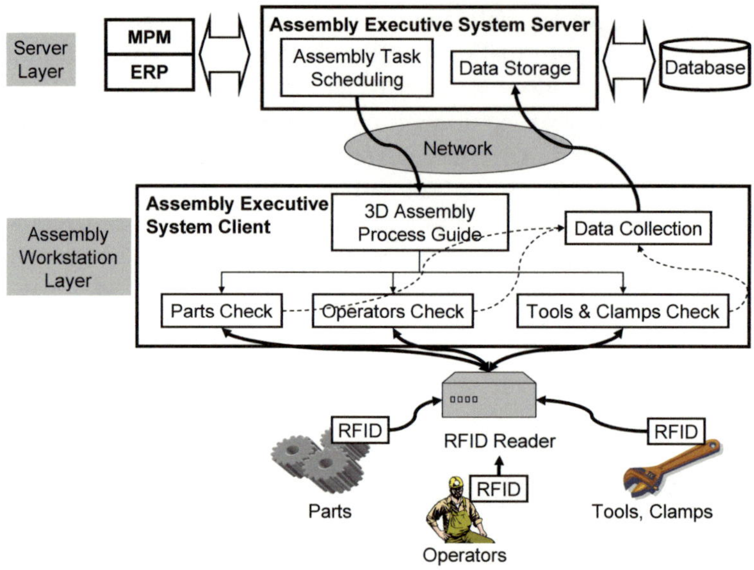

Fig. 9. The implementation framework

4.3 An example

To illustrate the method discussed above, we developed a prototype system. As the core product in the Dassault Systemes' 3DVIA Composer solution, the 3DVIA Composer is chosen to implement extended assembly steps. It is a desktop application for the creation of highly compressed lightweight product documentation contents directly from 3D digital product data. The ThingMagic's M5E-MF4E RFID UHF reader and Psion Teklogix's 7527C mobile terminal are used to read RFID tags. Some metal proof RFID tags are used to identify physical parts.

As shown in figure 10, this example covers two scenes.

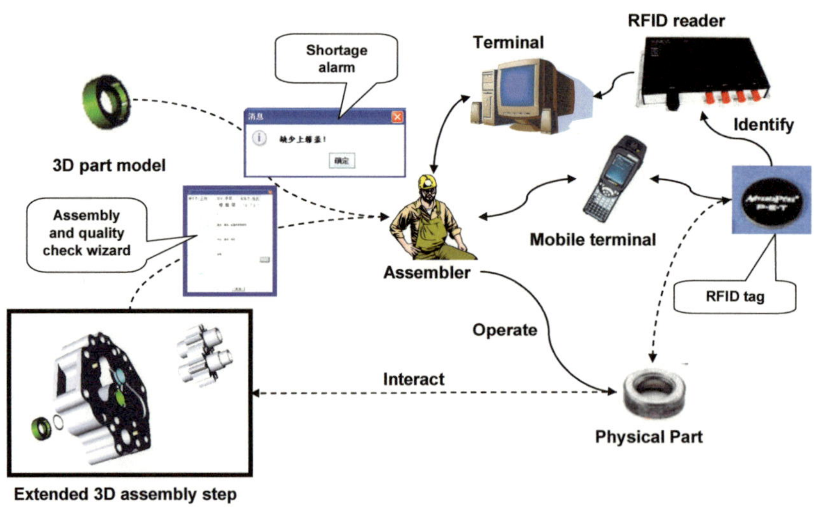

Fig. 10. An example of interactive 3D assembly operation guide

The first one is part check before the assembly operation. When parts related to an assembly task arrives, the assembly executive system triggers the RFID reader to identify all parts. After matching with the BOM tree, a result comes out. If a part is not included, a shortage alarm is displayed. Then the operator should start a shortage report workflow to ensure all parts are provided.

The second one is assembly operation guide during the assembly operation. Guided by extended 3D assembly steps, the operator carry out assembly operations. When a part is needed, its model is highlighted and the animation pauses. And RFID reader is triggered to wait for the operator to pick up the right part. When a part is provided, the assembly executive system get the model's unique ID by reading the RFID tag's storage space or get information from database. If the physical part matches with the highlighted model, the system record the part's ID and continues the extended 3D assembly step. Otherwise, the system keeps on waiting, until the queried part arrives. Under such circumstance, wrongness and missing in the assembly operation can be eliminated mostly.

5. Conclusion

This chapter presents two typical application cases of RFID Technology in the complex product assembly executive process. The first one solves the asynchrony problem between the logistics stream and the information stream in the complex product assembly executive process. The second one associates with the on-spot assembly operation guidance, and achieves dynamic matching mechanism between 3D models and physical parts. Both of them are discussed from methodology and implementation. These two cases illustrate potential of RFID technology's application in enhancing controlling and monitoring methods of complex product assembly executive process.

6. Acknowledgments

The research is under the support of the "National Natural Science Foundation of China" (NSFC, No.: 50805122) and the Science and Technology Innovation Foundation of Northwestern Polytechnical University (No. 2008KJ02017). Authors hereby thank them for the financial supports.

7. References

Fu-Shiung Hsieh (2006). Robustness analysis of Petri nets for assembly/disassembly processes with unreliable resources. *Automatica*, Vol. 42, pp. 1159-1166.

George Q. Huang, Y.F. Zhang, P.Y. Jiang (2007a). RFID-based wireless manufacturing for walking-worker assembly islands with fixed-position layouts. *Robotics and Computer-Integrated Manufacturing*, Vol.23, pp. 469-477.

George Q. Huang, YF Zhang, PY Jiang (2007b). RFID-based wireless manufacturing for real-time management of job shop WIP inventories. *The International Journal of Advanced Manufacturing Technology*, DOI 10.1007/s00170-006-0897-4

Guanghui Zhou, Pingyu Jiang. (2005). Using Mobile Agents to Encapsulate Manufacturing Resources over Internet. *The International Journal of Advanced Manufacturing Technology*, Vol. 25 No. 1, pp. 189-197.

Hossein Tehrani Nik Nejad, Nobuhiro Sugimura, Koji Iwamura, et al. (2008). Agent-based Dynamic Process Planning and Scheduling in Flexible Manufacturing System. *Manufacturing Systems and Technologies for the New Frontier*, pp. 269-274.

Huibin Sun, Zhiyong Chang and Rong Mo (2009a). Monitoring and controlling the complex product assembly executive process via mobile agents and RFID tags. *Assembly Automation*，Vol. 29, No. 3, pp.263-271.

Huibin Sun, Zhiyong Chang and Rong Mo (2009b). An interactive 3D assembly process model. *Applied Mechanics and Materials*, Vols. 16-19, pp. 1087-1090.

Jose Barata, Luıs Camarinha-Matos, Gonc-alo Candido (2008). A multi agent-based control system applied to an educational shop floor. *Robotics and Computer-Integrated Manufacturing*, Vol. 24, pp. 597-605.

Kyung-Hyun Choi, Dong-Soo Kim, Yang-Hoi Doh. (2007). Multi-agent-based task assignment system for virtual enterprises. *Robotics and Computer-Integrated Manufacturing*, Vol. 23, pp. 624-629.

Lu, B.H., Bateman, R.J.and Cheng, K. (2006). RFID enabled manufacturing: fundamentals, methodology and applications. *Int. J. Agile Systems and Management*, Vol. 1 No. 1, pp. 73–92.

Robin G. Qiua.(2007). RFID-enabled automation in support of factory integration. *Robotics and Computer-Integrated Manufacturing*, Vol. 23, pp. 677-683.

Tang Xinmin, Zhong Shi-sheng (2006). Petri Nets Based Air craft Maintenance Disassembly and Assembly Process Planning. *Journal of Civil Aviation University of China*, Vol. 24 No. 5, pp. 21-25. (in Chinese).

Weijun Zhang, Theodor Freiheit, Huashu Yang (2005). Dynamic scheduling in flexible assembly system based on timed Petri nets model. *Robotics and Computer-Integrated Manufacturing*, Vol. 21 No. 6, pp. 550-558.

Zhong Shisheng, Tang Xinmin, Chi Shanchun (2006). Conflict of Shared Resource Oriented Modelling and Scheduling of Aero- engine Assembly Using Petri Nets. *Aviation Precision Manufacturing Technology*, Vol. 42 No. 6, pp. 52-55. (in Chinese).

6

Using RFID Technology for Simplification of Retail Processes

Azra Bayraktar[1], Erdal Yılmaz[2] and Şakir Erdem[3]
[1]*Marmara University, School of Economics and Business Administration, Department of Marketing and Production Management,*
[2]*Marmara University, School of Economics and Business Administration, Department of Marketing and Production Management,*
[3]*Marmara University, School of Economics and Business Administration, Department of Marketing and Production Management,*
Turkey

1. Introduction

"Technology has made our lives more full, yet at the same time we've become uncomfortably "full". (Maeda, 2006, p.I). After considering these words within marketing context we can see that each day we are dealing with an information flow. Although there is a lot of information about almost everything (sold products, barcodes, invoices, information from the supplier, prices, customer data, competitors etc.) today's managers are more unsecure to take certain decisions. They also don't have enough time to pay attention for these controllable or uncontrollable forces. During a business process the way from production to wholesalers and than to retailers is very complicated. After products finally meet the customers, feedbacks are coming back to the companies and the cycle begins from the start. This product and information flow makes business processes very complex because different people (engineers, sales staff, consumers, managers) interfere this cycle. In this study our aim is - based on the simplicity theory of John Maeda – with the help of RFID technology to create a simple process model for retailers. By using RFID tags in their warehouses and stores they might be able to serve better and more efficiently to their customers and have a better overview in a short period of time. The information supplied via RFID allows corporations to plan their internal processes more efficiently. We also would like to analyze the pitfalls of RFID with a case from Turkish retail industry especially for In-Store usage of RFID.

2. RFID implications and simplification of processes

In his theory Maeda emphasizes simplicity laws: (Maeda, 2006, p.I)
Reduce: The simplest way to achieve simplicity is through thoughtful reduction
Organize: Organization makes a system of many appear fewer
Time: Savings in time feel like simplicity
Learn: Knowledge makes everything simpler

Differences: Simplicity and complexity need each other ....

1. **Reducing:** Reducing the unnecessary information through RFID tacked products and shelves.
2. **Organize:** Organizing the sales person more effective so they can have necessary information about their customers with the help of wireless handheld devices and can read the personalized customer cards.
3. **Time**: With the help of RFID tagged products the store managers can gather information about stock levels and responds to customer requirements on time.
4. **Learn:** Store managers and marketing managers can learn detailed and up to minute product information during the day.
5. **Differences**: Not every customer is the same; marketers should approach different customers in different ways.
6. **Shrinking the time**: With the help of RFID- you can get crucial and accurate information about your customer preferences on time.

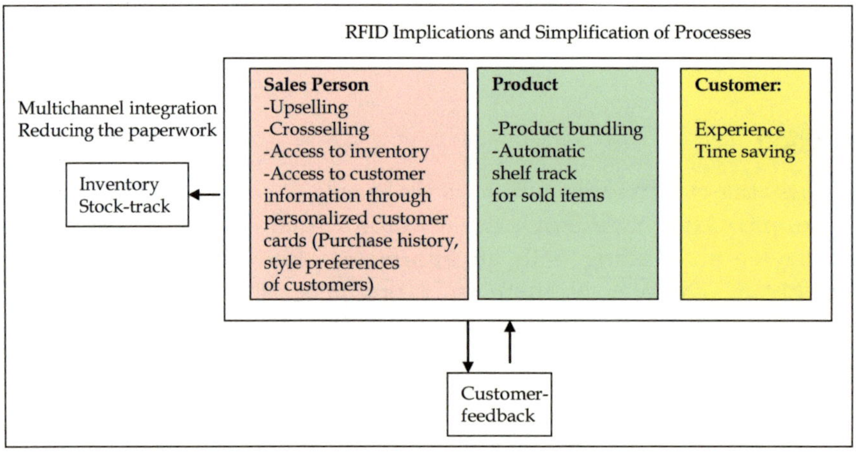

Fig. 1. RFID Implications and Simplification of Processes

2.1 The definition of RFID

RFID is a form of automatic identification and data capture (AIDC) technology that uses electric or magnetic fields at radio frequencies for identification, authentication, location, or automatic data acquisition and transmit, and support a wide range of applications—everything from asset management and tracking to access control and automated payment. RFID systems have the capability of sharing information across organizational boundaries, such as supply chain applications (Sabbaghi and Vaidyanathan, 2008, p.73).

Radio Frequency Identification (RFID) is a generic technology concept that refers to the use of radio waves to identify objects (Auto-ID Center 2002). The core of RFID technology is the RFID transponder (tag) - a tiny computer chip with an antenna. Consumer good suppliers attach these tags to logistic units (palettes, cases, cartons and hanger-good shipments) and, in some cases, to individual items. Logistic units and individual items are identified by the Electronic Product Code (EPC). An RFID reader is used to identify the EPC stored on the RFID tag. The antenna enables the microchip to transmit the object information to the reader, which transforms it to a format understandable by computers (Angelles, 2005, p. 52).

Empowered by the capability to identify uniquely and automatically provide continuous, accurate and real time information on the position and the status of product instances, RFID offers a great improvement opportunity to the shelf replenishment process (Bardaki, Pramateri,2 008; p:4)

2.2 Components of RFID systems

RFID Technologies support a wide range of applications—everything from asset management and tracking to manufactured products and related customer services to access controls and automated payments. Each RFID system has different components and customizations so that it can support a particular business process for an enterprise. Depending on the application in an industry and the enterprise within an industry, A RFID system can be very complex, and its implementations may vary greatly. Conceptually, RFID system may be composed of three subsystems as shown in the figure below (Sabbaghi and Vaidyanathan, 2008, p.73):

1. An RF subsystem, which performs identification and related transactions using wireless communication,
2. An enterprise subsystem, which contains computers running specialized software that can store, process, and analyze data acquired from RF subsystem transactions to make the data useful to a supported business process, and
3. An inter-enterprise subsystem, which connects enterprise subsystems when information needs to be shared across organizational boundaries.

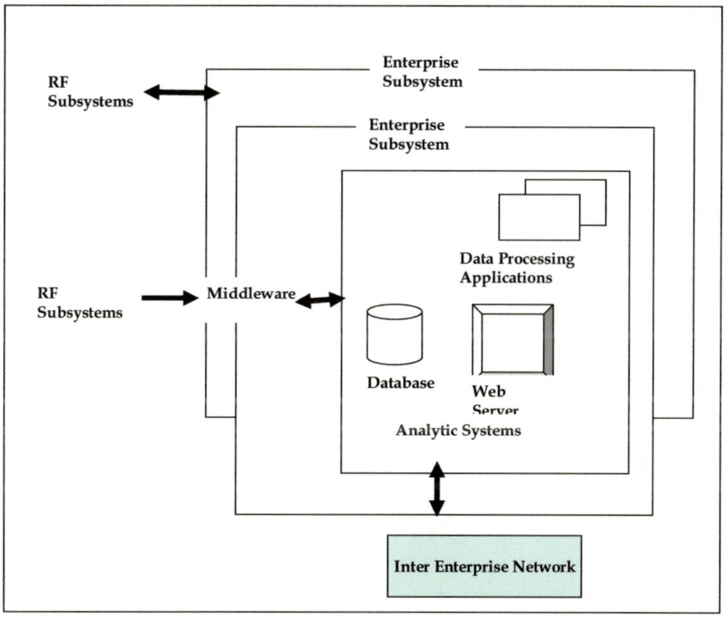

Fig. 2. Inter-Enterprise Architecture

Asghar Sabbaghi and Ganesh Vaidyanathan "Effectiveness and Efficiency of RFID technology in Supply Chain Management: Strategic Values and Challenges", Journal of Theoretical and Applied Electronic Commerce Research ISSN 0718-1876 Electronic Version Vol. 3 / Issue 2 / August 2008 / 71-81

Every RFID system contains an RF subsystem, which is composed of tags and readers. In many RFID systems, the RF subsystem is supported by an enterprise subsystem that is composed of middleware, analytic systems, and networking services. However, in a supply chain application, a tagged product is tracked throughout its life cycle, from the manufacture to final purchase, and sometimes even afterwards (e.g., to support targeted product recalls or related service), and thus its RFID systems has to share information across organizational boundaries. Thus, the RFID systems supporting supply chain applications have also an inter-enterprise subsystem (Sabbaghi and Ganesh, 2008, p 71-81).

3. Warehouse applications for RFID

RFID can be used for many warehouse inventory management operations, including receiving, storage, picking and shipping procedures. With RFID system, items can have a unique and secure serial numbers and it became so visible in inventory and supply chain operations. This visibility brings several benefits that can eliminate current disadvantages and this wireless system can be more efficient.

Warehousing Operations:

Receiving; when pallets are unloaded from the truck, they are automatically identified with fixed position or mobile RFID readers. Fixed position RFID readers can be mounted at the dock door. Mobile readers can be designed as a PDA or they can be mounted on a forklift. Mobile RFID readers can be more effective because they can be used throughout the facility and they require less investment. While RFID system is integrated with Warehouse Management System (WMS), data read from the pallet's tags are transferred into WMS and updating inventory files. This warehouse process reduces the labor needs. If bar codes were being used in this process, all received pallets would have to be scanned by workers and on the contrary RFID, bar codes needs clearly visible labels.
Storage; in the conventional warehouse systems, different items should be storage different locations. Whereas RFID readers can scan locations and read RFID tags from anywhere. As a result of this, items do not have to be storage in specific locations. In this way, many different storage location alternatives can be used for fast replenishment and picking.
Picking; when RFID system integrates with order management system, the order is checked by the WMS to confirm the picked item belongs with the order.
Shipping; the order management system can confirm pallet loads and improve the accuracy of the shipping process with RFID readers. A RFID tagged pallet can be identified a fixed position RFID reader (as vehicle mounted) or a mobile reader (handheld device). RFID allows for an automatic check of the items loaded into the trailer against the customer order. (Jones and Chung, 2008, pg. 325)
Mandates from the large retailers (Wal-Mart, Target, Albertson etc) and government agencies in USA have increased the awareness of RFID. But still companies that have not been affected by the mandate requirements prefer to wait until the technology matures so that they have adequate knowledge about its potential benefits, especially many companies are concerned about the ROI (Return on Investment) models of RFID (Bhattacharya et al. 2007, p. 1; Jabjiniak and Gilbert, 2004). Determining accurate measures for RFID ROI is very important in order to convince managers. Developing a comprehensive framework for all short term and long term benefits will contribute toward the development of ROI measures (Bhattacharya, 2007, p.2).

4. Benefits of RFID system in retailing

RFID technology can track inventory more accurately in real time resulting in reduced processing time and labor. There are many applications and possibilities for RFID/EPC as these objects in motion are traced throughout the supply chain. The complete visibility of accurate inventory data throughout the supply chain from manufacturer's shop floor to warehouses to retail stores brings opportunities for improvement and transformation in various processes of the supply chain. RFID technology can help a wide range of organizations and individuals such as hospitals and patients, retailers and customers, and manufacturers and distributors throughout the supply chain to realize significant productivity gains and efficiencies (Sabbaghi and Vaidyanathan, 2008, p.72).
RFID usage in retailing has taken a lot of attention recently (Bhattacharya et al.2007) because the retail industry is one of the most aggressive supporters of this technology; In comparison, a 2005 report by Frost & Sullivan determined the revenue in the RFID retail market to be $400.2 million in 2004, a figure expected to grow to $4,169 million by 2011 (Bacheldor, 2006, p.1).
During the last decade several research studies have focused on RFID and its benefits and challenges in retail sector (Bhattacharya, 2007, p.3) :

Larsson and Qviberg (2004)[1]	Justification of RFID implementation
Jones et al. (2004)[2]	Potential benefits and challenges of RFID throughout the supply chain for retailers in UK.
Koh, Kim and Kim (2006)[3]	Issues and critical factors of RFID in retail industry
Vijayaraman and Osyk (2006) [4]	Empirical study of RFID implementation in warehousing industry
Karkkainen (2003) [5]	Analysis of RFID benefits obtained by increasing supply chain efficiency for short shelf life products.

Table 1. Studies about RFID benefits in Retail Sector

Bhattacharya, Mithu; Chu Chao-Hsien; Mullen Tracy (2007). RFID Implementation in Retail Industry: Current Status, Issues and Challenges; *Decision Science Institute (DSI) Conference*, Phoenix Arizona AZ, p.3
According to IdTechEx (Bhattacharya, 2007, p.6) the retail industry will comprise %44 of the global RFID market value system including tags by the year 2016. In retail industry RFID is

[1] Larsson, B. And Qviberg, O. (2004): Evaluation and Justification of an RFID implementation, Master Thesis, Department of Management and Economics Industrial Engineering and Management Institute of Technology, Linköping University
[2] Jones, P., Clark-Hill, C. Shears, P.,Comfort, D., and Hillier, D. (2004).Radio Frequency Identification in the UK: Opportunities and Challenges, *International Journal of Retail & Distribution Management*. Bradford, Vol..32, Iss. 2/3; pp.164..
[3] Koh, C.E.,Kim,H.J., and Kim, E.Y (2006)."The Impact of RFID in Retail Industry: Issues and Critical Success Factors. Journal of Shopping Center Research, Vol. 13, Iss.1,pp.107-117.
[4] Vijayaraman, B.S., and Osyk, B.A.(2006). An Empirical Study of RFID Implementation in the Warehousing Industry. International Journal of Logistics Management. Ponte Vedra Beach, Vol.. 17, Iss.1;pp.6.
[5] Karkkainen, M. (2003)."Increasing Efficiency in the Supply Chain for Short Shelf Life Goods Using RFID Tagging. International Journal of Retail and Distribution Management, Vol. 31, ISS.10, pp.529-536.

expected to replace the barcode technology as it provides more benefits. The most important benefits for the future is integrated supply chain management, which enables availability of products, inventory management and decreasing of the costs (Bhattacharya, 2007, p.6). After conducting a content analysis in 2006 about RFID and retrieved 362 articles which have been published between 2002 and 2006 Bhattacharya *et al.* have categorized and summarized the challenges, drivers and benefits of RFID technology for retail industry along with the frequency of articles that support the analysis (Bhattacharya, 2007, p.9- 15)

Drivers for RFID in Retail Ind.	**Percentage**
Benefits*	92,68%
Wal-Mart Mandate	2.85%
Decreasing Cost of tags and readers	2.03%
EPC Global initiatives for Standardization	2.03%
Anti-Counter feiting	0,41%
TOTAL	100%
Benefits* from RFID Implementation	**Percentage**
Operational Efficiency - Reduced out of stock - Accuracy, speed and efficiency of process - Automated shipping/receiving - Reduced Inventory - Improved efficiency of store operations - Improved labor productivity - Streamlined process achievement/Leaner manufacturing	36.84%
Improved Visibility - Real-time Visibility - Tracking and Tracing - Improved visibility of orders and inventory - Asset Management - Return/Recall Management - Tracking shopping behavior - Streamlined reverse logistics	24.12%
Reduced Costs - Reduced labor requirements/costs - Reduced overall costs	10.53%
Improved Security Security against theft/fraud Reduced shrinkage Improved supply chain security Eliminates return merchandise fraud	9.21%
Improved Customer Service Levels	7.89%
Better Information Accuracy Improved packing and shipment accuracy Business Intelligence	7.46%
Increased Sales	3.95%

Table 2. Benefits from RFID Implementation in Retail Industry

Bhattacharya, Mithu; Chu Chao-Hsien; Mullen Tracy (2007). RFID Implementation in Retail Industry: Current Status, Issues and Challenges; *Decision Science Institute (DSI) Conference*, Phoenix Arizona AZ, 2007, (pp 9-11).

Bhattacharya *et al.* has also emphasized that all these retailer specific benefits have a potential to impact customer service levels positively. Although RFID imbedded loyalty programs can add extra values for the customers, store managers and shopping mall management, many companies are acting reluctantly still to use RFID for their Customer Relationship Strategies (Bayraktar; Yilmaz, 2011).

Privacy issues are one of the main concerns of RFID. A balance should be kept between the benefits that consumers can get in terms of better service, time saving and protection of their privacy. Also many business managers would like to see a detailed cost and benefit analysis of RFID implementation.

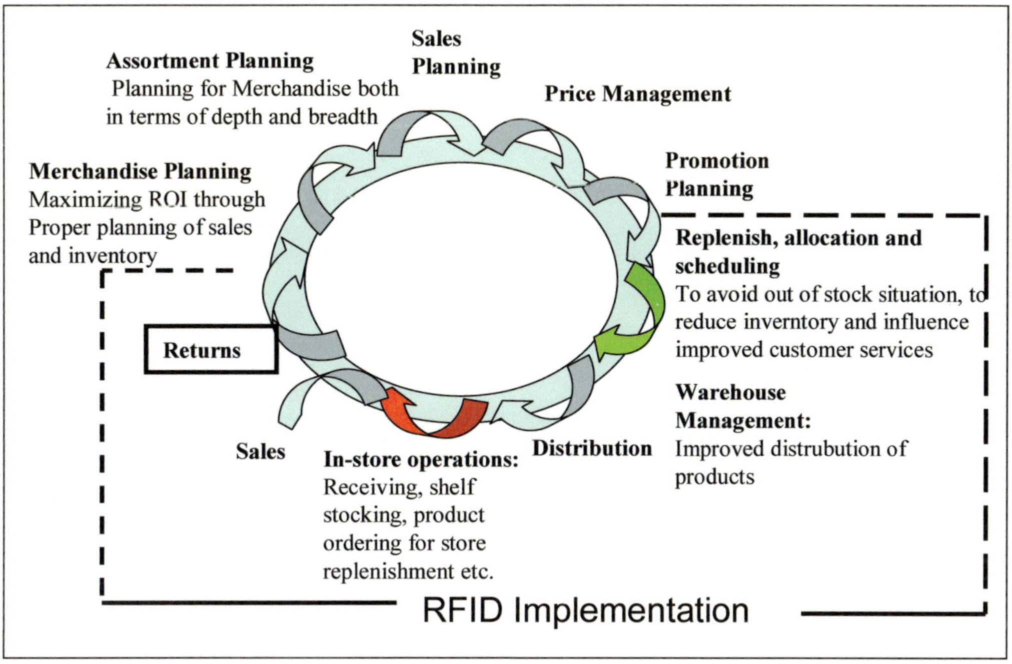

Fig. 3. Integrated Supply Chains (adapted from Callana, 2006; Bhattacharya, 2007 p. 14-15)

We see that most of the dominant RFID benefits are focusing on the lower side of the supply chain. RFID technology has the ability to provide up-to-minute information on sales of items, thus can give accurate information about inventory levels. With this accuracy managers may hold their inventory levels at minimum and this may cause to reduce their inventory costs. RFID technology at the pallet level has the potential to automate the distribution of goods between manufacturing plants warehouses and retail stores of different organizations. Companies can cut their costs also down from lost/misplaced inventory (Sounderpandian et al., 2007, p. 105).

There are evidences which prove a positive ROI for warehouse application but in our study we also would like to consider the advantages of in-store applications for store management and customers.

4.1 In-store applications for RFID

In a retail store RFID tag information is generated based on events: A product is leaving a shelf or a product being checked out by a customer at a checkout corner (Sounderpandian et al., 2007, p. 105). The tag readers should be deployed in a shelf; these tag readers are responsible for reading RFID tags of items on the shelf. Items read by the tag at the checkout generate messages for the host system. After processing these messages the host system informs other partner in supply chain. In addition the host system may send some of the RFID transaction data to the enterprise system of the retailer. The host system is connected to the enterprise information system via a virtual private network (Sounderpandian et al., 2007, p. 106).

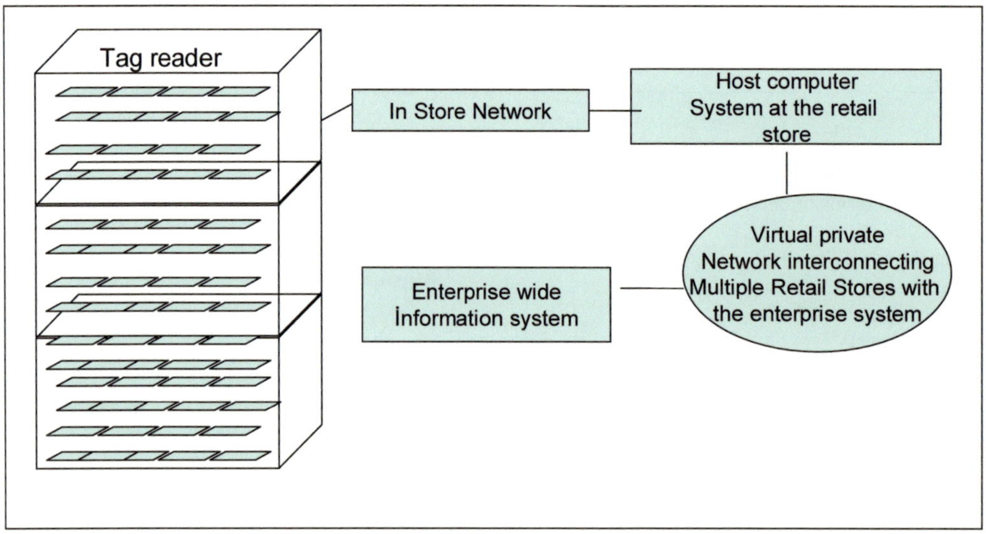

Fig. 4. RFID infrastructure in a retail store

Sounderpandian Jayavel; Boppana Rajendra V ;Chalasani Suresh ; Madni Asad M. (2007). Models for Cost-Benefit Analysis of RFID Implementations in Retail Stores. *IEEE Systems Journal*, Vol..1, No,2 December. p. 106.

In Europe, Metro Group has been using In-store RFID implications in its Future Stores since 2004. At the end of 2008 Metro group has brought 200 sales points in action, included all Metro Cash&Carry big supermarkets, , in 9 central distribution-warehouse of Metrogroup and also most of the Real Warehouses. They also started in 2008 France with RFID applications of 1.3 billion palettes for 89 Metro Cash and Carry Stores (http://www.future-store.org/fsi-internet/get/documents/FSI/multimedia/pdfs/broschueren/RFID%20und%20MG-D-271108-Internet.pdf, p. 21)

There are new usage areas for RFID- Instore applications that have been applied by Metro Ag. at Galeria Kaufhof Essen. Applications are following (Metro Group-Future Store-Guided Tour):

4.1.1 Personal digital assistants/smart shelf

Retailers have limited shelf space available. The choice of which items to stock and the allocation of scarce shelf space among the stocked items are relevant issues for the retailer. For individual SKUs these decisions are important determinants of sales and marketing effectiveness. At the aggregate level, shelf allocation is an important factor in the revenue, cost, and eventual profit of a product category. Complementary to the amount of space to allocate to an item, there is the problem of the location of the item on the shelf. For example, items on the lower shelf usually get less consumer attention than items on upper shelves. The items on the lower shelves may therefore have lower sales and may also benefit less from promotions.

Finding the profit-maximizing shelf arrangement while, at the same time, meeting manufacturers requirements is far from easy. A prerequisite to actual shelf optimization is a proper measurement of the effect of shelf layout on sales and marketing effectiveness (Nierop amd Franses, 2008, p. 1).

Retailers are testing the Smart Shelf, where an RFID reader is incorporated into the shelf and stocked with tagged product. The Smart Shelf monitors its rate of depletion, provides an alert when stock runs low and automates reordering to minimize out-of stocks. Gillette is testing smart shelves in an attempt to minimize theft. Because store personnel program the system with store sales data, the system detects behavior outside the norm and can alert store personnel by transmitting information to a personal digital assistant. By identifying the nature of the stock loss and mapping and addressing points of vulnerability, losses in some stores have been reduced by 70% to 80% (Thompson, 2004; p.3).

Advantage to the company with RFID systems: RFID also provides transparency on the sales floor. Every item of clothing is assigned a certain position on hangers or shelves. This data is saved in the outlets database. Employees record these items on shelves with "Personal Digital Assistants (portable RFID readers)". Incorrectly stacked goods could easily be found and resorted (Metro Group-Future Store- Guided Tour; Yalçınkaya; 2007). Advantage to the Customers: In store located RFID readers constantly detect item transponders. Current stock is shown detailed (according to color and size) on the screen. This way customer can check if desired items are in stock on the hangers or shelves (Metro Group-Future Store- Guided Tour).

4.1.2 Check out

Today, staff at the check out scan the barcodes on items to calculate the total amount purchased and deactivate the EAS (electronic article surveillance).

Advantage to the company with RFID systems: During the payment process, data is removed from the RFID system that operates parallel to the merchandise management system. Advantage to the Customers: No links are made between the purchased items and personal data - regardless of whether payment is with an EC Card or credit card. Customer can ask staff to remove the transponders completely if they wish (Metro Group-Future Store- Guided Tour).

4.1.3 Smart mirror

The advantage to the Customers: Customers could check the Smart Mirror to see if the selected garments fit. Integrated RFID readers detect the transponders that are fitted to the clothes being tried on. Detailed information (washing instructions, price etc.) on a chosen product then appears on the mirror's surface if requested (Metro Group-Future Store- Guided Tour).

Also another brand for luxury goods, Prada is using the smart mirrors in its stores in New York Epicenter Store. The mirrors in dressing rooms become magic mirror with a combination of a touch activated display and cameras, so customers may see what they try on from various angles. All articles in this shop have been tagged with RFID transponders which enables this technology (Spektrum RFID, 2011).

4.1.4 Smart dressing rooms

Advantages to the Customers: There were also touch screens in the cubicles in Gardeur shop. Smart Dressing rooms identify which item customers try on and shows product details on the screen. It also gives tips on accessories and possible combinations. Customers can also access details on suggested items by touching screen (Metro Group-Future Store-Guided Tour).

4.1.5 Handheld reader for salespersons

Advantages to the company with RFID systems: Within the store, which has been tagged with RFID transponders a handheld reader tells employees which items are still in stock. This is also an enormous advantage for customer service and availability of goods (Spektrum RFID, 2011).

4.1.6 Anti-theft system

Advantages to the company with RFID systems: RFID readers were also installed at transition points around escalators and lifts. In the future Metro AG is also planning to use passive RFID transponders (Metro Group-Future Store- Guided Tour).

4.2 Cost and benefit analysis of RFID systems in retailing

Benefits of the RFID systems in various industries (defense, healthcare, entertainment ets.) have been widely discussed, but managers still have some concerns about using this technology especially for in-store applications.

The advantages of In-store RFID systems are automatic check-out and reducing inventory costs due to the efficient shelf replenishment. Another issue is also the reduced losses due shoplifting. The points for these concerns are:

- Tag readers cost's, infrastructure costs (hardware and software costs, including the communication network required for RFID implementations)
- Yearly operational or maintenance costs (Sounderpandian, 2007, p. 106).
- Educating sales personnel about the new technology,
- Ethical issues and security concerns from the customers (Bayraktar, Yilmaz, 2010).

Sounderpandian *et al.* (2007) have calculated the formula for a retail store to evaluate whether or not an RFID implication is beneficial for a retail store. Then they showed with the help of a numerical example a retail store which uses part-time employees and implemented RFID system in its stores (Sounderpandian et al, 2007, p. 112): The formula is:

$$FRFID < F + BSL + BPOS + TIC * \sqrt{(1 - VRFID/v)}$$

FRFID = Fix costs: The costs at the maximum number of shelf replenishments $312.000

V= Variable costs: Wages of part-time employees who have been hired depending on the workload and the cost of consumables, $0,75. Variable costs depend on the number of replenishments in a year.

FRFID: RFID implementation costs: Amortized cost of computer hardware, RFID Related Equipment, RFID Tags, salaries of full-time employees, and the wages of (fewer) part-time employees, variable costs are the costs for the consumables.

BSL: Benefit from shoplifting: Using RFID reduces the loss due to shoplifting (for example $10 a day makes $3650 a year)
BPOS= Savings from the use of RFID at POS terminals so the checkout will be faster. Store needs fewer checkout counters and fewer cashiers. BPOS:$ 24.000 (in a year)
TIC: Total Inventory costs: Total costs of carrying and ordering inventory. Carrying costs: The costs of the shelf space: Rent, utilities maintenance of the area where the shelves are kept. TIC: $980.000
Sounderpandian and his colleagues have made the calculation for a retail store and discovered that RFID installation is beneficial.

$$\$750\,000 < \$312\,000 + \$3650 + \$24\,000 + \$980\,000 * \sqrt{(1 - 0{,}10/0{,}75)} =$$

$$\$\,750\,000 < \$961\,805.$$

However in countries where labor costs are relatively cheaper, managers may prefer to pay wages for part-time workers instead of paying for RFID implementation.

5. Concerns for global brands

Another problem for Global Brands is that the franchisees may not agree to tag their products with RFID transponders unless the main company (franchisor) agrees to place the RFID tags in the products before arriving at the retail store. (Number 7 in Figure 5) .

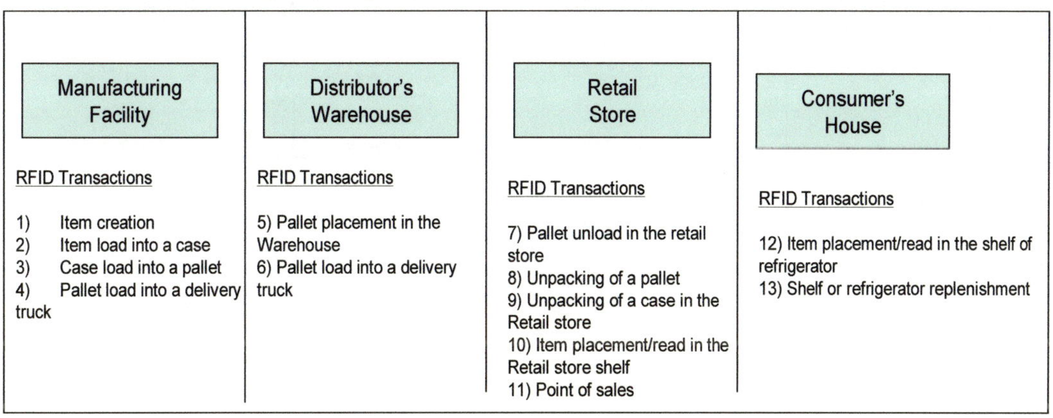

Fig. 5. Transition of an item from the manufacturer to the consumer in the supply chain and the relevant RFID transactions.

Sounderpandian Jayavel; Boppana Rajendra V ;Chalasani Suresh ; Madni Asad M. (2007). Models for Cost-Benefit Analysis of RFID Implementations in Retail Stores. *IEEE Systems Journal*, Vol.1, No,2 December, pp. 107.
It is not feasible for a large retailer, one with a daily turnover of 40,000-60,000, to place RFID tags on products before entering the store.
Another issue is that they have to change all labels and price tags three times a year according to the sales etc. Usually on product labels there is information about the country of origin, washing instructions etc. Therefore an extra RFID tag is sometimes too much for an item.

Other problem will be in point 8 and 9 in Figure 5. If the retail store is in a shopping mall, they have to unload their pallets in certain hours, and they can use loading elevators in shopping malls during limited hours. Until the stores open at 10.00 am, all the articles have to be unpacked, tagged, labeled and placed in the store shelves. Therefore with the increase the item numbers have been sold in the store this system could be complicated and costly.

6. Case Study: RFID Application in a Turkish Retail Company

Turkey's ready-to-wear clothing industry is one of the major industries of its economy and international trade. According to Sevim and Emek (2006) clothing and textiles have annual sales of $30 billion and a 26 percent share of total export Vol.ume in 2005. Turkey is the fourth largest clothing supplier in the world and second largest supplier to the European Union. Under the World Trade Organization Agreement on Textiles and Clothing this sector continues to maintain and enhance its competitiveness despite the abolition of quotas (Sevim and Emek, Turkish Clothing Industry Report, Export Promotion Center of Turkey, 2006). Germany, the UK and the U.S. are the most important markets for Turkish exports, with export shares of 27%, 18%, and 8% respectively. However, compared to 2004 data, exports to the U.S. have declined 21 percent. Hence, understanding the causal factors has potential for reversing this decline and growing apparel and textile exports to the U.S (Seitz, Neace, Razzouk, Keyfli, Tung,2008; p:173)
Throughout the theoretical findings we prefer to make a exploratory study about using RFID technology. In general, exploratory research is appropriate to any problem about which little is known. Exploratory research then becomes the foundation for a good study (Churchill, 1999;p:103).

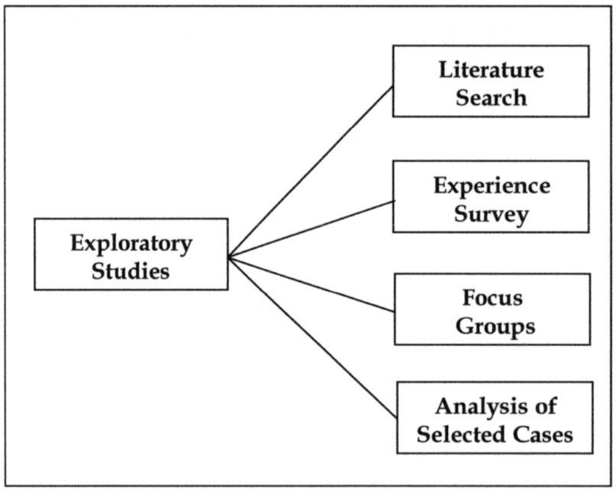

Fig. 6. Types of Exploratory Studies

Seltiz, C., Wrightsman, L.S.,Cook, S.W. (1976). Research Methods in Social relations, 3rd ed. New York. pp.90-91.
As shown in figure 6, "analysis of selected cases" is a type of exploratory study. The analysis of selected cases is sometimes referred to as the analysis of "insight-stimulating examples." By either label, the approach involves the intensive study of selected cases of the phenomenon under investigation. Examination of existing records, observation of the occurrence of the phenomenon, unstructured interviewing, or some other approach may be used. The focus may be on entities (individual people or institutions) or groups of entities as sales representatives or distributors in various regions (Churchill, 1999; p:113).

For the case study we have contacted a Turkish retailer and focused on a company. The company (mentioned as X to protect the confidentiality of the company) is one of the fastest growing retailers in Turkey. The company is active in the textile business and its product categories consist of casual children's, men's and women's articles. Their price ranges are between low and medium. After discussing with operations and logistics manager, we gathered enough information regarding RFID.

The company has been using the RFID technology for 3 three years. Contrary to other companies we have seen in the literature, the company X made some tests for its warehouse about the feasibility of the process three years ago but then decided to use RFID technology in its five selected (two stores in Istanbul, three stores outside of Istanbul) stores but not in its warehouse and supply chain system.

6.1 RFID test project

Products should be placed with RFID tags by the suppliers first. For the textile products they have ordered samples for washable, sewn-in RFID tags and made reliability tests for stock control. Companies who tried the string RFID tags had problems because they can break off and get lost easier; therefore, they have built antennas in the warehouses and made RFID track tests for the sample products with sewn RFID tags.

Results of the test: For stock follow ups and inventory levels they have nearly 100% reliability. Defection of the tags and the possibility of not reading the RFID codes within the cases was nearly 0%. In product incoming process into the warehouse:

Defection rate because of the inability of tag reading was again nearly 0%.

For the calculating stock level's accuracy rates were nearly 100% because of the durability of RFID tags. Live time value of RFID tags are 5 years but they should not come near the magnetic areas like near the transformer room etc.

Following up customer complaints: Even if a complaint comes from the customer for a specific product the company can match the Product ID with factory code. Especially The sewn-in tags are more durable than the thermal ones during warehouse processing. This system also facilitates the quality control process.

6.2 Advantages of RFID tags within the supply chain for company X

Unique ID Number: The advantage of unique RFID tags compared to Product ID is each product with the same color, the same size and the same model will have the same Model ID. Even 100 products will have the same Model ID. But with RFID each product will have a unique product ID and company X will follow the products from supplier to consumer with this unique ID number.

Sorting system: Before products have been distributed from the main warehouse, RFID can also make it easier during the preparation stage of delivering goods. Even within the stores, if there is an elevator in the store, building RFID antennas at elevator doors can help sorting the system between the stages and make the follow-up process easier for the employees. Company X delivers 3600 cases/hour daily from its main warehouse to its stores.

Quality of RFID Tags: Company X uses thermal etiquettes now and they can be damaged because of dust, tearing off, getting lost etc. but sewn RFID etiquettes can avoid all these disadvantages.

Collecting of Off-Season Articles: If a company use a trolley with an RFID antenna at the bottom, they can collect all the RFID tagged off-season articles in the store and when the

trolley leaves the store and goes to the store warehouse, items can be deleted from the system automatically.
Despite all these advantages the company X decided not to implement RFID technology for their supply chain system at this time. Reasons of these decisions are following:

6.3 Disadvantages of RFID tags for company X in their supply chain

Problems for different Product types: Company X needs different types of RFID tags (for belts, for the coats for the shoes, etc.)
Implementation costs: Each year 160 million products are sold. For each sewn-in RFID tags they had to paid (in 2007) 11-14 cents. If they implement this strategy they would like to eliminate the barcode system.
The cost for 160 million item delivery per year and implementation cost for 320 stores in Turkey (including software costs, RFID tags of all Products, handheld computers for employees, antennas, etc.) is 37 billion dollars.
As fixed cost the company calculated for the stores' antennas at the door and at the check out point.
Variable costs are handheld terminals, RFID tags.
ROI time : In retailing ROI time is approximately one year. According to their calculation company X can get this investment in 27 months.
Production in different locations: The Company has 300 producers and 50% of them are outside of Turkey. Their production facilities are in Bangladesh, Egypt, Sri Lanka and in China. After sewing RFID tags into the products, suppliers should also match the RFID tags with the unique products and load information in to the Network system.
Problems with sewn-in Tags: These tags should be sewn separately and carefully without damaging the antenna. For this sewing process they have to pay 1 cent extra and it makes the RFID tag cost 15 cent per unit. Because the retail prices for their products are not so high, company X can not add these extra costs to their prices.
Different system in the Warehouse and in the stores: If one company decides to use RFID in their warehouses and it they should also implement this strategy in their stores to make a hypenate for the information cycle. If the company uses barcodes in the stores and RFID tags in the warehouses, the information system should always transfer data between Model ID and Unique Product ID. Now in their system, stores are transferring RFID data as a barcode data to the warehouse.
Therefore company X did not implement RFID technology in its supply chain and warehouse system. They are still using barcodes for the product groups in factories and warehouses.

6.4 RFID applications for company X within the stores

Company X selected two test stores for RFID tags in two different areas in Istanbul. They have contacted a Turkish IT company, who invented a special alarm tag EAS (electronic article surveillance) with RFID for them.
Process for the incoming products: Before placing the incoming products on the shelves, sales personnel have to match each product with the barcode and the RFID transponders. In order to simplify this process company X placed a horizontal antenna. Since products are shipped from the factories with a barcode inside, employees should match barcodes with

RFID transponders. After the matching process is completed, employees place the products on the shelves.

For missing products in the stores: Company X is not using an intelligent shelf system. If a customer searches for a product which is in store network system but not on its shelf, employees are entering its RFID code in the handheld computers and walking within the store until transponders give a signal for the lost item.

Deactivating RFID tags: After customers finish their payment process, employees tear off the EAS tags (within RFID transponders). These tags and their information are deleted at the end of the day, so one EAS (RFID transponders) can be matched with another barcode and used 10.000 times. With this system customers can be sure about the privacy issue, because they wont carry the RFID tags on the products (Orel, 2006).

6.5 Advantages of RFID tags for company X within the stores

Reducing check out time: Employees are also using this system for the item counting. The main benefit of RFID transponders is reducing check-out time and line. Since the company has low prices, customers usually buy three or more items. Employees are putting all purchased products on the antenna, and the RFID transponder's, system can read them within seconds.

RFID Basket:

The main advantage of RFID is using RFID reader imbedded baskets in the store warehouse. Employees can bring the items in/from the store with these baskets and RFID readers transport information of these items directly to make stock control processes more efficient.

6.6 Disadvantages of RFID tags for company X within the stores

After discussing with the store manager, she explained the difficulties of RFID tags, as follows;

Using RFID tags in the EAS alarm tags : Managers ordered these tags in order to decrease number of etiquettes which have been tagged on the items but these tags are really big and cause some difficulties for small items (like baby products, underwear). They damage the item or the item is so thick, alarm can fell down easily.

The impact of RFID for Incoming and Store Warehouse Process:

This test store is one of the most crowded stores of the company. Each day this store receives 120-150 big boxes of products.

For the incoming process employees should open the boxes. Since each product is placed in a bag they should open them before placing the incoming products on the shelves. Sales personnel have to match each product with the barcode and the RFID transponders. Than place the tag into the products. This process negatively effects the time for bringing the product on the shelves.
This process takes 4 seconds. If employees would eliminate the matching process (in case of using the barcodes only) they only have to read the barcodes per product and this reduces the process by 1.5 seconds. These 4 seconds with RFID tags could be also longer sometimes if one tag is defect or if RFID tag has been matched with the wrong barcode.
This store receives 1000 products daily and with RFID tags they need 4 hours to finish incoming process and two employees are full time dedicated to do the matching.

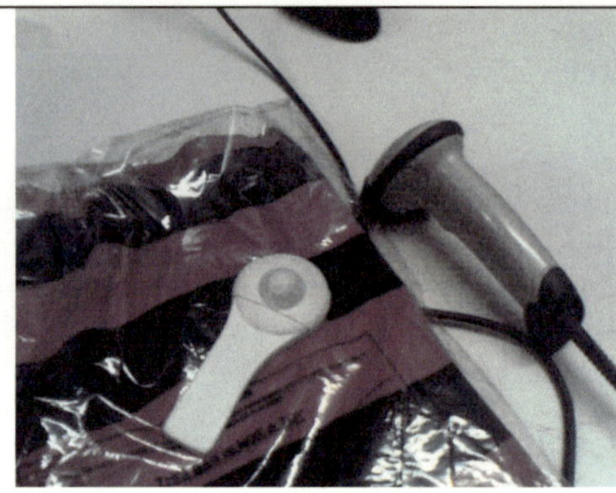

Check-Out Point:

Normally reading the barcodes of purchased items separately takes for multiple items 52 seconds in average. Reading these items with RFID readers decreases this time by 8 seconds. But RFID readers are located under the cash register (see picture at the right side) and the radius of these readers are 60 centimeters or more. Therefore sometimes the reader may not read only the tags of the purchased item but extra tags around the reader as well (like the tags which have been taken off and thrown in to the basket like in the picture).
In this situation the employee should repeat the purchasing process from the beginning and this decreases check-out point performance of the store.

Security Issue:
The needles of the alarms are very short and can be taken off very easily. This problem increases also shop-lifting per day (especially for expensive products like coats). To eliminate this problem Company X hired extra security personnel (4 full-time, 2 part-time). One person is responsible for the entrance and the others are responsible for inside the store and for dressing rooms. But still the Company can not prevent this situation.

Technological support service from the RFID Company:

In our case Company X has an outsourced RFID company, which supplies them the tags, portable readers, check-point readers etc. But this company was not able to give appropriate service for RFID. Tags, readers can be damaged easily and RFID company charged each time when they have to fix the devices extra cost to the head office of Company x.

Problems from the customer point of view:

Because of the reading problem at the check-out point customers may sometimes pay for the products that they had not bought. The RFID reader may read a barcode multiple times. The percentage of this problem is 5% to 6%. But if a customer has bought ten items, it can take half an hour for the employee to renew the process. Also for EAS alarms there is a last control device at the check-point before customer leaves the store. But for RFID there is no device and customers may face with alarm at the exit if one tag remained on the items.
These problems damage also the customer relations of the Company X.

7. Conclusion

For the new technological systems like RFID, mobile etiquettes and printers etc. retailers want the ROI less than one year. If the ROI time is longer than one year, they prefer not implementing these technologies.
To also consider the opportunity cost effect many retailers would like to invest this amount into advertising or other short term investments and gain the information instead of RFID with other cheaper technologies.
For in-store usage RFID Companies have to work especially on security issues. During the R&D process they have to work closely with the retailers and store managers in order to find the best solution, which will satisfy the need of each member in the supply chain.
If the retailers would like to implement this technology in their stores, it is appropriate only for stores which receive 100 items (on average) daily, have expensive merchandise and relative low in-store traffic.
But still the advantages are really very important. Also a lot of companies are ready to invest in this technology as soon as the cost for RFID tags decreases. Therefore for future research into new uses and of RFID tags need to be investigated.

8. References

Angelles, Rebecca (2005). RFID Technologies: Supply Chain Applications and Implementation Issues, Information System Management, Winter 2005, pp: 51-65

Bacheldor, Beth: (2006). Forecast Indicates Strong RFID Demand by Heavy Manufacturing. *RFID Journal*, Available at : http://www.rfidjournal.com/article/articleview/2652 accessed on 23.02.2011)

Bardaki, C., Pramatari, K. (2008). IP-Mapping a RFID-integrated Shelf Replenishment Information System for the Retail Industry to Assess Information Quality. *Proceedings of the 16th European Conference on Information Systems* (ECIS), Galway, Ireland, June 9-11, 2008.

Bayraktar Azra, Yilmaz Erdal, Yamak Oygur (2010). Implementation of RFID Technology for the Differentiation of Loyalty Programs. *Journal of Relationship Marketing*, 9,1:30-42.

Bhattacharya, Mithu; Chu Chao-Hsien; Mullen Tracy (2007). RFID Implementation in Retail Industry: Current Status, Issues and Challenges; *Decision Science Institute (DSI) Conference*, Phoenix Arizona AZ, Available at http://citeseerx.ist.psu.edu/viewdoc/download?doi=10.1.1.77.2715&rep=rep1&type=pdf (last accessed on 26.02.2011).

Callana, G. (2006). Improving Supply Chain Efficiency in Retail Sector. *Supply Chain Europe*, London: May/June. Vol..15, Iss.2; pp.22.

Churchill Jr, G.A. (1999). Marketing Research: Methodological Foundations. Dryden Press Seventh Edition. Orlando.

Jones, E.C. and Chung, C.A., 2008, RFID in Logistics: A Practical Introduction, CRC Press Taylor & Francis Group, USA.

Jabjiniak, B. and Gilbert, G, (2004), RFID Warrants a Strategic Approach, Business Integration Journal, pp.29-31.

Maeda John: (2006) Laws of Simplicity : The MIT Press, Cambridge Massachusetts London England. ISBN-10:0-262-13472-1; ISBN-13:978-0-262-13472-9

Metro Group-Future Store- Guided Tour: Available at: http://www.future-store.org/fsi-internet/html/de/23786/index.html;http://www.future-store.org/fsi-internet/html/de/1613/index.html, last accessed on March 2011.

Nierop,E.V., Fok,D., Franses,P.H. (2008). Interaction Between Shelf Layout and Marketing Effectiveness and its Impact on Optimizing Shelf Arrangements. *Marketing Science, Articles in Advance*, Published online ahead of print June 23, 2008 pp. 1–19

Sabbaghi,A., Vaidyanathan G. (2008). Effectiveness and Efficiency of RFID technology in Supply Chain Management: Strategic values and Challenges. *Journal of Theoretical and Applied Electronic Commerce Research* ISSN 0718–1876 Electronic Version Vol. 3, Issue 2, August, pp.71-81.

Seitz,V., Neace,M.B., Razzouk,N., Keyfli,E., Tung, C.W. (2008). Turkey: gaining market share in the U.S. ready-to-wear clothing market. *Review of Economic and Business Studies*, Vol.. 1, pp.171-178.

Seltiz, C., Wrightsman, L.S.,Cook, S.W. (1976). Research Methods in Social relations, 3rd ed. New York. pp.90-91.

Sevim and Emek, Turkish Clothing Industry Report (2006). Export Promotion Center of Turkey,

Spektrum RFID: Metro Group: Future Store Initiative, available at (http://www.future-store.org/fsiinternet/get/documents/FSI/multimedia/pdfs/broschueren/WISSB_Publikationen_Broschueren_SpektrumRFID.pdf)last accessed on February 2011.

Sounderpandian Jayavel; Boppana Rajendra V.; Chalasani Suresh; Asad M. Madni (2007). Models for Cost-Benefit Analysis of RFID Implementations in Retail Stores. *IEEE Systems Journal*, Vol..1, No,2 December, pp. (105-114).

Thompson,O. (2004). Supply chain payoffs with RFID:With advances in technology and increased Volumes, RFID will become cost effective for many applications. Food Engineering Magazine. April 6 2004 (http://www.foodengineeringmag.com/Articles/Column/718f8b49082f8010VgnVCM100000f932a8c0

Orel, Fatma Demirci (2006): Mağazacılıkta Ümit Veren Yeni Bir Teknoloji: RFID (A new technology which gives hope to retailing) , available at http://www.fatmaorel.net/bizim%20market/RFID.pdf , last accessed on 14 March 2011.

Yalçınkaya Levent (2007). RFID ve Hazır Giyim Sektörüne Katkıları (RFID and its contributions to ready to wear industry), March, available at http://sts-rfid.com/docs/doc1.pdf, last accessed on March 2011.

7

A Solution with Security Concern for RFID-Based Track & Trace Services in EPCglobal-Enabled Supply Chains

Wei He[1], Yingjiu Li[2], Kevin Chiew[2], Tieyan Li[3] and EngWah Lee[1]
[1]*Singapore Institute of Manufacturing Technology*
[2]*School of Information Systems, Singapore Management University*
[3]*Institute for Infocomm Research*
Singapore

1. Introduction

1.1 Overview

A supply chain represents the flow of materials, information, and finance as they move through supply chain partners such as manufacturers, suppliers, distributors, retailers, and consumers. The track & trace services in supply chains can help improve supply chain visibility and efficiency, and prevent counterfeiting and stealing of products thus enhance security. Track & trace services in supply chains require identification of items, capture of events as items move through supply chains, and query of events of items. RFID (radio frequency identification) is a technology that allows to identify objects simultaneously in a fully automated manner via radio waves. This advantage has enabled RFID technology to be used in many applications, including supply chain management (Angeles, 2005) and industrial production (Mintchell, 2002). RFID-based product track & trace in supply chains has attracted growing interests from both academic research and industrial practices.

1.2 RFID

The basic premise behind RFID systems is that each item in a supply chain is attached with an RFID tag. Such tag contains a transponder that emits radio waves of messages readable by specific RFID readers. Most RFID tags store identification codes such as customer number or product SKU (stock-keeping unit) code. The EPC (electronic product code) standard is a promising standard used for RFID identification codes. RFID tags may contain writable memories, which can be used to store extra information for sharing by various RFID readers in different locations. This information can be used to track the move of tagged items, and can be made available to each reader (RFID Journal, 1983). RFID tags can be classified in two general categories, namely active and passive, depending on their source of electrical power. Active RFID tags contain their own power sources, usually on-board batteries. Passive tags obtain power from the radio wave signals of external readers. RFID readers also come in active and passive varieties, depending on the types of tags they read.

1.3 EPCGlobal

Facilitating the use of RFID technology in global supply chains with low cost RFID tags and readers, the EPCglobal network is a platform to pass EPC numbers and leverage on the Internet to access large amount of associated information that can be shared among authorized users. Judging by interest in the global marketplace, EPCglobal is considered to be the next generation of automatic product identification system to facilitate object track & trace in real time throughout a supply chain (Tan, 2005). Its objective is to create a universal and open standard for identifying individual objects and sharing information as these objects traverse a supply chain. Besides a string of digits to identify manufacturer and product, EPCglobal adds another set of digits—serial number—which is unique to each object to identify and track a specific object as it moves through a supply chain. The EPCglobal number is stored on the microchip embedded in an RFID tag. An RFID tag reader sends out electromagnetic waves that can power up an RFID tag, enabling it to transmit back the information stored on its microchip. The reader receives the EPCglobal number, queries ONS (object naming service) about where to find the information about the tagged product, and retrieves the PML (physical markup language) data about the product from specific EPCIS (EPC information services) in the network as defined by ONS. Access to an EPCIS server is subject to authorization and authentication based on specific business agreements and contexts.

1.4 Security

Security has become a major concern while product and information move through a supply chain. An example is the product diversion such as smuggling, counterfeiting and terrorism. Questions of concerns include whether a received item is valid, whether an RFID reader is authorized to read its information, and how to keep the information secure among partners in the EPCglobal network. To address these challenges, hundreds of papers have been published in research literature on solving various security or privacy issues (Avoine, n.d.). Many international organizations such as Customs Trade Partnership against Terrorism (C-TPAT), Container Security Initiative, and Auto-ID Center are formed to address security issues in various industries (Auto-ID Centre at St. Gallen, 2006). However, the research for protecting RFID information in global supply chains is still in its infancy stage, and there are many issues to resolve before we can achieve a fully collaborative system (Sheu et al., 2006). In particular, there is a lack of unified RFID track & trace scheme to provide authenticity, integrity, privacy and accuracy for syndicated applications in EPCglobal-enabled supply chains.

1.5 Contribution and organization

The major contributions of this paper are as follows: (1) We propose a solution for RFID based track & trace services in EPCglobal-enabled supply chain with authentication process. (2) We implement a prototype for our solution. (3) We highlight the functionality of EPCIS in our system. (4) We design the models for track & trace services. (5) We summarize the industry interests in our system prototype.

The remaining sections are organized as follows. The related work on this topic is surveyed in Section 2, followed by our RFID-based track & trace solution with security concern in Section 3. The prototype design and implementation in an OM (order management) scenario are discussed in Section 4. Finally, the conclusion is given in Section 5.

2. Related work on RFID applications in EPCglobal-enabled supply chains

EPCglobal is an R&D effort of several reputable universities and institutes led by MIT Auto-ID lab. It has been the standard for the global supply chain track & trace. Based on EPCglobal, some other R&D efforts have been undertaken for the development of RFID middleware platforms to facilitate RFID application development. For instance, the Accada software package is an open source EPCIS repository and EPC middleware developed by MIT Auto-ID lab and Institute for Pervasive Computing, Zurich (Floerkemeier, Lampe & Roduner, 2007; Floerkemeier, Roduner & Lampe, 2007). Some researchers have also investigated methods of storage and management of RFID data (Derakhshan et al., 2007). RFID data can be stored in the EPCglobal network, RFID tags, or both. For example, Diekmann *et al.* (Diekmann et al., 2007) focused on the study of managing data in a complex RFID environment to deal with frequent data acquisition processes and increased data granularity. They explored the strategy of data management in EPCglobal network vs. RFID tags (i.e., data-on-network vs. data-on-tag) to facilitate the process management and the track & trace services. Other researchers put their efforts on improving the performance in data query response time and data reusability in EPCglobal network. For example, Song *et al.* (Song et al., 2006) proposed a proxy-based EPC track & trace service architecture with a proxy layer inside the EPCIS.

For RFID applications in supply chains, Straube *et al.* (Straube et al., 2007) investigated how to enhance supply chain visibility, efficiency, and performance from various perspectives. The key challenges include the identification and track & trace of items in supply chains, and information management and sharing, as well as security. As an advanced automatic identification technology, RFID allows supply chain partners to have real time information of supplies and demands and to avoid bullwhip effect (Huo & Jiang, 2007). Based on the analysis of SCM (supply chain management) visibility requirements and general RFID visibility potentials, Melski *et al.* (Melski et al., 2008) proposed a four-step approach to show how visibility in supply chains can be improved with RFID-generated data. RFID-enabled SCM is expected to establish item-level tracking, introducing another level of efficiency never seen before (Michael & McCathie, 2000). Previous study shows that RFID can also be used for reducing retailer product shrinkage with greater supply chain visibility (Huber & Michael, 2007). RFID applications have enabled inter-company integration in supply chains; however, it also triggers a high degree of implementation risk (Chuang & Shaw, 2007). This is because it demands for robust IT infrastructure, high investment, accurate and efficient data management (Imburgia, 2006).

In industry, many cases on RFID applications in supply chains have been reported. WalMart is one of the pioneers incorporating RFID in its retailing and supply chain system, and Gillette is one of the first eight companies to participate in the initial RFID pilot with WalMart. They used RFID technology to track their inventories as items move through a supply chain, from a manufacturer to a distribution center, next to a retailer stock room, and then to a shelf on the sales floor. DOD of US is also an early adopter of passive RFID to solve US military's huge logistics challenge (Thornton, 2006). Tibco, IBM and VeriSign jointly developed demo systems to promote the use of EPCglobal standards. All these applications have been developed to enhance product authentication (supplier: Gillette; retailer: WalMart) and new product visibility (manufacturer: Procter & Gamble; retailer: WalMart) (EPCglobal network, n.d.).

Intel introduced the connected digital supply chain in 2005, in which RFID in EPCglobal is the evolutionary enabler for optimizing supply chains and facilitating the acquisition, filtering, aggregation, and distribution of supply chain data for goods movement visibility (Intel, 2005). Partnering with Intel, OAT also developed a supply chain solution. The solution targets on a high-resolution view of product movement across extended supply chains based on OAT EPCIS edge servers and Intel processors. On the other hand, Sun developed an architecture as part of the Sun EPC initiative to integrate real-time data flow from existing business processes and back-end enterprise systems. Oracle developed the Oracle sensor data

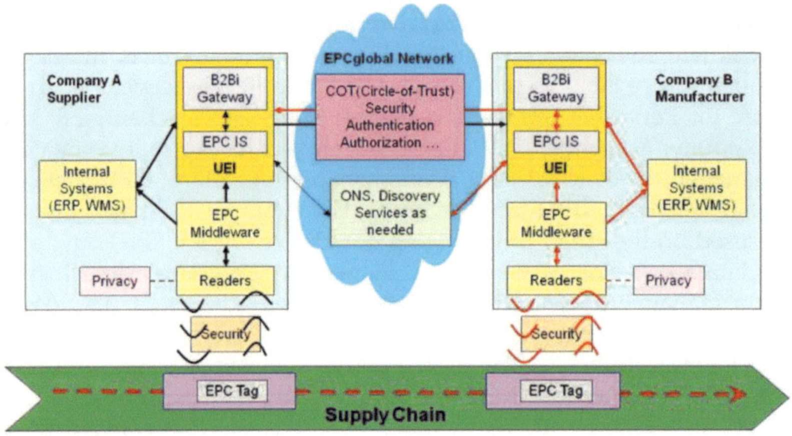

Fig. 1. The solution architecture

manager, in which EPCIS-compliant information service database and discovery services are used for searching for data in EPCglobal network. EPCglobal standards have also been used in developing Electronic Pedigree (E-Pedigree) in pharmaceutical industry against counterfeit drugs (E-Pedigree, n.d.).

3. Our RFID-based track & trace solution with security concern

3.1 The proposed solution

In general, the R&D efforts and solutions of RFID in the EPCglobal network reviewed in the previous section have their particular features and advantages in respective applications. However, there are some common limitations for these solutions, i.e., there are not specific security mechanisms applied to RFID tag authentication and data protection, neither proper security mechanisms at higher level for business information sharing and flow control. As discussed previously, secure and real-time track & trace, flexible business process, information flow control, and their synchronization are becoming increasingly important in supply chains. In view of the gap identified, we propose a solution with security concern for RFID-based real-time track & trace in EPCglobal-enabled supply chains. Figure 1 illustrates the architecture diagram of our solution.

In our proposed solution, track & trace is leveraging on the EPCglobal network. When a product bearing an RFID tag goes through its supply chain, an RFID reader reads the tag data at a reading point of a business step. The RFID data is then passed to EPC middleware for filtering and processing so as to create the EPC events which contain the information of what (the tag data is), when (it is captured), and where (it is captured). A UEI (unified EPCIS interface) designed in the solution captures EPC events and converts them to EPCIS events by adding why (it happened) information which is about the business context. Through the UEI, the B2Bi gateway system will query and retrieve the EPCIS events for business process control. The EPCIS RFID events stored on the EPCIS servers can be shared by other participants in the supply chain through EPCglobal network upon permission of access control.

The B2Bi gateway system (Tan et al., 2006) is a platform developed by SIMTech (Singapore Institute of Manufacturing Technology). It allows companies to participate in B2Bi collaborations to facilitate company collaboration in supply chains. This platform also provides configurable business templates with which users can customize the steps on each business transaction process to allow flexible process configuration. The B2Bi gateway system

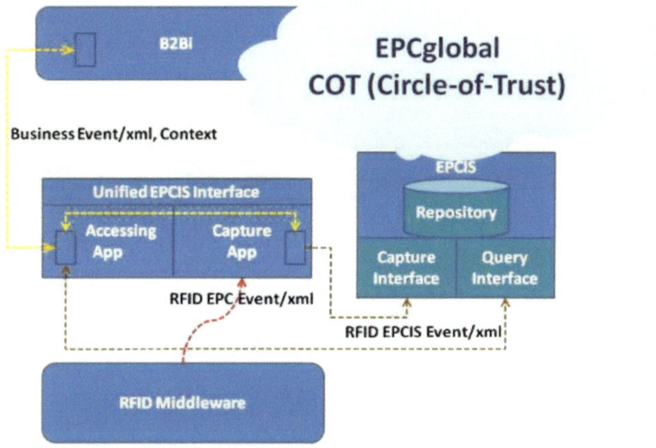

Fig. 2. UEI (Unified EPCIS Interface) architecture

currently only manages business information flow in business processes without involving the track & trace services for the physical items.

As aforementioned, security is important in RFID applications. There are two levels of security that are studied and designed in this solution. One is at lower level, i.e., data security between reader and tag. The other one is at higher level to control information sharing among participants through a COT (Circle-Of-Trust) model proposed for supply chains.

Our solution allows flexible business process configuration, secure information flow control and physical item track & trace in supply chains; and more importantly, it allows the synchronization of all of them. It enables the system and business processes to be fully automated and thus to improve collaboration efficiency. Some technologies developed in the proposed solution are elaborated in the following subsections.

3.2 UEI (Unified EPCIS interface)

As discussed earlier, the track & trace service in supply chains is based on RFID events capturing and querying through EPCIS (EPCIS, 2007), and sharing among participants in the EPCglobal network. In this solution, it is important to address how RFID events are captured into EPCIS and retrieved and used by B2Bi systems. The UEI is designed for this purpose as illustrated in Figure 2. It is one of the main components in the solution which facilitates the connection among EPC middleware, EPCIS, and the B2Bi gateway system.

The UEI consists of a CA (capture application) and an AA (accessing application). The CA serves capturing EPC events from RFID middleware and storing them to EPCIS as EPCIS events. The AA allows enterprise systems to query and retrieve EPCIS events by two ways, namely direct query and subscription. Web-services technologies are used in the UEI to enable loose-coupling among components and to make the design generic.

3.3 COT (Circle Of trust)

In a dynamic supply chain environment with track & trace services, multiple parties need to establish trust between each other to facilitate the secure exchange of sensitive information. In addition, a company may be participating in several supply chains at the same time as a partner collaborating with various companies. There are a lot of information/data sharing and exchange among them. It is critical to establish a trust relationship among partners for protecting business information/data flows in real-time in a dynamic environment. Currently,

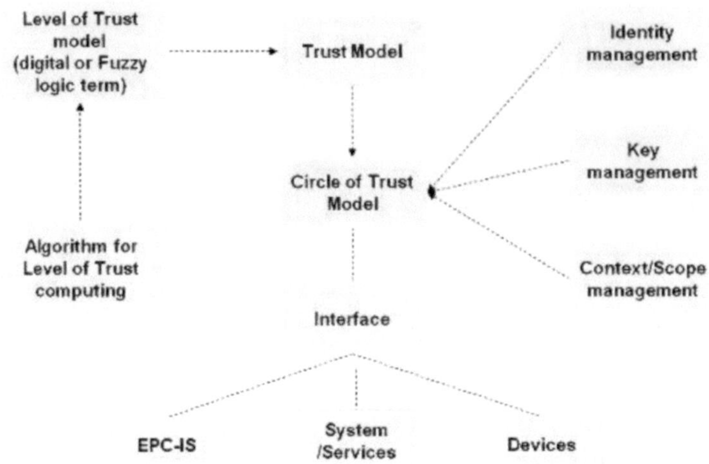

Fig. 3. COT (Circle Of Trust) model

there are mainly four kinds of existing models for trust establishment, namely centralized model, subordinate hierarchy, 2-party trust negotiation, and distributed trust evaluation (Maurer, 1996; Neuman & Ts'o, 1994; Xiong & Liu, 2004; Yu et al., 2000). However, they are not flexible enough to handle multiple parties in a dynamic supply chain environment yet lacking of security.

In our proposed solution, a COT (circle of trust) model is designed for the required purpose. The major components of the COT model consist of a trust algorithm, a trust model for two participants, a trust model of graph-based circle for multiple participants, and their control logic. Figure 3 illustrates the details of the COT model.

In short, the COT model can enable high level business information and low level RFID data to be shared and exchanged securely with different trust levels as specified among participants in a dynamically formed community in supply chains. Different trust levels will determine different levels of information access/exchange. Technical details of the COT design are not convenient to release here because it is under invention filing process.

3.4 Tag-reader security schemes

We design security schemes for protecting a tag at an end system level while it traverses a supply chain. In our schemes, a tag is marked initially at its manufacturer's site, whereas the mark is verified by the downstream partners of the supply chain. The mark is not a fixed one, but subject to changes (re-marked, and then re-verified) made by authorized partners. We adopt the standard security primitives (at the reader side) and tags that conform to specifications for EPC class 1 generation 2 RFID tags. At current stage, we developed three different protection schemes for protecting a tag, namely, a basic scheme, a batch scheme, and an undetachable scheme. The proposed schemes are secure, scalable, efficient, and easy to deploy. On one hand, it can resist un-authorized vendors from producing authentic tags quickly and massively thus raising the bar of difficulty for illegal behaviors. On the other hand, it stimulates the distributors or retailers of a supply chain on validating the goods/tags. This maintains the integrity of the tagged product in a supply chain within a complete EPCglobal network.

Our basic scheme is illustrated in Figure 4. When a tag is initialized at some partners in a supply chain, a security mark is generated and written into its user memory by a reader.

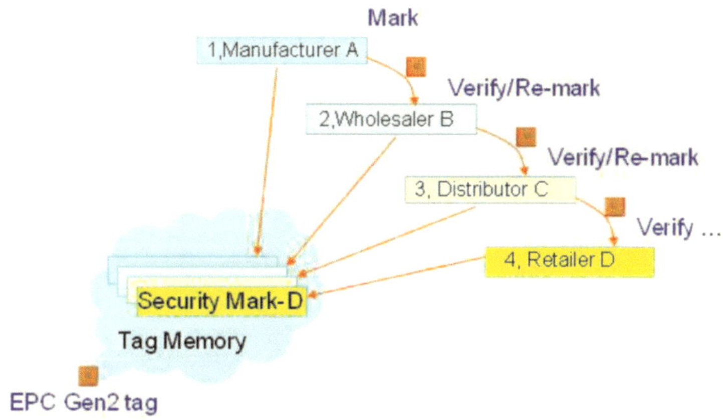

Fig. 4. Basic tag-reader security scheme (a tag is marked, verified, and re-marked when traversing different partners in a supply chain.)

The security mark is calculated based on the information such as tag identifier, reading point and time, and a secret key, all of which make it impossible to reversely disclose the relevant information from the mark. When the tag moves down to the next partner in the supply chain, an authorized reader can verify the mark and also leave its own mark. This process continues until it arrives in the destination of the supply chain. The tag-reader security scheme provides a secure, efficient and flexible verification in the track & trace process of a supply chain, as against risks such as counterfeiting.

Our batch scheme assumes a batch of tags attached to goods (e.g., packaged in a case), Instead of marking all the tags in a batch, the proposed method employs only a batch tag each time and marks it with our secure marking scheme. Moreover, besides the batch tag, an additional (randomly selected) tag, namely a pair-wise tag with the batch tag, is also securely marked. By pair-wising an additional tag at each step, we achieve efficient and secure tracing overall the supply chain.

Our undetachable scheme is suitable for the cases that require the presence of all tags for a complete verification. Our method makes these tags linked with each other so that any missing tag may cause a failed verification. To be efficient, we choose one tag in the set to be marked at each stage. Only by presenting all marks in the whole set can provide a complete verification.

In summary, all of the above methods make use of standard security primitives and conform to EPC class 1 generation 2 RFID tag specification. The proposed system is secure, scalable, efficient and easy to deploy. On the one hand, it can resist counterfeiting vendors from producing authentic tags quickly and massively. In other words, it raises the bar of hardness for the counterfeiting behaviors. On the other hand, it stimulates the distributors or retailers of a supply chain on validating the goods/tags.

3.5 Privacy-enhanced security scheme

In above schemes (subsection 3.4), when a participant of a supply chain leaves its mark on the tag, it also discloses its identity, which is not a good *privacy* property for this participant that may want to preserve its identity. We further devise a privacy enhanced tag protection scheme for marking tags and preserving the privacy of all participants in a secure RFID-based supply chain. This protection scheme provides participants with three privacy options, namely *public*, *limited*, and *private*. For the public option, the identity of a participant can be verified publicly

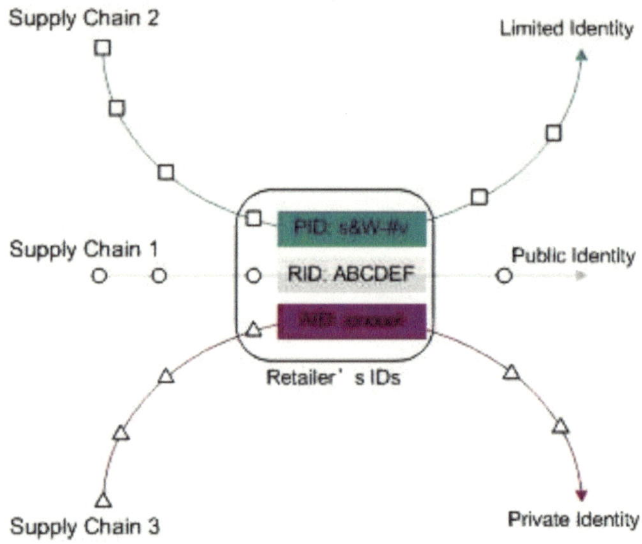

Fig. 5. Privacy-enhanced security scheme: a retailer assigns optional privacy (public, limited, or private) IDs for different supply chains

by any other participants in the supply chain; for the limited option, the verification can only be conducted by a small set of authorized participants; while for the private option, the identity of a participant cannot be disclosed unless being recovered by a legal authority in case of a dispute. Thus, a participant can freely choose a privacy option at any time for any goods in any supply chain. This makes the proposed scheme more secure, efficient and flexible.

For instance, as illustrated in Figure 5, we consider a participant (a retailer) traversing a number of supply chains with different privacy requirements. Firstly, upon receiving an item (with a tag attached) from Supply Chain 1, the retailer may consider it as a "non-privacy" case and continue using its original ID in the transaction documents as well as in our verification scheme. It chooses to put RID (e.g., $RID = ABCDEF$) into the tag in our tag marking equations. When the mark is recovered later on, this original ID will be posed to the verifier. Secondly, suppose that the item is received from Supply Chain 2 where the retailer's ID can be protected limitedly. The retailer uses a pseudo-ID (e.g., $PID = s\&W\text{-}\#v$) by employing the limited protection scheme. Thus, only authorized readers (in Supply Chain 2) can recover the original ID. Lastly, for Supply Chain 3 in which the retailer wants to hide itself against anyone but the trust authority, it can assign the AID (e.g., $AID = xxxxxx$) randomly to the tag protection scheme so that no one except the trust authority can recover its real identity.

4. Prototype implementation

A prototype is implemented for the above proposed solution. A sample application scenario of the prototype for OM (order management) business process is shown in Figure 6.

An OM business process involves two parties i.e., a manufacturer and a supplier. The higher lever business process between these two parties is managed by the B2Bi gateway system and includes the following steps:

- A manufacturer creates a PO (product order) and sends the PO to a supplier;
- The supplier acknowledges and confirms the PO from the manufacturer;

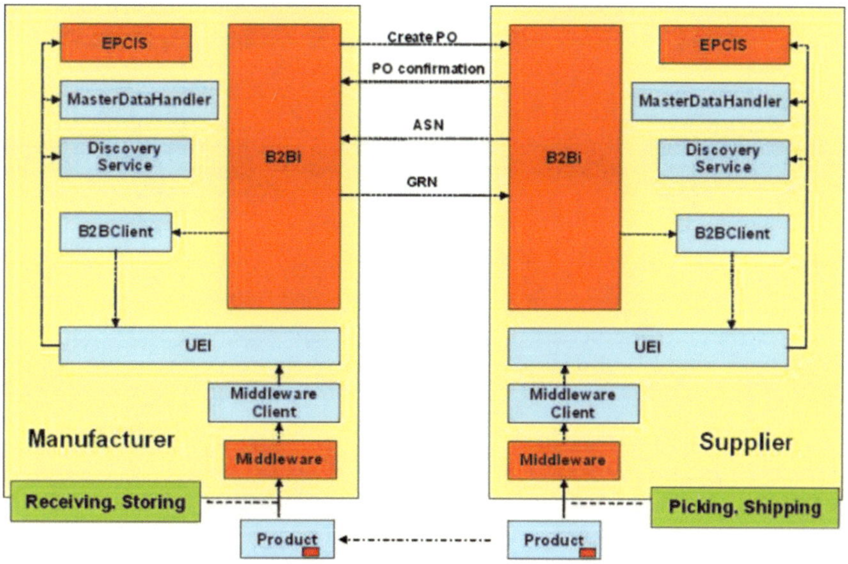

Fig. 6. An application scenario for OM processes

- The supplier sends an ASN (advance shipment notice) to the manufacturer;
- The manufacturer sends a GRN (goods receiving notice) the supplier upon receiving the products.

The physical product flow at the lower level is tracked and traced by RFID. The product states in the process are captured by a supplier when this product is picked up from a warehouse or shipped out, and by a manufacturer when it is received and stored into a warehouse.

In the solution prototype, the information flow and physical product flow are synchronized. The B2Bi process initiates the physical product flow which triggers and automates the business process. For example, when a PO is confirmed, the picking of products from a warehouse to fulfil the order is initiated at the supplier side; whereas at the manufacturer side, the reception and accept of products automatically trigger the GRN sending to the supplier.

In the prototype development and implementation, Accada EPCIS (Accada EPCIS, n.d.) is adopted as the EPCIS server. The EPCIS repository is deployed in MySQL database, while its capture and query interface services are deployed in Tomcat. The capture operation to Accada EPCIS repository is an HTTP-POST action where the data is in XML format (Accada EPCIS User Guide, n.d.). Because Accada's EPCIS repository implements the SOAP/HTTP binding for the query interface, it needs to construct a query for wrapping it into a SOAP request and sending it to the repository.

The data capture and configuration for EPCIS events in the OM process are such defined as shown in Figure 7. In real applications, GLN (global location number) can be used to identify RFID reading points. GRAI (global returnable asset identifier) and SSCC (serial shipping container code) are used for carton boxes and container tagging. The prototype is equipped with four logical readers, i.e., two readers named *s_picking* and *s_shipping* are used to simulate the picking and shipping out RFID gantries of suppliers, indicating business steps of "picking" and "shipping"; while the other two logic readers named *m_receiving* and *m_storing* are used to simulate the receiving and storing RFID gantries of manufacturers, representing business step of "receiving" and "storing". Figure 7 shows some other details, in which the

	Supplier		Manufacturer	
LogicReader	s_picking	s_shipping	m_receiving	m_storing
eventType	ObjectEvent	ObjectEvent	ObjectEvent	ObjectEvent
actionType	OBSERVE	OBSERVE	OBSERVE	OBSERVE
bizStep	urn:sg:pp:bizStep:picking	urn:sg:pp:bizStep:shipping	urn:sg:pp:bizStep:receiving	urn:sg:pp:bizStep:storing
readPoint	urn:sg:pp:rdPoint:001	urn:sg:pp:rdPoint:002	urn:sg:pp:rdPoint:101	urn:sg:pp:rdPoint:102
bizLocation	urn:sg:pp:bizLocation:SupplierWarehouse	urn:sg:pp:bizLocation:SupplierWarehouse	urn:sg:pp:bizLocation:ManufacturerWarehouse	urn:sg:pp:bizLocation:ManufacturerWarehouse
disposition	urn:sg:pp:disp:InProgress	urn:sg:pp:disp:InTransit	urn:sg:pp:disp:InProgress	urn:sg:pp:disp:InProgress

Fig. 7. Data capture in OM process.

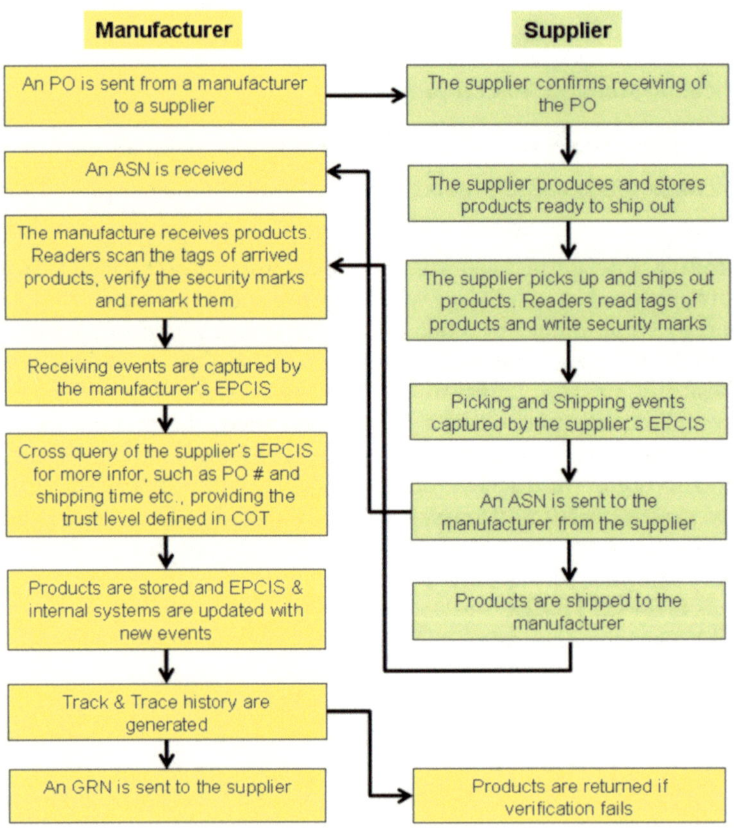

Fig. 8. Logic flow of the OM process enabled by the secure RFID-based track & trace solution

RFID tags used in the prototype are EPC class 1 generation 2 passive tags, and the readers and antennae are from Intermac and Symbol.

Figure 8 shows the logic flow of business information and physical products enabled by the secure RFID-based track & trace solution in the OM process of the prototype. The process

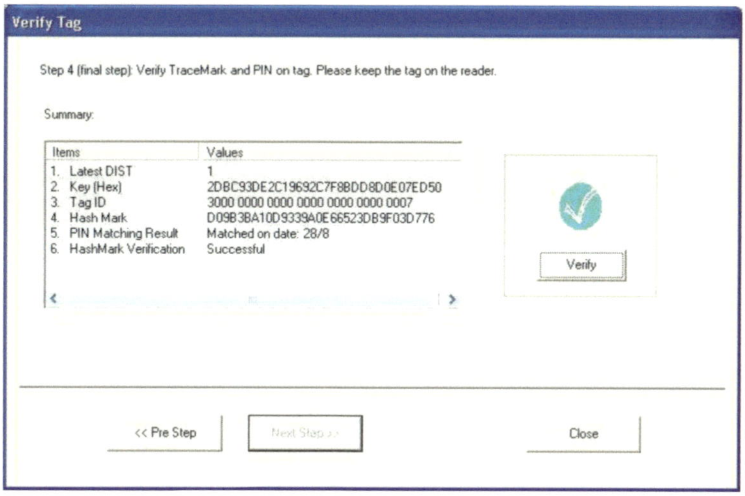

Fig. 9. Tag verification result

begins with a PO initiated and sent from a manufacturer to a supplier. After the supplier confirms the order, the products can be picked up from a warehouse to fulfill the order. After the products are shipped out, an ASN will be sent to the manufacture. During the processes off picking up and shipping out, each RFID tag attached on a product is read by a reader, at the same time a security mark is generated and written into the tag.

When the manufacturer receives the products, an RFID reader reads the tags on the products and checks the authentication of the products by verifying the security marks written in these tags by the supplier. Note that there are a number of ways to mark a tag and verify it later on. Without losing of generality, we hereby briefly introduce a basic scheme to illustrate the marking and verifying processes.

We assume a collision-free hash function $H(\cdot)$ that outputs an m-bit string on an ℓ-bit input message M, i.e., $H(M) : \{0,1\}^{\ell} \rightarrow \{0,1\}^{m}$. This hash function is to be implemented at the reader side. A mark is generated based on the identification information of the reader and the tag (RID and EPC, respectively), and the former mark. Thus, we can calculate the new mark as: $Mark_{new} = H(Mark_{old}||H(RID||EPC))$. In our prototype, we use SHA-1 as the secure hash function which generates a 160-bit hash value. However, we only keep the least significant 32 bits of the hash value as the mark to be stored in the user memory of the tag. We then generate the new PIN code for the tag to protect the memory[1] as: $PIN_{new} = H_k(Mark_{new})$, where k is a secret key only known to an authorized party and $H_k(\cdot)$ is a secure keyed hash function. To make the paper compact, we omit the details on key management, flexible marking processes and online/offline verification mechanisms. Thus far, the tag is assigned with a new PIN and a new mark before being shipped out.

At the receiving side, the verification is actually a reverse process of the generation process. The latest mark is read and checked against the PIN. If they are equal, the product is then accepted as successful verification and stored into the warehouse for storage; otherwise, the verification fails and the product is rejected. Figure 9 shows a case of successful verification result.

---

[1] Given a valid 32 bits access PIN, the tag is transitioned into what is called a "secured" state. Only at a "secured" state, the tag can be accessed for some restricted functions like read from the reserved memory bank and write to the user memory bank.

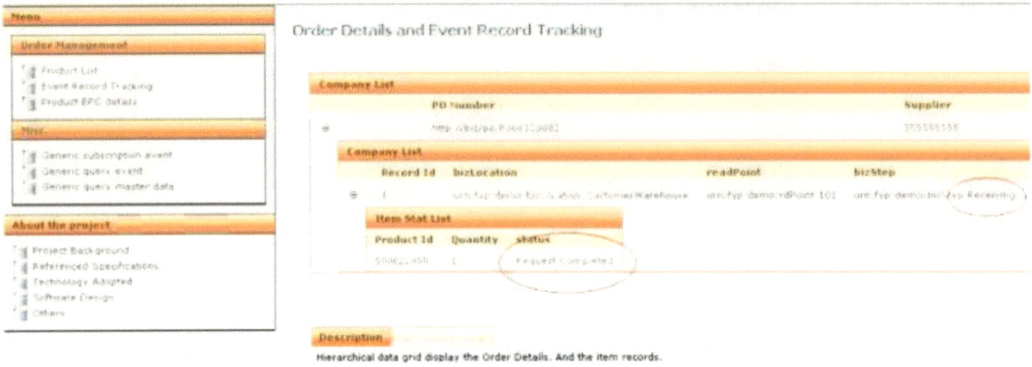

Fig. 10. Web GUI showing order details and RFID event tracking

The track & trace history can be generated through the query of EPCIS servers of both the manufacturer and supplier for further verification. The security ensures that the product manufacturer is from the right source. Additionally, to protect the privacy of the participants (e.g., manufacturers or suppliers) of a supply chain, we build an added optional privacy for those participants to choose. As discussed in subsection 3.5, a current participant can choose to use its real ID (RID) in the marking process, or to use a pseudo-ID (PID) or anonymous ID (AID) to hide its real identity. Thus, only authorized participants can recover the real identifier of a former participant.

Track & trace information and RFID events captured in EPCIS can also be used by an internal enterprise system of a company for decision making and business process control. For example, when the prototype is integrated with the inventory management system at the manufacturer side, the inventory level of the products can be automatically updated. Not only manufacturers but also suppliers can have in-time information of the inventory level of particular product, so that PO can be automatically generated at the manufacturer side or shipment can be automatically triggered at the supplier side when the inventory is below a certain critical preset level. Figure 10 shows a sample of Web GUI displaying the information details and RFID events tracking for an PO.

The system has undergone rigorous tests with some testing cases. The testing results show that products can be properly tracked and traced in an OM process. The security applied can eliminate the chances of counterfeiting products. The solution has significantly improved operation efficiency through the automated processes. Some companies have shown strong interests in our solution.

5. Conclusion

Given the current industrial demand for efficient and secure supply chain management, we have analyzed the issues on how to enable track & trace services (e.g., item identification, event capture and management, information storage, and information sharing among authorized parties) for RFID applications in an efficient and secure manner. We propose an RFID-based track & trace solution with security concern in supply chains based on EPCglobal standards. In this solution, a B2Bi gateway system is designed to manage high-level business information flows and processes, and EPCglobal network is leveraged on to manage the physical product flows. To address the security concern, we have proposed two levels of protection in our solution, namely (1) the COT for high-level business information sharing, and (2) the security schemes at reader-tag level for preserving participants' privacy. We have

shown that our solution can achieve secure information flow control for product track & trace services. We have also implemented a prototype of our solution for OM processes.
The working prototype of our solution has demonstrated high feasibility and efficiency in industrial scenarios under rigorous testing. It has attracted significant interests from industrial participants. While we are patenting the solution at this stage, we plan to commercialize the solution and make it a product of application package in the future.

6. Acknowledgment

This work is partly supported by A*Star SERC Grant No. 082 101 0022 in Singapore.

7. References

Accada EPCIS (n.d.).
URL: *http://www.accada.org/epcis/*

Accada EPCIS User Guide (n.d.).
URL: *http://www.accada.org/epcis/docs/userguide.html*

Angeles, R. (2005). RFID technologies: supply-chain applications and implementation issues, *Information Systems Management* 22(1): 51–65.

Auto-ID Centre at St. Gallen (2006). *Anti-counterfeiting and secure supply chain.*

Avoine, G. (n.d.). *Security and Privacy in RFID Systems.*
URL: *http://lasecwww.epfl.ch/~gavoine/rfid*

Chuang, M.-L. & Shaw, W.-H. (2007). RFID: integration stages in supply chain management, *IEEE Engineering Management Review* 35(2): 80–87.

Derakhshan, R., Orlowska, M. E. & Li, X. (2007). RFID data management: challenges and opportunities, *Proceedings of the 2007 IEEE International Conference on RFID,* Gravevine, TX, USA, pp. 175–182.

Diekmann, T., Melski, A. & Schumann, M. (2007). Data-on-network vs. data-on-tag: managing data in complex RFID environments, *Proceedings of the 40th Hawaii International Conference on System Sciences 2007*, Hawaii, USA.

E-Pedigree (n.d.).
URL: *http://www.axway.com/solutions/healthcare/epedigree.php*

EPCglobal network (n.d.).
URL: *http://www.epcglobalinc.org/about/media_centre/EPCglobal_Network_Demo.pdf*

EPCIS (2007). *EPCglobal EPC Information Services (EPCIS) Version 1.0 Specification.*
URL: *http://www.epcglobalinc.org/standards/epcis/epcis_1_0-standard-20070412.pdf*

Floerkemeier, C., Lampe, M. & Roduner, C. (2007). Facilitating RFID development with the accada prototyping platform, *Proceedings of the Fifth Annual IEEE International Conference on Pervasive Computing and Communications Workshops(PerComW'07),* While Plains, NY, USA, pp. 495–500.

Floerkemeier, C., Roduner, C. & Lampe, M. (2007). RFID application development with the accada middleware platform, *IEEE Systems Journal* 1(2): 82–94.

Huber, N. & Michael, K. (2007). Minimizing product shrinkage across the supply chain using radio frequency identification: a case study on a major australian retailer management, *Proceedings of the International Conference on Mobile Business 2007 (ICMB'07),* Toronto, Canada, pp. 41–45.

Huo, Y. & Jiang, X. (2007). Research on CPFR and warehousing management: A method to enhance supply chain visibility, *Proceedings of the 2007 International Conference on*

Wireless Communications, Networking and Mobile Computing (WiCom 2007), Shanghai, China, pp. 4645–4648.

Imburgia, M. J. (2006). The role of RFID within EDI: building a competitive advantage in the supply chain, *Proceedings of the 2006 IEEE International Conference on Service Operations and Logistics, and Informatics (SOLI'06)*, Shanghai, China, pp. 1047–1052.

Intel (2005). Building the digital supply chain: an intel perspective, *Intel Solutions White Paper: Supply Chain Technology* .

Maurer, U. (1996). Modelling a public-key infrastructure, *Proceedings of 1996 European Symposium on Research in Computer Security (ESORICS '96)*, Rome, Italy, pp. 325–350.

Melski, A., Muller, J., Zeier, A. & Schumann, M. (2008). Improving supply chain visibility through RFID data, *Proceedings of the IEEE 24th International Conference on Data Engineering Workshop (ICDEW'08)*, Cancun, Mexico, pp. 102–103.

Michael, K. & McCathie, L. (2000). The pros and cons of RFID in supply chain management, *Proceedings of the 4th International Conference on Mobile Business (ICMB'05)*, Sydney, Australia, pp. 623–629.

Mintchell, G. (2002). It's automatic: automation shifts transmission assembly into high gear, *Control Engineering* 49(6): 12.

Neuman, B. C. & Ts'o, T. (1994). Kerberos: an authentication service for computer networks, *IEEE Communications Magazine* 32(9): 33–38.

RFID Journal (1983). *A Guide to Understanding RFID.* URL: *http://www.rfidjournal.com/article/gettingstarted/*

Sheu, C., Lee, L. & Niehoff, B. (2006). A voluntary logistics security program and international supply chain partnership, *Supply Chain Management: An International Journal* 11(4): 363–374.

Song, S., Shim, T.-K. & Park, J.-H. (2006). Proxy based EPC track & trace service, *Proceedings of the 2006 IEEE International Conference on e-Business Engineering (ICEBE'06)*, Shanghai, China, pp. 528–531.

Straube, F., Vogeler, S. & Bensel, P. (2007). RFID-based supply chain event management, *Proceedings of the 1st Annual RFID Eurasia 2007*, Istanbul, Turkey, pp. 1–55.

Tan, J. S. (2005). ISO focus, *The Magazine of the International Organization for Standardization* 2(2): 19–25.

Tan, P. S., Goh, A. E. S., Lee, S. S. G. & Lee, E. W. (2006). Issues and approaches to dynamic, service-oriented multi-enterprise collaboration, *Proceedings of 2006 IEEE International Conference on Industrial Informatics (INDIN '06)*, Singapore, pp. 399–404.

Thornton, F. (2006). RFID security, *Syngress* pp. 46–48.

Xiong, L. & Liu, L. (2004). PeerTrust: supporting reputation-based trust in peer-to-peer communities, *IEEE Transactions on Knowledge and Data Engineering (TKDE), Special Issue on Peer-to-Peer Based Data Management* 16(7): 843–857.

Yu, T., Ma, X. S. & Winslett, M. (2000). PRUNES: an efficient and complete strategy for automated trust negotiation over the internet, *Proceedings of the 7th ACM Conference on Computer and Communications Security (CCS 2000)*, Athens, Greece, pp. 210–219.

8

Discovery Services in the EPC Network

Martin Lorenz, Jürgen Müller, Matthieu-P. Schapranow,
Alexander Zeier and Hasso Plattner
Hasso-Plattner-Institute
Germany

1. Introduction

Recent advances in Auto-ID technology, especially RFID, provide great potential for the innovation of existing processes in Supply Chain Management (SCM). Accompanied with item level identification using the EPC, companies are able to capture product lifecycle information at unprecedented levels of detail. RFID readers placed at strategic points in the supply chain automatically capture information about passing objects while they move along their way from the manufacturer to the consumer. Modern RFID tags can be equipped with sensors for temperature, humidity or other physical conditions, providing information systems with instant data on the current location and status of objects. Auto-ID bridges the gap between the physical and the digital world, providing real-time information about current supply chain operations. It provides companies with increased supply chain visibility [Melski et al. (2008)], resulting in reduced uncertainty, regarding operational and tactical supply chain planning. Overall, Auto-ID supports companies by providing higher information quality and quantity.

While most of the aforementioned aspects concern company internal processes, an even greater potential is being anticipated for company-overlapping supply chain collaboration. The possibility to provide real-time information about intra-company operations to trading partners, up- and downstream the supply chain, allows companies to increase value creation over all levels of the supply chain. In particular, planning activities of adjacent trading partners can be performed with a higher degree of certainty, reducing the need for high safety stock levels, which in turn reduces inventory costs [Simchi-Levi et al. (2003)]. On the other hand, many industries struggle with volatile demands, leading to the risk of running out of stock in times of higher demand. Real-time information can help to detect critical stock levels early. Sharing that information instantly with suppliers allows them to take immediate action such as rescheduling of shipments or increasing production rates to cope with temporary increased demand. Section 2 of this chapter will go into the details of two selected industry use cases that outline the benefits of company-overlapping collaboration.

The existence of practical scenarios for supply chain collaboration based on Auto-ID data demands for an infrastructure of information systems to support these use cases. EPCglobal, a joint venture between GS1 (formerly known as EAN International) and GS1 US (formerly the Uniform Code Council, Inc.), introduced the EPCglobal Architecture Framework, which is suppose to increase visibility and efficiency throughout the supply chain as well as to guarantee higher quality information flow between companies and their trading partners [EPCglobal (2007a)]. The EPCglobal Architecture Framework, for the rest of this chapter named EPC Network, is derived from the concept of the "Internet of Things" (IoT). The IoT

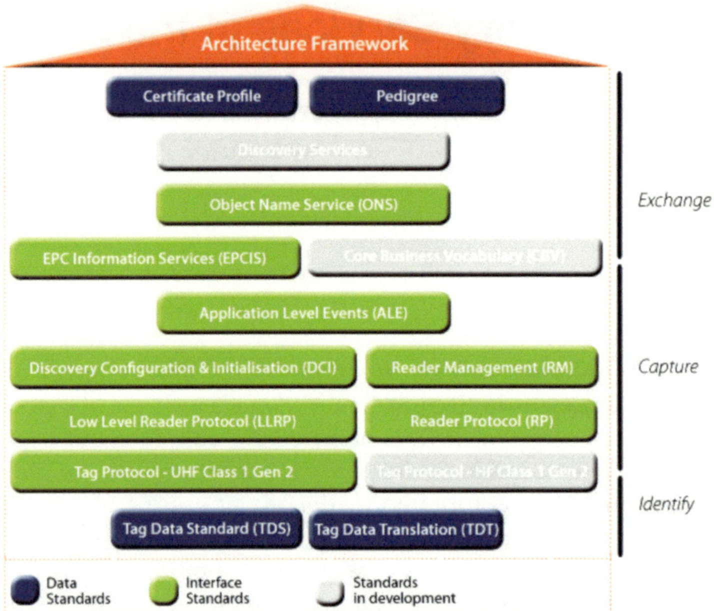

Fig. 1. EPCglobal Architecture Framework

is a concept that describes a self-configuring wireless network of sensors whose purpose is to provide objects with a means to interconnect and to interact [Polytarchos et al. (2010)]. Based on this idea, the EPC Network defines information systems, communication protocols, and data types that support capturing, storage, and exchange of EPC data among participants of a supply chain network. Figure 1 depicts the different standards defined for the EPC Network. The architecture includes specification for low level communication protocols such as the air interface between tag and reader as well as high level aggregated business information such as the EPC Information Services (EPCIS) and the EPC Discovery Service (EPCDS). Especially the latter play key roles for the company-overlapping exchange of information.

The diagram depicted in Figure 1 shows the discovery service component in a pale green color, indicating that it is still question to research how such a discovery service has to be designed. The purpose of this chapter is to elaborate on the complexity of this issue and introduce scientific work related to the definition of a discovery service component for the EPC Network. There are numerous functional and non-functional requirements that make the definition of an application layer protocol for a discovery service a difficult task. In Section 2, we present real world use cases that require the existence of a discovery service, to substantiate the necessity for such a component. In Section 3, we take a closer at the EPC Network components that are needed to support the use cases described in 2. Subsequently, we enumerate requirements for a discovery service to support the presented use cases. Based on these requirements, we propose a discovery service design for the EPC Network in Section 5. Section 6 gives an outlook on future work.

2. Industry use cases

To stress the need of a discovery service for the EPC Network, we present two real world industry use cases in this section. We do this for two reasons. First of all, practical use cases proof the necessity of a research topic, regarding its significance to economic interest for industries. Secondly, use cases can be used to derive concrete requirements for the design and the implementation of an information system. For this purpose, we introduce an anti-counterfeiting scenario in the context of the European pharmaceutical supply chain in Section 2.1, and we describe the process of product recalls in Section 2.2, focusing on the localization of effected products to provide effective recall management, keeping the financial impact as low as possible.

2.1 Use case 1: Anti-counterfeiting

As production in low-wage regions and global trade increases, opportunities for producing and selling counterfeit products also arises. The Organization for Economic Co-operation and Development (OECD) conducted a comprehensive study in 2008 [OECD (2008)], which was updated in 2009 related to the economic impact of counterfeiting and privacy [OECD (2009)]. It estimates that the trade volume of pirated and counterfeit goods could sum up to $250 billion excluding domestically produced and consumed products and pirated digital products. This is an equivalent of 1.95% of the world trade volume.

This poses a financial risk to companies because fake or smuggled goods reduce their sales volume. The pharmaceutical industry moved to public focus by the operation MEDI-FAKE, conducted by custom authorities in all EU members states. More than 34 million fake drug tablets were detected at customs control at the borders of the European Union in a two month period [Group (2009)]. This can put lives in danger as pharmaceuticals might not contain active pharmaceutical ingredients, wrong ingredients, a wrong dosis or other harmful substances.

To increase process efficiency and fight smuggling as well as counterfeiting, companies more and more inspect the concept of "unique identification", meaning that not only the product manufacturer and the product type is encoded but that each and every single item receives a unique serial number. That is the point where EPC an RFID comes into play. With the ability of unique identification using EPC and ubiquitous data capturing using RFID, it is possible to track items along their way from the point of production to the consumption. A major component in such a scenario is the company's read event repository, which stores the events captured by the RFID readers. Each company in the supply chain that captures Auto-ID data from their processes, needs to operate such a read event repository, to persist its data. Combining the information distributed over all repositories of the companies that are part of the manufacturing and/or distribution process, allows to reconstruct a complete trace of each individual item. Such a trace can be used to verify the origin and the distribution path of an item, providing customers only with pharmaceuticals from licit supply chains.

The problem is that a retailer needs to determine all resources of information, i.e., the addresses of the read event repositories that contain information regarding the particular EPC. Globalized trade, dynamic business relations, re-importing, and multiple levels of wholesalers and distributors, require a dynamic aggregation of information from a number of potentially unknown resources. To gather all this information, a component is needed that, given an EPC, provides pointers to the resources that contain the read events created during the travel of the item through the supply chain. Such a component is the EPC Discovery Service.

2.2 Use case 2: Product recall

The second use case that we want to present is product recalls. Product recalls usually occur due to safety or quality issues. They require a higher planning effort than most other return types. Key to a successful management of recalls is information technology and effective communication. Product recalls can be voluntary or mandated by legal obligations. A recent example is Toyota's production problems in October 2010 [Ohnsman & Kitamura (2010)]. They had to recall 10 million vehicles globally, because particular models might have brake system and gas pump issues. For many industries that are susceptible to recalls, like the automotive or food industries, a poorly managed recall can create a tremendous negative impact on the economic side of the company. Even more problematic is the accompanying damage in reputation, which can become a threat to existence.

In such a scenario like in the case of Toyota, it is most important to determine the exact number of affected products to act fast and target-oriented to contain the potential financial damage. In most cases not all of a company's products need to be returned. Temporary production problems in one of the production plants might have caused a subset of all products to be erroneous. Consequently, the company needs to find out where these products have been and who they have been sold to. That way it is possible to keep the number of recall products as small as possible, recalling only the ones that have been identified as potential defects.

Using RFID and EPC, it is possible to trace the distribution of each individual product. In case of food or life stock, it is also possible to determine all products that the item has been in contact with during storage or transportation, eliminating the possibility of collateral damage due to dispersion of poison or illness.

Again, this information is distributed over a number of independent read event repositories, which are operated by the companies that traded the goods. To perform effective product recall, we need to aggregate and analyze all the information distributed among the resources. Just like for the anti-counterfeiting scenario, a discovery service needs to be present to enable such kind of innovative process.

Now that we presented industry scenarios where Auto-ID technologies can help a great deal to improve current processes, we want to take a closer look at the EPC Network and the components that are needed to support our ideas.

3. EPCglobal architecture framework components

The previous section described practical use cases for a discovery service for the EPC Network. In this section, we go into the details of the EPC Network to understand the interconnection between the individual components and their relation to the use cases. We need to do this because most of the requirements for a discovery service are based on the existing components, the data that is available in the network, and the interfaces used to access the data. We will not go into the details of low-level physical data access and tag encodings, instead we restrict our discussion to the components above Application Level Events (ALEs), see Figure 1.

3.1 Read events

The primary type of data exchanged in the EPC Network are read events. read events are business-level events, which represent a scan of an RFID tag or 2D barcode associated with business context. There are five types of events: EPCISEvent, ObjectEvent, AggregationEvent, TransactionEvent and QuantityEvent. Figure 2 depicts an UML class diagram, showing the relation between the different types of events.

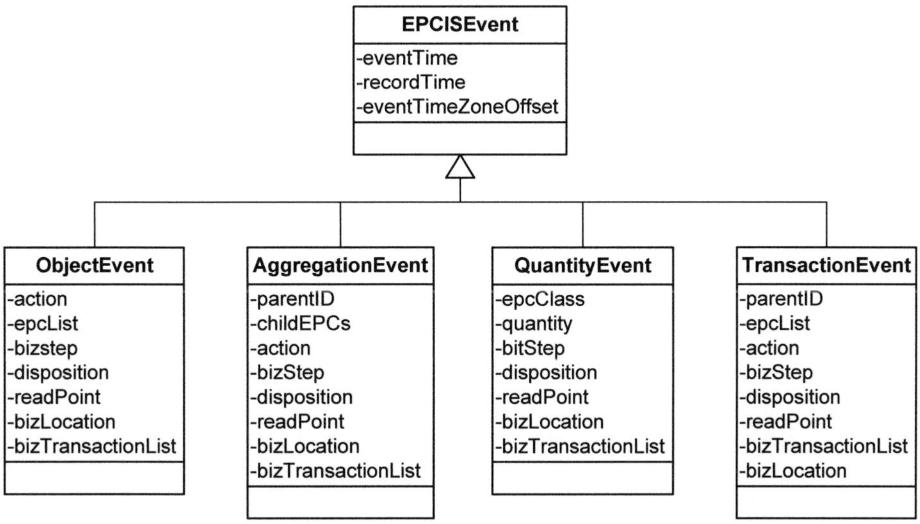

Fig. 2. Class Diagram of EPC Event Types

These events answer the questions *What*, *Where*, *When*, and *Why*. The EPCglobal standard allows to extend these data into each direction to provide companies with the ability to adapt the data to their special needs. For a detailed discussion on the meaning of the individual attributes of the events, we point the interested reader to the EPCglobal EPCIS standard [EPCglobal (2007b) (Section 7)]. With these read events, it is possible to identify location and business context of items during their travel through the supply chain.

3.2 EPC information services

Once these events are created, they need to be stored persistently at some point, to provide other applications with the ability to use these events. For this purpose, the EPC Network defines the EPC Information Services. The EPCIS provides a repository to store the information about read events that is why it is also called read event repository. Furthermore, it provides a capture interface to provide a way to store the events, as well as a query interface to query for stored events. Each company, which captures Auto-ID data is supposed to operate an EPCIS to be able to store and to exchange the information with internal and external applications. Figure 3 illustrates the process of information storage and exchange with the EPCIS. However, the EPCIS is nothing more than a repository for read event data. It solely serves as a resource of information and does not implement any business logic. In order to be able to leverage the full potential of the information distributed among the EPCIS servers of different trading parties, it is necessary to derive the exact addresses of the EPCIS servers that posses information about a particular item, i.e., EPC. The EPC Network defines two

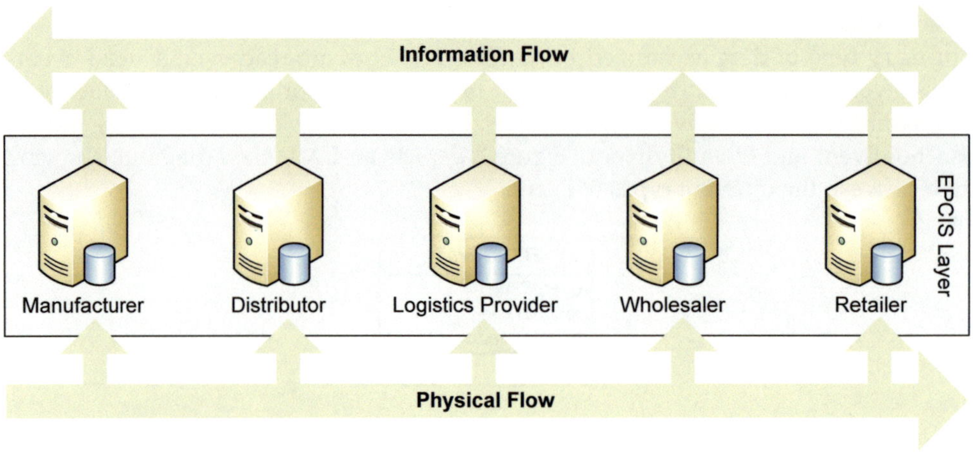

Fig. 3. EPC Information Services Data Flow

information systems that provide such kind of functionality, namely the Object Name Service (ONS) and the EPC Discovery Service.

3.3 Object name service

The ONS is a DNS-based service, whose purpose is to resolve information resources to an EPC. Information resources in the context of ONS can be websites, web services, or an EPCIS repository. However it is important to note that the ONS does not process the serial version of the EPC. Figure 4 depicts the EPC numbering scheme. It consists of a header, defining the version of the EPC, an EPC manager, identifying the authority that assigned the EPC to the object, an object class, which identifies the type of object, and a serial number, used to identify a particular item among a number of items of the same class and manager. The ONS neglects the serial number of an EPC [(EPCglobal, 2008, Section 5.2.1)]. The granularity of ONS resolution is currently limited to product type, rather than serial-level lookup. i.e. an ONS is not expected to retain distinct records for two objects of the same product type that only differ in their serial numbers. The only EPCIS server address that is being stored by the ONS is the manufacturers EPCIS, where the EPC has been assigned to the item. So if we want to store a list of different EPCIS server addresses for an individual item, we need another information system.

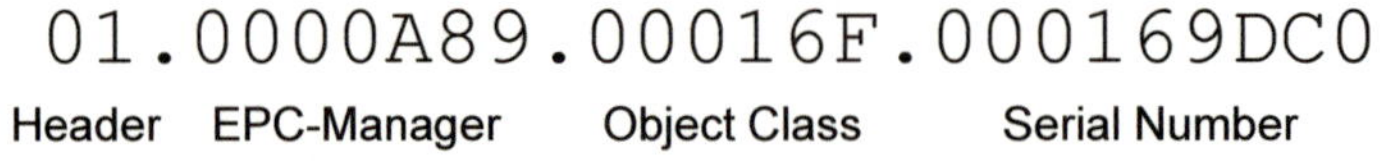

Fig. 4. Structure of the EPC Numbering Scheme

3.4 EPC discovery service

The EPC Discovery Service standard is currently in development by the EPCglobal Data Discovery Working Group. Its main purpose is "Finding and obtaining all of an item's relevant visibility data, of which a party is authorized, when some of that data is under the control of other parties with whom no prior business relationship exists" [EPCglobal (n.d.)]. The EPCDS can be seen as a search engine for EPC-related information. Given an EPC, it returns a list of URLs of the query interfaces of EPCIS servers, which are in possession of information related to the particular EPC. With this functionality, authorized and authenticated clients are able to reconstruct traces of items and to track the current location of items. Figure 5 illustrates the semantic difference between ONS and EPCDS. Looking at the use cases from Section 2, only the EPCDS provides enough functionality.

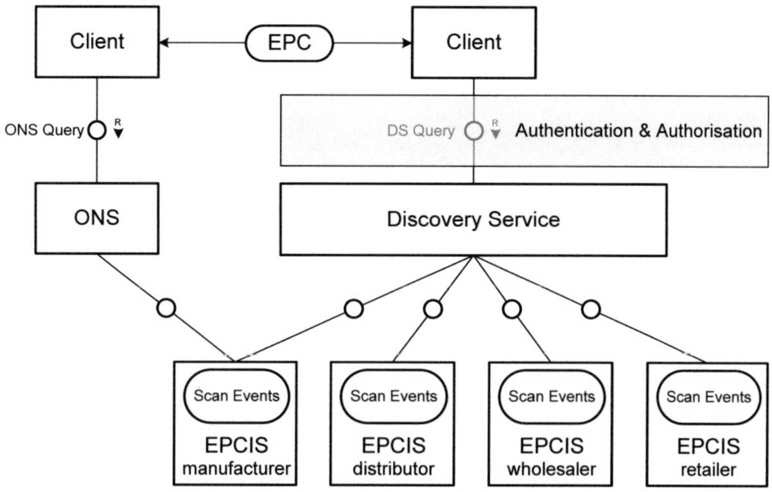

Fig. 5. Object Name Service vs. EPC Discovery Service

In this section, we looked at the individual EPC Network components and defined their particular roles, regarding information storage and exchange. We introduced the concept of the EPC Discovery Service, which is the central component to support the use cases from Section 2. The following section takes the prerequisites from this section and the use case definitions and derives a list of basic requirements for a discovery service for the EPC Network.

4. Discovery service requirements

In order to create an architecture design proposal for a discovery service, we need to define a set of requirements. This section enumerates requirements, which are used in Section 5 to reason on the design of our proposed discovery service architecture. The requirements have been gathered and consolidated from a number of resources. First and foremost, we used the results of the BRIDGE project, which is an integrated Project addressing ways to resolve the barriers to the implementation of the EPCglobal Network in Europe. Work package two of this project accessed requirements and designs for a serial-level lookup service for the EPC Network. Furthermore, we have taken argumentations from Müller et. al. [Müller, Oberst, Wehrmeyer, Witt & Zeier… (2009)] and Kürschner et. al. [Kürschner, Condea & Kasten… (2008)], which contribute to create a comprehensive set of discovery service requirements. These requirements include the main topics core functionality, data ownership, security, business relationship independent design, organic growth, scalability, quality of service, client complexity, and bootstrapping.

4.1 Core functionality

At its core, a discovery service needs to store the EPC, which has been observed, the URL of the EPCIS server that stores the actual event and a timestamp. In order to store this data, the EPCDS needs to offer a notification interface that can be used by resources to publish their information. Additionally, there needs to be a query interface, which allows clients of the discovery service to request the stored information. Parallel to this query interface, which allows ad hoc queries, there should be a way to register standing queries, which provide companies with the ability to get instant information on incoming notifications. Since the information, stored at the discovery service, is highly sensitive to companies, there should also be a security component in place that implements authentication and authorization functionality, to protect the data. The following enumeration summarizes the core functional requirements of the EPCDS labeled from **RQ1** to **RQ5**.

RQ1: A discovery service needs to provide a way for resources to publish their information, i.e., EPC and corresponding EPCIS server address.

RQ2: It needs to store the EPC/URL mappings and the according timestamps persistently.

RQ3: It needs to provide a way for clients to execute ad hoc queries for EPC-related information.

RQ4: It needs to provide a way for clients to register/unregister standing queries to provide instant information on incoming notifications.

RQ5: It needs to provide authentication and authorization mechanisms to protect the stored data.

4.2 Data ownership

According to Kürschner et al., data control aspects have to be considered by any discovery service approach. Their investigations showed that there exist companies that are not willing to share their EPCs or EPCIS addresses with other companies. The reason for this is self-interest, i.e., system owners have greater interest in system success than non-owners. The issue of data ownership is considered to be a major reason for managers to decline the participation in supply chain overlapping business collaboration. Neglecting this fact will lead to a reduced adoption rate of the particular discovery service approach among supply chain partners. Based on their findings, Kürschner et al. defined two requirements for the discovery service design, regarding data ownership.

RQ6: Companies shall be in complete control over their data including EPCIS addresses, read events, business data as well as setting of detailed, fine-grained access rights.

RQ7: Companies shall be able to track the usage or the requests upon their data. Particularly, publications of data at the discovery service level should be avoided.

4.3 Security

Security is a vital factor in any enterprise application. In case of the discovery service this issue becomes even more relevant due to the fact that it operates on public networks, keeping sensitive information potentially necessary for business success. Kürschner et al. derive a set of characteristics from the overall topic of security. These are availability, reliability, safety, confidentiality, integrity, and maintainability. Although all of the above mentioned characteristics are essential features of a discovery service, only three of them are regarded as outstanding for the design of a discovery service.

RQ8: The confidentiality of both the publisher data and client query shall be ensured by the discovery service design.

RQ9: The discovery service architecture shall ensure a high overall system availability and reliability.

Additionally to the security requirements **RQ8** and **RQ9**, we consider data integrity as a fourth characteristic. Business relations on the level of supply chain management rely on trust. In a collaborative effort to increase the efficiency of modern supply chains, managers base their decisions on data delivered by supply chain participants that have been categorized as trusted partners. To keep these trust relations valid over digital cooperation, there need to be mechanisms to verify the correctness of origin and integrity of the data.

RQ10: The discovery service design shall ensure the correctness of origin and the integrity of the shared data.

4.4 Business relationship independent design

Different customer demands, globalization, discovery of uncharted market opportunities, outsourcing, innovation and competition are some of the major factors that determine significant partner changes in supply chains. From a strategic point of view, for some companies, changing supply chain partners is simpler and cheaper than changing internal processes. Section 4.2 adduced the need for information ownership and fine-grained access control for information sharing. Companies that modify their trading partners relations frequently need to define access rights, reflecting these new business relationships. Having this in mind, it is important to minimize the access control maintenance effort for companies.

RQ11: Changes in business relationships shall not affect the way in which a company interacts with the discovery service.

4.5 Organic growth

Organic growth as used by Kürschner et al. is derived from a definition by Rogers in [Rogers (1995)], where he categorizes adopters of new ideas into innovators (2.5%), early adopters (13.5%), early majority (34%), late majority (34%), and laggards (16%). As a result of this development there will be only few companies initially joining the network in the beginning. However, the actual value of the EPC Network depends on the number of participating companies. Consequently, it is of high importance to lower the threshold for joining the network for less innovative companies, fostering the adoption of the EPC Network.

RQ12: The discovery service architecture shall encourage participation in the EPC Network.

Although this requirement is somewhat straightforward for any new technology, it is worth special consideration, because the value of the network and therefore the acceptance of the EPCglobal idea, to support supply chain innovation for all industries, depends on the fast adoption of discovery services.

Low threshold in this context can be related to technical, financial and political obstacles. In order to push the desire to participate in an innovative idea such as the EPC Network, it is important to keep a positive relation between opportunity and risk. An economically expensive solution, creating large administrative overhead, leads to a low adoption rate, resulting in an EPC Network with low attraction to potentially interested parties.

4.6 Scalability

Another very important requirement is scalability. Müller et al. have already been aware of the problem of handling large amounts of requests and data. The issue of information production in RFID-enabled supply chains has been topic to a number of research works all aiming to understand the nature and behavior of these RFID enabled supply chains [Ilic, A. Groessbauer and & Fleisch (2009)]. Depending on the industry and application scenario, a discovery service can become a bottleneck or, even worse, a single point of failure when scalability becomes an issue.

RQ13: The discovery service architecture shall be highly scalable to be able to handle both, data volume and number of participants.

4.7 Quality of service

From a client's perspective, quality of service means the discovery service needs to provide data that is accurate, complete, and delivered within acceptable time frames. In this context, the predicate "accurate" means that the response of the discovery service contains all information, necessary to perform the desired queries on the individual EPCIS servers. Response time is also an important characteristic. Research showed that the acceptable time for an ad hoc query is only a few seconds [of Cambridge et al. (2007)]. Completeness of the result means that the discovery service's response contains all information available in the network and accessible to the client with regard to access control rights. From these findings, requirement number fourteen is derived.

RQ14: The query result shall be complete and correct, respecting the clients' access rights defined separately by each information provider.

Since the original requirement does not contain the dimension time, we propose a second version of this requirement.

RQ14a: The query result shall be complete, correct and within an acceptable time frame, respecting the clients' access rights, defined separately by each information provider.

4.8 Low client complexity

Client complexity is an important requirement because it has a great impact on the usability of the discovery service. It directly determines the interaction behavior between client and discovery service, potentially aggravating possible use cases. This can lead to a low adoption rate.

RQ15: Client complexity for discovery services shall be as low as possible, without loosing functionality.

4.9 Bootstrapping

One of the major concerns for the implementation of successful discovery services is the bootstrapping process. The bootstrapping process enables an interested and authorized client to locate an object's discovery service, using only the object identifier, i.e., the EPC. For many reasons, this is a serious problem. First of all, the plain amount of data produced by RFID-enabled supply chains and the number of queries, requires to operate a number of distributed discovery services, to share the work load [Ilic, Groessbauer, Michahelles & Fleisch (2009); Müller, Pöpke, Urbat, Zeier & Plattner (2009)]. Secondly, there are political problems that prevent the successful operation of a single global discovery service. Companies from many different countries and industries would have to agree on publishing their data to a discovery service, operated by some authority organization. It is most likely that there are countries and individual organizations that are not willing to publish their data to such a discovery service for a number of political reasons. Thirdly, the operation of a global discovery service would require processing power and storage space similar to the data processing centers of the major search engine providers. However, search engine providers are able to be financed via advertisements and additional services. An organization running a global discovery service would have to be financed by its users, who might not be willing or able to pay for the service. This issue directly influences requirement seven.

In the above paragraph, we identified technical, political and economical problems that lead to a distributed network of independent, collaborating discovery services. These discovery services will be operated by different providers such as legal authorities, companies themselves, or third-party profit organizations. In [A. Rezafard (2008)] Rezafard assumes that there will be globally operating communities (supply chains) that commit to a discovery service of choice.

It has been suggested that the ONS could be used for the bootstrapping process. However, the ONS is authoritative in that the entity that has change control over the information about the EPC is the same entity that assigned the EPC to the item to begin with. This means that the entity that assigned the EPC has to determine the discovery service that each company, which gets in contact with the object, has to publish its information to. This procedure may be feasible for supply chains, completely owned by a single company, but it is not possible to force all supply chain participant in global dynamic supply chains to publish their information to a particular discovery service. We already mentioned the issue of information ownership. Each company, producing RFID data is in full control of the data and decides autonomously about the publication of this information. That way it might be possible that the information about an EPC is distributed over a number of different discovery services.

Until now there is no accepted network architecture for discovery services. The reason for this is the fact that there is no common understanding about the distribution of EPCs among discovery services as introduced above. An industry-wide agreed distribution schema for EPCs is the basis for the design of a network architecture for discovery services. Once there is an agreed network structure, it is possible to develop bootstrapping mechanisms that enable supply chain partners to determine suitable discovery services just by means of the given object identifier, e.g., EPC. Recent research proposes Peer-to-Peer overlay networks as a promising way for discovery services to collaborate [A. Rezafard (2008); Shrestha et al. (2010)].

RQ16: The discovery service architecture should support communication of independent discovery services, serving distinctive concerns of individual companies.

RQ17: The network of discovery services should provide a bootstrapping strategy for clients to approach the correct discovery service, only by having the EPC at hand.

5. Discovery service architecture design

In this section we summarize and evaluate existing theoretical and practical discovery service approaches. We reason on their suitability for the EPC Network. Afterwards, we present a discovery service architecture that we designed and implemented prototypically, followed by a comparison of the different approaches with our new design.

5.1 Existing discovery service approaches

Here, we present research related to the definition of a discovery service design for the EPC Network. The different theoretical and practical approaches described below, contribute to our own approach, presented in 5.2.

5.1.1 Beier et al.: Discovery services

In [Beier et al. (2006)] a first implementation of a discovery service is presented. It can be summarized as Directory Look-up approach. In their paper, Beier et al. analyze the appropriateness of the Object Name Service [EPCglobal (2008)] and come to the result that this approach is improper for building discovery services. The developed Directory Look-up approach works as follows: real-world items attached with an EPC travel through the supply chain. At each company the item passes, the EPC is read and a read event is stored in the company's EPCIS. For each EPC that is stored in an EPCIS for the first time, the discovery service is notified and stores the EPC, the URL of the submitting EPCIS, a timestamp, the certificate of the submitter, and a visibility flag in its repository. The discovery service can then be queried with an EPC of interest. It replies with a list of relevant EPCIS URLs. Finally, the requester can query all relevant EPCIS servers by himself and aggregate the respective information. The underlying assumptions are that all participants of the EPC network are authorized by EPCglobal and equipped with a certificate by a trusted third party.

According to Beier et al. [Beier et al. (2006)], access to a company's EPCIS should be implemented role-based and policy-based with cell-level data disclosure control. At the discovery service level, row-level data access control should be enforced and, using the visibility flag, the owner of the data decides whether the record is shared among all authorized participants of the EPC network or access is restricted to companies, which have information about the same EPC. To retrieve EPCIS addresses confidentially, Beier et al. propose the usage of EPCIS proxy servers by storing not the real but the proxies URL at discovery service level.

5.1.2 BRIDGE project: High-level design for discovery services

BRIDGE is an acronym for Building Radio frequency IDentification for the Global Environment. The objective of this EU-funded project is to "research, develop and implement tools to enable the deployment of EPCglobal applications in Europe" [of Cambridge & UK (2007)].

In the report [of Cambridge & Research (2007)] the authors propose eight discovery service approaches, evaluate them, and finally judge four as promising candidates for large scale discovery services. It is important to understand that EPCIS servers can serve two different types of queries: ad hoc queries and standing queries. One-off queries are performed by a client once and no further communication between client and EPCIS is planned. Standing queries are subscriptions, which can be time-controlled using a query schedule (e.g., a client wants to be informed every hour) or trigger-controlled (e.g., a client wants to be informed if new information about an EPC of interest is available) [EPCglobal (2007b)].0

We will now briefly describe the four candidates identified by the authors. The first candidate is called Directory-of-Resources and equals the Directory Look-up approach by Beier et al. [Beier et al. (2006)]. The second candidate is called Notification-of-Resources and works as the Directory-of-Resources except that a client shows interest about certain information by creating a subscription at the discovery service. Once an EPCIS notifies the discovery service about an EPC, which matches the criteria defined in the subscription, the discovery service informs the client that the respective EPCIS is in possession of information related to the subscription. The third candidate is called Notification- of-Clients: EPCIS servers notify the discovery service for each new EPC they own information about. Once a client shows interest in an EPC the Discovery Server informs all relevant EPCIS servers. The servers send an availability notification to the client, which then queries the respective EPCIS servers. The last candidate identified by the authors is called Query Propagation and acts like Notification-of-Clients except the information is sent to the client by the EPCIS servers immediately without the availability notification. The authors summarize the first two candidates as Directory Service approach and the last two as Query Relay approach.

5.1.3 Kürschner et al.: Discovery service design in the EPCglobal network

In their related work, Kürschner et al. describe that the concepts of the Domain Name Service and the Service Location Protocol are not appropriate to solve the discovery service problem [Kürschner, Condea & Kasten... (2008)]. The authors present the Directory Look-up approach by Beier et al. and criticize that the EPCIS address of companies having information about the EPC of interest might be revealed if there are no access control policies in place. Otherwise, if these policies were established, the maintenance effort and complexity would rise because fine-grained access rights would have to be defined at discovery service level and policies would have to be synchronized between companies' EPCIS servers and the discovery service. To fulfill the requirement of low access right maintenance effort, they present the Query Relay approach developed in [of Cambridge & Research (2007)] in detail. The idea is to use the discovery service as a relay by forwarding the respective client queries to all relevant information holders. The EPCIS servers reply directly to the requester. Therefore, the EPCIS address is not revealed to the requester if the respective company decides not to reply to the query at all. Finally, both presented approaches are discussed and evaluated.

5.2 Our new design - an aggregating discovery service

The idea of the *Aggregating Discovery Service (ADS)* is to forward client queries to relevant EPCIS servers, aggregate their responses and synchronously respond to the client request. This reduces client complexity, brings low response latency, delivers complete and correct information for the requester, ensures data ownership for the information holder, avoids the need for fine-grained access control replicated at discovery service level, and guarantees confidentiality of clients and information holders. The ADS is a centralized service, which offers two interfaces (see Figure 6).

The *query interface* is used to gather information about an EPC of interest from the EPC Network. The ADS links EPCs to supply chain partners, which can provide detailed information about those EPCs. Certificates are used to provide authentication as proposed in [Beier et al. (2006); Kürschner, Condea, Kasten & Thiesse (2008)].

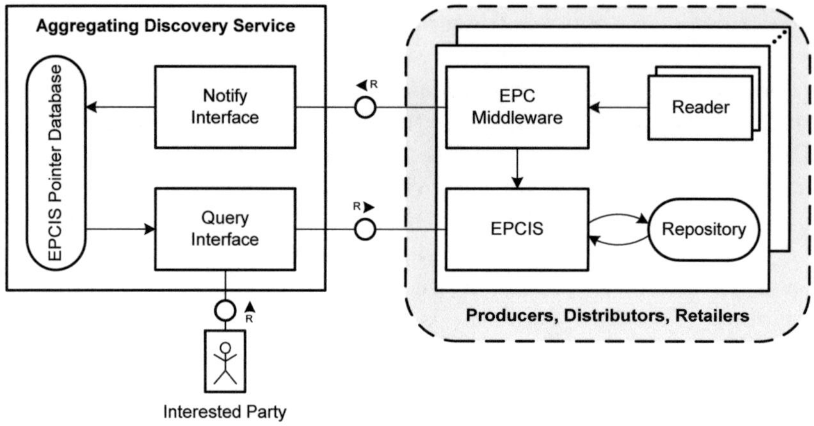

Fig. 6. Aggregating Discovery Service Architecture

The *notify interface* is used to inform the ADS about read events to be shared within the EPC Network. The ADS receives the EPCIS URL of the submitting partner and one or more EPCs that have been handled by this entity. Submitting multiple EPCs at once allows the client to batch notify requests and improves performance by lowering connection overhead. We propose a simple, XML-based format for this message to be submitted via HTTP POST.
The ADS maintains an association between submitting EPCIS servers and submitted EPCs. This allows the ADS to determine all EPCIS servers that hold more information about an EPC. The query relay provides an EPCIS-equivalent query interface [EPCglobal (2007b)] as proposed in [Kürschner, Condea, Kasten & Thiesse (2008)]. Additionally to a *full query* the client also can identify relevant EPCIS servers using a *resource query* [of Cambridge & Research (2007)]. For both types of queries the execution is as depicted in Figure 7.

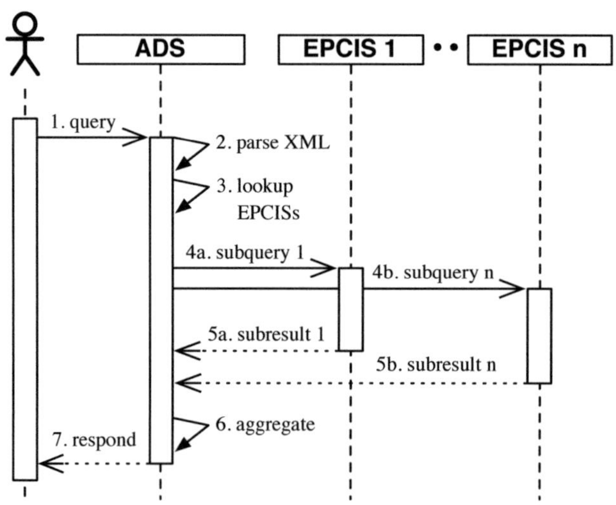

Fig. 7. Client Query Execution

The ADS waits for an incoming client query (*1.*) and parses the query to extract relevant EPCs (*2.*). The ADS then queries its internal database to look up the URLs of EPCIS servers,

which are relevant for this query (*3.*) and forwards the original query to those EPCIS servers (*4.*). After subresponses returned from the EPCIS servers (*5.*), they are parsed and the read events are extracted and combined (*6.*). The aggregated result is then returned to the client (*7.*). Effectively, this means the ADS acts as a proxy.
When querying EPCIS servers, problems might occur. Subqueries might time out, EPCIS servers may be temporarily unreachable or may refuse to answer the query. To prevent timeout of its client connection, the ADS will return a possibly incomplete result set marked as such, distinguishing between temporary problems (indicating that the client should try again later) and permanent reasons that prevent returning a complete response.
The ADS query interface should also support standing queries. The ADS needs to store all standing queries it received in order to forward them to an EPCIS when the EPCIS sends an event notification containing an EPC that matches a standing query of a client. This approach has the advantage that other EPCIS servers are not burdened with irrelevant standing queries, as they would be if the ADS was to distribute all standing queries to all EPCIS servers in order to achieve complete coverage.

5.3 Comparison of the different approaches

The fulfillment of the requirements stated in Section 4 and referenced literature in that section is substantial for a well-designed discovery service architecture that can easily be integrated in the EPC Network. In this section, we compare the existing approaches, introduced in Section 5.1, categorizing into the concepts Directory Service (DS) and Query Relay (QR), with our new Aggregating Discovery Service (ADS) approach. To do so, we elaborate on the different concepts, regarding the requirements, defined in Section 4. A summary of the comparison is depicted in Tables 1 and 2.

5.3.1 Directory Service (DS) approach

The Directory Service approach represents the most basic way to provide discovery service functionality. Given an EPC, a query to the discovery services would return a simple list of EPCIS server addresses that are in possession of read events for this particular EPC. Even though the design is simple, it has all means to provide functionality for the core requirements **RQ1** through **RQ4**. Discovery services are still question to research. However, security considerations such as for **RQ5**, **RQ8** and **RQ9** can be addressed by introducing existing authorization and authentication mechanisms, such as Public Key Infrastructures (PKIs). The integrity of the data (**RQ10**) can be ensured using digitally signed messages.
Data ownership (**RQ6**, **RQ7**) is a weak spot of this concept. The information residing at the discovery service comprises only of EPCIS server addresses. The actual read events are still stored at the respective EPCIS servers. However, according to [of Cambridge et al. (2007)] even such information is considered to be sensitive to some companies. To protect this information a discovery service following this concept would need to implement a role-based access layer. Such a layer is difficult to maintain, because of dynamic business relationships. Information about these business relations would need to be copied from the EPCIS servers to the discovery service, resulting in redundant information storage.
Business relationship independent design (**RQ11**) is directly related to the role-based access layer. Changing business relations are reflected by changing access permission for the particular trading partner. That means a company needs to update its access policies every time it adds, modifies, or removes permissions for trading partners.

Organic growth (**RQ12**) is a requirement that is hard to quantify in terms of good or bad. The DS concept provides a low technical boundary for potential users. However, due to the complex access control mechanism, it might be a problem for some companies to guarantee seamless collaboration with trading partners and at the same time protect their own interests. Frequently changing business relations have to result in frequent access policy updates at the discovery service. That is why we rate the support for organic growth for the DS concept rather small.

Looking at network traffic and produced data volume, we can state that the DS concept stores only a minimal set of data (EPC, EPCIS server address, timestamp). Messages between discovery service and client are also restricted to that type of information, leading to a small message size. Modern Database Management Systems (DBMS) are able to handle data volumes of many TB. The bigger problem is the potential request load, which increases with the number of clients and resources. These large data volumes produced by RFID enabled supply chains need to be searched very fast. This problem even aggravates when the number of parallel requests increases. We are currently conducting further research to analyze the impact of increased request load and data volume on the scalability of the different discovery service concepts. However, we expect the DS approach to be able to scale well by applying conventional scalability mechanisms such as load balancing and clustering. This assumption is based on the observation that most processing steps for the notification and the query of a discovery service can be parallelized very well.

RQ14 focuses on the quality of information. Assuming a suitable role-based access layer and a correct working query algorithm, the information returned by the discovery service is complete and correct.

We rate the client complexity (**RQ15**) for the DS concept high in comparison to the other two concepts. The DS approach only returns the URLs of the EPCIS servers' query interface. The client is responsible for invoking the individual EPCIS servers, to parallelize the different requests, to aggregate the information and to invoke successive request, related to different packaging hierarchies.

5.3.2 Query Relay (QR) approach

The Query Relay approach implements an asynchronous request/response paradigm, where the client submits a query for an EPC to the discovery service. The discovery service determines all potential resources for that EPC and propagates the query to these resources, which in turn answer directly to the client. The client needs to implement a callback interface, which is used to aggregate incoming EPCIS responses.

Just like the DS approach, the QR concept provides functionality for requirements **RQ1** through **RQ4**. An authorization and authentication layer needs to be implemented to restrict the number of authorized clients (**RQ5**).

Security (**RQ5**, **RQ8** and **RQ9**) can also be covered by introducing a PKI. By the same token, information integrity can be ensured using digital signatures based on certificates of PKI (**RQ9**).

In contrast to the DS approach, data ownership (**RQ6**, **RQ7**) is a strong feature of the QR concept. The discovery service relays the actual client query to the respective resources, which decide for themselves if they answer the request. That way, the resources are in full control of their data. The advantage is, that the discovery service does not need to provide a role-based access for the actual data. Redundancy and complex access right management are not necessary, because the responsibility to determine whether a client is allowed to receive the information is shifted to the resources.

Business relationship (**RQ11**) independent design is not as critical as for the DS approach, since the resources directly manage the access to their data. The interaction between client and discovery service is not affected by changing business relations. Clients need to negotiate with the operators of the resources, to get the desired information, but the discovery service is not involved in this process.

Organic growth (**RQ12**) is encouraged by the QR concept, because it requires less administrative overhead at the discovery service level to manage access policies. Therefore, it is easier for information providers and information consumers to join the network.

The discussion on scalability (**RQ13**) can be analogous to the DS approach in terms of network traffic and data volume. However, the QR approach has the advantage that it is relieved from complex role-based access policy checking, which can become a problem when the number of concurrent requests rises. So comparing the QR approach to the DS approach, the QR concept has a slight advantage.

Quality of information is a critical point in the QR concept. A client requesting information has no information about the number of potential EPCIS servers that are in possession of information regarding the queried EPC. Consequently, it has no information how many answers it needs to expect. A client querying a discovery service implemented according to the QR approach does not know how many answers from resources it has to expect. EPCIS servers might have slow response times, deny a response to his query or be temporarily not available. This leads to a situation that the client has to wait for a substantial amount of time to be sure that each EPCIS that replies to his query had the chance to do so. Therefore the client has to wait until a timeout is reached. This stands in contrast to a low response time. The result of a client query is complete and correct (**RQ14**) if and only if the client waits long enough to assure that no more replies are still underway. The client has no indication if EPCIS servers are temporarily unavailable.

The asynchronous communication inherent in the QR concept directly leads to an increased client complexity (**RQ15**). In the QR approach the client must be able to receive data from multiple previously unknown sources without knowing the exact number of responses. This results in the need for a complex software design that has to handle multiple incoming connections for a single request. Furthermore, the client has to aggregate the EPCIS responses by itself. Given the fact that client queries are forwarded to respective EPCIS servers immediately, the client is not in full control of its query. It cannot cancel the request or deny that his query is forwarded to a specific EPCIS, which might be a competitor's EPCIS.

5.3.3 Aggregating Discovery Service (ADS) approach

The ADS approach combines the advantages of the DR and the QR. The ADS shifts the complexity (**RQ15**) of query parallelization and the aggregation of EPCIS responses from the client to the discovery service and creates a view of the relevant information for the client. Hence, a query is immediately forwarded to all relevant EPCIS servers. The client is no longer in control of the query once it submitted it. If EPCIS severs enforce role-based access control this is not an issue because only the client role is revealed to the information holder, not the client identity.

Similar to the previous two approaches the ADS supports the four core functionalities (**RQ1-RQ4**). Security measures (**RQ5**, **RQ9**) and **RQ10**) can also be taken from the DS and the QR approaches.
The first major improvement compared to DS and QR is data ownership (**RQ6** and **RQ7**). The discovery services relays the client query to the respective EPCIS servers, providing complete privacy for the resources (**RQ8**), but in contrast to the QR approach, the ADS can control the query process, enabling it to take remedial action upon non-responding resources. The ADS is able to provide the client with a complete and correct set of information (**RQ14**), under the assumption that the client is allowed to see all the information. Otherwise, the result would contain a hint that there is additional information, the client is not allowed to see.
To show the scalability (**RQ13**) of our approach we discuss relevant aspects in Section 5.4.1. Since the ADS does not need to implement fine-grained role-based access control, to protect the companies' information, there is no need for adaptation when companies change their trading partners (**RQ11**). Closely related to this topic is the issue of organic growth. Low technical boundaries and flexibility regarding trading partner management encourage companies to add value to the EPC Network, to provide a beneficial environment for all participants. Table 1 and 2 summarize our evaluation of the presented discovery service concepts, regarding the requirements, defined in Section 4.

	Core Functionality				
	RQ1	RQ2	RQ3	RQ4	RQ5
DS	•	•	•	•	•
QR	•	•	•	•	•
ADS	•	•	•	•	•

Table 1. Fulfillment of Core Operational Requirements

	Selected Requirements											
	RQ6	RQ7	RQ8	RQ9	RQ10	RQ11	RQ12	RQ13	RQ14a	RQ15	RQ16	RQ17
DS	•	–	•	•	•	–	–	•	•	–	–	–
QR	•	•	•	•	•	•	•	•	–	–	–	–
ADS	•	•	•	•	•	•	•	•	•	•	–	–

Table 2. Fulfillment of Selected Requirements

5.4 Evaluation

One of the most important criteria for the successful operation of a discovery service for a larger community of collaborating trading partners is its ability to scale with an increasing number of participants. In detail, we need to focus on increasing network traffic, request load, and data volume. In this subsection, we draw a light on the scalability aspects of our ADS approach.

5.4.1 Scalability

The ADS provides additional functionality, which requires more computing power than the Directory Service or Query Relay approach. Like stated before the ADS has to wait for all responses of the subqueries, thus maintaining a connection's state for the request-response

cycle with the client. In this section we show that it is possible to implement a scalable discovery service following the ADS approach.
We exemplify the potential load for a discovery service in the U.S. pharmaceutical supply chain by a back-of-the-envelope calculation. Following the supply chain network model of Williams et al. [Williams et al. (2008)] a discovery service has to deal with 1,000 notifications per second at peak times and 200 queries per second in average. We assume the worst case scenario that supply chain partners conduct a query for each item they notify as indicated by [of Cambridge & UK (2007)]. The ADS therefore has to deal with the same amount of queries to the discovery service. As the authors additionally state a supply chain does not exceed 15 partners.

5.4.2 Load balancing and data partitioning

Distributing incoming notification messages and client queries to many self-contained application servers allows the ADS to scale very well. HTTP load balancing can be performed in both, hardware and software for very high connection speeds. Additional servers can be added at any time allowing the system to grow in size.
HTTP reverse proxy servers balance incoming HTTP queries. They accept incoming HTTP connections and are able to act based on the queried URL or even arguments in the HTTP request. Implementations like the event-driven *nginx*[1] can help to lower the CPU load on application server machines by mapping requests to a specific EPC to one specific server. Each server is then responsible for a range of EPCs, implementing partitioning at the application server and database tier. Client queries always refer to one or more EPCs. No single database query will ever need to JOIN any data with rows for other EPCs. Database queries will only perform index lookups for EPCs and return the corresponding EPCIS URLs. This allows the database to scale by horizontally partitioning all data by EPC. Every database server is then responsible to serve requests for a range of EPCs. The database lookup only consists of small queries requiring basic database functionality. There is no need for complex locking mechanisms because all data has to be stored persistently and no tuples will be deleted or updated. Furthermore there is no need for synchronizing database partitions.

5.4.3 Open connections

As depicted above it is assumed that the Discovery Server has to handle about 1,000 client queries per second. For a supply chain with $n = 13$ enterprises in average this results in $\frac{n-1}{2} = 6$ relevant EPCIS servers in average per query. This results in 6,000 subqueries per second. Assuming a query to an EPCIS is replied to in one second on average the system has to hold about 6,000 connections simultaneously.
We tested how many connections one commodity PC is able to hold. To simulate a real-world scenario we requested 30,000 random Websites we gathered by querying a search engine with random keywords. All DNS resolving was done before starting the test at a limited upstream speed of 1 Mbit/s and 2.4 GHz CPU speed. Using asynchronous I/O processing we were able to have a single-threaded Python script sustain 3,000 connections (1,100 active) while using 22 to 24% CPU power. A low number of commodity-level servers can easily handle the total amount of connections.

[1] `http://nginx.net`

5.4.4 Bandwidth

In the basic 'Query Relay' architecture, the queried EPCIS servers reply directly to the client. In comparison, the ADS is the single response endpoint for all subqueries. Like described before, during peak hours the ADS has to be able to cope with 1,000 incoming client requests per second. For 6 relevant EPCIS servers on average, it has to send 6,000 subrequests and receive 6,000 subreponses per second. We expect each (sub)query to be 1 KB, each subresponse to be 2 KB in size, and each aggregated response to be 12 KB in size.
Receiving 1,000 queries/s at 1 KB per query and 6,000 subresponses/s at 2 KB per subresponse comes out to an inbound bandwidth of $\frac{(1{,}000 \cdot 1)+(6{,}000 \cdot 2) \cdot 8}{1000} = 104\ Mbit/s$. On the other hand, sending 6,000 subqueries/s at 1 KB per subquery and 1,000 aggregated responses/s at 12 KB per response equals an outbound bandwidth of $\frac{(6{,}000 \cdot 1)+(1{,}000 \cdot 12) \cdot 8}{1000} = 144\ Mbit/s$. Both throughputs are perfectly feasible using available internet connections.

5.4.5 XML handling

All replies sent back from EPCIS servers to the ADS use the XML format standardized by EPCglobal. It wraps all `ObjectEvents` in a single `EventList` [EPCglobal (2007b)]. XML parsers optimized for high throughput provide efficient functionality for aggregating these XML responses. SAX or Pull parsers have proven their efficiency in SOAP environments where a large number of small XML queries have to be processed [Chiu et al. (2002)]. While receiving the XML data stream from a responding EPCIS every parsed tag inside the `EventList` can instantly be created on the output stream that, after all EPCIS servers replied, will be sent back to the client. This eliminates the need to add further buffers for XML objects and reduces XML rendering time.

6. Summary and future work

We started out by motivating the necessity of a discovery service for the EPC Network by introducing real world use cases that require the presence of such a component. In Section 3, we looked at the components of the EPC Network, discussed their particular roles within supply chain collaboration scenarios, and defined their relation to the discovery service. Section 4 introduced requirements for the implementation of a discovery service, followed by a description of existing theoretical and practical discovery service approaches. We also proposed a new design for an Aggregating discovery service, which we compared to the existing concepts in Section 5.
We see two major directions for future work. First of all, it is clear that there will not be a single discovery service, serving all industries. Scalability and political issues require to run a number of independent discovery services. Future research should include the investigation of inter discovery service communication and the definition of a communication protocol to support the exchange of information among independent discovery services. Secondly, we need to quantify the impact of an increasing number of clients onto a single discovery service or a network of interconnected discovery service, to support design decisions for the architecture of discovery services.

7. References

A. Rezafard, A. C. (2008). Extensible Supply-Chain Discovery Service Problem Statement, *IETF Proposal*.

Beier, S., Grandison, T., Kailing, K. & Rantzau, R. (2006). Discovery Services – Enabling RFID Traceability in EPCglobal Networks, *Proc. of the 13th International Conference on Management of Data (COMAD)* .

Chiu, K., Govindaraju, M. & Bramley, R. (2002). Investigating the Limits of SOAP Performance for Scientific Computing, *Proceedings of the 11th IEEE International Symposium on High Performance Distributed Computing* pp. 246 – 254.

EPCglobal (2007a). Architecture Framework Version 1.2.

EPCglobal (2007b). EPC Information Services Version 1.0.1.

EPCglobal (2008). Object Name Service Version 1.0.1.

EPCglobal (n.d.). Discovery Services Standard (under development).

Group, I. C. (2009). Ip crime report 2008-2009. IP Crime Report.

Ilic, A., A. Groessbauer and, F. M. & Fleisch, E. (2009). Understanding Data Volume Problems of RFID-enabled Supply Chains, *Business Process Management Journal*, Vol. 16.

Ilic, A., Groessbauer, A., Michahelles, F. & Fleisch, E. (2009). Estimating Data Volumes of RFID-enabled Supply Chains, *15th Americas Conference on Information Systems (AMCIS)*.

Kürschner, C., Condea, C. & Kasten…, O. (2008). Discovery Service Design in the EPCglobal Network, *The Internet of Things* .

Kürschner, C., Condea, C., Kasten, O. & Thiesse, F. (2008). Discovery service design in the EPCglobal network: towards full supply chain visibility, *IOT'08: Proceedings of the 1st international conference on The internet of things*, Springer-Verlag, Berlin, Heidelberg, pp. 19–34.

Melski, A., Müller, J., Zeier, A. & Schumann, M. (2008). Assessing the effects of enhanced supply chain visibility through rfid, *14th Americas Conference on Information Systems (AMCIS'08), Toronto, Canada*.

Müller, J., Oberst, J., Wehrmeyer, S., Witt, J. & Zeier…, A. (2009). An Aggregating Discovery Service for the EPCglobal Network, *hicss* .

Müller, J., Pöpke, C., Urbat, M., Zeier, A. & Plattner, H. (2009). A Simulation of the Pharmaceutical Supply Chain to Provide Realistic Test Data, *Advances in System Simulation, International Conference on* 0: 44–49.

OECD (2008). *The Economic Impact of Counterfeiting and Piracy*.

OECD (2009). *Magnitude of counterfeiting and piracy of tangible products*.

of Cambridge, A. U. & Research, S. (2007). High Level Design for Discovery Services. BRIDGE project.

of Cambridge, A. U. & UK, G. (2007). Requirements document of serial level lookup service for various industries. BRIDGE project.

of Cambridge, U., wireless, A., Research, B., Research, S., Zurich, E. & UK, G. (2007).

Ohnsman, A. & Kitamura, M. (2010). Toyota Recalls Increase on Brake Flaw Shared by Honda.

Polytarchos, E., Eliakis, S. & Bochtis, D. (2010). Evaluating Discovery Services Architectures in the Context of the Internet of Things, *Unique Radio Innovation* .

Rogers, E. M. (1995). *Diffusion of innovations*, Free Press, New York.

Shrestha, S., Kim, D. S., Lee, S. & Park, J. S. (2010). A Peer-to-Peer RFID Resolution Framework for Supply Chain Network, *Future Networks, International Conference on* 0: 318–322.

Simchi-Levi, D., Kaminsky, P. & Simchi-Levi, E. (2003). *Managing the Supply Chain : The Definitive Guide for the Business Professional*, McGraw-Hill.

Williams, J. R., Sanchez, A., Hofmann, P., Lin, T., Lipton, M. & Mantripragada, K. (2008). *Modeling Supply Chain Network Traffic*, p. 242.

9

Advantages and New Applications of DHT-Based Discovery Services in EPCglobal Network

Juan Pedro Muñoz-Gea, Pilar Manzanares-Lopez and Josemaria Malgosa-Sanahuja
Department of Information Technologies and Communications
Polytechnic University of Cartagena
Spain

1. Introduction

Radio Frequency IDentification (RFID) technology can track the movement of products throughout an entire supply network by giving a Globally Unique Product Identifier (GUPI) (Främling et al., 2007) to every product. The identification is used to connect the physical object to an information service run by a host on the Internet. It is there that all the information associated to that physical object is stored. Nowadays, several approaches are used to implement this connection, such as the EPCglobal Network (Armenio et al., 2009) developed by the Auto-ID consortium, the DIALOG system (Främling et al., 2006) developed at the Helsinki University of Technology and the World Wide Article Information (WWAI) system[1] proposed by Trackway. The EPCglobal Network stands out among the rest because in 2003 it was authorized as a Global Standards I (GS1). The GS1 system of standards is the most widely-used supply-network standards system in the world, the traditional barcode being its most widely used standard.

1.1 Product identifier proposals

The DIALOG system was developed at the Helsinki University of Technology. In this approach an ID@URI notation is used to create a GUPI, where the ID part identifies the product item located at the URI. If the URI is an URL, it is a straightforward task to link it to an information service. For an ID@URI to be a GUPI, the ID part should be unique for the corresponding URI. In the DIALOG system every product is implemented as a software agent (Nwana, 1996), and the information of each of them is accessed and updated through methods in the product agent interface (Främling et al., 2006). These interface methods are as follows: *update()* and *getProductInformation()*, which are used to append and retrieve information, respectively, and *getCompositeInformation()*, which relates to managing component hierarchies. The World Wide Article Information (WWAI) protocol developed by Trackway (formerly known as Stockway) is based on P2P principles. The manufacturers form a network of nodes which are identified by company numbers. When a node has joined the network, it can autonomously issue identifiers for individual products, the product GUPI consisting of a concatenation of the company prefix and item-specific suffix. The service provided by

[1] http://www.trackway.eu

Electronic Product Code (96–bit Version)			
02 .	0000A79 .	00013D .	000154ECD
Header	General Manager Number	Object Class	Serial Number
8 bits	28 bits	24 bits	36 bits

Fig. 1. EPC tag data structure.

the WWAI network is the mapping of the WWAI identifiers to the URL of the information provider, using a DHT (Distributed Hash Table). A lookup is only needed when a product identifier for a given company is seen for the first time. After that, URLs of known nodes are cached so that new node lookups do not need to be performed unless the cached address fails or changes for some reason.

The common characteristic of the two previous architectures is that all the data related to a specific product is available in only one information service. This means that every organization with information about a specific product has to publish it in the corresponding information service and then, the queries have to be addressed to a specific information service depending on the specific GUPI. On the other hand, one of the fundamental design principles for the EPCglobal Network is that each company retains control over the data that they collect or generate within their own organization, i.e. information about a specific product is decentralized across multiple organizations.

The EPC[2] serves the role of a GUPI in the EPCglobal Network. There are two different lengths of EPC tag data that have been ratified by the EPCglobal board: 64 bits and 96 bits. Figure 1 shows the EPC tag data structure of one 96-bit version: Header identifies the version of EPC itself; General Manager Number identifies an organization that maintains the numbers in the field of Object Class and Serial Number; Object Class refers to a unique type of products produced by an EPC general manager; Serial Number uniquely identifies each item within an object class. The EPC serial number allows individual items tracking, which is the key feature and advantage that distinguish EPC from a bar code standard. This feature enables RFID technology to capture the information about the movement of individual products in supply networks. Next section explains the rest of components of the EPCglobal Network architecture in more depth.

1.2 EPCglobal Network

One of the fundamental design principles for the EPCglobal Network is that each company should be able to keep control of the data that they collect or generate about a specific product within their own organization, i.e. information is decentralized across multiple organizations. With a suitable service-oriented architecture the EPC can be used both to locate a source of information, via lookup services, as well as for extracting relevant information about a particular product from each source, by using the EPC as a lookup key within a database.

The EPCglobal Network architecture includes specifications that deal with the collection of captured data and their distribution across organizations. However, the initial standards development has focused more on the EPCglobal Network's lower levels than on data exchange across supply networks. Figure 2 represents the EPCglobal Network architecture framework.

[2] http://www.epcglobalinc.org/standards/tds/tds_1_4-standard-20080611.pdf

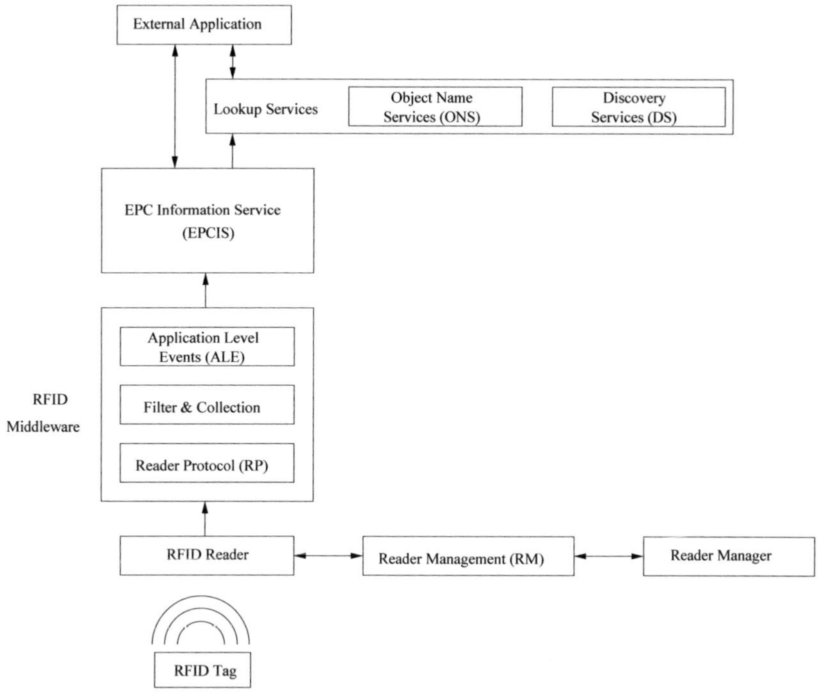

Fig. 2. EPCglobal Network Architecture.

The reader protocol (RP) is an API that abstracts from the underlying RFID hardware readers. RP is complemented by the reader management (RM) protocol, which facilitates the naming of readers and locations. In this context, it is necessary to take into account the fact that the enormous quantity of data generated by even a few readers can easily result in an unacceptable load. For this reason, an additional middleware layer is needed that filters the data collected from readers. In the EPCglobal Network, this task is supported by the application-level events (ALE) interface. ALE allows for the conversion of a series of tag reads into a single event message with a time interval attribute. In addition, the ALE standard also supports the declaration of logical readers that collect the data streams from multiple physical readers.

RFID readers are distributed across factories, warehouses and stores. In this context, EPCIS provide a repository of historical events and related information, and they are fed by an EPCIS capturing application, such as a management system connected to an ALE middleware. To find EPC-related information, business applications use the Object Name Service (ONS) to provide an EPCIS URL when queried with a tag's EPC. The granularity of ONS resolution is currently limited to product type, i.e. an ONS is not expected to retain distinct records for two objects of the same product type. Another point to note is that ONS is currently implemented using the Domain Name System (DNS) (Mockapetris, 1987), using Type 35 Naming Authority Pointer (NAPTR) (Mealling & Daniel, 2000) records to return the information. Queries to ONS are therefore performed by means of a DNS query for a hostname derived from an EPC. However, the ONS is allowed only to point to the manufacturer's EPCIS repository. For this reason, the EPCglobal Network is being extended to include Discovery Services (DS)[3]. These

[3] http://www.epcglobalinc.org/standards/discovery

services will allow applications to find third parties' EPCIS repositories with events related to a specific EPC. A critical component of the DS is the data storage component, which stores information about the list of EPCIS instances that have information about a particular EPC. There currently exist several options to implement this component, although two in particular are worthy of note, LDAP (Lightweight Directory Access Protocol) and DHT (Distributed Hash Table).

2. Distributed Hash Tables (DHT)

DHTs are decentralized distributed systems that distribute the management of a key table between the participating nodes. Each node maintains a routing table with a list of its neighbors, so that it can route messages to the unique owner of a given key. DHT is designed to scale to a large number of nodes and is prepared to deal with continuous node arrivals and failures by constructing a structured P2P (peer-to-peer) network.

There currently exist several structured P2P overlay network proposals such as Chord (Stoica et al., 2003), Pastry (Rowstron & Druschel, 2001), Tapestry (Zhao et al., 2004) and Kademlia (Maymounkov & Mazières, 2002). The difference among them is the way in which nodes are organized in the overlay network. In Chord, nodes are organized in a ring. However, in Pastry, Tapestry and Kademlia nodes are organized in a tree structure (Balakrishnan et al., 2003).

As an example, we are going to present an overview of the Pastry operation. The node identifiers and the routing procedure in Pastry depend on the base used to represent both node and key identifiers. For example, if a hexadecimal base is used and the system requires five digits to represent the full node space, an example of a node identifier could be 65A1F. Each node maintains a routing table formed by as many rows as digits needed to represent the full node space and as many columns as possible digit values (in our example each routing table is formed by 5 rows and 16 columns). Row 0 maintains the IP addresses corresponding to nodes without a common prefix with the local node identifier. Row 1 contains the IP addresses corresponding to nodes with a 1 digit-long prefix common to the local node identifier, and so on. As we have seen, a row is formed by some columns. The *i*-th column of a row stores, if it exists, the IP address of a node whose identifier is determined by the associated row prefix followed by the *i* digit. Please note that there could be more than one node that satisfies the requirements of an entry. If so, the Pastry implementation will use some predefined criterion to select an entry (for example, the node with lowest physical delay). In our example, row 0 stores the IP addresses of nodes without a common prefix with 65A1F: The first entry at row 0 stores the IP address of a node whose identifier is 0**** (if it exists), the second entry stores the IP address of anode whose identifier is 1**** (if it exists), and so on. Row 1 stores the IP addresses of a maximum of 16 nodes, with one digit-length common prefix: the first entry stores the IP address of a node whose identifier is 60*** (if it exists), the second entry stores the IP address of a node whose identifier is 61***, and so on. It must be emphasized that if the local node does not know of a suitable node to fill a concrete entry in the routing table, that entry is left empty.

A requester node elects from its routing table the node matching the longest prefix with the key. The lookup message is sent to that node, and then it repeats the same algorithm until the searched node is found. During this process, a required entry in the routing table might be empty. In this case, the node uses the closest numerical non-empty entry. The routing table definition assures the convergence of the lookup algorithm. In our example, if the 65A1F node is searching for the key 654B2, the longest common prefix in the routing table is 654** (that is

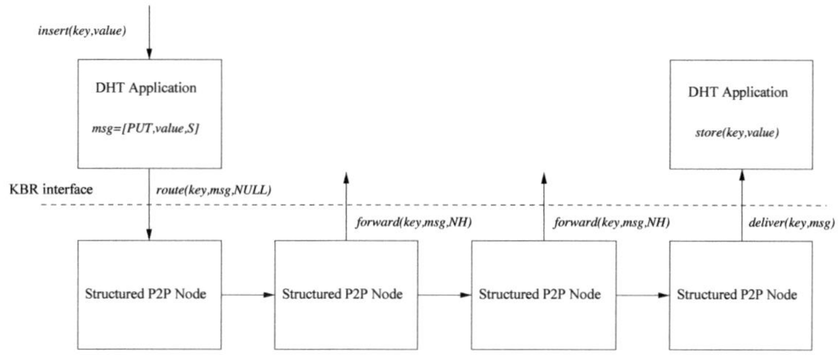

Fig. 3. *insert()* operation.

to say, the entry associated to row 2 and column 5). The local node uses the IP address stored in this entry (or the numerically closest node if the entry is empty) to forward the lookup message.

In general, structured P2P networks offer a routing service to send a message, with an associated key, to a single node, responsible for that key. For this reason, this service is called Key-Based Routing (KBR). There exists a common API (Dabek et al., 2003) with the ability needed to interact with the KBR service offered by any structured P2P network. These are as follows: *void route(key k, msg m)*, which forwards a message, *m*, towards the responsible key node *k* (it is implemented in the structured P2P network layer); *void forward(key k, msg m, nodehandle nexthopnode)*, which is invoked from the structured P2P network layer of each node that forwards message *m* during its routing (this function is implemented in the application layer); *void deliver(key k, msg m)*, which is invoked from the structured P2P network layer of the node that is responsible for key *k* upon the arrival of message *m* (this function is also implemented in the application layer).

Over the previous interface (in the application layer), the services offered by a DHT application can be implemented. The DHT abstraction provides the same functionality as a traditional hash table, by storing the mapping between a *key* and a *value*. That is, it implements a simple store and retrieve functionality, where the value is always stored at the overlay node to which the key is mapped by the KBR layer. This application provides two operations: *insert(key, value)*, and *value = lookup(key)*. A simple implementation of *insert()* routes a PUT message containing value and the *local node?s nodehandle* (S), using *route(key, [PUT,value,S], NULL)*. The node responsible of that *key*, upon receiving the message, stores the *(key, value)* pair in its local storage. On the other hand, the lookup operation routes a GET message using *route(key, [GET,S], NULL)* to the node responsible of that *key*, and it returns the associated *value* in a single hop using *route(NULL, [value], S)*. The single hop routing option uses the *nodehandle* of the source node (S) to send the message directly to that node, without hoping along other nodes in the network. Figures 3 and 4 represent the *insert()* and *lookup()* operation.

There exist several free software implementations of structured P2P networks, like Chimera[4] (a C implementation of Tapestry), the Chord project[5] (a C implementation of Chord) and FreePastry[6] (a Java implementation of Pastry). From among these FreePastry has been selected because it is implemented in Java and because it is an active project in development at the MPI-MPG[7], although it initially started at Rice University in USA.

[4] http://current.cs.ucsb.edu/projects/chimera
[5] http://pdos.csail.mit.edu/chord
[6] http://www.freepastry.org/FreePastry/
[7] http://www.mpi-inf.mpg.de

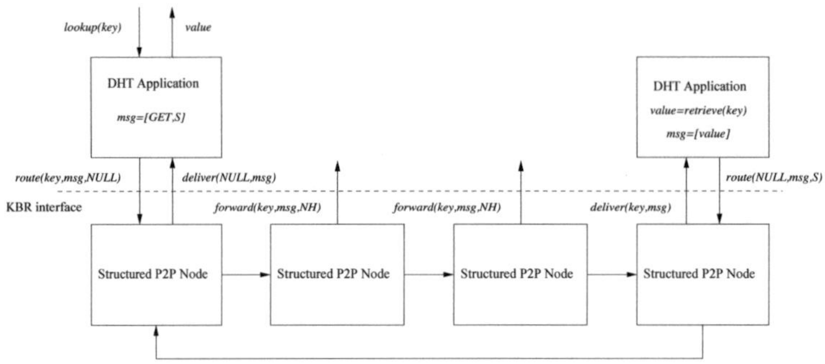

Fig. 4. *lookup()* operation.

3. Advantages of DHT-based discovery services

Recently, there have been several proposals for the implementation of the DS, like the DS prototype of the Bridge project[8], or the Extensible Supply-chain Discovery Service (ESDS) developed by Afilias (Young, 2008). In both the DS has been implemented as a centralized database, to be more specific in the Bridge Project it has been developed as a centralized database based on LDAP (Lightweight Directory Access Protocol) (Zeilenga, 2006). In this work, we are going to develop a DS prototype based on DHT.

LDAP is a networking protocol for querying and modifying directory services running over TCP/IP. In this context, a directory is a set of information with similar attributes organized in a hierarchical manner. LDAP could be thought of as type of database, different from a relational database. Databases are usually designed to perform many changes to their data whereas LDAP directories are optimized to read performance, as such, it is particularly useful for storing information that needs to read from many locations. On the other hand, LDAP permits secure delegation of reading and modification authority based on specific needs using ACLs (Access Control Lists); although this is not part of the LDAP protocol, many implementations offer this feature. The integration of the DS records into the data structure of the LDAP is based on the decomposition of the EPC tag into the LDAP tree. That is, fields such as Company Prefix, Item Reference or Serial Number can be used to distribute the EPC among the tree structure. In addition, security can be integrated with the fine grained access control policies of the LDAP implementation and can be used to limit the access to a particular record.

On the other hand, one of the characteristics of the DHT model is its decentralization, meaning that there is no central coordinator in the nodes. For this reason, the failure of a node never compromises the whole system. Another advantage is that nodes are organized in a way that a node only needs to coordinate with a few other nodes in the system (usually *log N*, with *N* participants). Therefore, if a node joins or leaves the system, this does not affect the whole system. These features provide the system with high scalability and fault tolerance. The integration of DS records into the data structure of the DHT is based on the use of a SHA1 hash of the EPC as the key for the hash table. However, in these systems, security options like authentication or fine grained access control have to be implemented.

As a conclusion, DHTs might be said to have several advantages over LDAP: in LDAP there is a bottleneck in the root of the architecture, whereas in the DHT the failure of one node does not affect the whole system; LDAP is not optimized for massive update operations, while DHTs are optimized for massive search and update operations. On the other hand,

[8] http://www.bridge-project.eu

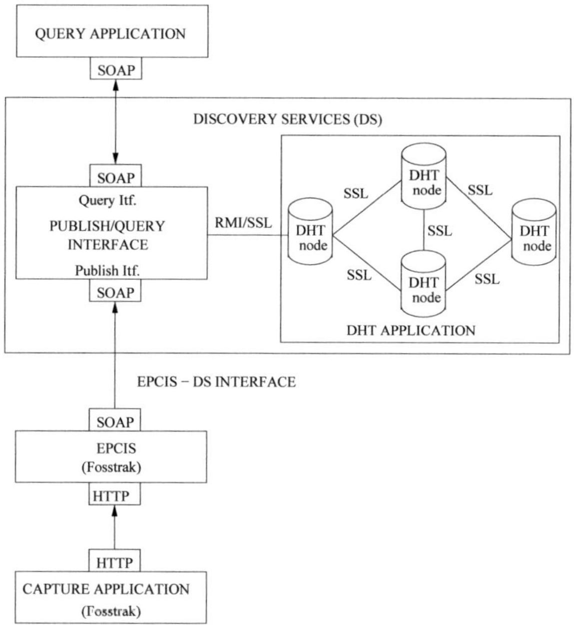

Fig. 5. Discovery Services architecture.

DHTs have an important drawback with respect to LDAP: while in LDAP the access control is already implemented by Access Control Lists (ACLs) in DHT fine grained controls have to be implemented.

4. DHT-based discovery services prototype

The objective of the designed DS is to return a time-ordered list of links to multiple ECPIS instances which hold information related to a specific EPC. Therefore, the DS is designed to create this list of links. The architecture of the full prototype has been divided into a set of logical components represented in Figure 5.

4.1 Discovery services

The DS implementation is composed by two components: the DHT application and the Publish/Query interface.

4.1.1 DHT application

The public methods (*insert()* and *lookup()*) of every DHT node in our implementation can be accessed by a remote application using Java RMI[9], which is a Java API which performs the object-oriented equivalent of remote procedure calls (RPC). The prototype of the previous methods is as follows: *public Boolean insert(MessageDigest key, String url)* and *public String[] lookup(MessageDigest key)*. The insert method has two parameters: *String url* represents the URL of the EPCIS query interface with information about a specific EPC; *MessageDigest key* represents the SHA1 hash (Eastlake & Jones, 2001) of the related EPC. This method returns a *Boolean* which indicates if the insert operation has been properly finished. On the other hand,

[9] http://java.sun.com/javase/technologies/core/basic/rmi/index.jsp

the lookup method has a single parameter: the SHA1 hash of an EPC, and it returns an array with all the associated URLs.
The *insert()* method of the DHT node creates a content object with the *key*, the URL and the its own nodehandle. After that, it inserts the previous content in a message with the following fields: the identifier of the origin node, the identifier of the destination node (which corresponds with the *key* of the message), the message type (for an INSERT message corresponds type 0) and the previously created content. Once the message has been created, the insert method calls the *route()* function with the previously created message.
The previous message is routed to the node responsible of the message's key. Then the FreePastry software will call the *deliver()* implementation of our application. This method extracts the type of the message, and if it is an INSERT message it extracts the *key* (the SHA1 hash of the EPC) and the URL and it inserts both of them in a MySQL[10] database. The communication between our Java application and the database is possible thanks to the JDBC[11] driver. Our database only has one table, with two columns: one for the key (SHA1 hash of the EPC) and the other for the URL. Multiple insertions of the same key, each one of them with a different URL, are allowed. By this way, the database stores the different URLs of the EPCISs with information about a specific EPC. The operation of the *lookup()* method is equivalent to the *insert()* operation, but in this case the destination node extracts from the database all the URLs associated to the key of a specific EPC.

4.1.2 Publish/query interface

This interface has been implemented as a Web Service[12] using two methods available thorugh SOAP[13] operations: *publish()* and *query()*, and it has been deployed in a Glassfish application server[14]. There are currently several frameworks to program Java Web Services, but the two most important are Apache Axis2[15] and Sun JAX-WS[16]. The Java API for XML based Web Services (JAX-WS) is the successor of the JAX-RPC specification. Its configuration is managed by annotations[17], therefore Java 5 or higher is required. With JAX-WS it is relatively easy to write and consume web services. The default values of numerous parameters are comfortable from the point of view of the programmer and simple methods declared with a @WebService annotation can be used as a service. A suitable WSDL document can also be generated from the class.
The *publish()* method of the web service implementation has two parameters: the SHA1 hash of the EPC and the URL where the information related to the EPC is available. The web service has previously been configured with the URL where the RMI stubs of all the DHT nodes are available. Therefore, the web service selects one of the DHT nodes of the network, it obtains its RMI stub and, after that, it calls the *insert()* method of the DHT application. After calling the *insert()* method, the web service receives if the insert operation has been performed in a correct way or not. Therefore, this interface is only a proxy between the EPCIS repository of the query application and the DHT nodes.

[10] http://www.mysql.com
[11] http://dev.mysql.com/downloads/connector/j/5.1.html
[12] http://www.w3.org/standards/webofservices/
[13] http://www.w3.org/TR/soap/
[14] https://glassfish.dev.java.net/
[15] http://ws.apache.org/axis2/
[16] https://jax-ws.dev.java.net/
[17] http://java.sun.com/j2se/1.5.0/docs/guide/language/annotations.html

The *query()* method is similar to that previously mentioned, but in this case it only accepts one parameter: the SHA1 hash of an EPC. This method selects one of the DHT nodes and it calls the RMI *lookup()* method of the DHT application. After that, the web service receives the URLs associated to all the EPCISs with information about that EPC. All this information will be returned to the application which called the *query()* method.

4.2 EPCIS-DS interface

In the first place, we have deployed the Fosstrak EPCIS software[18] in a host with a Glassfish application server. The Fosstrack EPCIS software is a complete implementation of the EPCIS standard specification (Version 1.0.1 of September 21, 2007) and it allows users to deploy an EPCIS repository. In addition, Fosstrak developers also provide an interactive EPCIS capture application which allows users to fill the EPCIS repository with EPC data using a graphical user interface. In our prototype, we have used this graphical tool to insert events into the EPCIS Repository.

The current EPCIS standard does not include a specific communication mechanism between an EPCIS and a Discovery Service. For this reason, we have developed a module to be integrated into the Fosstrak software. Its goal is to send the association between an EPC, with information registered in that EPCIS repository, and the URL to query that information to the DS, but only once. That is, it is possible that the EPCIS registers several events associated to the same EPC because it is possible that it has several associated RFID readers. Therefore, our module has to assure the DS stores only one association between that EPC and the public URL of the EPCIS. We have studied the EPCIS software implementation and we have detected that it has two independent modules: the EPCIS capture interface and the repository, implemented in a MySQL database. The communication among them is by means of the TCP protocol, and the SQL queries are sent to the 3306 TCP port of the localhost. SQL queries are sent in plain text; therefore, we decided to develop a traffic sniffer to detect the SQL queries sent to the MySQL database.

The sniffer module has been developed in Java language, using the jNetPcap[19] software development kit. The basic function of jNetPcap is to provide a java wrapper to popular libpcap library[20] for capturing network packets. Our module captures the packets in the loopback interface, because the TCP communication between the EPCIS capture interface and the MySQL database is in the localhost. But it does not capture all the traffic; we have implemented a filter in order to detect the SQL queries with EPCs. After the filter detects one of these SQL queries in a specific packet it will be redirected to a special method to process it. This method inspects the full SQL query to obtain the associated EPC. After detecting the EPC, our program checks if the detected EPC has been observed previously. In the case that it has not been detected, that is, it is the first time that information associated to that EPC is registered in the EPCIS repository, the program calls the *publish()* method of the publish/query web service interface.

4.3 Query application

This application allows a user to make real queries to the DS deployed. It accepts an EPC number and returns necessary information to reconstruct the supply network. In order to do

[18] http://www.fosstrak.org
[19] http://jnetpcap.com
[20] http://www.tcpdump.org

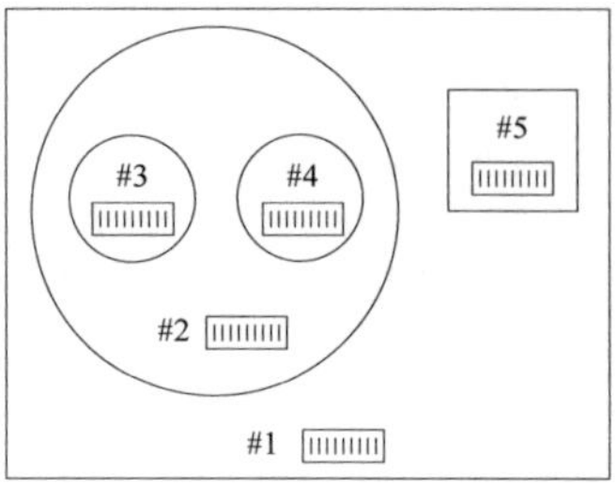

Fig. 6. A product structure.

that, the application implements a web service client which interacts with the *query()* method of the publish/query web service interface.

The last step in the supply network reconstruction process is the graphic representation of the received data, so that the information can be easily understood and interpreted by the final user. This functionality has to be added to the query application. It will depends on the final application developed using the DHT-based Discovery Service Architecture.

5. New Applications of DHT-based discovery services

5.1 Automatic traceability

5.1.1 Introduction

Suppliers, manufacturers, distributors and retailers are typically interconnected within networks and it is for this reason that the relationships among them are represented as supply networks (Wareham et al., 2005),(Phillips et al., 2006),(Poulin et al., 2006),(Li & Chandra, 2007). In this respect, supply chains are special types of supply networks in which organizations are organized in linear chains. However, many organizations do not have sufficient information about the full supply networks in which they are involved, for example, they do not know who supplies their suppliers because they are not directly connected to them. This information would be very useful in planning strategies or in assuring product quality. For example, in order to solve some problems of delayed shipments from suppliers, the organization might need to analyze its full supply network to identify which supplier is most to blame for the delay. Unfortunately however, many organizations do not have access to this information.

In this work, we propose a mechanism for automatically obtaining the supply network associated to a specific product using the EPCglobal Network. In (Bi & Lin, 2009) the authors proposed a methodology with the same objective but the main difference with the proposal set out in this article is that in (Bi & Lin, 2009) the client has to do all the operations to reconstruct the supply network. That is, initially, it has to obtain the URLs (Uniform Resource Locator) of the information services with information about a specific product, then it has to access the corresponding information services to get the necessary information, and after that it can reconstruct the full supply network. However, in our proposal the client does not have to perform any operations, that is, it queries the supply network associated to a specific product and the results are obtained directly.

A concrete example (extracted from (Bi & Lin, 2009)) is used to demonstrate how a supply network can be mapped using our methodology. Suppose that a retailer A sells a product that is assembled by a manufacturer B. Every product item and each of its components are given an EPC tag. The structure of the product with five EPC tags is shown in Figure 6. Note that component #2 is a sub-assembly that consists of two smaller components.

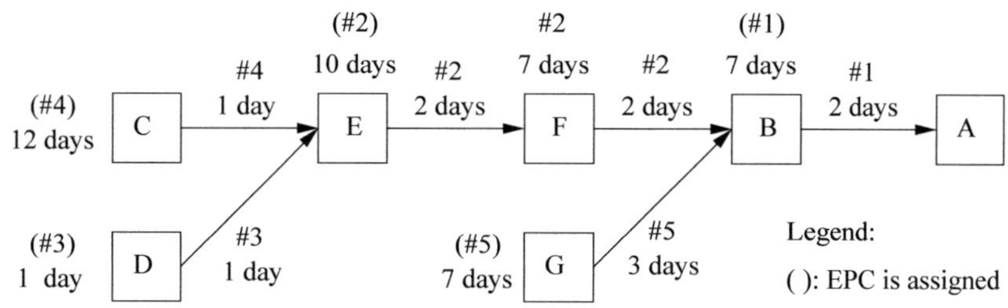

Fig. 7. Supply network associated to EPC #1.

After receiving a several requests to repair component #4, retailer A decides to map its supply network for this product in order to redirect the requests for repairs to the company that produces component #4. Retailer A is also interested in calculating the amount of time taken for products to flow between entities in this supply network. The full supply network that retailer A wants to reconstruct is represented in Figure 7.

Now, the necessary information to reconstruct the previous supply network needs to be deduced. Firstly, we concentrate on the events that assign an EPC to a new product (ASSIGN events, they are represent by ( ) in the figure). These events can be divided in two different sub-types. The first of which corresponds to the assignment of EPCs #3, #4, and #5 to the corresponding products. They are not all composed of other products, that is, they do not all have smaller components. The necessary information to represent the previous events is the following: #EPC, Company, Event=ASSIGN and Date. On the other hand, the assignment of EPCs #1 and #2 is different from the previous one, because in this case the new products are composed of several components with an assigned EPC. In this case, the necessary information to represent these events is similar to the previous one with the difference that the identity of the components of the product is added. That is, it will be: #EPC, Company, Event=ASSIGN, Date and #EPC of Components.

Next, we concentrate on the flow of products. To represent this we need to deduce the amount of time for the flow of products from one company to another company and the direction of the flow. The previous information can be deduced if we extract the exact date on which the product is sent from the company of origin (SHIP event) to a specific destination company, and the exact date on which the product is received by the destination company (RECEIVE event). In order to represent the SHIP event we need the following information: #EPC, Company, Event=SHIP, Date and Destination. On the other hand, to represent the RECEIVE event we only need: #EPC, Company, Event=RECEIVE and Date. In conclusion, in order to reconstruct a full supply network we need a time-ordered list of ASSIGN, SHIP and RECEIVE events with the necessary information to represent them all.

In order to obtain the previous information, it is necessary to send a query to the query interface of the Foostrak EPCISs with information about a specific EPC, asking for some information which depends on the event (ASSIGN, SHIP or RECEIVE) which we want to identify.

For an ASSIGN event without components, it is necessary to ask for the *date* of an *ObjectEvent* with *action=ADD* associated to a specific EPC. For an ASSIGN event with components, it is necessary to ask for the *date* and the *childEPCs* of an *AgregationEvent* with *action=ADD* associated to a specific EPC. For a SHIP event it is necessary to ask for the *date* and the *purchase order* associated to the *business transaction* of an *ObjectEvent* with *action=OBSERVE* and *businessStep=SHIPPING* associated to a specific *EPC*. In order to guess the destination company of a SHIP event we are going to use the purchase order associated to the business

Event	eventType	EQ_action	3rd Parameter
ASSIGN	ObjectEvent	ADD	MATCH_epc = urn:epc:id:sgtin:0034000.987650.2686
ASSIGN with comp.	AggregationEvent	ADD	MATCH_parentID = urn:epc:id:sscc:0614141.1234567890
SHIP	ObjectEvent	OBSERVE	EQ_bizStep = urn:epcglobal:cbv:bizstep:shipping
RECEIVE	ObjectEvent	OBSERVE	EQ_bizStep = urn:epcglobal:cbv:bizstep:receiving

Table 1. Parameters to query the ASSIGN, SHIP and RECEIVE events.

transaction. This purchase order includes the identities of the buyer and the supplier, the product related to the purchase order and the quantity of the ordered product. Finally, for a RECEIVE event, it is necessary to ask for the *date* of an *ObjectEvent* with *action=OBSERVE* and *businessStep=RECEIVING* associated to a specific *EPC*.
The query interface of the Foostrak EPCISs is provided by means of a web service; therefore, it is necessary to program a client web service to ask for the previous information.Table 1 represents the parameters to query the previous ASSIGN, SHIP and RECEIVE events.
The *poll()* method returns an array with the following information: Event, occurred, recorded, Parent ID, Quantity, EPCs, Action, Business step, Disposition, ReadpointId, Business location, Business transaction. Therefore, we only have to go to the corresponding column to obtain the necessary information.

5.1.2 Implementation

In this section, we are going to present the integration of a supply networks discovery mechanism within our DHT-based DS prototype. We have added a new public method to our DHT application, called *supply_network()*, which can also be accessed by a remote application using Java RMI. The prototype of the previous method is the following: *public Object[][] supply_network(MessageDigest key)*. This method has a single parameter, the SHA1 hash of an EPC, and it returns a bi-dimensional array of *Objects*. The Java *Object* class sits at the top of the class hierarchy tree in the Java development environment, that is to say, every class in the Java system is a descendent of the *Object* class. The bi-dimensional *Object* array has all the necessary information to represent the supply network. That is, the columns have these values: #EPC, Company, Event (ASSIGN, RECEIVE, SHIP), Data, Destination Company (for SHIP events) and Components (for ASSIGN events with different sub-products). In addition, every row represents an event (ASSIGN, RECEIVE or SHIP) associated to a specific EPC.
We have also added a new method to the web service publish/query interface called *network()*. The *network()* method is similar to the *query()* method. It only accepts one parameter: the SHA-1 hash of an EPC. This method selects one of the DHT nodes and it calls the RMI *supply_network()* method of the DHT application. After that, the web service receives the previous *Object* bi-dimensional array, with all the necessary information to represent the supply network. All this information will be returned to the application which called the *network()* method.
The *supply_network()* method of the DHT node creates a content object with the key and its own nodehandle. After that, it inserts the previous content in a message with the following fields: the identifier of the origin node, the identifier of the destination node (which corresponds to the key of the message), the message type (for a SUPPLY_NETWORK message corresponds to type 4) and the previously created content. Once the message has been created, the *supply_network()* method calls the *route()* function with the previously created message. The *supply_network()* method also uses the *continuations* functionality offered by FreePastry.

```
IF URL[].length == 1 THEN
    ASSIGN_function(URL)
ELSE
    FOR each component of URL[]
        IF URL is the last
            URL.poll(RECEIVE)
        ELSEIF URL is the first
            URL.poll(SHIP)
            ASSIGN_function(URL)
        ELSE
            URL.poll(SHIP)
            URL.poll(RECEIVE)
        ENDIF
    ENDFOR
ENDIF

FUNCTION ASSIGN_function(URL)
    COMPONENTS[]=URL.poll(ASSIGN with components)
    IF COMPONENTS[].length > 0
        FOR each COMPONENT
            supply_network(COMPONENT.\#EPC)
        ENDFOR
    ELSE
        URL.poll(ASSIGN without components)
    ENDIF
RETURN
```

Table 2. Algorithm of the *supply_network()* method.

The previous message is routed to the node responsible for the message's key. Then the FreePastry software calls the *deliver()* implementation of our application. This method extract the type of the message, and if it is a SUPPLY_NETWORK message it extracts the key of the EPC, and after that it extracts from the database all the URLs associated to the key of a specific EPC. It is necessary to take into account that the URLs are obtained in a time order. That is, the first extracted URL corresponds with the first EPCIS which registered information about the corresponding EPC. Then, the application follows the reconstruction algorithm presented in Table 2.

As an example, we present the operations performed by the DHT application in order to reconstruct the supply network associated to the previously presented product. In this process, we assume that the DHT node responsible for storing the URLs of the EPCIS with information about the EPC #x is the DHTx node (e.g., DHT1 has the URLs with information about the EPC #1). Therefore, the call to the *supply_network()* method with EPC #1 as a parameter, gets to DHT1. This node has two URLs associated to EPC #1, [URLB, URLA]. The application takes the last URL (URLA) and it sends a query to the query interface of the corresponding EPCIS asking for a RECEIVE event. After that, it receives the necessary information to construct an *Object* array. That is, it constructs [#1; RECEIVE; A; 10/02/09; NULL; NULL]. Then, the application takes the first URL (URLB), it sends a query asking for a SHIP event, with the received information it construct the *Object* array [#1; SHIP; B, 08/02/09; A; NULL] and it adds this array to the previously constructed *Object* array. Immediately, the application sends a new query to the same URL asking for an ASSIGN event with components, with the received information it construct the *Object* array [#1; ASSIGN; B; 01/02/09; NULL; #2, #5], and it adds this array to the previously array. Finally, the application calls the *supply_network()* method taking EPC #5 as a parameter. When it receives all the information related to this EPC, it will call the *supply_network()* method taking EPC #2 as a parameter. The operation mode of these two *suply_network()* calls is equivalent to the presented one and, as a result, both of them return an *Object* bi-dimensional array, with all the information associated to the corresponding sub-supply networks associated to EPC #2 and EPC #5. All this information is added to the previously constructed *Object* bi-dimensional array (with the information about EPC #1) and it is returned to the *network()* web service

EPC	EVENT	COMPANY	DATE	TO	CONTAINS
#1	ASSIGN	B	01/02/09	—	#2,#5
#1	SHIP	B	08/02/09	A	—
#1	RECEIVE	A	10/02/09	—	—
#2	ASSIGN	E	10/01/09	—	#3,#4
#2	SHIP	E	20/01/09	F	—
#2	RECEIVE	F	22/01/09	—	—
#2	SHIP	F	29/01/09	B	—
#2	RECEIVE	B	31/01/09	—	—
#5	ASSIGN	G	13/12/08	—	—
#5	SHIP	G	20/12/08	B	—
#5	RECEIVE	B	23/12/08	—	—
#3	ASSIGN	D	10/12/08	—	—
#3	SHIP	D	11/12/08	E	—
#3	RECEIVE	E	12/12/08	—	—
#4	ASSIGN	C	08/12/08	—	—
#4	SHIP	C	20/12/08	E	—
#4	RECEIVE	E	21/12/08	—	—

Table 3. Information to reconstruct the supply network associated to EPC #1.

method. The information contained in this *Object* bi-dimensional array corresponds to the information presented in Table 3. Figure 8 represents the interaction among the different DHT notes involved in the reconstruction of this supply network.

In addition, every DHT node which receives a SUPPLY_NETWORK query stores temporarily in a new database all the deduced information (contained in the *Object* bi-dimensional array), and it associates all this information to the corresponding EPC. By this way, subsequent calls to the *supply_network()* method will be resolved in a shorter period of time. Finally, the access control service is also used by this mechanism. The PEP located in the DHT node (responsible for storing the URLs associated to the EPC included in the *supply_network()* method) calls the *getDecision()* method of the PDP situated in the publish/query interface in order to get a read authorization decision. If it is allowed, the DHT application gets the necessary information to reconstruct the supply network. Otherwise, the DHT application returns an error message. This is possible because we have added a new field to the SUPPLY_NETWORK message: the identity of the query application which calls the *network()* web service method. Our access control mechanism implements one policy for the read operation implemented in the *supply_network()* method of the DHT application. The default behavior is to deny the access to the information except for those clients defined by the companies of the consortium as partners.

5.2 Nested package in supply chains

5.2.1 Introduction

In the EPC technology research, most of the work about track and trace corresponds to item level tracking (Bi & Lin, 2009), (Goebel et al., 2009), (Beier et al., 2006). To face this challenge, all of them assume that items are always visible along the whole supply chain. However, in many industrial fields this supposition does not reflect the reality. For example, clothing industry tags at item level, but products are distributed and move along the supply chain

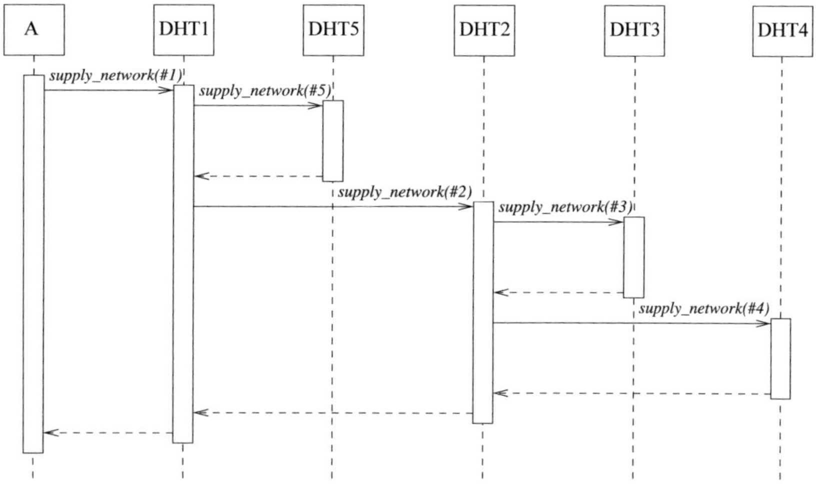

Fig. 8. Interaction among the DHT nodes involved in the reconstruction of this supply network.

within different storage systems (trays, packages, boxes, etc.) (*Charles Voegele Group Finds RFID Helps It Stay Competitive*, 2009). Although each item has its own EPC, items do not go independently towards the destination. In many cases, some items will be packaged together to facilitate transport and distribution.

Item package may produce scenarios with null (or limited) item level visibility. Each item is associated to a RFID tag but, if they are packaged together in a particular storage system, items are not visible to RFID readers. Only the RFID tag associated to the new package is visible. Storage process may be repeated as many times as needed (items into small boxes, small boxes into cases, cases into pallets, etc.). Usually, the EPC associated to the overall wrapper identifies the order.

This section presents an extension of the automatic traceability application to recover, in an efficient way, the complete supply chain of an item in a nested package scenario.

5.2.2 Description of the scenario

An example of nested package identification is shown in figure 9. There, four levels of package are carried out. $\#it_i$ tags identify the items, C_{1x}, C_{2x}, C_{3x} and C_{4x} tags identify the different level packages, being C_{41} the tag associated to the overall wrapper.

Figure 10 shows two supply chains corresponding to the previous nested package scenario: item 1's and item 10's supply chains. Because item tags are not always readable, supply chains will be reconstructed from the tag collections obtained by the RFID readers from the own item or the packages that, at different levels, contain the item. Each supply chain shows three possible actions: package actions, advance actions and unpackage actions. Associated to these actions, three zones can be identified from left to right:

- The first zone represents the package process. ($\#it_i$) indicates the assignment of an EPC to an item. ($\#C_{jk}$) indicates the assignment of an EPC to a level j package where $j = 1...4$. $\#it_i$ and $\#C_{jk}$ indicate the reading of RFID tags by a RFID reader. For the sake of simplicity and without loss of generality, it is considered that all package actions, from the item creation to the highest level package, are done at the same chain element.

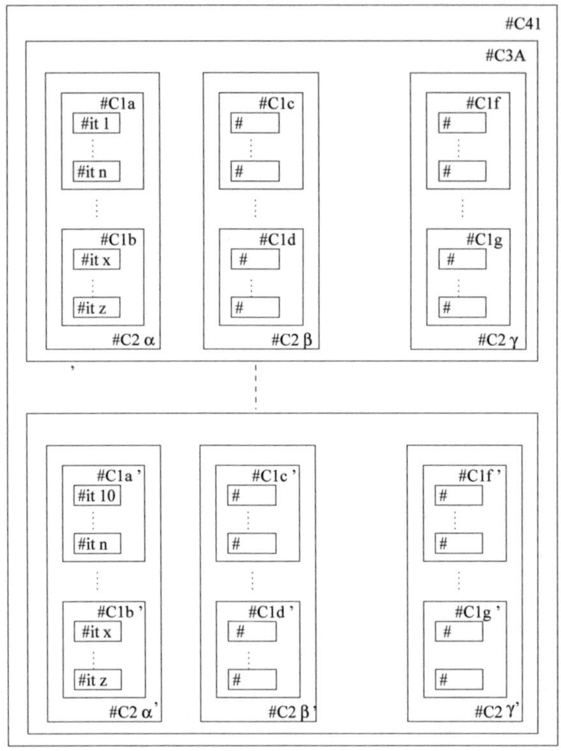

Fig. 9. An example of nested package identification, corresponding to four level package.

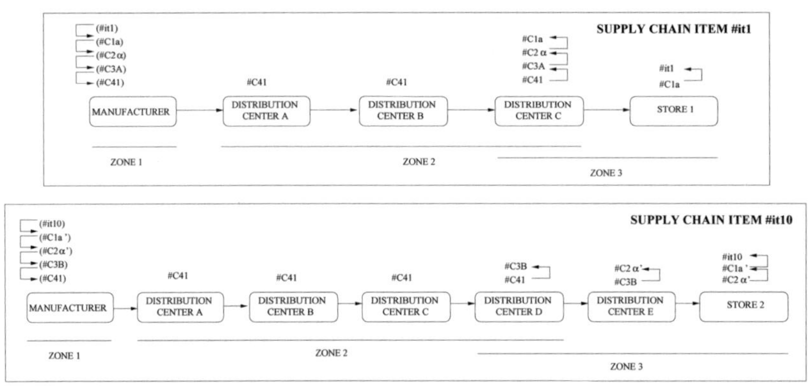

Fig. 10. Supply chains corresponding to items #it1 and #it10. $(\#it_i)$ and $(C\#_{jk})$ indicate the assignment of EPCs to items and packages. $\#it_i$ and $C\#_{jk}$ indicate the reading of tags by RFID readers. Package and unpackage actions are represented by arrows.

- When the highest package level is created (in our example it corresponds with C_{41}), it will pass through several organizations during the advance along the distribution channel.
- The third zone corresponds with the unpackage process. This task could involve several organizations, depending on the unpackage level reached at each organization. In item 1's supply chain, the unpackage process involves Distribution Center C (DC C) and Store 1. DC C unpackages to the lowest package level (C_{1a}), and after receiving the smallest

package unit, Store 1 unpackages the item. On the other hand, the unpackage process in item 10's supply chain involves three organizations: Distribution Center D, Distribution Center E and Store 2.

Although it is not represented in the supply chain example above, in some cases there is tracking information which is obtained during the second zone. In many industrial sectors, random verifications at item level are done at intermediate points of the supply chain. That means, a random lowest level package is chosen and then, all the items of this package (or just some of them) are examined. This action will generate additional information, corresponding to events happened in the central zone of the supply chain.

5.2.3 Supply chain recovery

The solution described in this section will allow the reconstruction of a supply chain at item level, in a totally distributed way. As it was described, the DHT-based DS architecture minimizes the number of messages that a requester organization must send (normally the requester will be an organization with a limited network connection). Most of the queries required to obtain all the information will be performed by DHT (Distributed Hash Table) nodes that, like EPCIS servers, will be components with better features in connectivity, reliability and security. On the other hand, the use of this architecture in nested package scenarios reduces the total number of messages required to obtain multiple supply chains corresponding to different items. Due to the distributed work during supply chain reconstruction, and only maintaining a cache at the DHT nodes, previously requested information can be reused during the reconstruction of new supply chains.

Figure 11 shows, by means of an example, the behavior of the distributed architecture during the storage and recovery of information required to reconstruct the supply chain of a nested package. In this example, the item whose EPC is #1 is created and packaged into three levels (the package systems are identified by $\#C_{11}$, $\#C_{21}$ and $\#C_{31}$ respectively) by company A. The highest level package passes through the distribution channel (companies B and C) to company D, where all the unpackage actions will be done. Interaction among the different elements involved in the reconstruction of the supply chain is represented in the figure. Messages 1 to 20 correspond with the storage in the EPCISs of the read EPC events and the registration of the EPCISs with the DHT network. After that, DHT node P -the responsible of key hash(#1)- stores the IP addresses of $EPCIS_A$ and $EPCIS_D$ (that is, the EPCISs storing events about EPC #1). DHT node Y -the responsible of key hash($\#C_{11}$)- stores the IP addresses of $EPCIS_A$ and $EPCIS_D$ (that is, the EPCISs storing events about EPC $\#C_{11}$). DHT node M -the responsible of key hash($\#C_{21}$)- stores the IP addresses of $EPCIS_A$ and $EPCIS_D$ (that is, the EPCISs storing events about EPC $\#C_{21}$). Finally, DHT node O -the responsible of key hash($\#C_{31}$)- stores the IP addresses of $EPCIS_A$, $EPCIS_B$, $EPCIS_C$ and $EPCIS_D$ (that is, the EPCISs storing events about EPC $\#C_{31}$). The rest of the messages are required to reconstruct the supply chain of item #1, process that is initiated by company D. Organization D will query its associated DHT node (node Z) to initiate the recovering of the supply chain (message 21). That DHT node will locate the DHT node responsible for the item EPC #1 (message 22). The node responsible for #1 (in the example node P) stores the addresses of the EPCIS that have data about the EPC ($EPCIS_A$ y $EPCIS_D$). Consequently, node P will be able to query them all the information about #1 (messages labeled jointly in the figure as 23). Information associated to EPC #1 indicates that the item was packaged into $\#C_{11}$. Therefore, instead of replying directly to node Z (the node that initiates the lookup), node P will locate the DHT node responsible for $\#C_{11}$ (message 24). Only when node P receives the data about $\#C_{11}$, it will

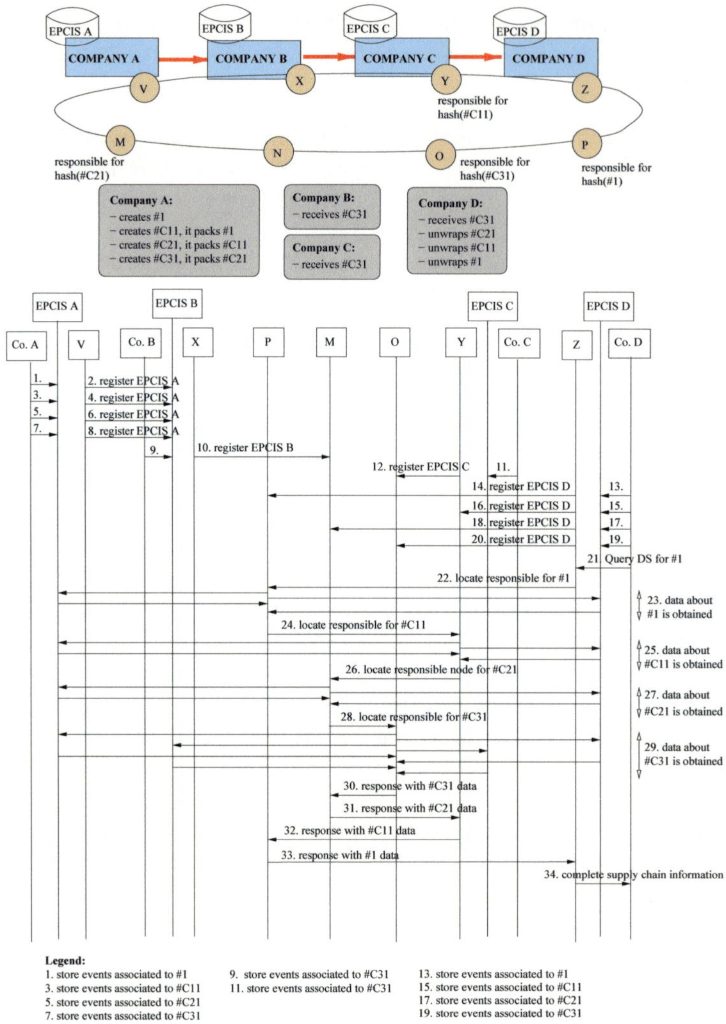

Fig. 11. Reconstruction of a supply chain in a nested package scenario.

reply to node Z with all the information. Due to $\#C_{11}$ is associated to $\#C_{21}$, node Y also locates the DHT node responsible for $\#C_{21}$ before answering to node P (message 26). Finally, node M, which is responsible for $\#C_{21}$, will locate the DHT node responsible for $\#C_{31}$ (message 28). Starting at node O, all the involved DHT nodes will respond successively and in the opposite direction to the lookups, with the requested information (messages 30 to 33). Finally, node Z will be able to answer the initial query performed by company D. This response will contain all the information required to reconstruct the supply chain of item #1.

This solution allows to recover, in a distributed way, supply chains in a nested package scenario. As it can be seen, each DHT node will find out a portion of the supply chain, and all these portions will form the complete supply chain. In fact, thanks to this feature the proposed solution offers a significant gain against the EPCglobal solution in terms of network usage. To obtain this gain, it would be sufficient if DHT nodes acted as proxies, that is, if they stored in a cache the information that was recovered during the reconstruction of a supply chain after answering the requesting node. In this way, when the supply chain of a new item is required, and this item coincides at any package level with another item whose supply chain has been obtained before, some portions of the supply chain have already been reconstructed and it is not necessary to query again. Depending on the level of coincidence between the items, the gain will be greater or smaller. On the other hand, depending on the cache maintenance (in terms of time-to-live of entries), the system gain will also vary.

5.2.4 Evaluation

5.2.4.1 Use of DHT nodes

The main contribution of this solution is the use of a DHT network to implement, in an efficient and distributed way, the item level track&trace service. Therefore, it is interesting to evaluate the consequences in the DHT nodes of storing the information associated to the supply chains. Specifically, the amount of involved nodes and information that each node must maintain is going to be evaluated. Remember that each node will store, for all the EPC under its responsibility, the URL of the EPCIS that have information about the EPC.
If an item package process reaches i levels, the number of DHT nodes which will be involved is $(i+1)$ (the nodes responsible for keys $\#it$, $\#C_{1k}$,..., $\#C_{im}$). Each node should store an entry, that corresponds to the Internet address of the manufacturer. On the other hand, the movement of the highest level package along the distribution chain does not involve other nodes, though it will add new entries on the node responsible for that package. That is, if j elements are gone through, the node responsible for $\#C_{im}$ should store j new entries. In the same way, the unpackage tasks will only affect to the number of entries stored on the nodes who are responsible for the assigned EPCs, but it does not modify the number of involved DHT nodes. Due to this last phase, a new entry for each unpackage level will be generated.
Usually, any industrial activity does not create a single item but a high number of them. If P items are produced, and they are grouped into i levels, the number of EPCs that are used by the manufacturer is:

$$P + \lceil P/n_1 \rceil + \lceil \lceil P/n_1 \rceil / n_2 \rceil + \lceil \lceil \lceil P/n_1 \rceil / n_2 \rceil / n_3 \rceil + ... \leq P + i + \sum_{j=1}^{i} \frac{P}{\prod_{m=1}^{j} n_m} \quad (1)$$

where n_i is the number of packages of $i-1$ level that are packaged into a i level package.
Because the number of involved nodes depends only on the number of EPCs, if N is the number of nodes that belong to the DHT network and considering that the amount of items is substantially higher than the number of nodes ($N << P$), due to the DHT network features, each node will be responsible for the following number of EPCs:

$$\frac{P + i + \sum_{j=1}^{i} \frac{P}{\prod_{m=1}^{j} n_m}}{N} < \frac{2P+i}{N} \sim \frac{2P}{N} \quad (2)$$

that is, a limited number of entries that can easily be managed by any current database system.

5.2.4.2 Network capacity gain

One of the most frequently used performance features of information system is the efficient use of network capacity. Here, this feature is measured in terms of absolute number of application messages exchanged to store and obtain the data related to an item supply chain. Application messages are the queries, responses, inserts, lookups, etc. generated in the evaluated systems. In fact, each application message could involve more than one network

message. For example, in the centralized EPCglobal proposal, a query to the Discovery Service will involve a certain number of messages partly due to the use of a DNS-based ONS. In the proposed distributed solution, an insert or a lookup message will involve a certain number of messages due to the structured overlay network mechanisms (Chord, Pastry, Tapestry,...), which is due to the use of a DHT-based ONS.
Both ONS architectures are compared by simulation in section 5.2.4.4. As it will be concluded there, the number of nodes that must be contacted to resolve a ONS query is always lower in the DHT-based solution than in the DNS-based solution. Therefore, an DHT-based application message implies a lower number of network messages.
In this section, it is considered that both solutions implement the same functionality. That is, DHT nodes in the distributed solution do not use cache tables.
To carry out the evaluation, the following parameters are considered: l is the number of elements of a supply chain ($l \in N^{+} \backslash \{1\}$), i is the number of package levels and p is the coincidence level between two items. If both items are packaged in the same level 1 package, p value is 1. However, if both items are packaged in different level 1 units, but in the same level 2 package, p value is 2, and so on.
For the sake of simplicity (like in the example in figure 11), it is considered that the package and unpackage actions take place in the beginning and at the end of the supply chain respectively, and therefore, the intermediate elements along the supply chain just read the highest level package. It is also considered that the item supply chain reconstruction is initiated by the last element.
According to a centralized solution, the number of messages required to store and later reconstruct the supply chain of an item is:

$$M_{centralized} = 2 \cdot (i+1) + (l-2) + 2 \cdot (i+1) + (l-2) + 2 \cdot (i+1) + 1 \cdot 2 \cdot l + i \cdot 2 \cdot 2 \quad (3)$$

The first four terms correspond with messages which are generated during the storage of data, that is, messages which are generated when the item is created and moves along the distribution channel until the destination. $2 \cdot (i+1) + (l-2)$ messages are sent to the EPCISs and $2 \cdot (i+1) + (l-2)$ are sent to DS. The rest of terms are created due to the recovering of the item supply chain. Let's identify each one: $2 \cdot (i+1)$ messages are exchanged between the element requiring the supply chain and the DS, and $1 \cdot 2 \cdot l + 2 \cdot 2 \cdot i$ messages are exchanged between the last element and each involved EPCIS.
According to a distributed solution, the total number of messages is:

$$M_{distributed} = 2 \cdot (i+1) + (l-2) + 2 \cdot (i+1) + (l-2) + 2 + 2 \cdot (i+1) + 1 \cdot 2 \cdot l + i \cdot 2 \cdot 2 \quad (4)$$

Like equation 3, the four first terms correspond to messages which are generated during the storage of data. If $2 \cdot (i+1) + (l-2)$ messages are sent to the DS in the centralized solution, in this solution all these messages are distributed amount $(i+1)$ nodes: The nodes that are responsible for the $i+1$ EPCs involved in the item supply chain. The rest of messages are generated during the supply chain reconstruction process. Although the amount of messages is almost the same, the sending and the reception of the messages is completely decentralized. Whereas in a centralized solution the last element is responsible for sending all the messages, in a distributed solution, all the DHT nodes that store data about the required supply chain will participate in the supply chain data recovery. In the distributed solution, the requester element initiates the supply chain reconstruction process by querying the DHT

node associated to its EPCIS. Finally, that element will receive from that DHT node all the data required to reconstruct the supply chain. That is the reason for the +2 term in equation 4.
It can be concluded that the gain between the centralized solution and the distributed proposal is not related to message consumption. In fact, the main improvement of the last solution is the distribution and decentralization of message sending and reception. In addition, as it will be described in the next section, the distributed behavior of the last solution will make possible to reduce the number of required messages when reconstructing more than one supply chain if they work as proxies.

5.2.4.3 Network capacity gain using cache

The DHT-based DS architecture offers an efficient and distributed solution to reconstruct supply chains. In fact, the proposed distributed solution also offers an additional advantage in nested package scenarios. To obtain this improvement, it is necessary that the DHT nodes maintain a cache table, where the information that they have obtained to respond to another DHT node's query during a supply chain recovery process will be stored. Thus, the total number of messages needed to reconstruct the supply chain of different items is reduced. For example, using the example shown in figure 11, if item #2 is packaged together with item #1 (and item #1 supply chain has been obtained), data about #C_{11}, #C_{21} and #C_{31} is already available to DHT node Y, which will respond to the lookup sent by the DHT node responsible for #2 during the supply chain reconstruction process.
In the previous section, equation 4 corresponds to the number of messages that are exchanged during the storage of supply chain information and the recovery of that information without using cache. If DHT nodes maintain a cache table, the number of messages required to store and retrieve the supply chain information of an item which does not coincide with other item at any level (level of coincidence p=0) is logically the same in that equation. However, if the required item coincides at any level (p) with a previous item whose supply chain was already reconstructed, the number of messages is reduced according to the next expression:

$$M_{distrcache} = 2 \cdot [2 \cdot (i+1) + (l-2)] + 2 + 2 \cdot (p+1) + 2 \cdot 2 \cdot p \quad (5)$$

As in equation 4, the first term corresponds with messages which are generated during the storage of data. Actually, both equations only differ on the number of message during the data recovering process. Therefore, to obtain the gain of the distributed solution using DHT nodes as proxies, only these terms are considered, as described in equation 6.

$$gain = 1 - \frac{2 + 2 \cdot (p+1) + 2 \cdot 2 \cdot p}{2 + 2 \cdot (i+1) + 2 \cdot 2 \cdot i + 1 \cdot 2 \cdot l} \quad (6)$$

Table 4 shows the gain values according to equation 6 corresponding to different package levels (i) and different levels of coincidence (p). l indicates the length of the supply chains. Next, the main conclusions of these results are exposed.
First, it can be noticed that for the same i and p values (that is, for the same number of package levels and the same level of coincidence between items), the gain increases as supply chain length is longer. This is because the information about the highest level package, which contains both coincidence items within any of its internal packages, has already been obtained during the first supply chain reconstruction. It will not be necessary to find out again what happened with the highest level package along the supply chain. Therefore, the greater the number of elements in the supply chain is, the greater the gain is.

		i=1	i=2	i=3	i=4	i=5
l=2	p=1	0.2857	0.5000	0.6154	0.6875	0.7368
	p=2	-	0.200	0.3846	0.5000	0.5768
	p=3	-	-	0.1538	0.3125	0.4211
	p=4	-	-	-	0.1250	0.2632
	p=5	-	-	-	-	0.1053
l=3	p=1	0.3750	0.5455	0.6429	0.7059	0.7500
	p=2	-	0.2727	0.4286	0.5294	0.6000
	p=3	-	-	0.2143	0.3529	0.4500
	p=4	-	-	-	0.1765	0.300
	p=5	-	-	-	-	0.1500
l=4	p=1	0.4444	0.5833	0.6667	0.722	0.7619
	p=2	-	0.3333	0.4667	0.5556	0.6190
	p=3	-	-	0.2667	0.3889	0.4762
	p=4	-	-	-	0.2222	0.3333
	p=5	-	-	-	-	0.1905
l=5	p=1	0.5000	0.6154	0.6875	0.7368	0.7727
	p=2	-	0.3846	0.5000	0.5789	0.6364
	p=3	-	-	0.3125	0.4211	0.5000
	p=4	-	-	-	0.2632	0.3636
	p=5	-	-	-	-	0.2773

Table 4. Network capacity gain when using caches. i is the number of package levels, p is the level of coincidence between the requested item and a previous item. l is the length of the supply chain.

Secondly, for items that move along the supply chain within the same highest level package, the lower p value is, the greater the gain is. That is, the best gain values are obtained when the coincidence between items happens in a lower level. Logically, if the number of common levels is higher, the number of queries is smaller because this information is already obtained. Finally, for supply chains of the same length and the same level of coincidence between items, the higher the total number of levels is, the greater the gain is. A bigger difference between p and i indicates that the number of common levels is higher. Thus, the amount of information to find out is reduced.

The above values represent the gain when recovering the information about an item using cache at DHT nodes in comparison to the recovering of a previous item if any cache is used. However, these values also represent the relation between the amount of messages generated during the first (all caches will be empty) and the second recovery in a distributed system with caches. Here, if a third item is considered, the gain is determined by the best level of coincidence with both previously requested items. Therefore, to obtain the network capacity gain of DHT-based DS architecture due to the use of cache tables, it is necessary to consider all the previously requested items.

Figure 12 shows the network capacity gain results in a nested package scenario. In this scenario, 100 items are packaged at level 1, 20 level 1 units are packaged at level 2, 10 level 2 units are packaged at level 3. Three level 3 packages have been created, which correspond to 100000 items. The length of the supply chain (l) is 5. The X axis corresponds to the amount of supply chains that have been resolved as time goes by. Items are randomly chosen. This figure represents the gain obtained for each requested supply chain and also the average network capacity gain. Since only three levels of package are used, the possible gain values are 0.6875, 0.5 and 0.3125, depending on the level of coincidence. Thus, three zones can be distinguished in the figure. In the first zone (until the value 100, approximately) the most common gain

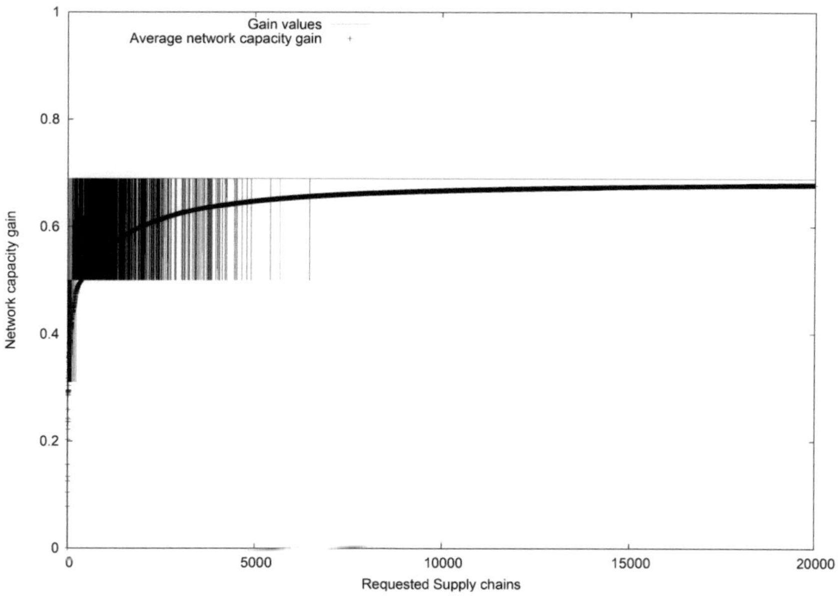

Fig. 12. Gain value of each requested supply chain and the average network capacity gain.

value begins 0.375, then changes up to 0.5 and even some 0.6875. In the second zone (between 100 and 2500), gain values move between 0.5 and 0.6875. Finally, in the last zone the most repeated gain value is 0.6785. Figure 12 shows how the average network capacity gain quickly increases until the constant value of 0.6743 is reached.

5.2.4.4 Application level messages

The parameter used in the previous sections to measure the network capacity is the number of application messages exchanged to store and obtain the data related to an item supply chain. In fact, these messages (queries, responses, inserts, lookups,...) involve more than one network message, depending on the number of nodes that it is necessary to contact in order to perform the required operation.

Here, this last parameter is going to be evaluated in two architectures: the first one implements a ONS based on the traditional DNS, and the second one implements a ONS based on a DHT network. The evaluation has been performed by simulation. The supply chains related to each EPC have been constructed using a C++ STL-like container class for trees, which depends of the probabilities of increasing the length and the width of the tree. The DNS-based approach has been evaluated using the OMNeT++ discrete event simulator, and the DHT-based approach has been evaluated using the OverSim P2P simulation framework for OMNeT++. All the simulation parameters are equally configured in both cases. For example, the number of nodes immersed in the ONS service is 65 (in the DNS-based implementation these nodes are organized in a tree hierarchy and in the DHT-based implementation these nodes compose a Chord overlay network (Stoica et al., 2003)).

Table 5 presents the simulation results. The first column represents the probability of increasing the length of the supply chain and the second one the probability of increasing the width. The third and fourth columns represent the average number of nodes that it is necessary to contact to reconstruct a full supply chain that has been built taking into account the previous probabilities. The average values are obtained from the reconstruction of 100000

p(length)	p(width)	Number of nodes	
		DHT	DNS
0.30	0.10	3.24	7.56
0.30	0.30	3.80	8.89
0.30	0.50	4.51	10.54
0.50	0.10	3.61	8.44
0.50	0.30	5.38	12.54
0.50	0.50	8.51	19.85
0.70	0.10	4.72	11.03
0.70	0.30	15.84	36.89
0.70	0.50	288.46	673.05

Table 5. Number of contacted nodes using DHT-based solution and DNS-based solution

supply chains, which is greater enough to guarantee stationary values. From the results it can be concluded that the number of nodes contacted is always lower in the DHT-based solution than in a DNS tree. On the other hand, it has also been noticed that the necessary number of contacted nodes in the DNS tree is roughly a 133% greater than the required number in the DHT-based solution. That is, the increment is approximately a constant value.

6. Conclusion

In this chapter we have analyzed the advantages of implementing the Discovery Services (DS) component of the EPCglobal Network architecture using a Distributed Hash Table (DHT) application. In addition, we have also showed that it is possible to develop new applications over the DHT-based discovery services.

Recently, there have been several proposals for the implementation of the DS, like the DS prototype of the Bridge project, or the Extensible Supply-chain Discovery Service (ESDS) developed by Afilias. In both the DS has been implemented as a centralized database, to be more specific in the Bridge Project it has been developed as a centralized database based on LDAP (Lightweight Directory Access Protocol).

DHTs might be said to have several advantages over LDAP: in LDAP there is a bottleneck in the root of the architecture, whereas in the DHT the failure of one node does not affect the whole system; LDAP is not optimized for massive update operations, while DHTs are optimized for massive search and update operations. On the other hand, DHTs have an important drawback with respect to LDAP: while in LDAP the access control is already implemented by Access Control Lists (ACLs) in DHT fine grained controls have to be implemented.

We have proposed a mechanism for automatically obtaining the supply network associated to a specific product. There are other systems with the same objective but the main difference with the proposal set out in this chapter is that in other systems the client has to do all the operations to reconstruct the supply network. That is, initially, it has to obtain the URLs (Uniform Resource Locator) of the information services with information about a specific product, then it has to access the corresponding information services to get the necessary information, and after that it can reconstruct the full supply network. However, in our proposal the client does not have to perform any operation, that is, it queries the supply network associated to a specific product and the results are obtained directly.

On the other hand, in the EPC technology research, most of the work about track and trace corresponds to item level tracking. To face this challenge, all of them assume that items are always visible along the whole supply chain. However, in many industrial fields this supposition does not reflect the reality. For example, clothing industry tags at item level, but products are distributed and move along the supply chain within different storage systems (trays, packages, boxes, etc.). This chapter has proposed a distributed architecture to recover, in an efficient way, the complete supply chain of an item in a nested package scenario. In addition, the proposed solution improves the EPCglobal Network features in terms of network usage and also in terms of distribution and decentralization of tasks associated to capturing and querying event information.

7. Acknowledgments

This research has been supported by the MICINN/FEDER project grant TEC2010-21405-C02-02/TCM (CALM) and it is also developed in the framework of "Programa de Ayudas a Grupos de Excelencia de la Región de Murcia, de la Fundación Séneca, Agencia de Ciencia y Tecnología de la RM (Plan Regional de Ciencia y Tecnología 2007/2010)".

8. References

Armenio, F., Barthel, H., Dietrich, P., Duker, J., Floerkemeier, C., Garrett, J., Harrison, M., Hogan, B., Mitsugi, J., Preishuber-Pfluegl, J., Ryaboy, O., Sarma, S., Suen, K., Traub, K. & Williams, J. (2009). Epcglobal architecture framework, Available online at: http://www.epcglobalinc.org/standards/architecture/architecture_1_3-framework-20090319.pdf.

Balakrishnan, H., Kaashoek, M. F., Karger, D., Morris, R. & Stoica, I. (2003). Looking up data in p2p systems, *Communications of the ACM* 46(2): 43–48.

Beier, S., Grandison, T., Kailing, K. & Rantzau, R. (2006). Discovery services - enabling rfid traceability in epcglobal networks, *Proceedings of International Conference on Management of Data (COMAD)*.

Bi, H. H. & Lin, D. K. J. (2009). Rfid-enabled discovery of supply networks, *IEEE Transactions on Engineering Management* 56(1): 129–141.

Charles Voegele Group Finds RFID Helps It Stay Competitive (2009). Available at: http://fridjournal.com/article/view/4836.

Dabek, F., Zhao, B., Druschel, P., Kubiatowicz, J. & Stoica, I. (2003). Towards a common api for structured peer-to-peer overlays, *Proceedings of the 2nd International Workshop on Peer-to-Peer Systems (IPTPS03)*, Berkeley, CA, pp. 33–44.

Eastlake, D. & Jones, P. (2001). RFC 3174: Secure hash algorithm 1 (sha1), Available online at: http://tools.ietf.org/rfc/rfc3174.txt.

Främling, K., Ala-Risku, T., Kärkkäinen, M. & Holmström, J. (2006). Agent-based model for managing composite product information, *Computers in Industry* 57(1): 72–81.

Främling, K., Harrison, M., Brusey, J. & Petrow, J. (2007). Requirements on unique identifiers for managing product lifecycle information: comparison of alternative approaches, *International Journal of Computer Integrated Manufacturing* 20(7): 715–726.

Goebel, C., Tribowski, C. & Günter, O. (2009). Epcis-based supply chain event management - a quantitive comparison of candidate system architectures, *Proceedings of the International Conference on Complex, Intelligent, and Software Intensive Systems (CISIS'2009)*, pp. 494–499.

Li, X. & Chandra, C. (2007). Efficient knowledge integration to support a complex supply network management, *International Journal of Manufacturing Technology and Management* 10(1): 1–18.

Maymounkov, P. & Mazières, D. (2002). Kademlia: A peer-to-peer information system based on the xor metric, *IPTPS '01: Revised Papers from the First International Workshop on Peer-to-Peer Systems*, London, UK, pp. 53–65.

Mealling, M. & Daniel, R. (2000). RFC 2915: The naming authority pointer (naptr) dns resource record, Available online at: http://tools.ietf.org/rfc/rfc2915.txt.

Mockapetris, P. (1987). RFC 1035: Domain name system, Available online at: http://tools.ietf.org/rfc/rfc1035.txt.

Nwana, H. S. (1996). Software agents: An overview, *Knowledge Engineering Review* 11(3): 1–40.

Phillips, W., Johnsen, T., Caldwell, N. & Lewis, M. A. (2006). Investigating innovation in complex health care supply networks, *Health Services Management Research* 19(3): 197–206.

Poulin, M., Montreuil, B. & Martel, A. (2006). Implications of personalization offers on demand and supply networks design: A case from the golf club industry, *European Journal of Operational Research* 169(3): 996–1009.

Rowstron, A. & Druschel, P. (2001). Pastry: Scalable, distributed object location and routing for large-scale peer-to-peer systems, *IFIP/ACM International Conference on Distributed Systems Platforms (Middleware)*, Heidelberg, Germany, pp. 329–350.

Stoica, I., Morris, R., Liben-Nowell, D., Karger, D. R., Kaashoek, M. F., Dabek, F. & Balakrishnan, H. (2003). Chord: a scalable peer-to-peer lookup protocol for internet applications, *IEEE/ACM Transactions on Networking* 11(1): 17–32.

Wareham, J., Mathiassen, L., Rai, A., Straub, D. & Klein, R. (2005). The business value of digital supply networks: A program of research on the impacts of globalization, *Journal of International Management* 11(2): 201–227.

Young, M. (2008). Extensible supply-chain discovery service (esds) concepts, Available online at: http://tools.ietf.org/id/draft-young-esds-concepts-04.txt.

Zeilenga, K. (2006). RFC 4510: Lightweight directory access protocol (ldap), Available online at: http://tools.ietf.org/rfc/rfc4510.txt.

Zhao, B. Y., Huang, L., Stribling, J., rhea, S. C., Joseph, A. D. & Kubiatowicz, J. (2004). Tapestry: A resilient global-scale overlay for service deployment, *IEEE Journal on Selected Areas in Communications* 22(1): 41–53.

10

Application of RFID and Mobile Technology to Plaster Board Waste in the Construction Industry

Lizong Zhang, Anthony S. Atkins and Hongnian Yu
Staffordshire University,
United Kingdom

1. Introduction

Protection of the environment is one of the most sensitive topics of recent years and recognition of its importance is resulting in more effective recycling and reuse of material. Simply disposing of waste through landfill or burning, as was historically the case, will damage our environment. Nowadays, a number of government initiatives and environmental pressure groups are seeking more environmental, socially responsible methods of disposal. Undoubtedly, recycling is the best solution for protecting our environment by 'waste management'.

In general, waste management is the administration of collection, transport, processing, recycling and/or disposal of waste materials(McBean et al., 1995). Currently, it usually includes reduction and auditing activities for the protection of human health and the environment by producing less waste and by using it as a resource wherever possible. This type of waste management is known as 'sustainable waste management'- reduction, re-use, recycling, composting and using waste as a source of energy (DEFRA, 2007).

Today, waste generation and disposal has become a serious problem. Each year, approximately 335 Mt of waste are produced in the UK most of which is produced in England. In 2005, 272 Mt were produced in England, much more than Scotland, Wales and Northern Ireland, which produce less than 25% of the total UK waste (DEFRA, 2007). Information from the Environment Agency (EA) and SEPA (Scotland's Environmental Regulator and Adviser) indicated there were 4.12 Mt of hazardous waste produced in England and Wales (EA, 2007), with an additional 4,430 tonnes produced in Scotland (SEPA(Scotland), 2007). In 2006, this increased to 6 Mt in England and Wales (EA, 2008a).

In general, waste is a combination of many types of material, and most of them are harmful and polluting. Waste takes many years to break down, and can pollute water courses and the land even when it is carefully disposed of, especially the hazardous or controlled waste. However, fly-tipping and incorrect land filling of these types of waste will cause heavy pollution and damage to the environment. The UK produces over 1 million tonnes (Mt) of plasterboard waste per annum, of which only up to 7% is being recycled whilst the majority has been land filled which causes a potential environmental problem. Plasterboard waste contains a high percentage of gypsum which has resulted in serious problems because of the emission of hydrogen sulphide gas (H_2S) once it is land filled with organic waste. According to DEFRA (Department for Environment, Food and Rural Affairs), it can be anticipated that the volume of plasterboard waste will increase over the next 15 years due to the expansion of usage and the rise in construction projects. Another example is medical waste; it usually contains infectious materials, drugs and sharp objects such as syringes, which are undoubtedly harmful waste that contains a large number of viruses, bacteria and harmful chemical reagents.

In the UK, only a few recycling facilities are available in England, and it is usual for the recycling companies to take plasterboard waste from construction or demolition sites themselves, and charge a transportation fee. Depending on the distance to the recycling facilities, transportation fees can be expensive, and exceed the landfill mono-cell costs. Consequently, increasing recycling is a viable route to prevent environmental problems. To improve the recycling rate, a tracking and auditing system is needed to prevent fly tipping and other illegal disposal. A novel waste management prototype is outlined based on identification technology, current waste management process and reasoning techniques.

In this chapter, RFID (Radio Frequency Identification) technology is introduced which can potentially improve the waste management efficiency. Radio Frequency Identification (RFID) is an automatic identification technology used in assets tracking and logistics support in supply chain management by substituting barcodes with RFID tags.

2. Plasterboard landfill and fly-tipping problems

Demolition waste and construction waste are becoming an increasing problem, particularly due to the re-development of urban areas. This type of waste typically contributes more than 30% of the total waste in the UK. For instance, in 2004, as Figure 1 demonstrates, construction and demolition waste accounted for 106.1 Mt, which is 31.7% of the total UK waste (DEFRA, 2006).

Plasterboard waste is a major contributor to the problem of construction and demolition waste. The application of plasterboard in construction began in the late 1960s, and its lifespan, which averaged 30 years, depended on the type of plasterboard used (MTP, 2007a), therefore current refurbishment and demolition activities will increase the amount of waste plasterboard created in this sector.

Information from Waste and Resource Action Programme (WRAP) in 2006 and the UK Department for Environment, Food and Rural Affairs (DEFRA) in 2007 indicated that more than 1 Mtpa (Million tonnes per annum) of plasterboard is generated by the construction and demolition sector, and only 70,000 tonnes are being recycled each year. It is anticipated that over next 15 years the volume of plasterboard waste will increase because of expansion in its use and the rise in construction projects (DEFRA, 2007).

Plasterboards are made from fibre materials and gypsum. Normally, the middle layer of the plasterboard is gypsum ($CaSO_4{\cdot}2H_2O$) which is sandwiched together by two pieces of fibre material such as paper or cloth. Plasterboard is relatively quick to fit in buildings, requiring a low level of skill; it is widely used in domestic and commercial situations, and usually used in the interior of buildings to provide a high quality finish.

The amount of plasterboard produced and consumed in the UK is about 3 Mt (million tonnes), and between 1 and 1.3 million tonnes of plasterboard waste is generated annually which is a significant amount (DEFRA, 2007). This waste comes from demolition and refurbishment activities, and also from new construction sites, which typically waste 12% of the total raw plasterboard materials (Lund-Nielsen, 2007).

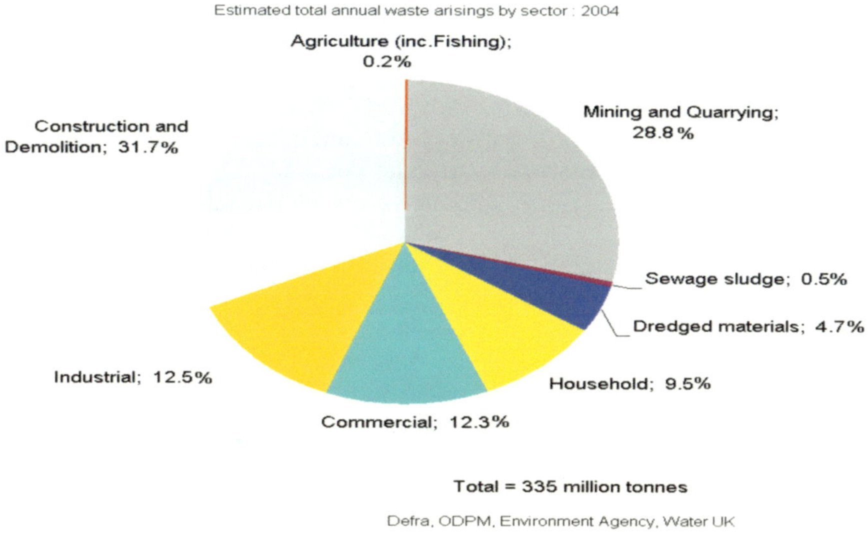

Fig. 1. Estimated Total Annual Waste by Sector in the UK 2004(DEFRA, 2006) (modified by authors)

Traditionally, landfill is the main method of UK waste disposal, and this includes plasterboard. However, plasterboard is made from more than 90% gypsum, and some fibrous materials, with less than 1% of other materials (PbrUK, 2007, Kleist et al., 2004). Plasterboard waste is therefore considered to be high sulphate waste, which can break down with other organic waste and emit H_2S gas, as represented in the following reaction:

$$SO_4^{-2} + 2\,CH_2O \text{ ---> } 2\,HCO_3^{-1} + H_2S$$

Sulphate (SO_4^{-2}) is the main component of gypsum, (which is $CaSO_4{\cdot}2H_2O$), and CH_2O is the organic carbon in organic waste material (McBean et al., 1995). Hydrogen sulphide (H_2S) gas is produced by the sulphate-reducing bacteria under anaerobic conditions in the landfill sites. 100 tonnes of landfilled sulphate can potentially produce 35 tonnes of H_2S (Heguy and Bogner, 2004). Plasterboard waste in landfill sites therefore poses a serious health and safety risk.

In December 2002, the European Council established criteria and procedures for the acceptance of waste at landfills referred to as 'European Council Decision 2003/33/EC'. This document clearly indicates that 'Non-hazardous gypsum-based materials should be disposed of only in landfills for non-hazardous waste in cells where no biodegradable waste is accepted'(European_Council, 2002).

In England and Wales, this decision has already been implemented in 'The Landfill (England and Wales) (Amendment) Regulations 2004'. This act extended the range of 'gypsum-based materials' to include 'Gypsum based and other high sulphate bearing materials', (Bradshaw, 2004) and a definition was necessary to determine what percentage of sulphate indicated 'high sulphate bearing materials'. In 2005 The Landfill Regulations were further amended, detailing the 'Criteria for stable non-reactive hazardous waste and non-hazardous waste deposited in the same cell with such waste', and stipulating that the sulphate level should be less than 20,000mg/kg dry substance or less than 10,000 mg/m^2(Bradshaw, 2005).

It relation to the revised regulations, the Environment Agency (DEFRA) has published at least two documents to interpret the definition of high sulphate waste. It 'considers 'gypsum-based and other high sulphate-bearing materials' to be waste with more than 10% sulphate in any one load' (Bourn, 2005), and that 'gypsum based and other high sulphate bearing materials relates to both gypsum and other forms of sulphate containing waste with a content of more than 10% sulphate per load' (Guidance for waste destined for landfill). In addition, Regulatory Guidance Note 11 indicates that this type of waste should be landfilled in specially designed cells with their technical requirements (Bourn, 2005). This 'guidance' or 'requirement' is referred to as the '10% Rule' for plasterboard disposal (James et al., 2006).

The '10%' rule has however resulted in a problem, in that it seems to encourage 'diluting' plasterboard with other waste under the 10% threshold rather than segregating the plasterboard from other waste for recycling purposes (James et al., 2006).

In November 2008, The UK government removed the 10% regulation, and consequently the plasterboard waste needs to be separated from other waste (EA, 2008b). The changes to the regulation represent awareness of public interest in plasterboard waste issues, and since that time, plasterboard waste has to be landfilled in a mono cell, which almost doubles the normal waste disposal cost. Additionally, there is an increase in transportation cost to these special mono-cell sites for the waste producer. Consequently, fly-tipping or illegal disposal is becoming a problem and a specialist management system for the plasterboard waste will be required.

3. Landfill sites and fees

According to the regulations, landfill sites which can accept plasterboard waste should be specifically designed, with a 'mono-cell'. This type of 'mono-cell' is expensive to construct in terms of high investment and the duration of construction period (Bourn, 2005). Consequently, due to the limited capacity and high cost of the mono-cell, this method of disposing of low-value and large volume waste such as plasterboard is not the preferred solution.

	Pure (100%) Plasterboard Waste using Mono-cell	Recycling Plaster-board Waste	Mixed (<10%) Plasterboard Waste using normal Cell (NOT ALLOWED since 11/2008)
Transport Fees £/t	Average £ 20 /t	Average £20 /t	Average £20/t
Landfill Tax £/t	£ 24 /t	N/A	£ 24 /t
Operation fees £/t	£95 /t	Unknown (from free to £75/t)	£24.5 /t
Total £/t	Average 139£/t	Minimal cost £75/t	Average £68.5/t

Fig. 2. Estimated Fees for Disposal of Plasterboard Waste (Source: (MTP, 2007a, James et al., 2006, Turban et al., 2005, Pal. and Shiu, 2004)).

Information from WARP indicates that there are two sites which have been identified as accepting high sulphate waste: Winterton Landfill in North Lincolnshire and Harmondsworth Landfill in Middlesex (James et al., 2006). However, there appears to be no information showing how many sites in England & Wales are licensed to accept high sulphate waste. Most applications for construction of mono-fills are currently for asbestos, because high sulphate waste is relatively new requirement (MTP, 2007b).

Figure 2 compares the estimated fees for the different disposal methods of plasterboard waste. In 2008, there were still large amounts, more than 1 Mt per year, of plasterboard waste being landfilled in the UK, and there is no information that can confirm the landfill amount for the most recent two years; butundoubtedly, the removing of the '10% rule' from the regulations has been playing a significant role for improving the percentage of plasterboard which is being recycled.

4. Plasterboard recycling

Recycling plasterboard waste is an important way to reduce the amount of plasterboard entering the waste landfill system and damaging the environment. There are different processing techniques for recycling plasterboard, but in general, they all include two major objectives: separation of gypsum from the paper, and crushing to produce a gypsum powder that is mainly used to make new plasterboard (Hamm et al., 2007, Hirz and Sterr, 1995, John and Knez, 1993, Tudahl and Bush, 2000).

Normally, the plasterboard waste usually contains impurities such as metal, plastic and other debris, which need to be removed before going to the feed hopper of the recycling equipment. This process is usually manual, but the removal of ferrous metal can be performed automatically using electromagnets (Turban et al., 2005, GRI, 2011).

Fig. 3. Locations of Current and Reported Potential Recycling Facilities/Transfer Stations for Gypsum Waste within the UK (excludes new NWG Plants as Location not Revealed)(James et al., 2006)

Figure 3 shows the current location of plasterboard recycling facilities, and demonstrates that in some parts of the UK it is not feasible to collect and transport the waste, either because of issues of transport, or cost implications. Depending on the distance to the

recycling facilities, transportation fees can be expensive, and exceed the landfill mono-cell costs. Therefore, logistics is an important area affecting recycling throughput.
A typical method of transportation is implemented by Gypsum Recycling International Company, which compacts the plasterboard waste and transports it using specially designed vehicles (Turban et al., 2005). This requires a site to be selected and waste plasterboard is placed in a special container which is provided by or rented from this company. It then regularly sends a specially designed vehicle with an on-board crane and weight system to collect the waste, and transport it to the recycling plant (Turban et al., 2005, GRI, 2011).

5. Introducing of RFID

Radio Frequency Identification (RFID) is an automatic identification technology which has received much attention recently because of its application in assets tracking and logistics support to supply chain management from the substitute barcodes with RFID technology. In this section, RFID technology will be introduced from its history, principle, standards, system and application scale viewpoints.

5.1 History of RFID

RFID started in the Second World War, but the real progress was not made until mid-1960s, when it became practical and in a form that would be recognized today(Roussos, 2008). During the 1960s and 1970s, RIFD became a more widely practical reality and turned from military to commercial usage. The Electronic Article Surveillance (EAS) is the first system used in commercial application that utilizes 1-bit tags to prevent theft of merchandise (Landt, 2005, Roussos, 2008). Although the system only has the feature to detect the presence or absence of a tag, it did not prevent EAS from becoming the first and widespread commercial use of RFID technology.
The first truly passive tags appeared in 1975, when an early and important development was made by the Los Alamos Scientific Laboratory and presented by Alfred Koelle, Steven Depp, and Robert Freyman (Landt, 2005), which indicates the complete first truly passive tag using backscattering development(Miles et al., 2008). During the 1970s and early 1980s, the large scale commercial usage began, but it was still limited by the electronic component and circuit technologies, most of the tags could only hold a few bits in that period. At the end of the 1980s, beside the significant and rapid development in electronic technology, which provided lower cost and higher performance, RFID technology became more practicable. In the following decades, many contactless applications using RFID appeared and RFID has become popular, particularly in access control and ticketing (Landt, 2005, Roussos, 2008).
The important expansion of RFID application happened in the early 2000s, when the enterprise information system had developed into the backbone of globe trade, and the supply chain introduced to the business process, which involved partnerships with millions of goods across the world (Landt, 2005, Lehpamer, 2008). Due to the precision and low cost of RFID technology, the large scaled RFID application was then expanded, and applied to many global companies during this period, such as DHL, Wal-Mart.

5.2 Principle of RFID technology

The basic principle of RFID technology seems simple and easy, and just two sentences can describe the whole procedure of RFID how works: 1) transmit adequate energy to power up the tag, and 2) communicate with the tag to request and receive the identifier (Roussos, 2008). In fact, RFID technology has involved many electromagnetic theories and electronic technology. In addition, there are many totally different principles/ theories that could be applied to RFID technology and this results in the diverse performance of RFID (Chawla and Dong Sam, 2007). For example, most low frequency tags use Inductive Coupling phenomenon (like a power transformer) using 'Load Modulation' to communication with reader, thus the communication range is limited to a few centimetres (Miles et al., 2008, Roussos, 2008). However most large communication range RFID equipment have adopted Capacitive Coupling, that provides more than 9 meters range, from result in laboratory testing using Alien RFID ALR-8800.

RFID tags are usually categorized in two groups, active and passive system. 'Passive' usually indicate two meanings: it can be used to describe tag's communications context and may also be used to refer to the tag's power. In communications method, a tag can be called 'active' if it contains its own transmitter, which can broadcast data even when no reader is present, it may be referred to 'transmitter tag.' Where as passive tag relies solely on backscatter modulation of the reader's signal for communication with the reader and usually there is no transmitter onboard. In a power context, the term 'passive' and 'active' are often used to mean 'beam powered' and 'battery powered' as well referring to the tag-reader communications methods.

However, passive or active is only a method to generally describe a tag, there are many categorisation methods available, for example, tags can also be sorted by 1-bit or Multi-bit System, by their working band (Ultra High Frequency (UHF: 865-954 MHz) / High Frequency (HF: 13.56MHz)/ Low Frequency (LF: 125-134 KHz) tag), and even by their memory (writeable or read-only tag) etc. The following section introduces some global standards for categorized tags from different manufactures.

5.3 The standard of RFID technology

A major RFID standard is the International Organization for Standardization (ISO), which issued about 50 standards related to RFID technology. However, to discuss the RFID standard, another important organization - Auto-ID lab must be mentioned, which was founded in 1999 for developing Electronic Product Code (EPC) technology for marking and identifying the goods in a global supply chain (Landt, 2005). At the same time, they started to develop a low-cost RFID system, which was already widely used in industry known as EPC G2C1, a part of 'EPCglobal' standard. In this section, some major RFID standard including LF, HF, and UHF will be discussed as follows (Landt, 2005, Lehpamer, 2008, Miles et al., 2008, Roussos, 2008):

EPC Standards: EPCglobal is an organization founded by Auto-ID in 2003 for pushing its low-cost RFID system utilization in global supply chains. The Auto-ID Centre developed its own protocol and licensed it to EPCglobal on the condition that it is royalty-free to manufacturers and end users who use the EPCglobe system. The aim is to construct the 'Internet of Things' which must based on a open and shared infrastructure for auto-identifiable networked objects (Roussos, 2008). Therefore, the structures and aims of the EPCglobal and ISO systems are fundamentally different. EPC standards were the competitor with ISO, which working in a similar system but much slower than EPC. After several years of conflict, they have now collaborated with EPC class 1 Generation 2 tags, and

introduced ISO standard as ISO18000-6 with a slightly modification(Landt, 2005). Currently, there are three major RFID EPC standards ('G' for generation, and 'C' for class):
EPC G1C0 - published in September 2003, working in UHF band, a backscatter read-only tag that was programmed at the time the microchip was made (Landt, 2005, Roussos, 2008).
EPC G1C1 - published in September 2003, working in UHF band, a passive read-only backscatter tag with one-time, field-programmable non-volatile memory (Landt, 2005, Roussos, 2008).
EPC G2C1 –Much more powerful and modern tag published in December 2004. Working in UHF band, passive backscatter, and multiple read and write tag. Four memory banks are included on the chip: which holds two 32bit passwords in reserved memory bank, EPC information on the EPC memory bank(Barber and Tsibertzopoulos, 2005). Tag identification bank contains tags own information such as the serial number, and the fourth user bank can be used freely by applications (Landt, 2005, Roussos, 2008).
ISO Standards: In recent years, ISO has approved serial RFID standards working in different band and for different applications, also including the UHF band. This consequently conflicts with the EPC Generation 2 class 1 standards. The UHF band standards took more time to be published (latest version published in 2006), that has a better interoperability in air interface with EPC G2 standards than the first version published in 2004. Excepting the conflict of UHF band RFID standard, ISO has also published several RFID standards(Kitsos and Zhang, 2008), and they are outlined below:
ISO 14443- A standard for payment systems and contactless smart cards, this is relatively complex as it involves payment function, working in 13.56 MHz band, short range communication usually in few centimetres (Landt, 2005, Roussos, 2008, Kitsos and Zhang, 2008).
ISO 15693- A standard for vicinity tags similar to ISO 14443, also working in 13.56Mhz band, communication range is larger than the ISO 14443, but only support simple data exchange (Landt, 2005, Roussos, 2008, Kitsos and Zhang, 2008).
ISO 18000- A standard for air interface including the LF, HF, UHF and microwave for active and passive RFID systems. It describes the physical layer specifications for communications between reader and tag.

5.4 The performance of RFID technology in laboratory environment

The experiments presented in this section aim to prove a concept for the design of 'Plasterboard Auditing and Tracking System'. The term 'passive' and 'active' used in this section refer to both communication and power context, i.e. passive tag are working in passive communication mode and with no battery on board.
The active tags such as the product of GAO RFID usually provide a large communication arrange and can be up to 150m. In this application the large communication range may result in conflict and incorrect reading. Battery maintenance and cost are also issues with active tags compared to passive tags, which only cost less than 10p and there is no need for any maintenance (no battery onboard). Consequently, from both economy and efficiency aspects, passive system is the most appropriate for this application.
However, the actual performance of RFID passive tags is difficult to confirm, because the literature shows a large difference in some cases between the communication range and performance. Some specialists claim that passive UHF RFID tags only provide a less than 0.5 meters range, but other investigators claim the range should be 1-2 metres, indeed, longer

communication range such as 2 metres to 5 metres can also be found in the literature (Landt, 2005, Roussos, 2008, Chawla and Dong Sam, 2007, Lehpamer, 2008, Min et al., 2007)
A laboratory based experiment is discussed to examine the RFID performance, using Alien RFID equipment for these experiments. ALR-8800 from Alien Technology was used and the UHF reader is operated in 865.7-867.5 MHz range. The reader supports two communication interfaces which are RS-232 and RJ-45 for TCP/IP communication. The Reader is also connected to a pair of antennas' one antenna is for transmitting the signal (write operation) and the other is used to receive the signal (read operation). The first experiment involves 7 different models of passive EPC G2C1 tags, 6 of them are the product of Alien Technology and one is BAP (Battery Associated Passive) tags from Power-ID, which has the battery to associate with. The equipment and tags are shown in the Figure 4.

Fig. 4. The Alien RFID Reader, Antenna and Tags

This experiment was conducted as illustrated in Figure 5, and setup for three different environments as follows:

1. To determine the maximum communication range in ideal environment (i.e. limited interference using paper rather than 'hand' to hold the tag.
2. To determine communication range in 'metal affected environment' from small metal objects were placed around the tags and the antennas (at least 1m from the equipment).
3. To determine communication range using hand held reader device in ideal environment similar to test 1.

Each scenario was tested 5 times and the result gave average range accuracy to 10 cm which is satisfactory for logistic tracking purposes. The experiment were conducted by using far to near movement to test the range i.e. the tags were moved from far to near until the tags were detected and then the measured distances recorded.
The results in the second column of Figure 6 indicates that, some of the passive tags performances are satisfactory for logistic tracking giving 7.5m on average, except the AL-9629, which is used for only item level application. The BAP (Battery Associated Passive) tags provided the best result and exceeded a 10m range.
The results indicate that the metal environment dramatically affected the tags' performance, and the read range was reduced to 3.4m for the AL-964X Tag and 3.5m for BAP Tag. The results showed that the BAP tag was only slightly better than the passive tags. In addition,

the handheld reader performance for each tag was also tested, and the results are shown in the last column of Figure 6. The results show that the range is reduced and only provides a maximum of 1.2 m and a few of the tags read < 10 cm.

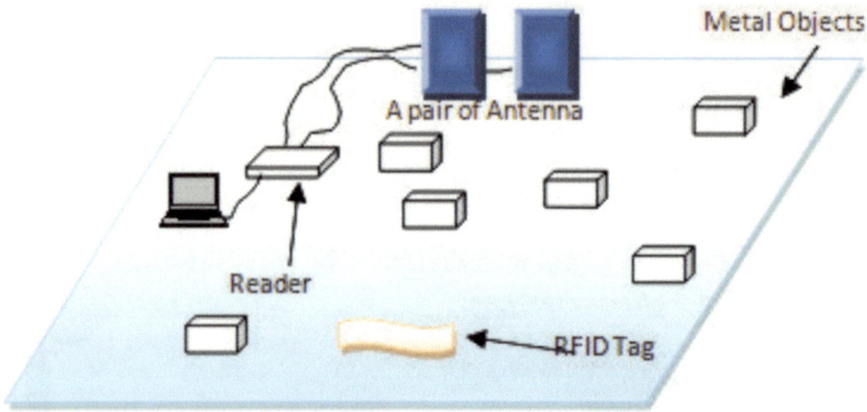

Fig. 5. Testing Setup for Environment Effects

Distance / TAG Name	Free Testing (No attachment)	Metal Environment (Free Testing)	Hand-held Device (Free Testing)
AL-9540	6.7m	3.2m	1m
AL -964X	8.2 m	3.4m	1.2m
AL-9654	7m	3.3m	50cm
AL-9634	7.8m	80cm	30cm
AL-9562	7.9m	80cm	20cm
AL-9629	1m	0.2m	<10cm
PowerID BAP Tag	10.2m	3.5m	1.3m

Fig. 6. RFID Test Results

6. Waste auditing system and knowledge hub

The plasterboard waste problem outlined would require an auditing and tracking system to provide verification and logistical support for authenticating environment disposal. Consequently, a prototype system was designed to ensure that plasterboard waste containers go to their correct destination, and provide verifiable evidence of each stage of the operation for auditing purposes and independent scrutiny.

6.1 System structure

The design will use RFID technology and digital imagery to integrate records including location, volume and weight, container movement, delivery tracking inventories and scheduling etc (Atkins et al., 2008). It works with the support of a knowledge management system which helps managers to make decisions of scheduled logistics of waste to treatment plants and also provides the instruction for operating staff to deal with the plasterboard waste and also other kinds such as medical waste etc.

The design of the prototype system can be viewed from two aspects: firstly, providing the evidence of plasterboard waste being sent to the correct treatment facility and preventing fly-tipping during transportation. This relies on comparison of the information from destination and the source sites, including RFID records, image or video records, operators checking and the possible use of built-in weight systems. The second aspect is the logistic / instruction support that helps management to choose the appropriate treatment facility to dispose of the waste and real-time instructions to the operating staff.

In addition, UHF RFID technology is introduced to the system which could provide automation on appropriate range and low cost by using passive tags. In the early stage of system design, Alien Technology AL-8800 reader and AL-9654 passive tags have been chosen for prototype design and feasibility evaluation. The system will use UHF tags which work in 865.7-867.5 MHz range, and the reader supports two communication interfaces which are RS-232 for serial connection and an RJ-45 for TCP/IP communication. Two antennas are linked to a reader, one for transmitting the radio power to the tags, and another one for receiving the feedback signal.

The prototype system was designed to consider three auditing aspects: 1, Auditing plasterboard material onto the construction site. 2, Monitoring the plasterboard waste removed from site 3, Checking that the wastes plasterboard has been sent to the correct destination.

An RFID tag is attached to the plasterboard stacks/pallet for auditing of new material moved into the construction site. The main gate is equipped with an RFID reader then can monitor the transporting of plasterboard material. Once having successfully scanned a valid RFID tag, a record will be generated and sent to the central server with date, time, and ID number. These records can be checked to show how much plasterboard is delivered to the site.

The plasterboard waste container (skip or compactor skip) is marked with a unique ID and RFID tag in the demolition or construction site. The ID relates the information of the waste container, such as the total load, unloaded weight, location etc. This information is located in the central information server, and can be checked or updated by a hand-held RFID device. The operator can input real time information about the waste as appropriate using a mobile device equipped with RFID module.

When the waste containers are fully loaded, they are transported through a special gate to the recycling company or appropriate licensed landfill site. The gate is equipped with RFID sensors and digital imagery to create records which could be supplemented with mobile imagery and logging devices on site. The record is uploaded to the central information server and shows the logistics of the containers and the appropriate tonnages of plasterboard waste being transported or delivered to recycling and/or landfill sites

Figure 7 is a Framework for Plasterboard Waste Management system using RFID technology and knowledge hub for tracking and verification purposes. The design will use RFID technology and digital imagery to integrate records including location, volume and weight, container movement, delivery tracking inventories and scheduling etc. It works with the support of a knowledge management system which helps managers to make decisions on scheduled logistics of waste to treatment plants and also provides the instruction for the operating staff dealing with the plasterboard waste and also other kinds such as medical waste etc. All the RFID fixed readers are associated with imagery equipment, digital imagery could be automatically taken when a valid tag successfully scanned by RFID reader. These digital imagery records will be well documented as the evidence to verify the transportation.

Figure 7 also illustrates the system of a 'main construction demolition site' and near 'smaller construction demolition site' which are the two typical source sites. The plasterboard waste is designed to be bagged in the source sites during the demolition/building process and a RFID tag is then attached to the container (bag, box, or bins etc.) immediately.

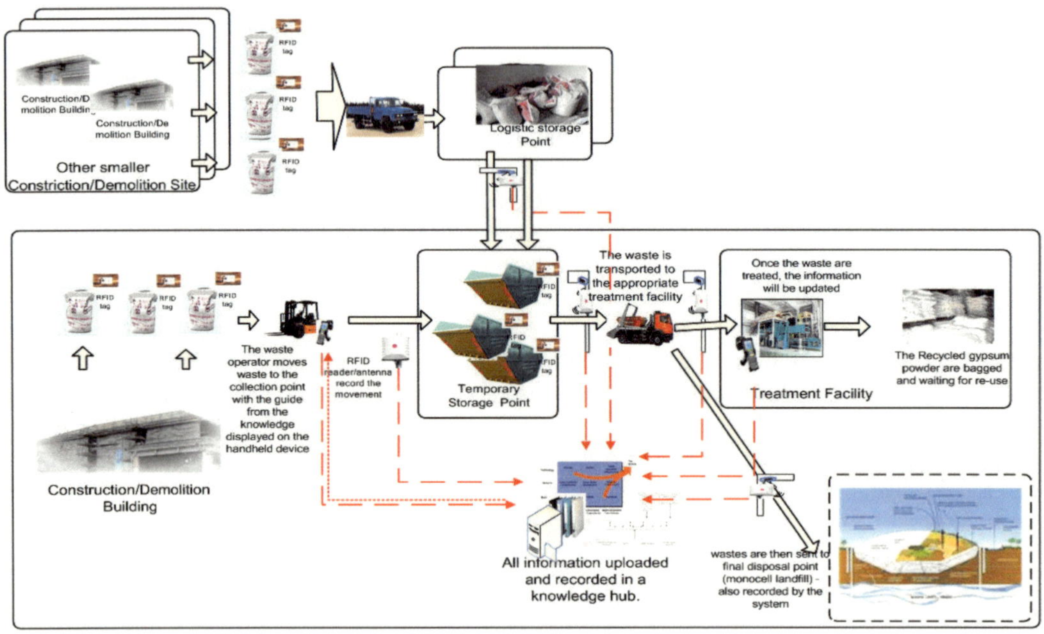

Fig. 7. Frameworks for Plasterboard Waste Management System

Plasterboard waste can go directly to the landfill with mono-cell. If the construction or waste company wishes to land fill them, the prototype system can fulfil the function of providing the evidence by records and image. The RFID equipment and RFID reader is set on the entrance of the landfill site to verify the arrival of the waste. When the containers pass this gate, a record will automatically be created and uploaded to the central server to show the logistics of the containers and the appropriate tonnages of plasterboard waste being transported or delivered to recycling and/or landfill sites.

Hand-held devices are used by the operating staff involved in the system, including vehicle drivers, cleaners, demolition operators and waste managers etc. The device is a small sensor that links to the central server, and can display information from the system. The instruction and logistical support information will be automatically downloaded from the knowledge management system when it is required. The information notifies the operators which container should be transported or moved to the correct location in a specific time, and also notifies the procedure of transporting this type of waste and any particular cautionary instructions.

6.2 Knowledge hub design

The prototype system is designed using a knowledge hub as the back end support, which includes a knowledge based system and reasoning to provide the logistical support for the waste management. The reasoning system is designed using Rule-based Reasoning, and the structure of the knowledge base is illustrated in Figure 8.

Figure 8 illustrates the structure of the knowledge hub system that is designed in four layers. The lowest layer is the hardware layer, called Data processing layer, which is the route for acquiring the data and information from the RFID and imagery equipment into the system (Zhang et al., 2008). The data gained from the equipments are separately sent to data bases, located in the second lowest layer.

The second lowest layer is the knowledge storage layer, called data integrate layer. This layer contained two databases which stores the RFID data and imagery data from lower layer, and another database is responsible for integrating the two types of information and prepares them ready for the next layer usage. In fact, this database is a 'fact' storage that used for the reasoning. In addition, the database can output 'fact' to a long term data storage data warehouse, and an OLAP (Online Analytical Processing) function can introduced into the system for better performance.

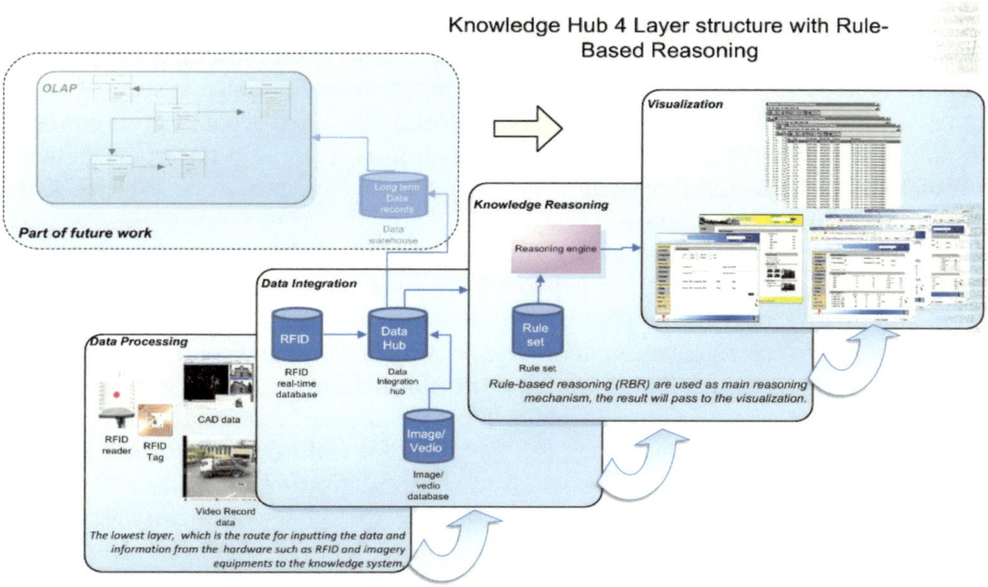

Fig. 8. 4-Layer Structure with Rule-Based Reasoning

The next higher layer is the core layer, which is called the knowledge reasoning layer. Rule-based reasoning is the main reasoning mechanism for generating the best solution for logistical and tracking support. The inference engine is the core of this layer that works with the rule base and the fact uploaded from lower layers.

The knowledge is stored in productive rule (IF…THEN…) format at the rule base. The three components compose the full Rule Based Reasoning system. The result of this layer is a suggestion solution' that is generated by the previously inputted rules, the reasoning aspect including the logistic suggestion and also the guidance for the waste operators, depending on the users requirements. Finally, the result is then passed to the highest layer - visualisation to provide the resolutions for decision support.

The highest layer bears the communication function between the system and users. This layer is called visualisation layer, which is designed to represent the logistical solution and the guidance in suitable client machine, either the desktop computer or hand held device. The visualisation layer can be associated with web-based application to represent data for easy access and flexible monitoring, and alternatively may use as individual programme to improve the security and more trustable evidence. The visualisation layer is also responsible for the user's command input; the command will pass to the lowest layer through the kernel module.

6.2.1 Adopting of rule-based reasoning

The rule-based system is usually called an expert system, and is the most popular choice for knowledge-based applications. A simplified definition of rule based reasoning is a technology in which knowledge is represented by a set of IF…THEN... production rules and data is represented by a set of facts(Giarratano and Riley, 2005). The rule will be executed when the fact matches the condition of a rule, and it may add or modified to fact for a new rule execution until the final result is determined(Giarratano and Riley, 2005).

Rule-based reasoning has some advantages compared with other reasoning technology and has been generally accepted as the best option for a knowledge-based system. It typically features natural knowledge representation, uniform structure, separation of knowledge from its processing and has the ability to deal with incomplete and uncertain knowledge. Some features of rule-based reasoning are suitable for the prototype system, and are discussed as follows(Giarratano and Riley, 2005).

Rule-based reasoning technology stores knowledge in IF…THEN structure meaning each piece of knowledge is relevantly independent from other knowledge. This structure is efficient for finding out the target knowledge when the waste regulation is amended. Secondly, the waste management system requires that knowledge should be easy to adopt into the reasoning system without complex transformation. In fact, it is better to input knowledge without any programme skills for ease of use and maintenance/updating purposes.Individual knowledge storage is a key required feature that separates knowledge from the system and thus it could be removed without affecting the system design and a new knowledge base which contains the knowledge for other waste management areas could be supplemented.

6.2.2 Optimization module design

The reasoning layer is responsible for the optimized schedule plan, generates the real time guidance and reports on the current situation function, but the optimized schedule plan is the major task of the knowledge reasoning layer.

Normally, schedules include two aspects: the time plan and the route plan. However, considering the application is designed for a waste recycling company and most waste collection times are contracted, therefore the prototype system only needs to generate the route plan and the time schedule has been assumed to be initially confirmed by contract between the waste company and the construction company.

The routing plan of the transportation can be seen as a classic TSP (Travelling Salesman Problem) question, which has the same requirement: the vehicle departs from the recycling facility, visit each site one time, and finally returns to the recycling facility. The major task of the reasoning layer is planning and finding an efficient route. It is also responsible for real-time planning in case of an emergency where a new route needs to be planned.

The requirement of the prototype system's application area restricts the route plan algorithm to matching the following features: 1) Inherent parallelism, which needs to consider more than one route at the same time 2) Efficient to solve TSP and similar problems. 3) Can be used in dynamic applications. Therefore, for this application, ACO (Ant Colony Optimization) will be introduced in the system that is responsible for generating the route plan (Colorni et al., 1991, Dorigo and Gambardella, 1997, Dorigo et al., 1999, Qiang and Qiuwen, 2008).

The ACO module is only dealing with the vehicle routing plan, therefore it needs to be independent from the main rule-base to reduce complications, and thus it does not need to be converted in production rule format. It only works when the vehicle type and target site has been decided by the rule based reasoning system; the vehicle and site information will be passed to the ACO module as the initial parameters, then the acceptable result can be generated in limited iterations and this is illustrated in Figure 9.

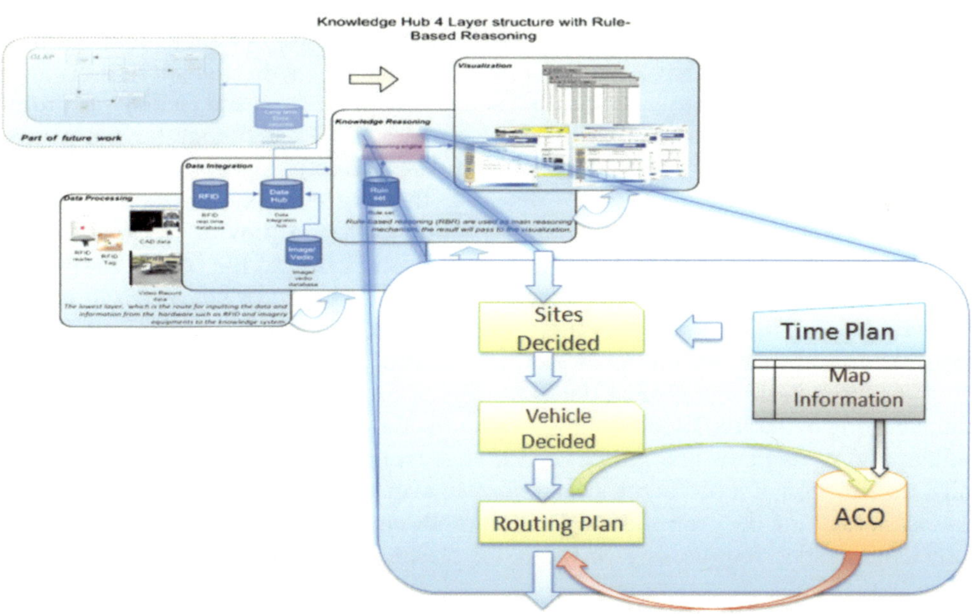

Fig. 9. The 4-layer Structure with ACO Module

The work procedure of the reasoning layer starts from the time schedule and routing plan. Firstly, the system will check the current time and query the database if there are any sites which need to be visited in this time (day, week or month) and also query the last operation

on that site to roughly estimate the tonnage of the waste. The estimating also takes into account the site project, construction progress and even its financial situation.
The next step is to decide the vehicle type and the number. After the site which must be visited in the next period has been decided and the waste tonnage of each site is estimated, obviously the total amount of waste will be known. The vehicle type can then be decided based on this information; the capacity of the vehicle should be larger than the tonnage and depends on the containers used on the sites. The rule-based system will be based on these 'facts' to reason out the vehicle type and number. Planning the details of vehicle routing is the function of the ACO, which firstly decides the routes to be calculated and the sites for a single trip. Then the exact route will be calculated by the ACO, in the prototype of the waste management system, only the original ACO will be introduced for evaluating purposes. After the routing has been decided, the details will be passed to the visualization layer for guidance.
Another important function of the prototype system is providing guidance to the operation staff to help them deal with the waste. It works as a handbook to remind them when, where and how to collect/transport the waste. The transport plan is part of the guidance information that can give clear instruction about route choice and waste collect procedure to the vehicle drivers.

7. Conclusions

This chapter introduced the current plasterboard disposal situation and addresses the logistical problem which is a barrier to an increased recycling rate. In the UK only four known recycling facilities are available, all of which are located in England, and two of them in the London area. This situation has caused difficulties with transportation, and the recycling fees are higher than landfill if the source site is far from the facility. A prototype system for waste management is outlined which uses RFID technology for the main data collection methods, and rule-based reasoning and Ant Colony Optimization for auditing/ tracking the plasterboard waste and detailing the reasoning system and optimization methods. It also has the function to make a schedule plan and provide the guidance to the operation staff to ensure that waste containers are transported to the correct locations. The system can also handle emergency changes such as traffic hold-ups etc, as it will re-arrange suitable routes that reduce potential loss. The structure of a waste management and work process are introduced, including the four layer structure showing the reliance of RFID technology for collecting logistical data and digital imaging equipments are used to give further auditing evidence. The reasoning core in the third layer is responsible for generating schedules and route plans and guidance, and the last layer delivers the results to the users. Finally, the function of a prototype system for waste management was discussed which uses RFID technology for the main data collection methods, and rule-based reasoning and Ant Colony Optimization for auditing/ tracking the plasterboard waste movement.

8. References

atkins, A. S., Zhang, L. & Yu, H. (2008) Issues in Environmental Recycling of Plasterboard Waste and Application of RFID and Knowledge Technology. International Conference on Software, Knowledge, Information Management and Applications

(SKIMA). March, Kathmandu, Nepal Co-sponsored by IEEE pp 54-59 ISBN 9781851432516.

Barber, G. & Tsibertzopoulos, E. (2005) An analysis of using EPCglobal class-1 generation-2 RFID technology for wireless asset management. Military Communications Conference, 2005. MILCOM 2005. IEEE.

Bourn, M. (2005) Landfill Directive Regulatory Guidance Note 11. ENVIRONMENT AGENCY.

Bradshaw, B. (2004) The Landfill (England and Wales) (Amendment) Regulations 2004. Statutory Instrument 2004 No. 1375. Department for Environment, Food and Rural Affairs.

Bradshaw, B. (2005) The Landfill (England and Wales) (Amendment) Regulations 2005. DEFRA, Statutory Instrument 2005 No. 1640.

Chawla, V. & Dong Sam, H. (2007) An overview of passive RFID. Communications Magazine, IEEE, 45, 11-17.

Colorni, A., Dorigo, M. & Maniezzo, V. (1991) Distributed optimization by ant clonies. Proceedings of the 1st European Conference on Artificail Life, 8.

DEFRA (2006) Estimated Total Annual Waste Arising by Sector: 2004 Department for Environment, Food and Rural Affairs, Available from:http://www.defra.gov.uk/environment/statistics/waste/download/xls/wrfg02.xls, Cited: 02-Oct-2007.

DEFRA (2007) Waste Strategy for England 2007. Department for Environment, Food and Rural Affairs.

Dorigo, M., Di Caro, G. & Gambardella, L. M. (1999) Ant algorithms for discrete optimization. Aritificial Life, 5, 137-172.

Dorigo, M. & Gambardella, L. M. (1997) Ant colony system: a cooperative learning approach to the traveling salesman problem. Evolutionary Computation, IEEE Transactions on, 1, 53-66.

EA (2007) Hazardous waste deposits in England & Wales 2005. Environment Agency, http://www.environment-agency.gov.uk/commondata/103601/05_tables_240707_1787522.xls, Cited 11/July/2008.

EA (2008a) Hazardous waste deposits in England & Wales 2006. Environment Agency, http://www.asiantaeth-yr-amgylchedd.cymru.gov.uk/commondata/103601/ew_haz_waste_2006_1902503.xls ,Cited 11/July/2008.

EA (2008b) Position Statement MWRP 007. Environment Agency.

EUROPEAN_COUNCIL (2002) Establishing Criteria and Procedures for the Acceptance of Waste at Landfills Pursuant to Article 16 of and Annex II to Directive 1999/31/EC. COUNCIL DECISION 2003/33/EC. European Council

Giarratano, J. C. & Riley, G. (2005) Expert systems : principles and programming, Boston, Thomson/Course Technology.

GRI (2011) "Gypsym Recycling International ". http://www.gypsumrecycling.biz, Cited 28/Jan/2011.

Hamm, H., Huller, R. & Demmich, J. (2007) Recycling of plasterboard. Zkg International, 60, 68-74.

Heguy, D. & Bogner, J. (2004) Cost-Effective Hydrogen Sulfide Treatment Strategies for Commercial Landfill Gas Recovery: Role of Increasing C&D (Construction and Demolition) Waste. Municipal Soild Waste Management.

Hirz, H. & Sterr, H. (1995) Recovery of components of waste plasterboard. IN PATENT, U. S. (Ed. B02C 2318 ed., Gebruder Lodige Maschinenbangesellschaft mit, beschrankter Haftung.

James, P. R., Pell, E., Sweeney, C. & John-Cox, C. S. (2006) Review of Plasterboard Material Flows and Barriers to Greater Use of Recycled Plasterboard. The Waste & Resources Action Programme.

John, S. & Knez, J. (1993) Method for recycling wallboard. B02C 2300 ed., Knez Building Materials Company.

Kitsos, P. & Zhang, Y. (2008) RFID security : techniques, protocols and system-on-chip design, New York ; London, Springer.

Kleist, R. A., Chapman, T. A., Sakai, D. A. & Jarvis, B. S. (2004) RFID Labeling-Smart Labeling Concepts & Applications for the Consumer Packaged Goods Supply Chain, Printronix.

Landt, J. (2005) The history of RFID. Potentials, IEEE, 24, 8-11.

Lehpamer, H. (2008) RFID design principles, Norwood, MA ; London, Artech House.

Lund-Nielsen, H. (2007) Experience in gypsum recycling on three continents. Global Gypsum MAGAZINE. Surrey,UK, PRo Publications International Ltd.

Mcbean, E. A., Rovers, F. A. & Farquhar, G. J. (1995) Landfill Gas Collection and Recovery. Solid Waste Landfill Engineering and Design. London, Prentice-Hall,Inc.

Miles, S. B., Sarma, S. E. & Williams, J. R. (2008) RFID technology and applications, Cambridge, Cambridge University Press.

Min, Z., Wenfeng, L., Zhongyun, W., Bin, L. & Xia, R. (2007) A RFID-based Material Tracking Information System. IEEE International Conference on Automation and Logistics.

MTP (2007a) BNPB2 Plasterboard - Waste Management. The Market Transformation Programme.

MTP (2007b) BNPB3 Plasterboard - Legislation and Policy Drivers. The Market Transformation Programme.

Pal., S. K. & Shiu, S. C. K. (2004) Foundations of soft case-based reasoning, Hoboken, N.J., Wiley-Interscience.

PBRUK (2007) Estimated cost model for plasterboard waste. Plasterboard Recycling UK, Available from:http://www.pbruk.co.uk/faq.htm, cited: 19-Sep-2007.

Qiang, Z. & Qiuwen, Z. (2008) An Improved Ant Colony Algorithm for the Logistics Vehicle Scheduling Problem. Intelligent Information Technology Application, 2008. IITA '08. Second International Symposium on.

Roussos, G. (2008) Networked RFID : systems, software and services, London, Springer.

SEPA (SCOTLAND) (2007) Waste Data Digest 7 - WDD7, 2005 and 2005/2006 data. Scotland's Enviromental Regualtor and Adviser.

Tudahl, D. L. & Bush, G. R. (2000) Apparatus and method for recycling gypsum wallboard IN PATENT, U. S. (Ed. B02C 1800 ed.

Turban, E., Aronson, J. E. & Liang, T.-P. (2005) Decision support systems and intelligent systems, Upper Saddle River, NJ, Pearson/Prentice Hall.

Zhang, L., Atkins, A. S. & YU, H. (2008) Application of RFID Technology and Knowledge Hub for Logistic Support in Scrapped Type Recycling. 9th Informatics Workshop for Research Students. June, Bradford UK, pp.190-195, ISBN 978 1 85143 251 6.

11

RFID-Based Equipment Monitoring System

Mohd Helmy Abd Wahab, Herdawatie Abdul Kadir, Zarina Tukiran,
Nor'aisah Sudin, Mohd Hafiz A. Jalil and Ayob Johari
Universiti Tun Hussein Onn Malaysia
Malaysia

1. Introduction

Automated monitoring systems are becoming trends, creating easier method to identify item, tracking, monitoring and add on security values. In places where there are lots of items accessed by many users, the tendency of loss is high due to weakness in items monitoring. Here, we briefly describe our research on the university's laboratory perspectives. The main aim of the research is to work out a generic approach of monitoring items in a place with several rooms. For example, there are laboratories with expensive equipments are available in a university to support teaching and learning session. Conventional approach of checking items for every session is difficult for lab administrator as most libraries are being used by more than 20 students per session.These leads to a challenge for lab administrator to monitor the flow of these items are always in place. Currently, the monitoring of laboratory equipments is performed manually by the lab administrator during each laboratory sessions. For every loan of equipment, a log book needs to be filled up in order to keep track the transaction information.This system was found to have a lot of weaknesses such as misuse of the equipment log records, losses of equipment, no in-out transaction record and misplace of equipments. To automate the process, Radio Frequency Identification (RFID) is identified as one of the most practical and applicable in real time implementation in-line with the nature where most of the systems are made computerized. In this paper, a solution has been provided for the problem encountered in laboratory equipment monitoring system using RFID technology. Therefore RFID-based monitoring system has been designed and developed to solve the problem associated with the handling of laboratory equipments. This chapter is organised as follows. Section 2 describes related works on RFID-based monitoring system. The architecture of the system is mentioned in section 3. Application scenario and the implementation are briefly explained in section 4 and 5 respectively. Finally, the chapter is concluded in Chapter 6.

2. Related work on RFID in monitoring

RFID is a wireless automatic identification that is gaining attention and is considered by some to emerge as one of the pervasive computing technologies in history (Roberts, 2006). As the technology grows very rapidly, RFID has received considerable worldwide attention and widely used in monitoring and tracking ranging from human identification to product identification. Previous research has successfully indicated that RFID has been increasingly expanded in various fields such as retail supply chain, asset tracking, postal and courier services, education, construction industry, medical, and etc.

The work presented by Tan and Chang (2010) who had developed an RFID-based e-restaurant system to change the traditional restaurant services which is considered as passive. The utilization of RFID is to improve the service quality which is customer-centered that enable waiters to immediately identify customers via their own RFID-based membership card. It can also provide customized services such as enhanced dining table service; pay the bills, instant transmission of customer orders to kitchens and flexibility of managing payments of bills and discounts. However, in Ngai et. al. (2008), designed and developed RFID-based sushi management system to help a conveyor belt sushi restaurant to achieve better inventory control, responsive replenishment, and food safety control, as well as to improve its quality of service.

In the perspective of animal tracking or livestock monitoring management system, Vouldimos et. al. (2010) developed FARMA project which combined with RFID technology and mobile wireless networking to track animal and the data in repository which contains animal data records. The purposes of the system are to identify animal in case it gets lost and identify some basic information about particular animals. A similar work done by Nor Suryani Bakeri et. al. (2007) and Ahmad Rafiq Adenan et. al. (2006) developed a livestock monitoring system using RFID. An RFID tag is used and attached to each livestock to monitor its movement in and out as well as the basic information about any particular animals.

The use of RFID also could assist in customs clearance process by reducing the delay time. According to Hsu, Shih and Wang (2009), the use of RFID can improve the efficiency of cargo process, and reduce the inventory and labor cost. The work presented based on the mathematical model of the customs clearance process-delay and the network of customs delay is reconstructed based on the use of RFID. RFID also has been successfully applied in global postal and courier services in monitoring the parcel delivery. One of the well known courier service company is DHL which has been using RFID in their services since 1988 and carried out 20 trials on active and passive technology and successfully proved it improved the service and reduce the costs (EPC Global, 2005). The application of RFID in global market in postal and courier services contribute 650 billion per year and Europe was the leader in utilizing RFID in postal and courier services (Zhang, et. al., 2006).

High quality service lead to customer satisfaction, increase market share, and enhance profitability of service organizations (Hoffman and Bateson, 1997). Oztaysi, Baysan, and Akpinar (2009) have done a study to investigate the possibility of using RFID as a tool for improving service quality in hospitality industry and primarily concern in tourism industry.

In monitoring of asset tracking, an effective and efficient managing the tracking of medical-assets in healthcare facilities can be performed by the means of RFID. Oztekin et. al. (2010) has done a study using enhanced maximal covering location problem along with critical index analysis metric to optimize the design of a medical-asset tracking system constrained by a limited number of RFID readers. Results indicate that the proposed technique has improved by 72% compared to the currently utilized expert placement strategy.

Yan and Lee (2009) developed RFID application in Cold Chain monitoring system to track the cold-chain product flowing in supply chain, ensure the products' quality and comply with relevant provisions during transportation. The system executes in real-time environment and can track the location and monitor the temperature of cold-chain products to ensure the quality. However, according to Loebbecke (2005) has done a research regarding the application of RFID in retail supply chain at a brink-and-mortar supermarket to investigate the advantages and challenges with the early RFID applications in terms of technological issue such as standardization, challenges on the data, network and application layers.

Haron, et. al. (2010), designed and developed of a context aware notification system for university students using RFID. The system aims to deliver urgent notifications to the intended students immediately at their respective locations. A quite similar work done by Herdawatie et. al. (2010) which integrates RFID and biometric sensor to track students in a boarding school of their location at the selected restricted area.

As summary, based on the successful of RFID applications in various fields as discussed above, it shows that its application is endless. This section onwards explains the RFID application in tracking of laboratory equipments movement to ensure its availability. It also aims at helping the lab administrator in monitoring the equipment from lost or misplaced. The monitoring of equipments movement is not only being monitored by the lab administrator but also by the top management through online databases.

3. System architecture

Building an automated tracking applications by integrating web services guarantee many benefits, such as reduce clerical task and ease the management burden. The RFID-based Equipment Tracking System is an integrated system that offers an effective solution of managing items especially for large scale environment. It combines the RFID technology and security devices to ensure the items are always been monitored and secured. The system enable the university to give admission to selected individual to access locations, permit movement of items, record the important data and also enable the viewing of record via internet.

A faculty usually has a number of laboratories. Faculties with technical courses such as Information Technology and Engineering usually have more laboratories. To implement the system, an appropriate design is required to make sure it is suitable for the number of laboratories and equipments in all laboratories. In this study, the design of the system which utilizes RFID is divided into two; hardware design and software design as shown in the architecture diagram in Figure 1.

There are six important components involved as illustrated in Fig. 1, (1) RFID Tag, (2) RFID Reader, (3) Personal Computer, (4) RS232 Cable, and (5) LAN HUB and 6) CCTV Camera The master server contains the database which is used to store all data collected from RFID reader where user can read or change information in the database. The RFID tags contain antennas to enable the receiving and transferring data. The passive RFID tag creates power from magnetic field and use it to energize the circuits of the RFID chip and sends information back to the reader in the form of radio-frequency waves. The physical layer of the system is depicted in Figure 2. It shows how the computers and the master server are connected. The software involved in developing the system is also outlined.

In the system, RFID technology were implemented to enable data to be automatically recorded where each tag is embedded in the metric card (working pass) for individual and attached to each equipments. The lab administrator will grant an access to selected individual to enter a laboratory and also enable selected individual to move items out from the lab and within the organization. Upon the individual is found attempt to force the process the camera is triggered and activated to document the image of intended person and buzz the alarm system and notify the person-in-charge.

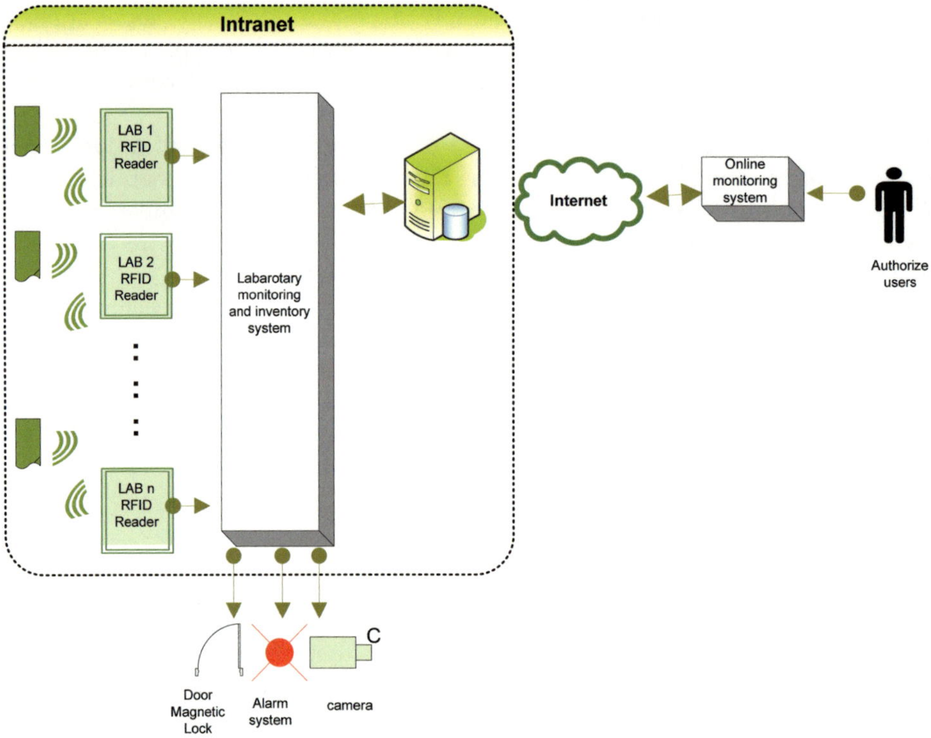

Fig. 1. System Architecture

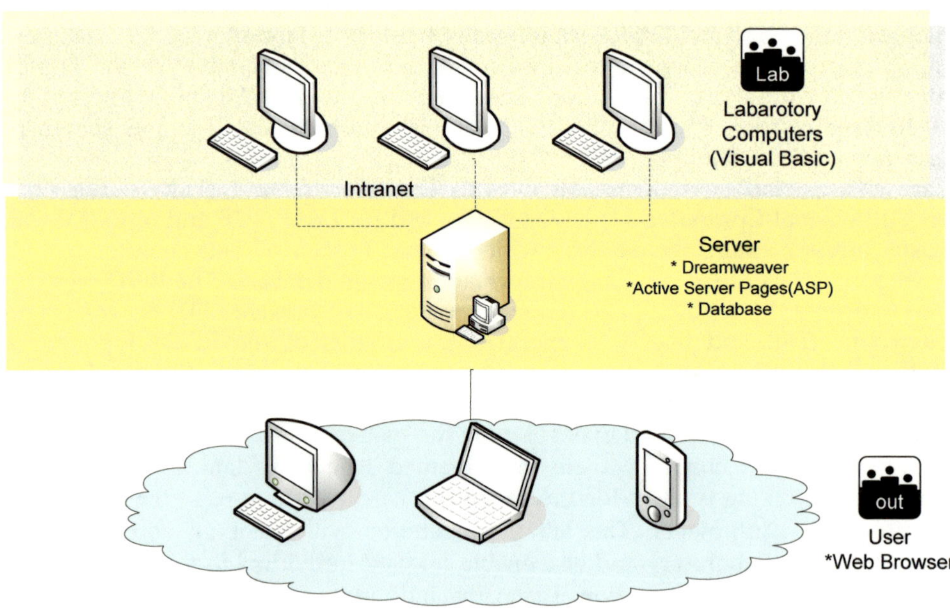

Fig. 2. Physical layer of the system

4. Application scenarios

The developed prototype is an online laboratory monitoring system that has three purposes; which typically composed of (1) Laboratory grant access (2) Inventory control, and (3) Online data viewing. The prototype has been applied at the UTHM research project lab. To illustrate the concept, a sample of layout of application was provided in Figure 3.

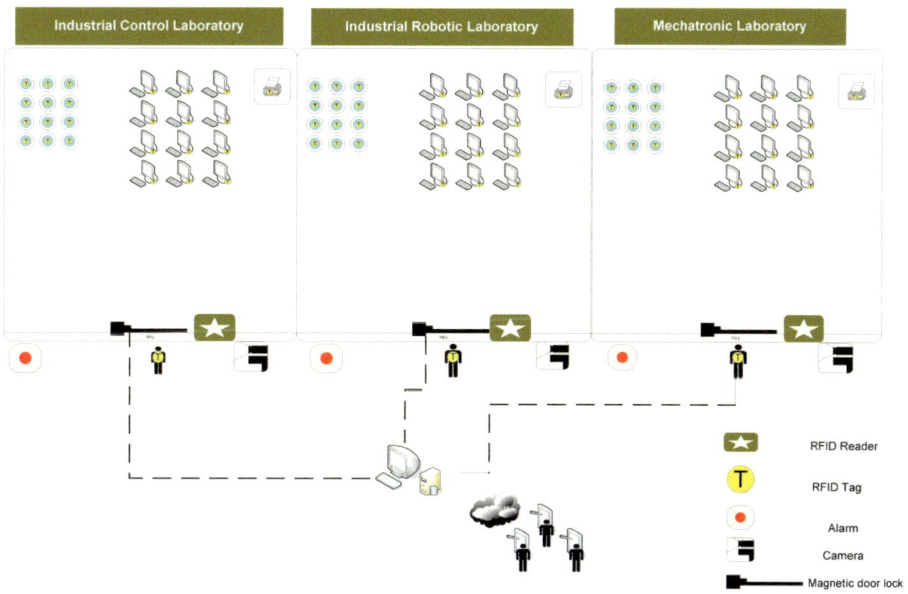

Fig. 3. Application scenario layout

In the system, RFID tag is attached to both users and equipments. The RFID reader is located at each Laboratory to record and verify the RFID tags in the area. Each laboratory is equipped with a surveillance camera and an alarm indicator to deal with unforeseen circumstance events. The recorded data is stored and managed by a central computer whereby each laboratory computer is connected via intranet connection to ease any information received from computer lab can be easily transmitted to central computer. The main purpose of data, which is stored at the central computer, is to ease the management to have a look the whereabouts of equipment and record of in-out information. The administrator will grant the personal level access, equipment status and also permit online monitoring to authorize individual.

Legally attempt to enter a laboratory with authorize RFID identification (id), lead the magnetic door to unlock (door open) and record the entry information. Illegally attempt to access the laboratory, the door keep locked and activate the camera and warning sign is indicated to the system.Once the system detects a forceful behavior such as shaking the door, the system triggers an alarm to notify the security. In side of inventory control, intended user must be registered with authorized id before granted to move or lend the equipments, once verified, magnetic door will unlock and information is recorded. Otherwise, the registered id requires re-verification.

5. Implementation

As mentioned in previous section, the RFID-based Equipment Monitoring System is used to keep track the record on laboratory equipment. Hence, the laboratory, its equipments and users who use the equipment in the laboratory need to be part of the system entities. This can be done by enrolling these entities in the system.

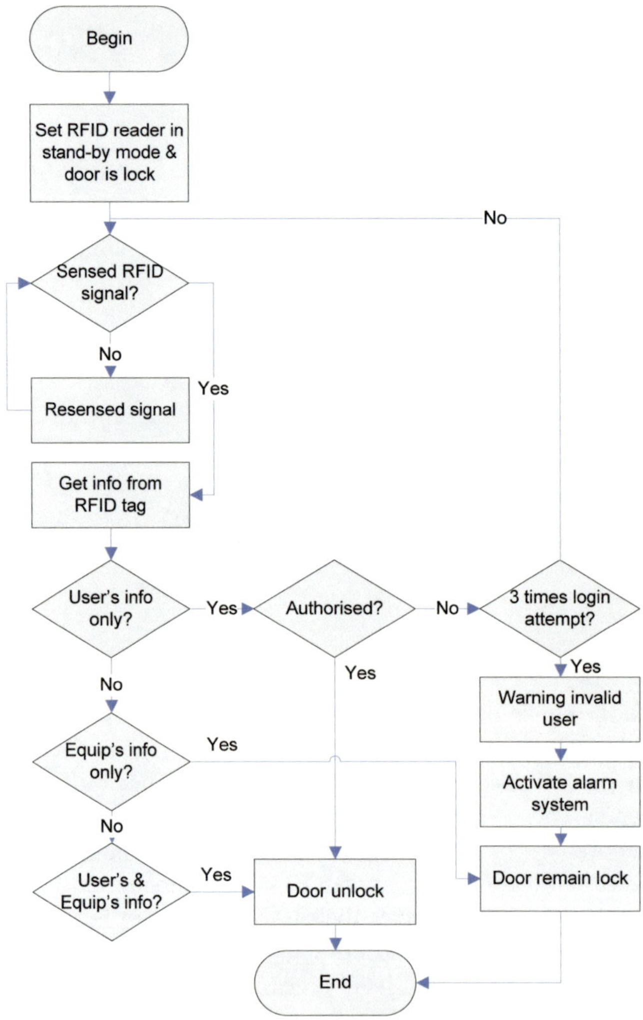

Fig. 4. Access laboratories and equipments flowchart

At this moment, only three (3) laboratories have used the system and only authorised personnel are allowed to access the labs. In order to ensure only the authorised user login to the labs, they need to present their RFID card. Each laboratory is equipped with magnetic door. Therefore, the RFID card acts as a key to unlock the magnetic door. All equipments placed in the labs are tagged with RFID and registered in the system. This equipment can be used and borrowed by the user either in the same laboratory or in other laboratories. For the latter the user's RFID card and the equipment's tag should be readable by the RFID reader.

Once the information is successfully matched by the system, the magnetic door will be unlocked. Otherwise, the door remain locked if only the reader able to read equipment's information but not the user's information. Figure 4 illustrates the flow to access laboratories and equipments.

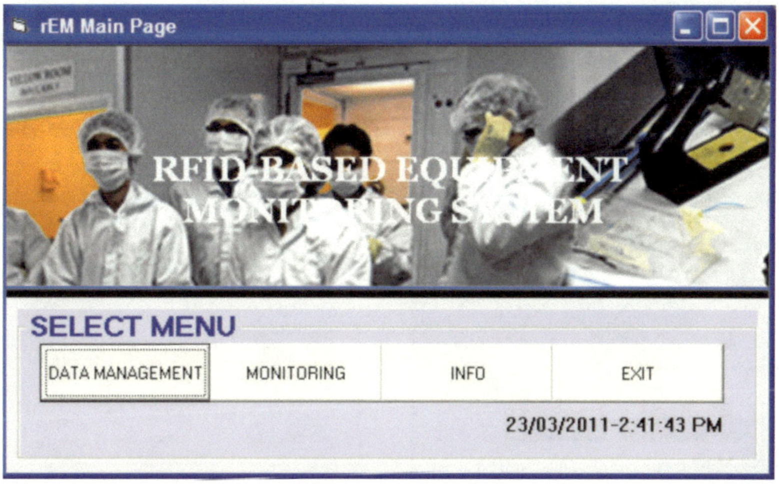

Fig. 5. The main GUI of RFID-based Equipment Tracking System

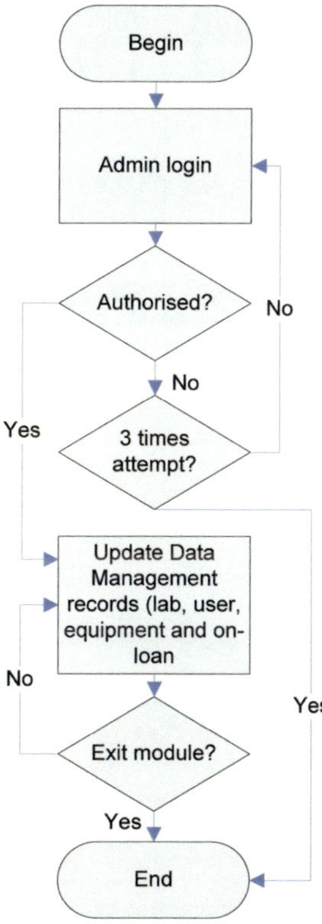

Fig. 6. The system flow of data management module

The system has two main purposes; first is to register user, equipment and laboratories to be part of the system entities. This is done by the system administrator through data management module. The second is to keep track the equipment and to monitor the activities of the user. The latter can be accessed through monitoring module. These two main purposes are presented in the form of graphical user interface as shown in Figure 5.
The data management module system flow is illustrated in figure 6. This module can only be accessed by authorised personnel to maintain the integrity of the data. Thus, system administrator needs to enter the correct password in login page as shown in Figure 7. Users are allowed to re-enter the password up to three (3) times for invalid password before the system activates the alarm system.

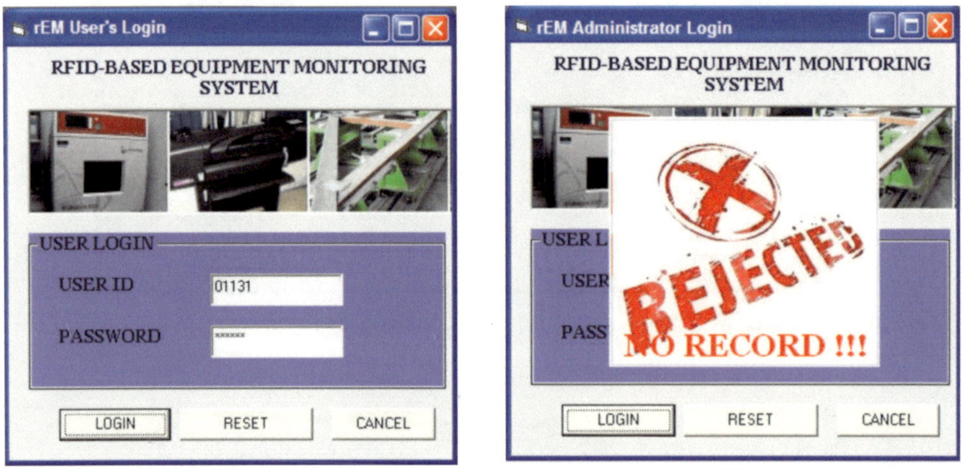

(a) (b)

Fig. 7. System administrator's (a) entering the password and (b) warning message for invalid access

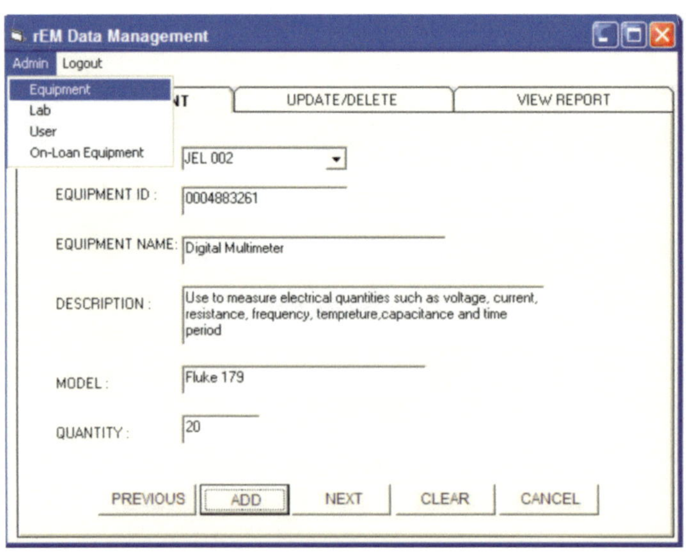

Fig. 8. Data Management Module

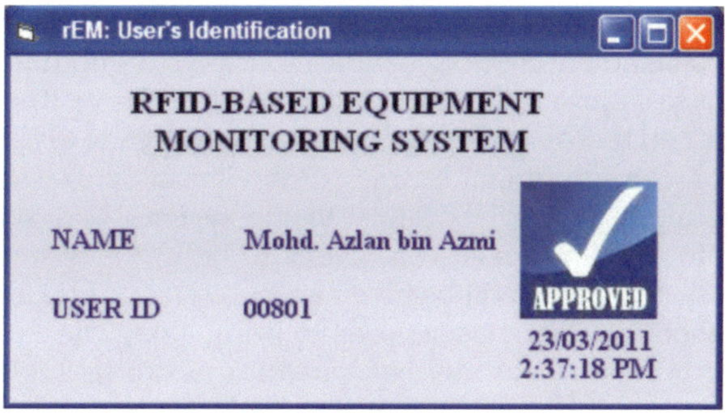

Fig. 9. Authorised user

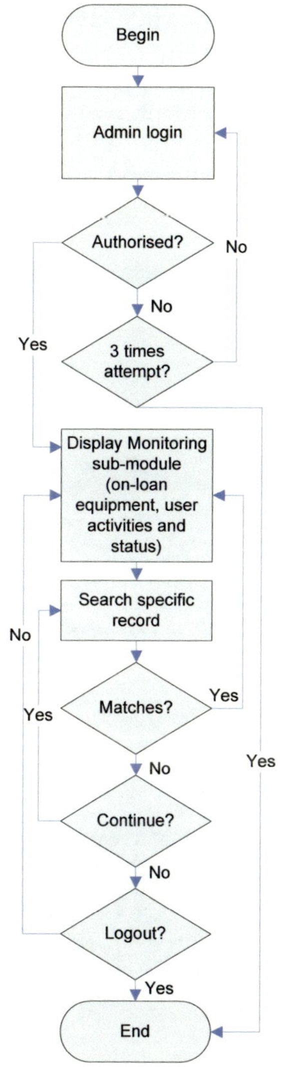

Fig. 10. The system flow of data monitoring module

Then system administrator is able to view the data if the administrator is the authorized personnel. The data management module handles four (4) sub-modules which are described as Figure 8. According to Figure 8, *Equipment* sub module allows the system administrator to add new equipment and maintain the equipment record that is assigned in the laboratory. RFID tag is attached on the equipment to track down its status. *Lab* sub module allows the system administrator to enroll any laboratory to the system. *User* sub module allows the system administrator to enrol any user that wants to be in the system. After the user has been registered in the system, each will be given an RFID card. This card is used to authorise access the intended laboratory. The user needs to bring along the card to enter or to leave the laboratory. If user brings in/out any equipment to/from its registered laboratory, the card and the equipment's tag should be read by the RFID reader without fail in order to unlock the magnetic door. Figure 9 shows an example of successful login to laboratory.

On-loan equipment sub module allows the system administrator to register the status of equipment; whether it is in place or it is circulated around laboratory under an authorised user.

The system flow of monitoring module is shown in Figure 10. The module allows the administrator to monitor on-loan equipment, users' activity and their status. This module is designed so that it can be viewed by the administrator internally (intranet access) or remotely (internet access). Here, the discussion is focused on remote access. In order to use this module either internally or remotely, the administrator needs to log-in to the system as shown in Figure 11.

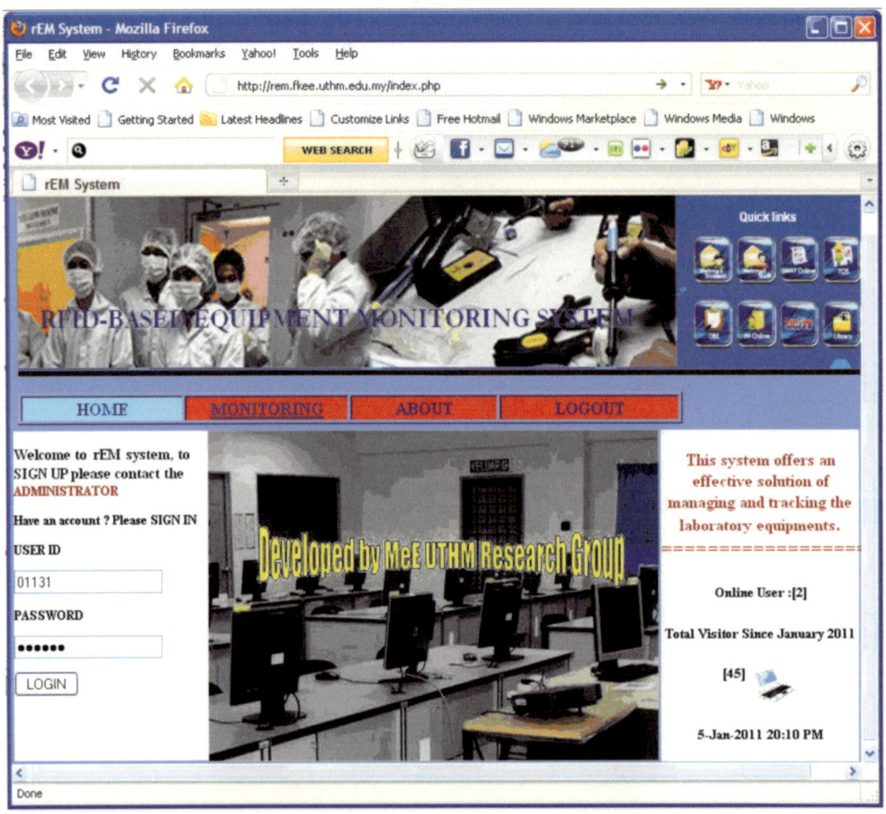

Fig. 11. The login page of system

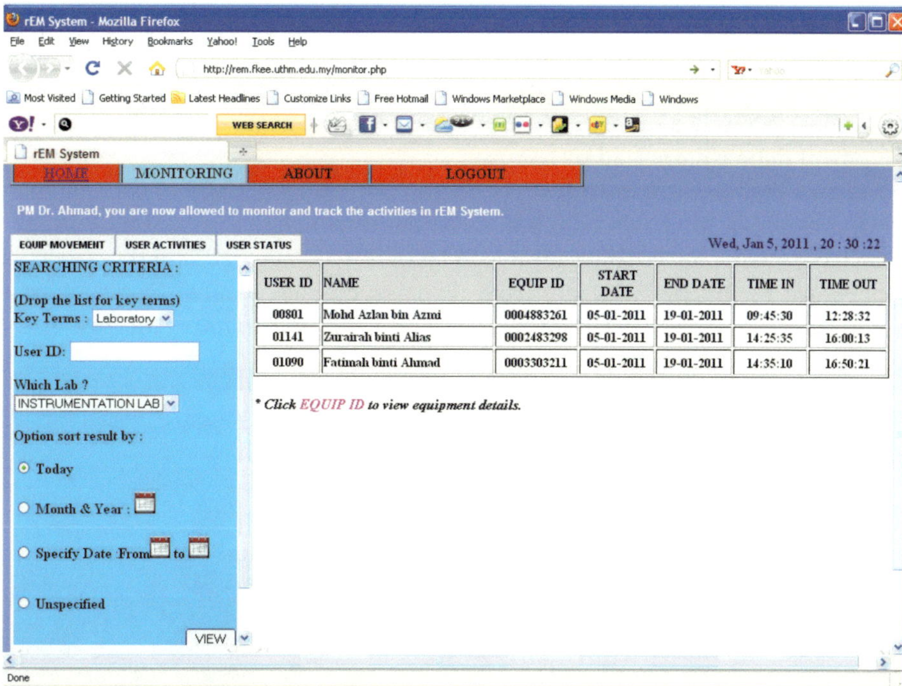

Fig. 12. On-loan equipment page

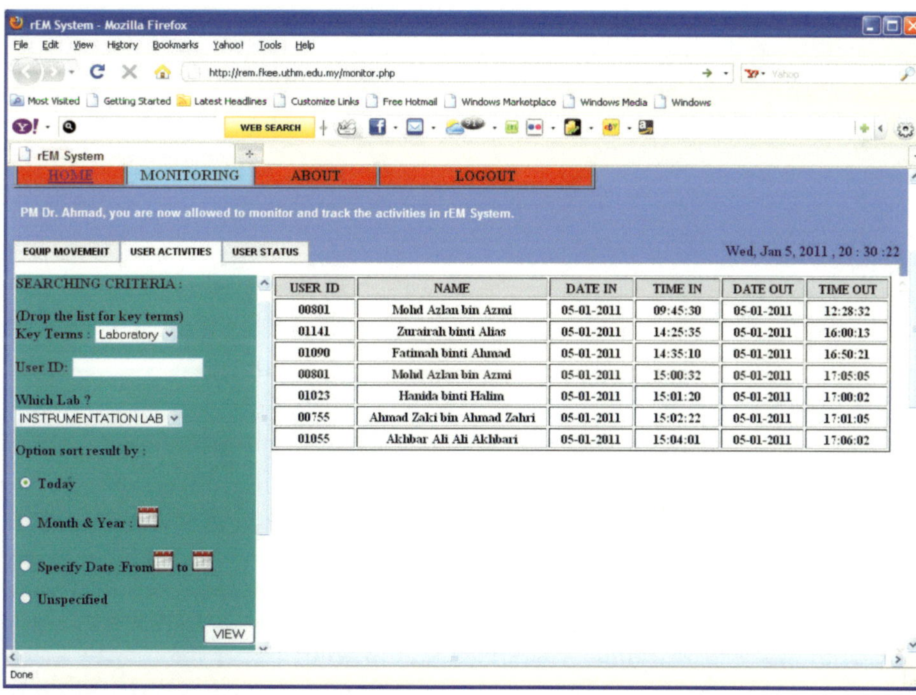

Fig. 13. Monitoring user's activity remotely

For successful login, the administrator is allowed to view and find a specific record on on-loan equipment. Figure 12 shows on-loan equipment based on laboratory and specific date. As shown below, the following on-loan information is taken from instrumentation lab for Jan 5, 2011. The system is also designed so that the administrator could click on *Equip ID to* view equipment details borrowed by the user.

Figure 13 shows that the administrator is able to view user's activity at each laboratory. In the following example, it shows who has used the instrumentation lab on Jan 5, 2011. The user status tab contains information on which laboratory is allowed and the valid period as shown in Figure 14. By default, this page displays the status of all users. It also could display the status of certain user by selecting specific information, for instance *UserID* keyword to perform the searching process.

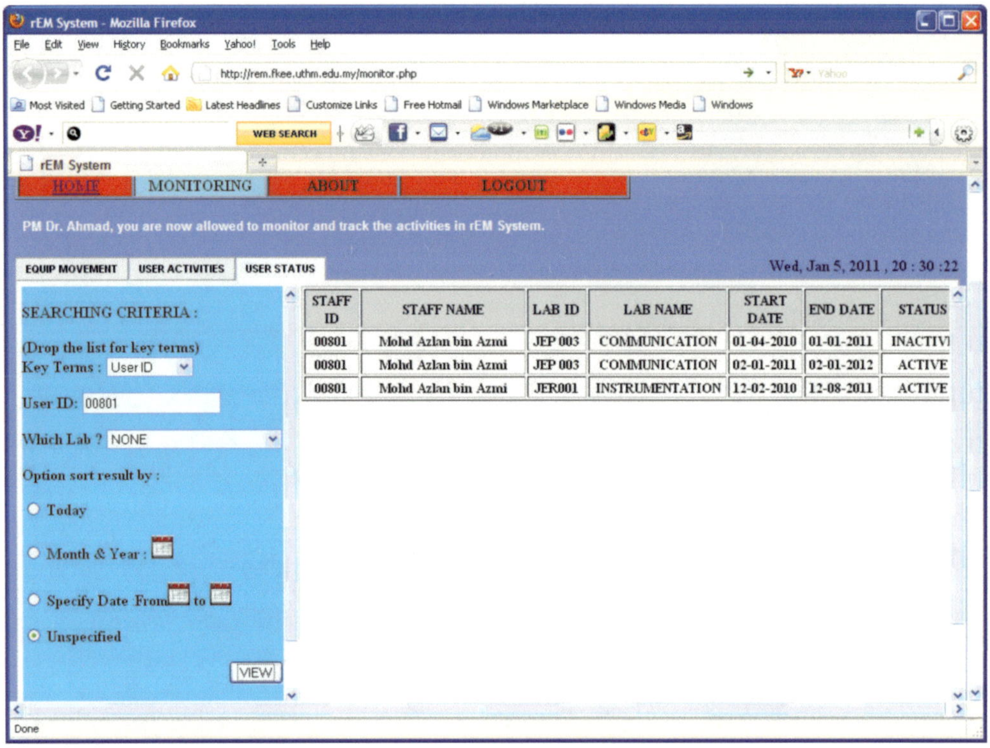

Fig. 14. Viewing user's status

6. Conclusion

Laboratory equipment monitoring system using RFID is proposed to effectively monitor the in-out equipment from the laboratory. Via this system, every activity involving laboratory equipment can be monitored and updated through web based environment. For security purpose, only authorized personnel have the permit to monitor the transaction activities of laboratory equipment in real-time. The adaptation of RFID-based Equipment Monitoring System also would promote diversity on laboratory management which previously are handled manually.

7. References

Ahmad Rafiq Adenan, Siti Zarina Mohd Muji, Mohd Helmy Abd Wahab. Automated Animal Tracking System using Radio Frequency Identification Tags. Proceeding of Computer Science and Mathematics Symposium 2006. KUSTEM, Kuala Terengganu, Terengganu, Malaysia, 8 - 9 November 2006

EPC Global. RFID smart label practice experience. (2005-08-07) [2006-04-04], http://www.rfidi--nfo.com. cn/report/dissertation/2OoS08/1655.html

Haron, N. S., Saleem, N. S., Hassan, M. H., Ariffin, M. M. and Aziz, I. A. A RID-based Campus Context-Aware Notification System. Journal of Computing. Vol. 2. Issue 3.

Herdawatie Abdul Kadir, Mohd Helmy Abd Wahab, Zarina Tukiran Mohd Razali Mohd Tomari and Mohd Norzali Hj. Mohd. (2010). Fusion of Radio Frequency Identification (RFID) and Fingerprint in Boarding School Monitoring System (BoSs), Sustainable Radio Frequency Identification Solutions, Cristina Turcu (Ed.), ISBN: 978-953-7619-74-9, InTech, Available from: http://www.intechopen.com/articles/show/title/fusion-of-radio-frequency-identification-rfid-and-fingerprint-in-boarding-school-monitoring-system-b

Hsu, C-I, Shih, H-H, Wang, W-C. (2009). Applying RFID to Reduce Delay in Import Cargo Customs Clearance Process. Computers & Industrial Engineering. Vol. 57. pp. 506 - 519.

Loebbecke, C. (2005). RFID Technology and Applications in Retail Supply Chain: The Early Metro Group Pilot. 18th Bled eConference on eIntegration in Action, June 6 - 8, 2005, Bled, Slovenia.

Ngai, E. W. T. and Lo, S. Y. Y. (2008). Development of an RFID-based sushi management system: The case of a conveyor-belt sushi restaurant. International Journal of Production Economics, Vol. 112, Issue 2, pp. 630-645.

Nor Suryani Bakery, Ayob Johari, Mohd Helmy Abd Wahab, Danial, Md. Nor. RFID Application in Farming Management System. In Proceeding of 3rd International Conference on Robotics, Vision, Information and Signal Processing 2007 (ROVISP2007), Penang, 28 - 30 November 2007

Oztekin, A., Pajouh, F. M., Delen, D., and Swim, L. K. (2010). An RFID Network Design Methodology for Asset Tracking in Healthcare. Decision Support Systems. Vol. 49, pp. 100 - 109.

Oztaysi, B., Baysan, S., and Akpinar, F. (2009). Radio Frequency Identify (RFID) in hospitality. Technovation. Vol. 29. Pp. 618 - 624.

Robert, C. M. (2006). Radio Frequency Identification (RFID). Computers & Security, Vol. 25. Pp. 18 - 26.

Tan, T-H, Chang, C-S. (2010). Development and Evaluation of an RFID-based e-Restaurant System for Customer Centric Service. Expert System with Applications. Vol. 37 Issue 9.

Voulodimos, A. S., Patrizakis, C. Z., Sideridis, A. B., Ntafis, V. A., and Xylouri, E. M. (2010). A Complete Farm Management System based on Animal Identification using RFID Technology. Computers and Electronics in Agriculture. Vol. 70. Pp. 380 - 388.

Yan, B. and Lee, D. (2009). Application of RFID in Cold Chain Temperature Monitoring System. 2009 ISECS International Colloqium on Computing, Communication, Control, and Management. Aug. 8 - 9, 2009. Sanya, China.

Zhang, X., Yue, S., and Wang, W. (2006). The Review of RFID Applications in Global Postal and Courier. The Journal of China Universities of Post and Telecommunications. Vol. 13. Issue. 4.

12

Developing RFID-Based Instruments Maintenance Management in Construction Lab

Yu-Cheng Lin, Weng-Fong Cheung, Yi-Chuan Hsieh, Fu-Cih Siao and Yu-Chih Su
National Taipei University of Technology/ Civil Engineering
Taiwan

1. Introduction

Maintenance management is very important subject special in construction lab. To manage related information of equipments and instruments plays an important role in the view of construction lab management. Those equipments and instruments need high standard and requirement in precision and accuracy of tests. Managing maintenance work effectively is extremely difficult in construction lab owing to various equipments and instruments with different specification. Furthermore, it will take high cost to maintain those instruments in the good conditions for the test correctness. With the advent of the Internet, web-based information management solutions enable information dissemination and information sharing among related maintenance staff members. Generally, maintenance managers and staffs require access to the equipments and instruments location to handle inspection and maintenance work in construction lab. Usually, maintenance staffs generally use sheets of paper to handle various types of maintenance information, including checklists, specification, and maintenance procedure. Consequently, there is serious rework progress regarding the data capture and entry in maintenance progress. In order to enhance the effectiveness of inspection and maintenance work in construction lab, this study presents a novel system called Mobile RFID-based Maintenance Management (M-RFIDMM) system for the acquisition and tracing of lab equipments and instruments maintenance information on locations and providing an equipments and instruments maintains information sharing platform among all participants using web technology and RFID-enabled PDAs. Integrating promising information technologies such as RFID-enabled PDAs, Radio Frequency Identification (RFID) scanning and data entry mechanisms, can help improve the effectiveness and convenience of information flow in the maintenance management. The primary objectives of this study include (1) applying such a system that integrates RFID technology with RFID-enabled PDAs to increase the efficiency of equipments and instruments inspection and maintenance data collection, and (2) designing a web-based portal for equipments and instruments management and control, providing real-time information and wireless communication between offices and instruments locations. The M-RFIDMM is then applied in a construction lab in Taiwan to verify our proposed methodology and demonstrate the effectiveness of maintenance progress in construction lab. The combined results demonstrate that, an M-RFIDMM system can be a useful web-based lab maintenance management platform by utilizing the RFID approach and web technology. With appropriate modifications, the M-RFIDMM system can be utilized at any instruments inspection and maintenance service for maintenance management divisions or suppliers in support of the M-RFIDMM system.

2. Problem statement

Maintenance management performance can be enhanced by using web technology for information sharing and communication. Information acquisition problems in instruments management follow from most of the data and information being gathered from the instruments location in construction lab. The effectiveness of information and data acquisition influences the efficiency of maintenance execution. Usually, maintenance managers and staff members generally use sheets of paper and/or field notes for maintenance progress in Taiwan construction lab. Restated, existing means of processing information and accumulating data are not only time-consuming and ineffective, but also compromise maintenance management in information acquisition. Such means of communicating information between instruments location and office, and among all participants, are ineffective and inconvenient. According to the questionnaire survey, the primary problems in inspection and maintenance regarding to data capture and sharing are as follows: (1) the efficiency and quality are low, especially in the inspection and maintenance progress in instruments management through document-based media, and (2) there are serious rework progress regarding the data capture and input in inspection and maintenance progress. However, few suitable platforms are developed to assist maintenance staff members with capturing and sharing the inspection and maintenance information when maintenance staff members need to handle inspection and maintenance work. Therefore, to capture data effective and enhance information communication in construction lab will be primary and significant challenge in the study.

3. Research objectives

This study utilizes the RFID and web technology to enhance the maintenance progress and effectiveness in instruments management service. This system is controlled by the management division, and provides maintenance managers and maintenance staff members with real-time instruments-related information-sharing services, enabling them to dynamically respond to the entire maintenance management network. This study develops Mobile RFID-based maintenance management (M-RFIDMM) system to improve efficiency and cost-effectiveness of instruments management, improve practical communication among participants, and increase flexibility in terms of service delivery and response times. M-RFIDMM system is a web-based system for effectively integrating maintenance managers, maintenance staff members and relative members, to enhance the instruments maintenance management in the construction lab. PDAs can extend M-RFIDMM systems from offices to instruments locations. Data collection efficiency can also be enhanced using RFID-enabled PDAs to enter and edit data on the instruments location. By using web technology and mobile devices, the M-RFIDMM system for the management division has tremendous potential to increase the efficiency and effectiveness of information flow, thus streamlining services processes with other participants. Maintenance managers and staff members frequently waste time by travelling to obtain information in the absence of other efficient means of communication. The portal and PDAs enable maintenance staff members to update data from the instruments location and immediately upload it to the system; Maintenance managers can receive maintenance information and make better decisions regarding future instruments management and control.

The main purposes of this study include (1) developing a framework for a mobile maintenance management system for instruments in the lab; (2) applying such a system that integrates RFID technology with PDA technology to increase the efficiency of instruments inspection and maintenance data collection in the lab; (3) designing a web-based portal for maintenance management and control, providing real-time information and wireless communication between offices and instruments locations, and (4) Evaluating the effectiveness of the proposed system in construction lab. Figure 1 illustrates solutions used in a real case utilized M-RFIDMM system in Taiwan construction lab. With appropriate modifications, the M-RFIDMM system can be utilized at any instruments inspection and maintenance service for maintenance management divisions or managers in support of the M-RFIDMM system.

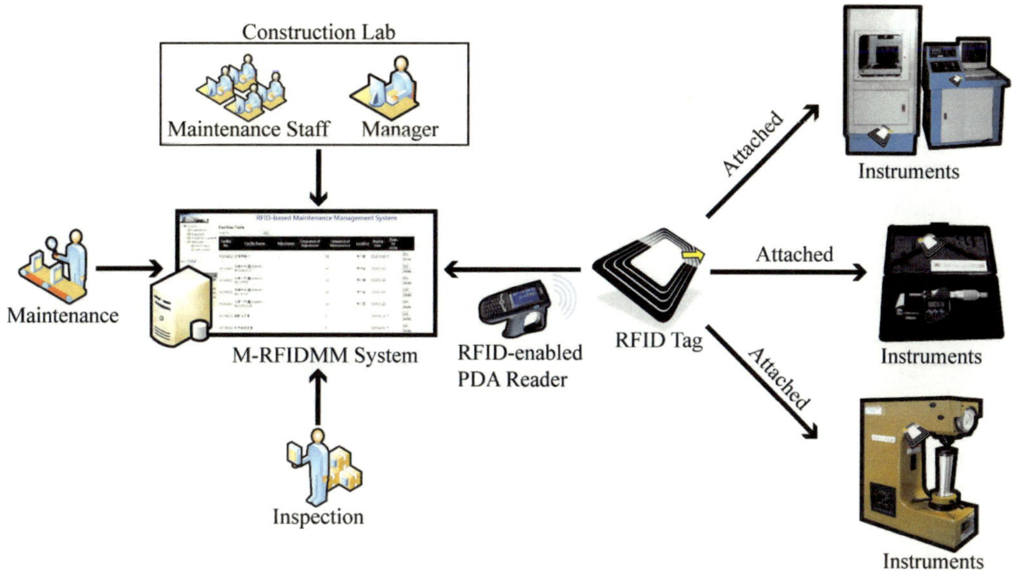

Fig. 1. M-RFIDMM System Framework Overview

4. Background research

RFID is an automatic identification solution that streamlines identification and data acquisition, operating similarly to bar codes. Automatic identification procedures have recently become very popular in numerous service industries for purchasing and distribution logistics, and in manufacturing companies and material flow systems. Jaselskis and Anderson (1995) investigated the applications and limitations of RFID technology in the construction industry, and attached read/write RFID tags to the surfaces of concrete test that were cast from the job site to test lab. This RFID technology has been widely applied in many areas in the construction industries for the following reasons: (1) to provide owners and contractors with information to enhance operation using RFID technology (Jaselskis and

Tarek, 2003); (2) to propose a novel concept of "parts and packets unified architecture" in order to handle data or information related to a product carried by product itself by utilizing RFID technology (Yagi et al., 2005); (3) to apply RFID technology as a solution to problems in pipe spools, and identify potential economic benefits from adopting RFID technology in automated tracking (Song et al., 2006); (4) to apply RFID combined with GIS technology in order to locate precast concrete components with minimal worker input in the storage yard (Ergen et al., 2006); (5) to improve the efficiency of tracing tools and tool availability using RFID (Goodrum et al., 2006); (6) to develop mobile construction supply chain system integrated with RFID technology (Wang et al., 2006); (7) to describe a prototype of an advanced tower crane equipped with wireless video control and RFID technology (Lee et al., 2006); (8) to improve tracing of material on construction using materials tagged with RFID tags (Song et al., 2006); (9) to present strategy and information system to manage the progress control of structural steel works using RFID and 4D CAD (Chin et al., 2008); (10) to enhance precast production management system integrated with RFID application (Yin et al., 2009), and (11) to present a new methodology for managing construction document information using RFID-based semantic contexts (Elghamrawy and Boukamp, 2010).

The use of technology to improve delivery process control is not a novel concept. Many industries have applied barcodes to track materials for many years. Construction companies began to examine the use of barcodes for tool management in the early 1990s. Although barcode is an established and affordable technology, it has presented problems in the construction industry due to the short read range and poor durability of barcodes — a barcode requires a line of sight, and becomes unreadable when scratched or dirty.

An RFID system is composed of an RFID tag and an RFID reader. The RFID tag comprises a small microchip and an antenna. Data are stored in the tag, generally as a unique serial number. The RFID tags can be either passive (no battery) or active (battery present). Active tags are more expensive than passive tags and have a read range of 10–100 meters. Passive tags have a read range of 10mm to approximately 5m (Manish and Shahram, 2005). The vast majority of RFID tags applied in the construction industries are passive.

The RFID reader functions as a transmitter/receiver. The reader transmits an electromagnetic field that "wakes up" the tag and provides the power required for it to operate (Lahiri, 2005). The tag then transfers data to the reader via the antenna. This data are then read by the RFID reader, and transferred to a Pocket PC or computer. Unlike barcodes, RFID tags do not require line-of-sight to be read; they only need to be within the reader's radio range. Additionally, RFID tags, unlike barcodes, can be read through most materials. RFID tags are shrinking, with some measuring only 0.33mm across. Although RFID systems can apply different frequencies, the most common frequencies are low (125KHz), high (13.56MHz) and ultra-high (UHF) (850–900MHz) (Lahiri, 2005).

Notably, RFID systems are one of the most anticipated technologies that will potentially transform processes in the engineering and construction industries. In the construction industry, RFID technology can be utilized with PDAs, thereby allowing staff members to integrate seamlessly work processes at labs and sites, due to the ability to capture and carry data. With a RFID scanner plugged into a PDA, the RFID-enabled PDA is a powerful portable data collection tool. Additionally, RFID readings increase the accuracy and speed of information communication, indirectly enhancing performance and productivity.

The advantages of using mobile devices in the construction industry are well documented (Baldwin et al., 1994; Fayek et al., 1998; McCullough, 1997). Moreover, mobile devices have been applied in numerous construction industries, to provide the following support: (1) providing wearable field inspection systems (Sunkpho and Garrett, 2003); (2) supporting pen-based computer data acquisition for recording construction surveys (Elzarka and Bell, 1997); (3) supporting collaborative and information-sharing platforms (Pena-Mora and Dwivedi, 2002); (4) using mobile computers to capture data for piling work (Ward et al., 2003), and (5) utilizing mobile devices in construction supply chain management systems (Tserng et al., 2005).

5. System implementation

5.1 System architecture

The M-RFIDMM system has three main components, a PDA, RFID and a portal. Significantly, both the PDA and RFID components are located on the client side, while the portal is on the server side. All instruments-related information acquired by maintenance staff members within the M-RFIDMM system is recorded in a centralized M-RFIDMM system database. All staff members can access required information via the portal based on their access privileges. Moreover, the portal is limited by design to thirty persons logging in when all participants acquire the same case information at same time. The M-RFIDMM system extends the RFID-based instruments management system from the office to instruments locations to assist with inspection and maintenance services, while the M-RFIDMM system primarily deals with data transactions in all departments or systems integration. When the data are updated on the M-RFIDMM system, e-mails are automatically sent from the server to the maintenance managers of the management division and to staff members involved in the relevant activity. The M-RFIDMM system consists of an inspection and management portal integrated with mobile devices and RFID technology (RFID-enabled PDA). Each module is briefly described below.

RFID Module of M-RFIDMM System

The RFID technology can be either a passive or active system. The major difference between an active and a passive RFID system is that an active tag contains a battery, and can transmit information to the reader without the reader generating an electromagnetic field. The case study uses UHF passive RFID technology due to budget restrictions and long distance read range requirement.

Mobile Device (PDA) Module of M-RFIDMM System

The M-RFIDMM system adopts a Unitech RH676 with an UHF RFID Reader as the RFID-enabled PDA hardware. Unitech RH676 PDA is operated on Windows CE. All data files in the PDA module are transmitted to the server directly through the web. The Internet explorer 6 was chosen as web browser in the RFID-enabled PDA hardware system.

Web Portal Module of M-RFIDMM System

The web portal is an information hub in the M-RFIDMM system for an instruments management. The web portal enables all participants to log onto a single portal, and immediately obtain information required for planning. The users can access different information and services via a single front-end on the Internet. For example, a customer can log onto the portal, enter an assigned security password, and access real-time inspection schedule information. A general contractor can check the test or inspection status, availability of reports and various other case-related data. The web portal of M-

RFIDMM system is based on the Microsoft Windows 2000 operating system with Internet Information Server (IIS) as the web server. The prototype was developed using ASP.NET, which are easily combined with HTML and JavaScript technologies to transform an Internet browser into a user-friendly interface. The web portal provides a solution involving a single, unified database linked to all functional systems with different levels of access to information, based on user role, both within an organization and across organizations and other members.

5.2 Modules of system functions

This section describes the implementation of each module in the M-RFIDMM system.

Test Report Module:

The report module provides maintenance staff members with a complete record of inspection and maintenance performed in the maintenance management.

Inspection and Maintenance Module:

Maintenance staff members can download the most up-to-date maintenance schedule from the Internet, and enter instrument maintenance results directly via a PDA. Additionally, PDAs display the checklist for every instrument maintenance task. Maintenance staff members can record instrument information for dates, conditions, inspection result, descriptions of problems and suggestions that have arisen during maintenance. Furthermore, maintenance staff members can also mark unacceptable tasks, and select relevant tasks from lists in the PDA. The module has the benefit that maintenance staff members can enter/edit inspection and maintenance test results, and all test records can be transferred between the PDA and portal by real-time synchronization, eliminating the need to enter the same data repeatedly.

Progress Monitor Module:

This module is designed to enable maintenance staff members to monitor the progress of inspections and maintenance. Additionally, maintenance managers and staff members can access the progress or condition of inspection and maintenance tasks. The progress monitor module provides an easily accessed and portable environment where maintenance staff members can trace and record all information regarding the status of inspections delivered to the maintenance or scheduled for repair.

E-Documents Module:

This module allows maintenance staff members to download manuals and specifications in advance, and reference them during inspection. This module also has a search function that enables the information to be found and retrieved easily, which is a valuable feature in dynamic environments. Moreover, maintenance staff members who do not need paper-based manuals or specifications can download e-manuals or e-specifications and access them directly using their PDAs.

6. Case study

This study is applied in Taiwan construction lab for the case study. This study utilizes an M-RFIDMM system in the instruments maintenance management in construction lab.

Existing approaches for tracking and managing instruments maintenance adopt manually updated paper-based records. The most of inspection in maintenance work were paper-based work by manual entry although instruments maintenance management system was developed for information management. However, information collected by staff members using such labor-intensive methods is rework and ineffective in the maintenance results entry. Therefore, maintenance management division and maintenance staff members utilized the M-RFIDMM system to enhance instruments inspection and maintenance management in the case study. UHF Passive read/write RFID tags were used in the case study. After the critical instruments were selected, each UHF RFID tag for the instruments was made, and the unique ID of the instrument was entered into the M-RFIDMM system database. After the instrument was assigned to be monitored for maintenance, the instrument was scanned with a RFID tag to enter the M-RFIDMM system.

Fig. 2. Displayed the UHF RFID tags using in the case study.

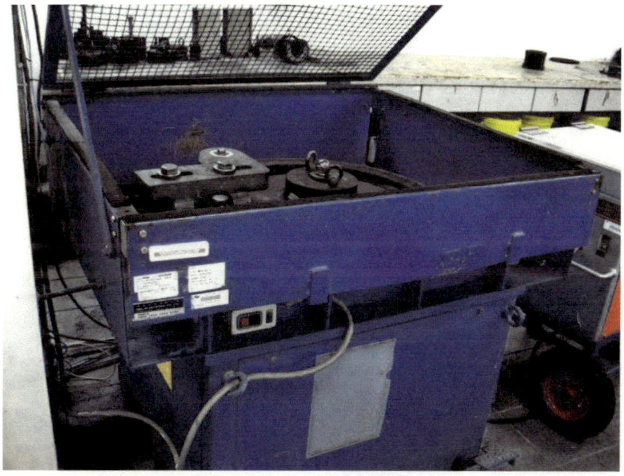

Fig. 3. Displayed the instrument attached UHF RFID tag in the case study (1).

During the setup phase, all the ID of instrument in the RFID tag had been determined and entered the database for system, and then the RFID tag was attached in the instrument. Finally, the tag will be scanned and checked before the maintenance work. Furthermore, the tag suggested to be attached on the non-metal surface (like wooden box) or placed a formcore (about 5 mm) for the interface for decreasing the influence because the RFID tag will be influenced by metal facilities (see Fig. 2-Fig.4).

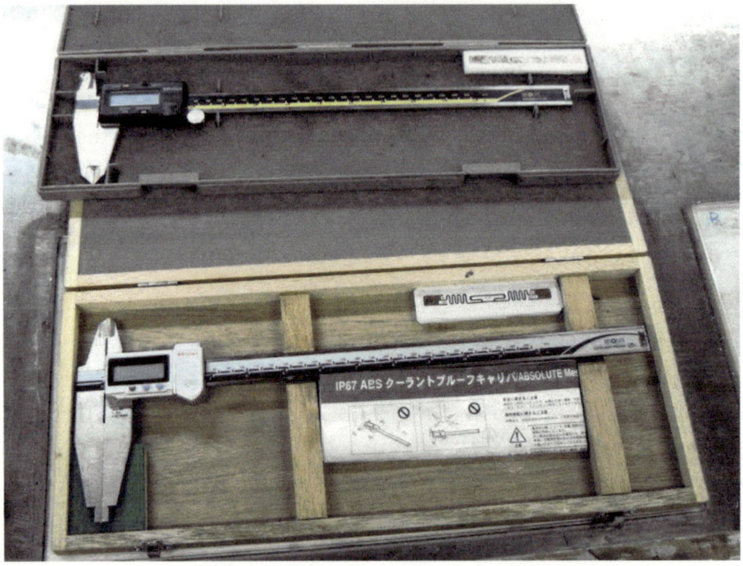

Fig. 4. Displayed the instrument attached UHF RFID tag in the case study (2).

Fig. 5. Displayed maintenance staff member used RFID-enabled PDA to scan RFID tags (1).

Fig. 6. Displayed maintenance staff member used RFID-enabled PDA to scan RFID tags (2).

Fig. 7. Displayed maintenance staff member entered the result of inspection, edited the description in the PDA.

Before the inspection/maintenance work, the instruments staff members can check the instrument list from PDA, refer the relative information and can make the preparation work without printing any paper document. During the inspection/maintenance progress, the instruments staff member scanned the RFID tag first and to confirm the instrument, then checked the further detail information like maintain procedure, notification, and fittings (see Fig. 5 and 6). The system would support any information of instrument via browser under wireless circumstance. After the instruments were inspected, staff members recorded the status and execute the work by procedure. After the operation, instruments staff member entered the result of inspection, edited the description in the PDA, and provided the updated information to the system (see Fig. 7). Once the instrument was break and need to be repaired, the system also can provide the supplier information and handle the problem immediately. Finally, the instruments manager and the authorized staff members accessed the updated information from office synchronously.

7. Field tests and results

Overall, the field test results indicate that UHF passive RFID tags are effective tools for instruments maintenance management in construction lab. All tags survived use in the instruments environment over one month testing period. The number of instruments for inspection and maintenance progress in field trials was around fifty. The M-RFIDMM system was installed on main server in the instruments management division of the construction lab. During the field trials, verification and validation tests were performed to evaluate the system. The verification aims to evaluate whether the system operates correctly according to the design and specification; and validation evaluates the usefulness of the system. The verification test was carried out by checking whether the M-RFIDMM system can perform tasks as specified in the system analysis and design. The validation test was undertaken by asking selected case participators to use the system, and provide feedback by answering a questionnaire. The case participators consisted of two maintenance managers with 6 years of experience; six maintenance staff members with 5 years of experience above in the case study. To evaluate system function and the level of system capability satisfaction, we distributed questionnaires, and the users of the system were asked to grade the conditions of system testing, system function, and system capability separately, compared with the typical paper-based maintenance method, on the five Likert scale. Some comments for future improvements of M-RFIDMM system were also obtained from the case participators through user satisfaction survey. Table 1 shows a comparison of the approximate time required for a typical instruments maintenance service using a traditional paper-based inspection approach and the proposed system. The next section presents the detailed results of the performance evaluation and the user survey conducted during the field trials.

Item	Paper-based Approach		Proposed Approach	
	Method	Average Time (Min)	Method	Average Time (Min)
Instruments located in the outside	Check instrument (using stair) for maintenance	5.2	Use PDA and read the tag attached in the instrument for maintenance	0.3
Find related maintenance information	Referring to maintenance menu	2.2	Automatic selection	0.1
Input maintenance description	Referring to maintenance item and checklist	2.5	Entry the PDA and store in the system	1.8
Check maintenance record	Paper forms	N/A	Read information directly from the system	0.2
Archive data	Re-entry at the office	6.0	Real-time Update database	0.2
Sharing maintenance information	Send the e-mail (at the office)	5.0	Access the system directly and share information	0.5

Table 1. System Evaluation Result

The 88% obtained from user satisfaction survey indicates that the M-RFIDMM system is quite adaptable to the current instruments maintenance management practices in construction lab, and is attractive to users. This result implies that the M-RFIDMM system was well designed, and could enhance the current time-consuming instruments maintenance process.

The 88 % obtained from maintenance staff members satisfaction survey indicates that the system automatically generated all documentation, and accumulated the related historical data in the central database server. The maintenance staff members could thus collect maintenance data, and send them electronically to the M-RFIDMM system. No additional work was required for any documentation or maintenance analysis after the data collection.

The 25% user shows the PDA is not so easy to operate because some of staff members are not used to use PDA in the beginning.

The advantages and disadvantages of M-RFIDMM system identified from the real case studies application are identified. However, over 80% of users obtained from maintenance staff members' satisfaction survey agree that the M-RFIDMM system is useful for improving the efficiency and effectiveness of automated data acquisition and information sharing in instruments maintenance service, thus assisting maintenance managers and maintenance staff members in managing and monitoring the maintenance progress of instruments in the building. UHF Passive tags are less expensive than active tags. Thus, UHF passive tags are suited to instruments maintenance management.

The use of RFID and web technology to collect and capture information significantly enhanced the efficiency of inspection and maintenance processes of instruments. RFID readers and tags are widely thought likely to improve in the future and significantly improving the maintenance processes efficiency.

In the cost analysis, the UHF tags adopted in this study cost under $0.2 US dollars each in 2010. The cost of these tags is decreasing every year. The total cost of the equipment applied in this study was $3250 US dollars (including RFID-enabled PDA reader and one server personal computer). Even the reader initial cost is higher, but it is function expandable and really decreases human work. Experimental results demonstrate that M-RFIDMM system can significantly enhance the instruments maintenance progresses. The use of RFID significantly decreases the overall maintenance operation time and human cost.

8. Conclusions

This study presents a Mobile RFID-based Maintenance Management (M-RFIDMM) system that incorporates RFID technology and mobile devices to improve the effectiveness and convenience of information flow during maintenance phase in construction lab. The M-RFIDMM system not only improves the acquisition of data on instruments maintenance efficiency using RFID-enabled PDA, but also provides a real time service platform during instruments maintenance progress. In the case study, plugging a RFID scanner into a PDA creates a powerful portable data collection tool. Additionally, RFID readings increase the accuracy and speed of information search, indirectly enhancing performance and productivity. Maintenance staff members use RFID-enabled PDAs to enhance seamlessly maintenance work processes at instruments locations, owing to its searching speed and ability to support any information during the process. Meanwhile, on the server side, the M-

RFIDMM system offers a hub center to provide instruments management division with real-time to monitor the maintenance progress. In the case study, the application of the M-RFIDMM system helps to improve the process of inspection and maintenance work for the construction lab in Taiwan. Based on experimental result, this study demonstrated that UHF passive RFID technology has significant potential to enhance inspection and maintenance work in instruments management. The integration of real-time maintenance information from instruments helps maintenance staff members to track and control the whole inspection and maintenance progress. Compared with current methods, the combined results demonstrate that, an M-RFIDMM system can be a useful web-based lab maintenance management platform by utilizing the RFID approach and web technology.

9. Recommendations

Recommendations for implementing the proposed system in the future are given below.

- Cost is a currently significant factor limiting the widespread use of RFID tags in the construction industry. Passive tags are cheaper than active tags. Therefore, passive tags are suited to the instruments management.
- If the RFID tag needs to be placed the interface of the metal instruments, the RFID tag should be isolated by formcore (over 3mm) or other non-metal formcore to avoid influence from metal instruments.
- The PDA screen is not large enough for operating the M-RFIDMM system fluently. The system should be redesigned and developed to be suitable for the PDA screen.
- It is necessary to consider the usage time of RFID. Currently, the average of longest time regarding to RFID tags is ten year. Therefore, if the instruments need to track over ten years then the RFID tag should be attached to replace easily and workable.

10. References

Baldwin, A. N., Thorpe, A. and Alkaabi, J. A. (1994), "Improved material management through bar-code: results and implications of a feasibility study," Proceedings of the institution of Civil Engineers, Civil Engineering, 102(6), 156-162.

Chin, S., Yoon, S., Choi, C., and Cho, C. (2008). "RFID+4D CAD for progress management of structural steel works in high-rise buildings,"Journal of Computing in Civil Engineering, ASCE, 22(2), 74-89.

Elghamrawy, T. and Boukamp, F. (2010). "Managing construction information using RFID-based semantic contexts," International Journal of Automation in Construction, 19(8), 1056-1066.

Elzarka, H. M. and Bell, L. C. (1997), "Development of Pen-Based Computer Field Application," Journal of Computing in Civil Engineering, ASCE, 11(2), 140-143.

Ergen, E., Akinci, B., and Sacks, R. (2006) "Tracking and locating components in a precast storage yard utilizing radio frequency identification technology and GPS," International Journal of Automation in Construction, doi:10.1016/j.autcon.2006.07.004.

Fayek, A., AbouRizk, S. and Boyd, B. (1998), "Implementation of automated site data collection with a medium-size contractor," in Proc. ASCE Computing in Civil Engineering, Boston, MA, 454-6.

Goodrum, P. M., McLaren, M. A., and Durfee, A. (2006) "The application of active radio frequency identification technology for tool tracking on construction job sites," International Journal of Automation in Construction, 15(3), 292-302.

Jaselskis, E. J. and Anderson, M. R. (1995). "Radio-Frequency Identification Applications in Construction Industry," Journal of Construction Engineering and Management, 121(2), 189-196.

Jaselskis, E. J. and El-Misalami, Tarek (2003). "Implementing Radio Frequency Identification in the Construction Process," Journal of Construction Engineering and Management, 129(6), 680-688.

Lahiri, Sandip (2005), RFID Sourcebook, Prentice Hall PTR.

Lee, Ung-Kyun, Kang, Kyung-In, and Kim, Gwang-Hee (2006). "Improving Tower Crane Productivity Using Wireless Technology." Journal of Computer-Aided Civil and Infrastructure Engineering, Vol. 21, pp.594-604.

Manish Bhuptani and Shahram Moradpour (2005), RFID Field Guide : Deploying Radio Frequency Identification Systems, Prentice Hall PTR.

McCullouch, B. G. (1997), "Automating field data collection in construction organizations," in Proc. ASCE Construction Congress V, Minneapolis, MN, 957-63

Pena-Mora, F. and Dwivedi, G. D. (2002), "Multiple Device Collaborative and Real Time Analysis System for Project Management in Civil Engineering," Journal of Computing in Civil Engineering, ASCE, 16(1), 23-38.

Song, J., Haas, C. T. and Caldas, C. (2006). "Tracking the Location of Materials on Construction Job Sites," Journal of Construction Engineering and Management, 132(9), 680-688.

Song, J., Haas, C. T., Caldas, C., Ergen, Esin, and Akinci, B. (2006). "Automating the task of tracking the delivery and receipt of fabricated pipe spools in industrial projects," International Journal of Automation in Construction, 15(2), 166-177.

Sunkpho, Jirapon and Garrett, J. H., Jr. (2003), "Java Inspection Framework: Developing Field Inspection Support System for Civil Systems Inspection," Journal of Computing in Civil Engineering, ASCE, 17(4), 209-218.

Tserng, H. P., Dzeng, R. J., Lin, Y. C. and Lin, S. T. (2005). "Mobile Construction Supply Chain Management Using PDA and Bar Codes." Journal of Computer-Aided Civil and Infrastructure Engineering, Vol. 20, pp.242-264.

Wang, L. C., Lin, Y. C. and Lin, P. H. (2006). "Dynamic Mobile RFID-based Supply Chain Control and Management System in Construction." International Journal of Advanced Engineering Informatics - Special Issue on RFID Applications in Engineering, Vol. 21 (4), pp.377-390.

Ward, M. J., Thorpe, A. and Price, A. D. F. (2003), "SHERPA: mobile wireless data capture for piling works, "Computer-Aided Civil and Infrastructure Engineering, 18, 299-314.

Yagi, Junichi, Arai, Eiji and Arai, Tatsuo (2005). "Construction automation based on parts and packets unification," International Journal of Automation in Construction, 12(1), 477-490.

Yin, Y.L., Tserng, H. P., Wang, J.C. and Tsai, S. C. (2011). "Developing a precast production management system using RFIF Technology," International Journal of Automation in Construction, 18(5), 677-691.

13

What are Authentic Pharmaceuticals Worth?

Matthieu Schapranow, Jürgen Müller, Martin Lorenz,
Alexander Zeier and Hasso Plattner
Hasso Plattner Institute, Enterprise Platform and Integration Concepts Chair, Potsdam
Germany

1. Introduction

Radio Frequency Identification (RFID) technology is named as a possible basis for future anti-counterfeiting by providing enhancements of existing business processes [Choi & Poon (2008)]. Hereby, the use of unique Electronic Product Codes (EPCs) [EPCglobal Inc. (2010)] for identification improves processing times during goods receipt and enables automated product tracking and tracing. The EPC is used to refer to a concrete item instance in a software system. For example, it identifies a concrete bottle of analgesic that was manufactured on May. 01, 2011 at 07:03 a.m. In contrast, currently used barcodes identify a class of pharmaceuticals, e.g. all analgesics of a certain manufacturer. RFID technology shows prevailing advantages in contrast to barcodes, RFID tags can be read without establishing a direct line of sight, multiple tags can be read simultaneously, and they can cope with dirty environments [Stiehler & Wichmann (2005); White et al. (2007)].

In the following, we refer to an RFID-aided supply chain when dealing with an supply chain solution that build on good's tracking and tracing functionality by integrating RFID technology [Schapranow et al. (2009)]. In context of the pharmaceutical supply chain, the integration of tracking functionality is widely considered, e.g. two-dimensional data matrix or RFID technology, since this specific industry is confronted with increasing counterfeit rates [European Commission Taxation and Customs Union (2009)]. However, advantages of using RFID technology only apply when all participants of the supply chain seamlessly integrate tracking solutions based on it.

Fig. 1 models components within an RFID-enabled company to support anti-counterfeiting using the Fundamental Modeling Concepts (FMC) [Knöpfel et al. (2005)]. These components can be established to track and trace goods on item level without media breaks. Since the depicted architecture switch is connected with high monetary investments, costs have to be accommodated by all participants of the supply chain [Schapranow, Nagora & Zeier (2010)]. Different levels of technology acceptance to transform towards an RFID-enabled company can result in exclusion of participants from the supply chain. We expect especially Small and Mid-sized Enterprises (SMEs) to be confronted with financial barriers to participate in global RFID-aided supply chains [Müller, Faust, Schwalb, Schapranow, Zeier & Plattner (2009)]. However, a gap-less integration of RFID technology at all supply chain participant sites is the basis for consistent tracking and tracing on item level in real time.

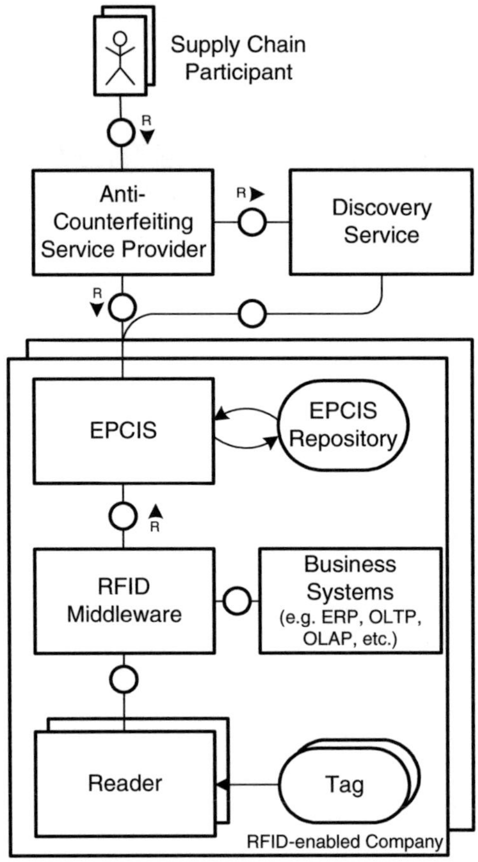

Fig. 1. FMC Block Diagram: Anti-counterfeiting Components of RFID-enabled Companies

We contribute by sharing our research results for enabling an integer RFID-aided supply chain. We focus on the business perspective and present concrete costs for RFID-enablement of supply chain participants and for operating a dedicated architecture for anti-counterfeiting. Our research activities are motivated by concrete requirements of the pharmaceutical industry. We present operating models to establish an RFID-aided supply chain while keeping initial infrastructure investments for involved supply chain parties at a moderate level. We discuss approaches for on-premise and on-demand operating models sharing hardware and software resources for cost-saving reasons. We identify cost-drivers for the proposed operating models, discuss cost-saving potentials, and define the amortization by product surcharges.

In the rest of our work we do not focus on how RFID technology may help to improve current pharmaceutical business processes, such as drug prescription, controlling of medication, or observation of patients. Instead, we stress on necessary adaptations to perform the transformation towards an RFID-aided supply chain. It is the key-enabler to observe product flows and to detect counterfeits by systematically analyzing the recorded movement of goods.

The rest of our contribution is structured as follows: Sect. 2 presents counterfeit challenges of the pharmaceutical industry from which we draw the motivation of our work. We define supply chain roles and their tasks within an RFID-aided supply chain to support automated anti-counterfeiting in Sect. 3. In Sect. 4 we perform a quantitative analysis of initial and operational investments for transforming towards an RFID-aided supply chain. Our work concludes in Sect. 5 by summarizing our finding and providing an outlook towards possible payment models.

2. Challenges in pharmaceutical supply chains

RFID technology is nowadays named to be the successor of existing tracking techniques such as scanning of one-dimensional barcodes [White et al. (2007)]. Making use of RFID tags results in various advantages. Tags can be read without establishing a direct line of sight, multiple tags can be read simultaneously, and they can cope with dirty environments. The logistics sector is currently one of the first implementers to guarantee traceability of fast-moving goods, e.g. life-saving pharmaceuticals, blood preservations, or organ donations. Tracking goods is an important factor for participants in global supply chains, i.e. RFID technology helps to keep goods moving on the road instead of keeping them in costly stocks [Schlitter et al. (2007)]. Compared to existing semi-automatic solution, e.g. scanning of barcodes, the implementation of RFID technology reduces time to process incoming and outgoing goods at all involved intermediate stations by enabling automatic product identification [Bovenschulte et al. (2007)].

Pharmaceutical counterfeits introduce the risk of harming human-beings, e.g. when applying wrong doses, invalid or missing active ingredients or poison combinations for people with certain risks [Bos (2009)]. In the context of global pandemic infections, such as pandemic influenza type H1N1 in 2009 or H5N1 in 2008, the impacts of counterfeits become visible [World Health Organization (2009)]. Illicit drug use is a major problem in the U.S. for years, e.g. approx. 20 million people used illicit drugs in 2007 and more than every fifth person between 18 and 20 contributed to this statistics [Barthwell et al. (2009)]. These drug-abusing people order prescription-based pharmaceuticals via the Internet without having a valid prescription of consulting a doctor. In case the expected medical effect does not occur, therapies are hard to develop, because pharmaceutical ingredients cannot be traced to an authentic manufacturer.

In terms of intellectual rights and property management new aspects of product tracking such as counterfeit detection become relevant. Upcoming regulations will force manufacturers, retailers, and pharmaceutical business partners to be reliable for products showing their company logo or involvement. Tracking of their products through the entire supply chain becomes necessary. A reliable tracking mechanism is the first step in fighting counterfeits of pharmaceutical products. Studies show that expensive products, such as cancer fighting drugs and drugs for AIDS therapies, suffer from product counterfeits with increasing rates. But also generic products are increasingly subject to plagiarism.

Pfizer reported experiences with RFID-based implementations to guarantee authenticity of its Viagra pills already in 2006 [U.S. Pharmaceuticals Pfizer Inc. (2006)]. These activities indicate ambitions of pharmaceutical manufacturers to validate the use of RFID technology as a possible way to protect their products.

Product counterfeits arrive in the United States (U.S.) of America and the European Union (EU) with steady increasing rates. A high level of integrity in the supply chain is the basis for reliable product tracking to reduce the amount of counterfeit cases. In the following, insights about the current pharmaceutical market situation in the European Union and the United

States are presented. They support the motivation to design innovative RFID implementations focusing on security aspects to be an integral aspect.

2.1 Threats in European Union

The EU consists of 27 member states since it has been extended lately in 2007 and the youngest member states Bulgaria and Romania joined. Its population covers approx. 500 million citizens, which is approx. 7.5 percent of the world's population. Yearly, approx. 30 billion packages of pharmaceuticals are manufactured for the entire European market [Müller, Pöpke, Urbat, Zeier & Plattner (2009)].
In 2007, a total of 43,671 reported counterfeit cases with approx. 80 million involved articles were reported. In contrast to 2007, a total of 49,381 counterfeit cases, i.e. an increase of 13 percent, with approx. 180 million involved articles, i.e. an increase of 125 percent, were reported in 2008 (European Commission Taxation and Customs Union). A fraction of 6.5 percent of all reported cases and approx. five percent of all articles were associated with the pharmaceutical sector. The European Commission reports an increase of 118 percent for pharmaceutical counterfeits detected at EU borders in 2008 compared to 2007. In addition to the categories CDs/DVDs and cigarettes, the pharmaceutical sector holds the third place according to growth rates of intercepted articles.
To stress the increase of detected pharmaceutical counterfeits, we provide the following quote:

> In a two-month period, more than 34 million tablets were seized, including fake antibiotics, anti-cancer, anti-malaria and anti-cholesterol medicines, painkillers and erectile dysfunction medication. [IP Crime Group (2008)]

The aforementioned quote underlines that by a single joined operation more than 30 million pharmaceutical counterfeits were detected at the borders of the EU. More than 90 percent of intercepted articles are suspicious in terms of trademark infringement. More than 50 percent of all articles were intercepted during import procedures, whereas most articles were detected in air transportation. The category of life-style drugs is reported to be number one regarding detected counterfeits [IP Crime Group (2008)].
India is named as the top source of counterfeit pharmaceutical products contributing more than 50 percent of all detected articles [Shukla & Sangal (2009)]. This development is constant for years. The example of India shows that counterfeiters in countries with low law regulation benefit from pandemic diseases, such as influenza H1N1 in 2009, because consumers buy medicines preventively via the Internet [World Health Organization (2009)].

2.2 Threats in the United States

The United States Federal Food and Drug Administration (FDA) detected more than 21 counterfeit cases between 2001 and 2003 [Food and Drug Administration (2004)]; in 2004 this number almost tripled with 58 confirmed cases [Food and Drug Administration (2005)]. In contrast to this development, in the years 1997 to 2000 the number of detected counterfeits did not exceed six per year. This outlines two aspects. On the one hand, the number of pharmaceutical counterfeits increases. On the other hand, counterfeit detection methods are continuously improved and former undetected counterfeits can be detected meanwhile.
An estimated number of 7,000 deaths are connected with counterfeit medicines in the United States per year [Jenkins et al. (2007)]. Health damages result in legal consequences for the manufacturer and loss of the company's reputation. To emphasize potential monetary impact,

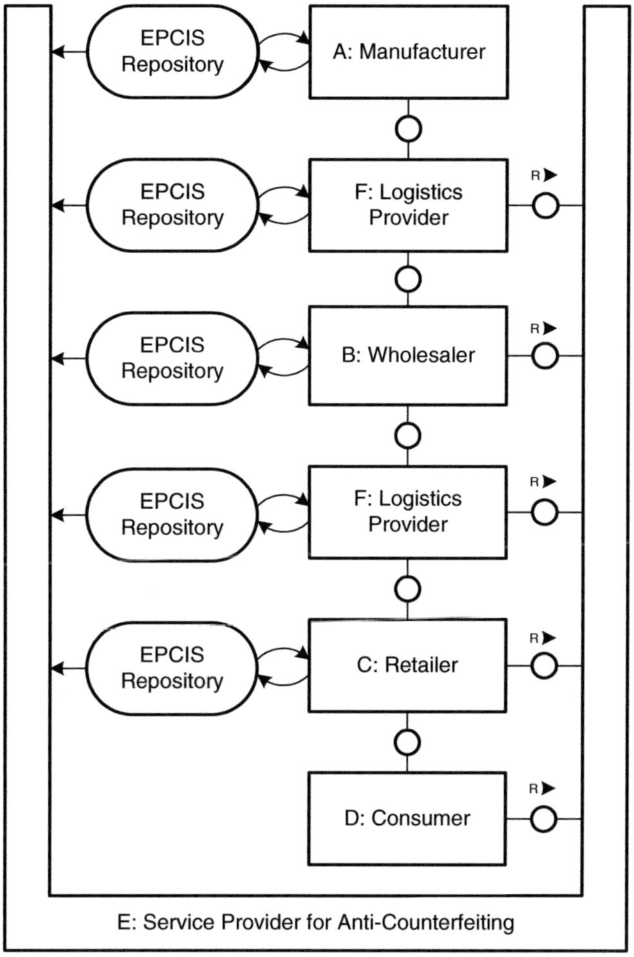

Fig. 2. FMC Block Diagram: Roles in the Pharmaceutical Supply Chain

Merck's medical vioxx evoked human damages and five billion USD were paid to avoid a lawsuit [Merck & Co. Inc. (2007)].

In 2004, it was estimated that more than 500 billion USD were traded in counterfeits, i.e. seven percent of the world trade in the same period [ICC Policy Statement (2004)]. It is stated, that this equals an increase of 150 billion USD in comparison to 2001 while the worldwide merchandise trade increased by approx. 50 billion USD in the same time, i.e. only one third of the increase traded in counterfeits [Staake et al. (2005)].

At this point, it is important to highlight that estimations about the monetary impact of counterfeits vary drastically. This fact underlines that only a small number of counterfeits can be detected nowadays and that the number of unreported cases is hard to derive. Technical improvements in counterfeit detection and goods protection help to increase the amount of detected cases by implementing new barriers to entrance counterfeits into large markets.

2.3 Sizing details for an RFID-aided pharmaceutical supply chain

The given case studies for the pharmaceutical industry in the U.S. and the EU highlight potential risks introduced by counterfeits and the need for active product protection. A high level of supply chain integrity is the basis for reliable product tracking and to support anti-counterfeiting. In the following, we focus on the European pharmaceutical supply chain, whereas similar conclusions can be drawn for the U.S. market. The European pharmaceutical supply chain consists of approx. 2,200 pharmaceutical manufacturers, 50,000 wholesalers, and 140,000 retailers [Müller, Pöpke, Urbat, Zeier & Plattner (2009)]. Every supply chain participant stores events e^* capturing the Electronic Product Code (EPC) [EPCglobal Inc. (2010)] of a certain item in an EPC Information Services (EPCIS) repository [EPCglobal Inc. (2007)] for all manufactured and processed goods.

A total amount of more than 30 billion pharmaceutical goods is manufactured in the pharmaceutical supply chain for the EU on yearly basis, whereas the half of them is available on prescription [Müller, Pöpke, Urbat, Zeier & Plattner (2009)]. As a result, we can derive an average daily production/handling rate of approx. 37,879 pharmaceutical goods that are produced per manufacturer, approx. 1,667 goods are handled per wholesaler, and approx. 595 goods are handled per retailers in the European pharmaceutical supply chain. To determine a lower threshold for the expected amount of captured EPC events for 30 billion pharmaceutical goods, we assume a minimal supply chain consisting of a pharmaceutical manufacturer with 360 production days per year and 24/7 manufacturing line, two wholesalers, a single retailer, and a customer. The manufacturer will capture at least a production and a shipping event for a certain pharmaceutical good. Both wholesalers will capture one event for goods receipt and goods shipment and two events that observe product movements within their stock locations. The retailer will capture a goods receipt event and a selling event, e.g. when a customer buys a medicine in the pharmacy. The customer will invoke an anti-counterfeiting check just in the pharmacy before buying the product, which results in a single check event. Ultimately, it sums up to eleven relevant captured EPC events distributed across the supply chain. As a result, a lower threshold of approx. 10,610 captured relevant events per second need to be expected across the entire global supply chain, each with an average size of 182 bytes [Schapranow, Müller, Zeier & Plattner (2010)].

A FMC model depicting the RFID-aided supply chain of the pharmaceutical industry is drawn in Fig. 2. It contains supply chain roles A to E and the involvement of a dedicated service provider for anti-counterfeiting. The service provider accesses individual EPCIS repositories of supply chain participants that are involved in handling a certain good to derive its virtual product history [Schapranow, Müller, Zeier & Plattner (2010)]. We agree that reliable product tracking and tracing across the entire supply chain can be implemented using RFID [Bundesverband Informationswirtschaft, Telekommunikation und neue Medien (2005)], but this technology is not designed to be immunized against threats, such as cloning, spoofing or eavesdropping [Schapranow et al. (2009)]. It is very important that customer profiles cannot be derived, because besides customers' privacy the entire supply chain would become vulnerable.

We agree that reliable product tracking and tracing across the entire supply chain can be implemented with the help of RFID solutions and open interfaces for supply chain participants [Bundesverband Informationswirtschaft, Telekommunikation und neue Medien (2005)]. However, RFID was not designed for secured data exchange of confidential details.

Hence, security threats exist, e.g. the possibility of cloning, spoofing or eavesdropping of tag reader communication to inject counterfeits [Schapranow et al. (2009)]. Possible measures against threats, e.g. mutual authentication, may reduce the probability for a certain threat [Schapranow, Zeier & Plattner (2010)].
We want to support the usage of RFID by introducing reliable IT infrastructure components that help to identify counterfeits by analyzing available event data, e.g. analysis of the goods path through the supply chain, suspicious ordering of actions, and semantic errors. We consider customer's privacy worthy of protection. From our perspective, customer profiling by combining checkout data with captured event data must be prevented to increase the acceptance of this anti-counterfeiting technique.
On the one hand, the presented scenario of the pharmaceutical industry underlines the need for reliable anti-counterfeiting mechanisms to prevent counterfeit injection. On the other hand, the pharmaceutical industry suffers from privacy concerns of end consumers while implementing RFID technology for tracking and tracing [Schapranow et al. (2009)]. We focus on the pharmaceutical industry in the following to support the fast adoption of RFID technology. We agree that this technology can contribute to establish an integer global pharmaceutical supply chain by establishing a permanent product trace. However, automated anti-counterfeiting is only feasible for expensive pharmaceuticals unless costs for RFID tags and components exceed an empiric threshold of less than about ten percent of the product's retail price.

3. Impacts of anti-counterfeiting for supply chain participants

The following section outlines our considerations for an anti-counterfeiting architecture based on location-based event data. The heart of the architecture builds a dedicated service provider for anti-counterfeiting as depicted in Fig. 1. It performs checks on event data for a given pharmaceutical good that is uniquely identified by its EPC. Furthermore, the service provider protects the privacy of inquirers and supply chain participants that handled a certain product as a kind of facade. On the one hand, queries are not directly sent to supply chain participants, i.e. the service provider prevents derivation of business relationships. On the other hand, supply chain participants cannot derive good's holder identity, e.g. to trace the good's complete path in the supply chain once it left the manufacturer.
Fig. 2 depicts the flow of data between roles involved in an RFID-aided supply chain to support anti-counterfeiting. Supply chain roles are described in further detail in subsequent sections focusing their business activities as participant of an RFID-aided supply chain.

3.1 Role A: Manufacturer

The manufacturer role consists of two separated tasks: product assembly and product creation. In terms of the product assembly it acts as an end consumer, i.e. consuming partly assembled products and removing them from the supply chain.
The task product creation is responsible to bring products alive. In this context the task of the manufacturer involves the creation of the product's meta data representation for the RFID-aided supply chain, which is covered by the following tasks.
The following steps are only required once a new product is created and can be considered as optionally if partially assembled product are consumed by the manufacturer.

1. Equip the product with a proper RFID tag to create the handling unit, i.e. the physical connection between the product and its tag. A handling unit can also be a transportation unit, such as a box or a container that groups multiple goods together.
2. Determine next available unique EPC for the created product. Therefore, the EPCIS repository of the manufacturer needs to be contacted.
3. Initialize the RFID with RFID-specific data, i.e. mandatory data, e.g. EPC, and optional data, e.g. authentication data.
4. Establish the virtual product history for the certain good by storing the creation event in the manufacturer's event repository.

Continue the business process on the manufacturer's site and capture events where the product's handling unit is involved. The following task is required for all kinds of products.

5. Capture all events defining the path at manufacturer's locations.

3.2 Role B: Wholesaler

The wholesaler's receives goods from various manufacturers, disassembles the handling units partially and reassembles them to new more specific handling units for certain retailers, such as hospitals or pharmacy chains.
The following tasks are required to contribute to the virtual product history.

1. Capture the goods receipt event.
2. Capture goods movement events within local storage, e.g. unpacking or new placing. All events are stored in the local event repository.
3. Equally to goods receipt processing the goods shipment is performed and. Corresponding captured events are stored in the local event repository.

3.3 Role C: Retailer

The retailer receives goods packed in so-called handling units, e.g. boxes or pallets. The latter are delivered by logistics provider from manufacturers, other retailers or wholesalers. Retailers use their local or more often a hosted event repository for storing captured events. The latter is available on subscription basis [Müller, Schapranow, Helmich, Enderlein & Zeier (2009)]. The retailer's task is to separate goods and sell them either to end consumers or to other retailers. When a product is sold to the end consumer the product history typically ends with the deactivation of the RFID tag at the point of sales [Schapranow et al. (2009)].

1. Receive handling unit and capture the goods receipt.
2. Unpack received handling unit recursively and process all contained goods individually, i.e. store events for all goods in the local event repository.
3. Capture the shipping event at the point of sales.

3.4 Role D: End consumer

The supply chain role end consumer occurs only once for a product and defines its sink. The end consumer in terms of the RFID-aided supply chain performs no additional tasks. In case of recall actions or warranty services, details about the product's path in the supply chain can be used to detect further cases. However, due to customer privacy concerns, we propose to deactivate tags after the end consumer passes the point of sales [Schapranow et al. (2009)].

3.5 Role E: Service provider

The service provider performs specific tasks not performed by other supply chain roles. It is responsible for counterfeit detection, e.g. by performing plausibility checks on the virtual product history with the help of the EPC of a certain product.

The service provider is provided by a trusted third party and can be contacted by any supply chain participant. It needs to contact the distributed discovery service to identify supply chain parties involved in handling a certain item [Müller et al. (2010)]. This requires a set of subsequent queries to retrieve event data from event repositories of involved parties. The service provider retrieves all event data via the discovery service and aggregates them to materialize the virtual product history [Schapranow, Müller, Zeier & Plattner (2010)].

In case of counterfeit detection the service provider returns the value $counterfeit$ and the product is removed from the supply chain for further investigations.

If the virtual product history is evaluated to be valid *authentic* is returned. If the outcome of the counterfeit detection cannot be derived automatically, e.g. in case of network partitioning or temporary failures, *unknown* will be returned to indicate the need for manual processing.

We define a function $service_{counterfeit}$ in Equation 1 performing checks with the help of a given epc_i. It returns either one of the results authentic a, counterfeit c, or unknown u.

$$service_{counterfeit} : epc_i \mapsto (epc_i, \{a, c, u\}) \tag{1}$$

3.6 Role F: Logistics provider

Logistics providers are responsible for transportation of handling units, i.e. moving them from a location to another location. On the route various intermediate locations are passed while the transportation is performed in a certain transportation time.

The logistics provider exposes details for transported goods, which involves capturing events at the start and end location of the transport. If the logistics provider additionally exposes details about intermediate locations, we refer to it as a logistics provider with real-time tracking capabilities [Zeier et al. (2008)]. A logistics provider in context of an RFID-aided supply chain performs the following tasks.

1. Capture all events at intermediate locations characterizing the path of a good in this part of the supply chain.

4. Quantitative analysis of EPC networks

After having discussed qualitative requirements in the prior section, we focus on quantitative considerations for supply chains based on RFID technology in the following. We present in detail the service provider for anti-counterfeiting and its impact on operative costs. A dedicated service provider performing anti-counterfeiting checks is anticipated. It provides a unified way for each supply chain participant to check authenticity of pharmaceutical goods based on their EPCs stored on RFID tags. From our perspective, an independent party should provide this service, which is not part of the pharmaceutical supply chain in order to guarantee trust for all participants. The operation of the service provider implies additional costs, i.e. surcharges for handled pharmaceuticals have to be considered. We present a model to identify costs by involving the amount of data transferred via the communication network.

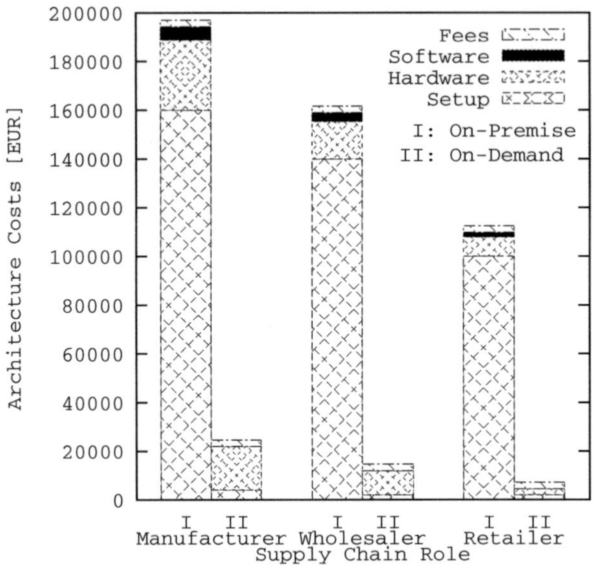

Fig. 3. Comparison of Costs for RFID-enablement per Supply Chain Role

Additionally, we compare the two operating models on-premise and on-demand for required RFID infrastructure components within participating companies.

4.1 Cost drivers for RFID-enablement in companies

For RFID-enablement of companies an initial monetary investment is required depending on the company's role within the supply chain. For example, a manufacturer requires both RFID reading and writing devices being capable to initialize RFID tags when new products are produced. In contrast, a retailer only needs to be equipped with RFID reading devices. Detailed results of our research for concrete costs are given in Sect. 6.

The initial investments for RFID-enablement are visualized in Fig. 3. It highlights the potential cost savings during enablement phase by using an on-demand operating model due to the reduced setup and implementation costs. In addition, it shows that costs for hardware components remain almost constant since this is required on-site equipment, e.g. RFID reader and writer devices. Tab. 1 contains the detailed criteria for comparison of investments for an on-premise solution with investments for a comparable on-demand solution. We divide costs accordingly to individual supply chain roles and categorize them using the following criteria [Schapranow, Nagora & Zeier (2010)].

- **Hardware** describes required investments associated with infrastructure components for establishing an RFID-aided supply chain, e.g. servers, RFID writing and reading devices, network components, etc.
- **Software** describes required investments for software operating the hardware, especially required licenses.
- **EPC Fees** describes required investments to operate as provider for certain EPC intervals, e.g. license fees for GS1 [GS1 Germany GmbH (2010)].

	Costs [EUR]	A: Manufacturer I	A: Manufacturer II	B: Wholesaler I	B: Wholesaler II	C: Retailer I	C: Retailer II
Hardware		28,906	17,988	15,339	9,880	7,929	2,470
RFID writers	3,526	3x		-		-	
RFID readers	913	6x		8x		2x	
Antennas	161	12x		16x		4x	
Workstations	3,261	2x	-	1x	-	1x	-
Servers	1,898	2x	-	1x	-	1x	-
Routers	300	2x	-	1x	-	1x	-
Software	908	6x	-	4x	-	2x	-
EPC Fees	2,650	1x		1x		1x	
Implementation	400	400x	10x	350x	5x	250x	5x
Total [EUR]		197,004	24,638	161,621	14,530	112,395	7,120

Table 1. Cost Distribution per Supply Chain Role: On-premise (I), On-demand (II)

- **Implementation** describes required investments for setting up the RFID infrastructure, e.g. costs for consulting, configuration of software and hardware respectively, implementation tasks, etc.

4.2 Role A: Manufacturer

Tab. 1 shows, that implementation costs contribute by approx. 80 percent to the total costs for supply chain role manufacturer, followed by hard- and software costs with approx. 15 percent. Applying a Software-as-a-Service (SaaS) solution results in reduction of costs for hard- and software components, such as workstations, servers, routers, and special software licenses, which have no longer to be paid by the manufacturer. Furthermore, the implementation effort for an on-demand solution is reduced since the configuration of existing hardware devices with the manufacturer, e.g. RFID reading and writing devices, is only required on-site. Although these on-site devices are required in a SaaS solution to scan items or write tags, they are reconfigured in a SaaS solution to transmit all incoming data directly to the on-demand solution hosted in the provider's cloud and to receive data from it.
Ultimately, this reduces initial investments for a SaaS solution by approx. 87 percent compared to an on-premise solution for the supply chain role manufacturer. Nevertheless, we expect that the SaaS approach to be uninteresting for manufacturers, because of related monthly rates for the on-demand solution. From our perspective, especially the manufacturer will benefit from an on-premise solution, because of the bulk amount of manufactured products per year, which need to be processed. Besides, the manufacturer typically owns a complex IT infrastructure consisting of enterprise applications, which have to be operated independently from participating in an RFID-aided supply chain. Its IT landscape consists of various enterprise systems, such as enterprise resource planning or customer relationship management systems, which are already administered by trained personnel. Thus, an initial investment with lower monthly fees will be more attractive for manufacturers.

4.3 Role B: Wholesaler

The effort of implementing RFID technology at the wholesaler's site when applying an SaaS solution equals less than 1.5 percent of the implementation costs required for an on-premise

solution as given in Tab. 1. By eliminating the need for huge on-site hardware investments in combination with the lower required administration effort, a SaaS solution helps to save more than 90 percent of the initial investment for the supply chain role wholesaler.
We believe, wholesalers will adopt a SaaS solution, because they primarily belong to the category of SMEs that these business models address. We expect the savings for initial investments also to be reflected by monthly saving, since the ratio of manufacturers and wholesalers in the European pharmaceutical supply chain is approx. 1:25. In other words, there are 25-times more wholesalers than manufacturers, which also reflect the amount of handled items.

4.4 Role C: Retailer

Comparable saving potentials exist for the supply chain role retailer. Approx. 93 percent of the implementation costs required for an on-premise solution can be saved when using an on-demand solution as shown in Tab. 1. Adding the savings introduced by eliminating investments for local hardware and the reduced on-site implementation effort, the total saving of initial investments are approx. 93 percent for the supply chain role retailer. This supply chain role belongs especially to the SMEs within the pharmaceutical supply chain, which are addressed by a SaaS solution. From our perspective, we expect monthly savings for a SaaS solution to be comparable to the savings for the initial investments, since the ratio of wholesalers and retailers in the European pharmaceutical supply chain is approx. 1:3.

4.5 Cost evaluation

Leveraging on-demand solutions reduces required implementation costs of a comprehensive and expensive on-premise solution. We compared the setup costs per supply chain role within the pharmaceutical supply chain between an on-premise and an on-demand solution. Independent from the role within the supply chain, costs savings for the initial investments of 80 percent and more can be achieved when applying an on-demand solution. Although the operation of an on-demand solution will be connected with monthly operational fees, we believe that the presented reduction for initial investments are the key enabler to increase the acceptance of RFID technology and supports SMEs to participate in RFID-aided supply chains without huge financial hurdles.

4.6 Amortization period

In the following, we derive required product surcharges to redeem initial investments for RFID-enablement in the European pharmaceutical supply chain. Let $p = 30$ billion products describe the annual manufacturing rate of pharmaceuticals available on-prescription, r describe the supply chain role, and x_r as defined in Equation 2. We assumed $a = 5$ years to describe the amortization period for all initial investments s_r.

$$x_r = \frac{s_r \cdot |r|}{a \cdot p} \tag{2}$$

Tab. 2 compares the required surcharges per product and role for an on-demand and an on-premise setup [Schapranow, Nagora & Zeier (2010)].
Based on the configuration of the supply chain as given in Sect. 4 the following total costs arise. Applying a SaaS solution for all roles of the pharmaceutical supply chain will result in

Supply Chain Role	$\|r\|$	On-demand x_r [EUR]	On-premise x_r [EUR]
Manufacturer	2,200	0.0004	0.0029
Wholesaler	50,000	0.0048	0.0539
Retailer	140,000	0.0066	0.1049

Table 2. Product Surcharges per Supply Chain Role for Amortization, a=5 years

a very low surcharge per item of

$$0.0004\ EUR + 2 * 0.0048\ EUR + 0.0066\ EUR = 0.0166\ EUR.$$

In comparison, an RFID-aided supply chain built on a purely on-premise solution requires a surcharge per item of

$$0.0029\ EUR + 2 * 0.0539\ EUR + 0.1049\ EUR = 0.2156\ EUR.$$

We expect to implement a combined solution of the given examples. A supply chain configuration consisting of manufacturers applying an on-premise solution and wholesalers as well as retailers accompanying an on-demand solution, the surcharge per item is given by

$$0.0029\ EUR + 2 * 0.0048\ EUR + 0.0066\ EUR = 0.0191\ EUR.$$

By applying this combined supply chain configuration it is possible to reduce the surcharge per item to less than 10 percent of the purely on-premise costs. Assuming an average pharmaceutical product price of 7.13 EUR. The expected surcharge per item for on-demand and the combined configuration are of 0.02 EUR resp. 0.22 EUR for on-premise, which equals 2.7 permille resp. 3.1 percent of the initial product price [European Commission (2008); Schapranow, Nagora & Zeier (2010)]. In all cases, the surcharge remains below our assumed empirical threshold of approx. 10 percent of the product's retail price as stated in Sect. 2.
The given surcharges are required to amortize the initial investment for RFID-enablement only. Regular costs, such as operational costs, maintenance costs for RFID devices, cost for RFID tags, monthly fees for subscription in an on-demand solution, etc. need to be added individually since they are not part of the given calculations.

5. Conclusions and outlook

In the given work, we considered RFID technology as the key-enabler for an integer and counterfeit-resistant pharmaceutical supply chain [Zeier et al. (2009)]. The pharmaceutical industry draws the motivation for our research activities due to the increasing number of detected pharmaceutical counterfeits within industry countries. We draw our business considerations for RFID-enablement of participants in an integer pharmaceutical supply chain. We shared our qualitative analysis of EPC networks architectures and compared operative factors. Based on our analysis, we derived costs for operating a dedicated service provider for anti-counterfeiting within an RFID-aided supply chain and compared possible models to operate this instance.
We expect that the acceptance of RFID technology depends on costs for RFID-enablement and its business advantages. For the given pharmaceutical case study, we expect RFID technology

to support authentic pharmaceuticals and automatic anti-counterfeiting by evaluating a good's product history. Ultimately, we compared required costs for RFID-enablement in an on-premise setup with an on-demand setup and derived per product surcharges for amortization of anti-counterfeiting in RFID-aided supply chains. The outcome of our research activities clearly depicts that initial investments for RFID enablement do no contribute to major product surcharges.
Our future research activities will focus on payment models for operation of the service provider for anti-counterfeiting. We will analyze the following payment models:

1. General post-payment models, e.g. once a month for large wholesalers,
2. Individual payment models, e.g. per anti-counterfeiting check for small wholesalers, or
3. Pre-payment models, e.g. for retailers when a predefined balance on an account can be used for checks.

6. Appendix A: Component Costs

Tab. 3 contains selected RFID components for RFID-enablement of a pharmaceutical company. We selected these components for pricing assumptions[1]. The given assumption can also be used for further industries. However, components may vary individually for specific industries and setups, which result in different costs per component and/or total costs.

[1] We assume USD 1.4184 = 1.0000 EUR

Component	Article	Costs [EUR]
Reader Equipment		
Device	Alien 9800 [RFIDSupplyChain.com LLC (n.d.b)]	913.00
Antenna	Alien 915 MHz Circular Antenna [RFIDSupplyChain.com LLC (n.d.a)]	101.00
Cable	Alien ALX-408 Extension Cable [RFIDSupplyChain.com LLC (n.d.d)]	44.00
Holder	Alien ALX-407 Mounting Bracket [RFIDSupplyChain.com LLC (n.d.c)]	16.00
Printer	Zebra R110Xi [IDAutomation.com Inc. (n.d.)]	3,526.00
Middleware		
Workstation	HP Workstation xw9400 [Hewlett Packard (n.d.)]	3,261.00
Software	IBM Websphere RFID Premise Server [IBM Corporation (n.d.)]	908.00
Internet Access		
Server	HP ProLiant DL380 G5 [macle GmbH (n.d.)]	1,898.00
Router		200.00
Network Cable		100.00
EPCIS		
Fosstrak	[Fosstrak (2009)]	open-source
Internet Access		2,298.00
EPC Fee	[GS1 Germany GmbH (2010)]	2,650.00
ONS		
Internet Access		2,298.00
Verification Server		
Internet Access		2,298.00
Tag	Thin Propeller Label[TAGnology RFID GmbH (n.d.)]	0.37
Consulting	Man-Day	400.00

Table 3. Costs Per RFID Component

7. References

Barthwell, A. G., Barnes, M. C., Leopold, V. R. & Wichelecki, J. L. (2009). National Survey on Drug Use and Health, *Technical report*, Center for Lawful Access and Abuse Deterrence.

Bos, J. V. D. (2009). Globalization of the Pharmaceutical Supply Chain: What are the Risks? – The FDA's Difficult Task, *Society of Actuaries*, pp. 23–26.

Bovenschulte, M., Gabriel, P., Gaßner, K. & Seidel, U. (2007). RFID: Prospectives for Germany – The State of Radio Frequency Identification-based applications and their Outlook in National and Internat. Markets.

Bundesverband Informationswirtschaft, Telekommunikation und neue Medien (2005). White Paper RFID Technologie, Systeme und Anwendungen, http://www.bitkom.org/files/documents/White_Paper_RFID_deutsch_11.08.2005__final.pdf[2].

Choi, S. & Poon, C. (2008). An RFID-based Anti-counterfeiting System, *International Journal of Computer Science* 35(1).

EPCglobal Inc. (2007). EPCIS Standard 1.0.1, http://www.gs1.org/gsmp/kc/epcglobal/epcis/epcis_1_0_1-standard-20070921.pdf[2].

EPCglobal Inc. (2010). Tag Data Standard 1.5, http://www.gs1.org/sites/default/files/docs/tds/tds_1_5-standard-20100818.pdf[2].

European Commission (2008). Pharmaceutical Sector Inquiry Preliminary Report.

European Commission Taxation and Customs Union (2009). Report on EU Customs Enforcement of IP Rights, http://ec.europa.eu/taxation_customs/resources/documents/customs/customs_controls/counterfeit_piracy/statistics/2009_statistics_for_2008_full_report_en.pdf[2].

Food and Drug Administration (2004). Counterfeit Drug Task Force Report.

Food and Drug Administration (2005). Counterfeit Drug Task Force Report.

Fosstrak (2009). Project License, http://www.fosstrak.org/epcis/license.html[2].

GS1 Germany GmbH (2010). Preise für die Nutzung des Leistungspaketes GS1 Complete, http://www.gs1-germany.de/service/gs1_complete/preisliste/index_ger.html[2].

Hewlett Packard (n.d.). HP Workstation xw9400, http://h20195.www2.hp.com/V2/GetDocument.aspx?docname=4AA0-4798EEE&doclang=EN_GB[2].

IBM Corporation (n.d.). Software Pricing, https://www-112.ibm.com/software/howtobuy/buyingtools/paexpress/Express?P0=E1&part_number=D0A4DLL,D0A4QLL,D0A4WLL,D0A4YLL,D0A5LLL,D0A5NLL,&catalogLocale=de_DE&Locale=de_DE&country=DEU&brand=ws&PT=html[2].

ICC Policy Statement (2004). The Fight against Piracy and Counterfeiting of Intellectual Property, http://www.iccwbo.org/home/intellectual_property/fight_against_piracy.pdf[2].

IDAutomation.com Inc. (n.d.). Zebra R110Xi RFID Printer Encoder, http://www.idautomation.com/rfid/Zebra-RFID-Printer.html[2].

IP Crime Group (2008). IP Crime Report, http://www.ipo.gov.uk/ipcreport08.pdf[2].

Jenkins, J., Mills, P., Maidment, R. & Profit, M. (2007). Pharma Traceability Business Case Report.

Knöpfel, A., Gröne, B. & Tabeling, P. (2005). *Fundamental Modeling Concepts. Effective Communication of IT Systems*, John Wiley.

macle GmbH (n.d.). HP Proliant DL380 R05, `http://www.macle.de/hp.html`[2].

Merck & Co. Inc. (2007). Settlement agreement, `http://www.merck.com/newsroom/vioxx/pdf/Settlement_Agreement.pdf`[2].

Müller, J., Faust, M., Schwalb, D., Schapranow, M.-P., Zeier, A. & Plattner, H. (2009). A Software as a Service RFID Middleware for Small and Medium-sized Enterprises, *Proceedings of the 5th European Workshop on RFID Systems and Technologies*, VDE.

Müller, J., Oberst, J., Wehrmeyer, S., Witt, J. & Zeier, A. (2010). An Aggregating Discovery Service for the EPCglobal Network, *Proceedings of the 43th Hawai'i Conference on System Sciences*, Koloa, Hawaii, USA.

Müller, J., Pöpke, C., Urbat, M., Zeier, A. & Plattner, H. (2009). A Simulation of the Pharmaceutical Supply Chain to Provide Realistic Test Data, *Proceedings of 1st International Conference on Advances in System Simulation*, IEEE.

Müller, J., Schapranow, M.-P., Helmich, M., Enderlein, S. & Zeier, A. (2009). RFID Middleware as a Service – Enabling Small and Medium-sized Enterprises to Participate in the EPC Network, *Proceedings of the 16th International Conference on Industry Engineering and Engineering Management*, Vol. 2, pp. 2040–2043.

RFIDSupplyChain.com LLC (n.d.a). Alien 915 MHz Circular Antenna (ALR-9611-CR), `http://www.rfidsupplychain.com/-strse-13/Alien-915-MHz-Circular/Detail.bok`[2].

RFIDSupplyChain.com LLC (n.d.b). Alien 9800 EPC Multiprotocol RFID Fixed Reader, `http://www.rfidsupplychain.com/-strse-98/Alien-9800-EPC-Multiprotocol/Detail.bok`[2].

RFIDSupplyChain.com LLC (n.d.c). Alien ALX-407 Antenna Mounting Bracket, `http://www.rfidsupplychain.com/-strse-195/Alien-ALX-dsh-407-Antenna-Mounting/Detail.bok`[2].

RFIDSupplyChain.com LLC (n.d.d). Alien ALX-408 Antenna Extension Cable, `http://www.rfidsupplychain.com/-strse-196/Alien-ALX-dsh-408-Antenna-Extension/Detail.bok`[2].

Schapranow, M.-P., Müller, J., Zeier, A. & Plattner, H. (2009). Security Aspects in Vulnerable RFID-Aided Supply Chains, *Proceedings of 5th European Workshop on RFID Systems and Technologies*, VDE.

Schapranow, M.-P., Müller, J., Zeier, A. & Plattner, H. (2010). RFID Event Data Processing: An Architecture for Storing and Searching, *Proceedings of the 4th International Workshop on RFID Technology - Concepts, Applications, Challenges*.

Schapranow, M.-P., Nagora, M. & Zeier, A. (2010). CoMoSeR: Cost Model for Security-Enhanced RFID-Aided Supply Chains, *Proceedings of the 18th International Conference on Software Telecommunications and Computer Networks*, IEEE.

Schapranow, M.-P., Zeier, A. & Plattner, H. (2010). A Dynamic Mutual RFID Authentication Model Preventing Unauthorized Third Party Access, *Proceedings of the 4th International Conference on Network and System Security*.

Schlitter, N., Kähne, F., Schilz, S. T. & Mattke, H. (2007). Potentials and Problems of RFID-based Cooperations in Supply Chains, *Innovative Logistics Management: Competitive Advantages through new Processes and Services*, Erich Schmidt Verlag GmbH & Co., Berlin, pp. 147–164.

Shukla, N. & Sangal, T. (2009). Generic Drug Industry in India: The Counterfeit Spin, *Journal of Intellectual Property Rights* 14: 236–240.

Staake, T., Thiesse, F. & Fleisch, E. (2005). Extending the EPC Network: The Potential of RFID in Anti-Counterfeiting, *Proceedings of the ACM Symposium on Applied Computing*, ACM, New York, NY, USA, pp. 1607–1612.

Stiehler, A. & Wichmann, T. (2005). RFID im Pharma- und Gesundheitssektor. Vision und Realität RFID-basierter Netzwerke für Medikamente, Berlecon Report.

TAGnology RFID GmbH (n.d.). Impinj Thin Propeller Label 3.875" x 0.5", `http://www.rfid-webshop.com/product_info.php/info/p482_Impinj-Thin-Propeller-Label.html`[2].

U.S. Pharmaceuticals Pfizer Inc. (2006). Anti-Counterfeit Drug Initiative Workshop and Vendor Display, `http://www.fda.gov/OHRMS/DOCKETS/dockets/05n0510/05N-0510-EC21-Attach-1.pdf`[2].

White, G. R., Gardiner, G., Prabhakar, G. & Razak, A. A. (2007). A Comparison of Barcode and RFID Technologies in Practice, *Journal of Information, Information Technology, and Organizations* 2.

World Health Organization (2009). Warning on purchase of antivirals without a prescription, including via the Internet, `http://www.who.int/medicines/publications/drugalerts/Alert_122_Antivirals.pdf`[2].

Zeier, A., Hofmann, P., Krüger, J., Müller, J. & Schapranow, M.-P. (2009). Integration of RFID Technology is a Key Enabler for Demand-Driven Supply Network, *The IUP Journal of Supply Chain Management* 6(3, 4): 57–74.

Zeier, A., Knolmayer, G., Mertens, P. & Dickersbach, J. (2008). *Supply Chain Management Based on SAP Systems*, Springer.

[2] All online references were checked on Apr. 28, 2011.

14

Security Control and Privacy Preservation in RFID enabled Wine Supply Chain

Manmeet Mahinderjit-Singh[1], Xue Li[1] and Zhanhuai Li[2]
[1]*The University of Queensland,*
[2]*Northwest Polytechnical University of China,*
[1]*Australia,*
[2]*China*

1. Introduction

Modern identification procedures such as radio frequency identification (RFID) are able to provide transparency in applications including supply chain, logistics and equipment management. The benefits of visibility and fast identification provided by RFID technology especially in supply chain management (SCM) reduce the risk of counterfeiting (Gao et al., 2004). There are two mainly ways in which RFID technology supports a visible and fast identification processes: 1) RFID allows for new, automated and secure ways to efficiently authenticate physical items; and 2) As many companies invest in networked RFID technology for varying supply chain applications, the item-level data can be gathered in any case (Lehtonen et al., 2006).

Despite these benefits, RFID technology is still not widely implemented. The main reasons are, firstly, the difficult are, firstly, the difficult technical aspects of implementation resulting in high setup costs, secondly, growing security and privacy concerns. Our focus in this paper is to discuss the second reason for the low take-up of RFID technology, that is, security and privacy concerns. We argue that without applying maximum security and privacy, trustworthiness between supply chain partners will be minimal. As a result, the effectiveness and collaboration of traditional supply chain environment with RFID technology cannot be achieved. Given that humans cannot read the RFID tags on items and the tags themselves maintain no history of past readings, the challenge of security and privacy in this technology is related to the nature of RFID tags and their functionality (Juels, 2005). A retailer inventory that is labeled with unprotected tags may be monitored and read by unauthorised readers. The inventory data holds significant financial value for commercial organisations and their competitors. Once data has been accessed by unauthorised users, it can be cloned on empty tags, giving rise to the counterfeiting issue. Counterfeiting in the form of cloned or fraudulent RFID tags is the consequence of a lack of security measures and trustworthiness among the supply chain partners when RFID technology is used to automate their business transactions.

Privacy violations stem from the fact that when goods are tagged, the manufacturers, retailers and consumers will be able to track the goods beyond point-of-sale (POS) because they have associated data. Even if the tags only contain product codes rather than unique serial numbers, a consumer's taste in brands "constellation" can betray their identity.

Moreover, even if the responses of the tags are encrypted, the owner can also be identified and tracked by the fixed encrypted code. While consumers fear omnipresent surveillance, organisations are primarily interested in protecting company-internal data from unauthorised access and potential manipulation. These problems are not, however, completely independent of one another, considering the fact that data security represents a prerequisite to guarantee data privacy.

Bottled wine counterfeiting is a multi-billion dollar industry, which has increased drastically since the early 1990s. A report produced by Australian IT (Mar, 2007) shows that counterfeit wines accounted for almost 10 percent of the global market. In terms of wines, most counterfeiters aim to counterfeit expensive wines by tampering the labels or marking of the bottles. Recently, the ability of RFID to identify, authenticate and track items and activities in the supply chain is seen as a possible solution to the counterfeiting damage occurring in the wine industry. RFID has been used in the wine supply chain to provide visibility at each step of the supply chain process and to provide unique identification for tracking not only lots and cases but also at the item-level. An example of RFID usage in the wine industry can be explored in the real-life business scenario of Domenitz.L and Kravitz.J (2008).

Even though RFID is seen as an anti-counterfeiting tool, the use of passive RFID tags is a significant problem for industry including the wine industry. The low-cost passive tags currently used in the wine industry may not be able to provide sufficient security compared to active tags. Passive tags have lesser storage and memory space and provide insufficient security against security threats such as RFID tag cloning and fraud which lead to counterfeiting. For example, the tags used by Domenitz.L and Kravitz.J (2008) for tracking purposes can be easily cloned and all the historical information can be stolen. If this occurs, a fraudulent batch of wines produced with similar historical data can hit the market without anyone noticing the lack of authenticity of the products.

In order to address wine bottle counterfeiting, three different modules need to be incorporated together. These three modules are: 1) prevention; 2) detection; and 3) privacy. Prevention techniques focus on preventing a clone or fraud attack. Meanwhile, detection techniques are used to notify or record an attack in progress. Both modules are equally important and are not interchangeable since we cannot prevent what we cannot detect. The prevention module aims to provide security to all the layers concentrating mostly on the application, communication and the physical layers. Prevention of cloning involves the design and development of tags from the physical layer up to application level. Since intrusion prevention systems (IPS) are able to prevent attacks in real-time and manage the supply chain functionality through traffic flows, each part of the RFID system such as the tags, readers, communication channel, middleware and database are able to be protected. We will propose a simple yet powerful method to prevent counterfeiting in a supply chain plant. The aim here is to shield the whole supply chain plant from security attacks such as skimming, eavesdropping and replay attacks from occurring in the first place.

In contrast to prevention, the detection module focuses only on the application level. An intrusion detection function can tackle a compromised system more precisely since the knowledge of how and what has attacked the system is more clear compared to a prevention system. Prevention techniques are not guaranteed and may let an attack through, but dealing with a compromised system by responding to suspicious behavior and generating an alarm is possible with a detection system. However, the issue we tackle here is beyond the effort to minimise the error rate: the aim is to improve the percentage of the incorrect prediction of class labels and to deliver higher detection accuracy. In real-world applications, cost is treated unequally and the misclassification cost can be significant. We argue that a cost sensitive approach is essential in reducing the risk of counterfeiting in a supply chain. For example, in medical diagnosis of a cancer disease, if the cancer is regarded as the positive class, and non-cancer (healthy) is regarded as the negative class, then missing a cancer (the patient is actually positive but is classified as negative), is called a "false negative" and is much more serious (thus expensive) than the false-positive error. The patient could lose his or her life because of the delay in the correct diagnosis and treatment. Similarly, in RFID clone and fraud detection, a false negative or failure to detect fraudulent tags could be very expensive with counterfeit items reaching the market and causing millions of dollars of loss. In this chapter we aim to construct cost model detection using supervised learners from available tools such as WEKA (Hall.M and Frank.E *et.al*, 2009). The objective of our study is to classify RFID tags using supervised learners to categorise RFID tags and detect the genuine (good) and fraudulent (bad) tags. Our RFID tag clone and fraud detector will employ RFID SCM tracking and tracing functions such as tag history attributes, event timestamp and time to live (TTL) (Li *et al.*,2009) as important factors. We believe this simple experiment using a cost sensitive detection method for RFID tags in a supply chain environment is the first of its kind.

Finally, the privacy module is useful to support the handling of security attacks such as cloning and fraud attacks. This is because tracking RFID tags is an essential step in cloning yet may compromise a partner's privacy (Mahinderjit-Singh & Li, 2009). As for certain applications which require tracking, such as supply chains and drug pedigree tracing, privacy is a sensitive issue since the tracing and tracking processes may violate privacy in the first place. Thus, ensuring privacy protection while dealing with cloning attacks is crucial. Our third objective will be to provide a comprehensive guideline in tackling privacy concerns in the counterfeiting issue.

This chapter is directed towards a problem-driven context. Counterfeiting in an RFID based-wine supply chain is considered as a problem. We will tackle the counterfeiting issue in this supply chain example by using three different modules, namely, prevention, detection and privacy. Our first contribution is to propose a prevention method which is simple yet affordable to be implemented in a supply chain environment. The second is to detect counterfeit tags attached to wine bottles by utilising the cost sensitive concept. Finally, we provide a comprehensive privacy guideline handling the counterfeiting issue in a supply chain plant.

The main significance of this paper is to demonstrate how privacy preservation and security protection through prevention and detection can be maintained in an open-loop RFID supply chain such as the wine industry. In addition, a complete methodology on the most optimal and easy to use technique, approach or guideline in dealing with counterfeiting is presented. This research will closely study the relationship between these three modules in an RFID-enabled supply chain. The information on handling the supply chain in the wine industry can be extended to other goods or other RFID applications. The solution will be supported by using our seven-layer trust framework. The rest of this chapter is constructed as follows. Section 2 gives a literature review on RFID security and privacy issues in the supply chain. It also demonstrates the proposed trust framework. Section 3 explains the RFID-based wine supply chain. In Section 4 we outline how all three modules of prevention, detection and privacy can be employed to tackle counterfeiting in the wine industry. Section 5 provides a discussion. Section 6 provides the conclusion and views on future work.

2. RFID security and privacy in supply chain system

In this section, we present a taxonomy of the security and privacy issues of RFID-enabled supply chain management. Firstly, we discuss the challenges and problems related to security in an RFID system. The discussions in this section are essential in understanding why the trust mechanism is important in addressing security and privacy issues in RFID.

Lack of standardisation among different manufacturers of tags and readers makes it harder for a sharable security mechanism to operate in an open system environment. The network issues include the insecure communication between tags and readers. The attacker is able to remove the tag from the product and the lack of sufficient pedigree security makes it much easier for an authentic product to be forged. In addition, the lack of communication bandwidth and management introduces the problem of key management in ubiquitous computing (Juels, 2005). The architecture deployment in a supply chain environment, which includes the position of tags and alignment of readers in a centralised server, could cause erroneous readings such as duplicate records in the system and a reduction in accuracy. In addition, the RFID tag scalability issue in the supply chain environment needs attention. The growth of tag and reader size over time according to the needs of the supply chain business shows the importance of designing an architecture that is able to cope with future advancement.

On the other hand, the simple middleware design currently used by the Electronic Product Code (EPC) global network (http://www.epcglobalinc.org) does not take into account the evolving RFID technology and meet the business owner's requirements. There is no dedicated middleware component for ensuring security needs such as authenticating (Lehtonen, Michahellas & Fleisch, 2007). These are among the issues concerning RFID in an open system environment as they affect the security and privacy of the data, which is the information on the tags linked to the enterprise database. This causes data inconsistency and leakage. In light of all these issues, the impact on human trust in the RFID technology is critical and contributes to the lack of data sharing mechanisms in SCM. The next sub-section will examine the RFID security taxonomy directed towards RFID security attacks.

2.1 Taxonomy of RFID security attacks in the supply chain

The taxonomy of RFID security attacks in the supply chain is based on three security mechanisms - authentication, authorisation and trust services. This section is structured as a discussion on each of these mechanisms.

A) Authentication

In a supply chain environment, there are several methods available to combat product counterfeiting. These methods include the use of electronic pedigree (Koh, 2003), serialisation (Johnston, 2007), product authenticity and RFID tag authentication. The electronic pedigree used for drug authenticity can trace and track items by checking and updating transactions in a sequence. This method will be invalid if any party, for instance, a retailer acting deliberately, does not update the database whenever a product leaves the store at the point-of-sale. Non-cryptography techniques that are simple and cost effective (Koh, 2003; Nochta, 2005; Johnston, 2007) were proposed to combat counterfeiting. But certain techniques such as the trace and track approach proposed by Koh (2003) do not solve cloning problems thoroughly due to the storage of the tag information in plaintext. However, the track and trace approaches which deal with the locality factor could bring additional information on when and where cloning would have taken place place

(Lehtonen, 2007). Other approaches are proxy-based, such as the RFID Authentication Processing Framework (APF) (Ayoade et al., 2005) and Certificate Authority (CA). Both these techniques can reduce the counterfeiting issue caused by an unauthorised reader by authenticating the reader and hindering the ability of a fake reader to access the tags. The CA functions in a similar but more systematic way by storing a list of good readers on the centralised server. Finally, techniques such as watermarking (Potdar & Chang, 2006) do provide some degree of protection against cloning. However, the protocols are not adequate for low-cost EPC tags and require higher numbers of bits to ensure ultimate security.

B) Authorisation

In order to realise the business benefits of RFID, trading partners must be able to exchange data. The manufacturer, distributor and retailer all share RFID data. Access control involves the process of determining whether a user can perform a specific operation on resources. Based on the policy introduced by Wang et al. (2008), we argue that RFID system attacks within a supply chain can be eliminated when policies are assigned at product-level and item-level. In addition, a Discovery Service (DS) which is still under development can be utilised as another registry where incoming and outgoing products are registered (Ranasinghe & Cole, 2007) and can function as an item-level tagging server. Another important point is using role-based policy for the RFID access control systems (Seong et al., 2006). Languages such as SAML, XML, XACML and even WS-Security are suitable and widely used in RFID especially in supply chain services. In addition, the concept of e-pedigree (Frey.M , 2008) in the pharmaceutical industry proves that the sharing and tracking of information within an EPC network is able to provide accuracy and eliminate security threats. Another approach to sharing and exchanging information in RFID is using the Electronic Product Code Information Services (EPC-IS) model (Ranasinghe & Cole, 2007) which is a component of the EPC global network. The method for exchanging EPC-IS events uses protected communication channels based on HTTPS and SSL. EPC-IS enhances data sharing and visibility and monitoring day-to-day RFID applications. Each local company will have its own local database and local EPC-IS.

C) RFID Trust Service

The importance of the trust role in dealing with the security threats in RFID is a novel approach and is the first of its kind. Trust in the adoption of RFID among business partners is likely to be affected by two main challenges in RFID technology. First, the security and privacy threats in the system reduce the confidence in the system especially when RFID tagging is used for counterfeiting purposes. Second, the lack of any attack detection model in the RFID network makes the security and privacy threats go unnoticed.

Among the other reasons which contribute to the decrease of trust in RFID are:

i. Open system environment - Supply chain management exists in an open system environment with various types of RFID system interface, organisational protocols and communication interface (Derakhshan, Orlowska & Li, 2007). As a result, with multiple data integrations models existing, it is harder to develop a standardised and common data exchange and integration model among them.

ii. Minimal authentication, authorisation and tracking services capabilities - RFID middleware only includes the common models for data exchange transactions. Models such as EPC-IS, ONS, EPC-DS only provide common transactions (http://www.epcglobalinc.org). The lack of capabilities in authentication models and tracking mechanisms built in to the RFID middleware supports the point that the design of RFID network infrastructure fails to address the security and privacy challenges. So far, e-pedigree, which is used for drug tracking, is the only model for tracking and tracing the whereabouts of products in the drug supply chain environment (http://www.rfidupdate.com/articles/index.php?id=1277).

iii. The existing EPC Trust Service – The current EPC trust model is the only trust model so far in RFID technology providing authentication and authorisation. EPC-Trust functions by using a third party (CA) in authenticating devices and users in a supply chain model (Verisign Inc, 2004). The trust model does not cover any security mechanism for RFID tags and readers and is without any detection model.

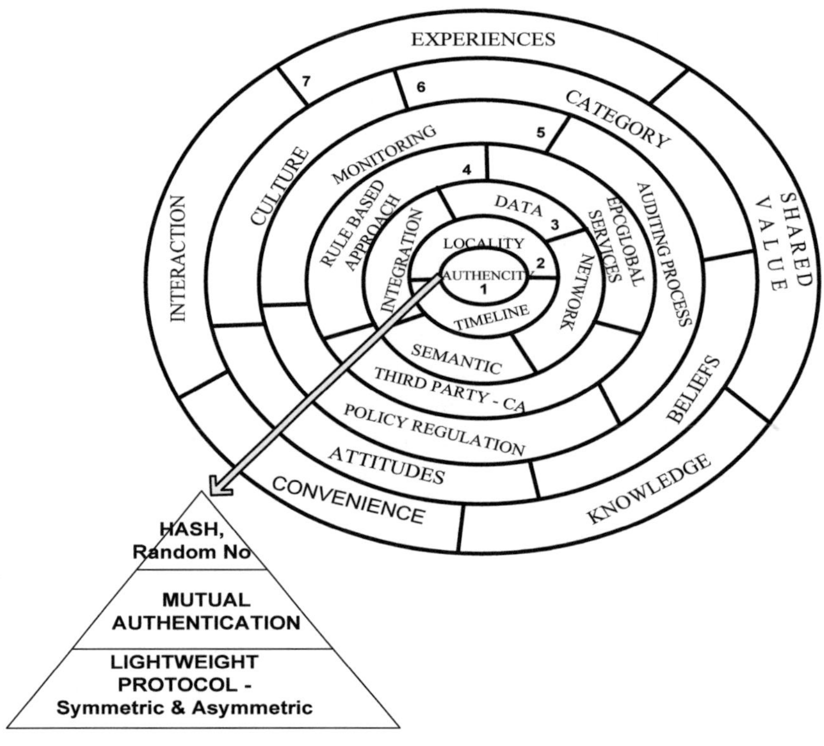

Fig. 1. Seven-layer trust framework (Mahinderjit-Singh and Li ; 2009)

Based on the factors that impact on the confidence rate in RFID adoption and the technology functionality, the base argument for this research is that both prevention and detection models are needed to tackle these issues. A better model of trust is needed especially one with the capability for preventing and detecting security and privacy threats. Further, the trust model must stand in an open system environment by supporting multiple protocols and communication interfaces between various organisations. This trust model could also be assimilated with supply chain service standards such as EDI/XML and SOA. Thus, in this chapter we carefully study the previous literature to determine the gaps in the research before proposing our idea and solutions. This extension of knowledge will be considered as a map for more secure and efficient business transactions in open system supply chain management.

Based on the seven-layer trust framework (Mahinderjit-Singh & Li, 2009), trust in an RFID technology system is defined as a "comprehensive decision making instrument that joins security elements in detecting security threats with preventing attacks through the use of basic and extended security techniques such as cryptography and human interaction with reputation models". In addition, a trust model for a technological system should always include human interaction through the use of a feedback and ranking model. Among the functions of the trust framework (Figure 1) is the provision of guidelines for designing trust to solve open system security threats. The next sub-section focuses on RFID privacy concerns.

2.2 RFID privacy taxonomy

An RFID system should consider both privacy and security in its design structure and the focus of the proposal should be on the information system and not the technology. Privacy is the ability of the RFID system to keep the meaning of the information transmitted between the tag and the reader secure from non-intended recipients. The main privacy challenge in RFID is due to the nature of the RFID tag operation. Tags are "promiscuous": they can be read by entities outside their owner's knowledge. Among the privacy concerns are tracing and tracking, profiling of products and secret tag reading (Ayoade, 2007). Approaches to deal with these concerns include: (i) tag killing (Sarma et al., 1999) in which the tags of sold items are disabled or removed at the point-of-sale; (ii) tag blocking (Juels et al., 2003) in which a blocker tag creates a radio frequency environment that prevents unauthorised scanning of consumer items; (iii) hash encryption (Juels, 2005) in which the information stored in tags is encrypted in a dynamic manner; and (iv) a rewriteable memory and random number approach (Gao et al., 2004) in which only authorised readers are able to access the tags.

In RFID applications such as a supply chain, an RFID tag may change its owner multiple times. To tackle this issue, a secure ownership transfer is essential. Ownership transfer means that once an RFID tag is transferred from two different owners, all information associated with the tag will need to be passed on as well. This should be done without compromising the privacy of either the old or new owner to ensure that tracing and retaining of the tag's information is not possible. Some ownership protocols that tackle ownership transfers are proposed by Osaka et al. (2006), Saito et al. (2005) and Song (2008). The Osaka-Takagi-Yamazaki-Takahashi (OTYT) protocol. (Osaka et al., 2006) uses symmetric encryption and hashing and provides privacy protection for both new and old owners. However, without any consideration of after-sale information recovery, this scheme is also prone to message manipulation attack since similar random numbers could be used to query a tag twice. The Saito protocol (Saito et al., 2005) makes use of properties such as three-way authentication using a TTP server but is prone to eavesdropping and only supports new owner privacy. This is because the fundamental approach of their scheme is to provide support for the backward channel without consideration of forward channel communication. Through security analysis done by Pedro (2010), the proposal by Song (2008) provides three important ownership transfers, which are new owner privacy, old owner privacy and authorisation recovery for transaction after POS. However, the mutual authentication method used is prone to many attacks such as tag and server impersonation, data leakage and denial-of-service attack. As a result, it is difficult to ensure privacy without

compromising security if only symmetric cryptosystem is used without any provisions made in terms of a secure server's communication setup.
Hargraves and Shafer (2004) suggested that identifiability, observe-ability and link-ability of RFID tags with associated data should be minimised and the RFID system should be developed with authorisation, authentication and encryption on a routine basis to ensure trustworthiness of the RFID system. In VeriSign (2008), an innovative way to minimise the sharing of information is by applying distributed network architecture. This type of networked RFID system ensures that partners only store their serialised information about each product in a database and this information is only accessible to authenticated and trusted partners. Another approach will be to apply policies (Garfinkel et al., 2005). Garfinkel et al. (2005) emphasise the need for guidelines which require human and technology intervention and the need to educate humans in accessing RFID technology and facilitate understandings of how privacy threat can be handled.

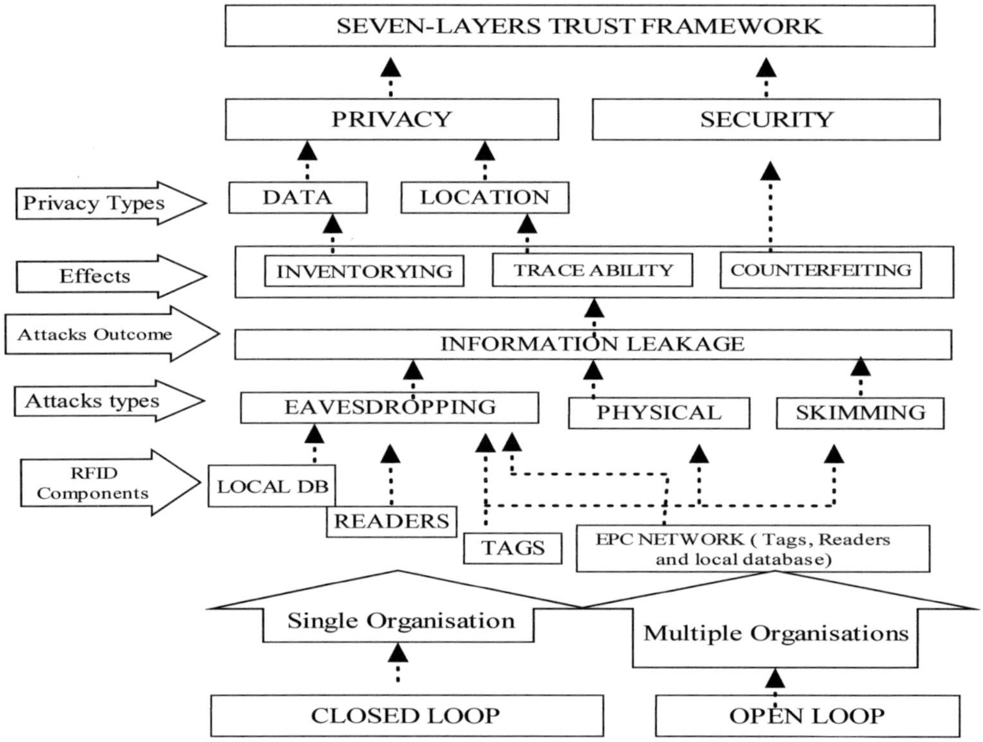

Fig. 2. RFID Privacy Concerns Categorisation

In the seven-layer trust framework (Mahinderjit-Singh & Li, 2009), both security and privacy are integrated in the first 5 layers. The trust framework could be applied to maintain an RFID system which is able to handle security threats without compromising privacy effects. Layer 2 - privacy looks into time and locality factors which are related to the privacy of data and location. Mahinderjit-Singh and Li (2009) argued that the privacy component is necessary to support the handling of cloning attacks because tracking of tags is an essential step towards cloning-detection and this may compromise a partner's privacy. Thus, this layer is to ensure the privacy protection while dealing with cloning attacks. We also believe trust management is the key for the overall protection of security and privacy in an RFID system. In Figure 2, we categorise privacy attacks in RFID within single and multiple organisation loops and show how both privacy and security are a part of any trust model, which in our case is the seven-layer trust framework.

3. An example of RFID SCM in wine industry

In this section, we present an example of the supply chain in the wine industry. This example is important for understanding the degree of the counterfeiting risk in RFID technology. The counterfeiting issue in this example will also be used to design an appropriate solution in terms of preventing counterfeiting, detecting the clone and fraud attacks and preserving the privacy of the users in this supply chain example.

The aim of counterfeiters is to counterfeit expensive wines by tampering with the labels or markings of the bottles. Among the anti-counterfeit techniques are the traditional method of tasting the wines, biochemical methods (http://www.enotes.com/forensic-science/wine-authenticity), and using hologram labels, tamper-proof security seals and smart corks (Sagoff, 2008). However, the easily tampered, unsecured holograms and lack of mechanisms for traceability offered by the above techniques have led to the problem of low visibility, non-authentic and inaccurate transactions for tracing and tracking the movement of wines in a supply chain. Instead of solving the counterfeiting issue, more vulnerability loopholes are presented to the counterfeiter to perform attacks. The challenges of RFID usage in the wine industry are as follows:

i. the identification of liquids
ii. the short lifespan of the passive tag battery currently used for RFID tracking and monitoring
iii. the lack of a preventive mechanism to cope with future counterfeiting once the tamper-proof seal on the wine is tampered with,
iv. the nature and limitation of the passive RFID tags.

The issue of identifying liquid is troublesome for the reason that liquid absorbs and reflects radio waves. The passive RFID tags for identification of the wines at e-Provenance are placed under the bottle and this reduces the read accuracy. According to Yeo (2006), the reading accuracy can be enhanced if the tag is placed on the top of the bottle. In order to be able to track and monitor purchased wine, the tags used for tracking must survive a life span of many years. However, the outcome of the RFID tags used currently is limited and only last for two years. The low-cost passive tags used currently may not be able to provide ultimate security compared to active tags. Passive tags have lesser storage and memory space and have insufficient security against security threats such as RFID tag cloning, fraud attack and counterfeiting. The tags used by e-Provenance (2008) for tracking purposes can easily be cloned and all the historical information can be stolen. A fraudulent batch of wines produced with similar historical data can hit the market without anyone noticing the lack of authenticity of the products.

3.1 RFID tagged wine supply chain management

Based on Report of Wine Traceability (2005), the function of each supply chain business partner in a typical wine production environment are as follows:

a. Wine Producer - The wine producer is responsible for receiving the grapes and for the production, manufacture and/or blending of wine products.

b. Transit / Cellar - The transit cellar is responsible for the receipt, storage, dispatch, processing, sampling and analysis of bulk wine, as well as record keeping of appropriate information about what is received and what is dispatched. The transit cellar can be part of the filler/packer company (geographically separate or not) or can be outsourced. What differentiates the bulk distributor from the transit cellar is that the former has a commercial role, whereas the latter has only a role of transit with no commercial and no invoicing goal.
c. Filler - The filler/packer is responsible for the receipt, storage, processing, sampling, analysis, filling, packing and dispatch of finished goods, as well as record keeping of appropriate information about what is received and what is dispatched.
d. Distributor - The finished goods distributor is responsible for the receipt, storage, inventory management and dispatch of finished goods, as well as re-packing and re-labelling.
e. Wholesaler / Retailer - The retailer receives pallets and cartons from the finished goods distributor and picks and dispatches goods to the retails stores. Figure 4 shows the flow of wine beginning from the grape grower up to the retailers.

Figure 3 shows the flow of supply chain business transaction between various partners in a wine environment. In addition, in this figure we are also able to pin-point the vulnerability points in which a counterfeit attack could takes place. Few scenarios of how the attack happens are also listed.

Besides the flow among normal supply chain partners, another process worth mentioning in the wine supply chain is the consolidation or merger of a few players in order to enhance profits and reduce the cost of labor and infrastructure. This process is critical if security measures are not taken upfront. The consolidation process could input counterfeit wines that are later sent to the distributor (licit chain) or the other retailers (illicit chain). The end process of the counterfeit wines here is the sale to the consumer. One more route of the counterfeiting process is the act of the thief in stealing information directly or indirectly. The direct stealing of information involves the help of a third party, someone who is the employer of the licit supply chain. An indirect attack is an attack done by using the internet such as eavesdropping, man in the middle and skimming. The function of the thief is critical. The thief can manipulate the information of the wines or even the wine bottles and input them into consolidation process or even sell the information to the retailer and consumer.

Based on the vulnerability points illustrated above in Figure 3, the following scenarios demonstrate typical cases of RFID tag cloning and RFID tag fraud:

- Bordeaux Corp produces 1000 cases of wines with each case containing 100 bottles. The cases are then sent to the distributor. Bob, an employee of the distributor, steals the EPC information of 100 wine cases and supplies it to Carol, the attacker. Carol then copies the EPC tag numbers into empty tags and tags fake cases of wines. These wines are later shipped to several states within the country to different retailers.
- Reagan Corp, a shipping company, is plotting to steal a bulk load of wines that it has been entrusted with transporting. These wines have tamper-proof bottles with passive RFID tags attached. Rather than trying to defeat the tamper-proofing of the bottles, Reagan creates fake cheaper wine bottles, and clones the associated passive EPC tags. It swaps the bogus bottles while it has custody of the real ones.
- An anonymous reader belonging to Carol (an attacker) was placed at the warehouse belonging to Alice. When the Cabernet Sauvignon wines transported by Suiko Corp reached the warehouse, Carol eavesdrops on the communication channel, actively

performs a relay attack (man in the middle attack) and records a series of messages exchanged between the genuine reader and the trusted local database. Based on the encrypted EPC data obtained, Carol's reader communicates with the database. As there is no reader authenticity needed at the database side, the encrypted key is exchanged by the database. Carol now uses this key information received and performs a brute force attack on other EPC tags tagged on the cases. The guess game was able to reveal the key used for all the EPC tags scanned. Carol now sells this information to Alex, Alice's competitor who injects the data into cloned EPC tags and tags them on to cheaper goods and sends the goods to another retailer.

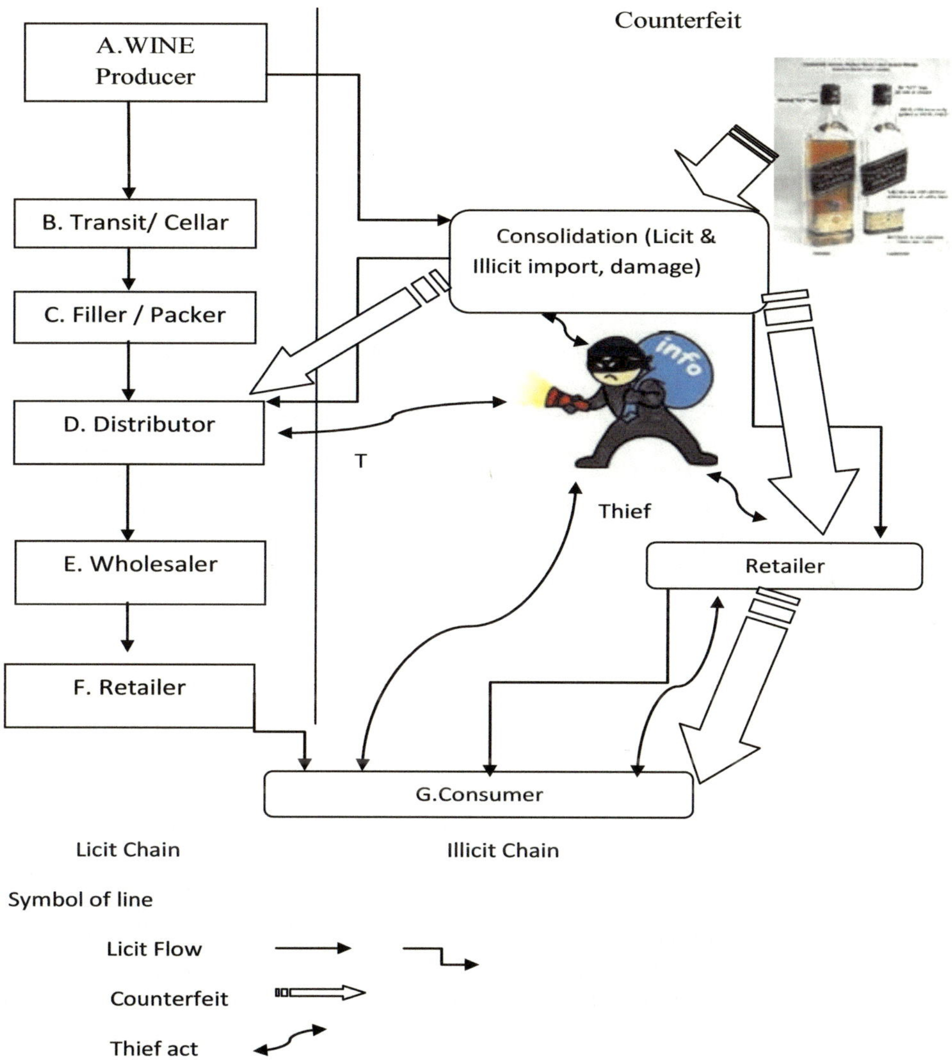

Fig. 3. Wine Vulnerability Points

Counterfeiting in the RFID-based system used in wine industry can be tackled using three categories: security, privacy and detection. The security solution looks into how we can protect the RFID tags on the wine bottles against cloning and fraud attacks. The privacy solution looks into how we can preserve the privacy of the partners and maintain the confidentiality of the information recorded by them and shared between them. Detection plays its role in detecting the cloned and fraud tags in an RFID-based system.

4. Clone/fraud handling through prevention, detection and privacy

4.1 Security - prevention of cloning in RFID-based wine system

The requirements of the cloning prevention system are data integrity and authenticity. In order to eliminate cloning, there is an essential need for complete authentication between all the RFID components. This includes providing integrity to the information within the tags. In addition, the need to sign the data is essential to show that the data has not been tampered with throughout the communication channel. The cloning prevention system must be able to prevent the skimming, eavesdropping and active attacks which are major security attacks that contribute to cloning in RFID systems. In addition, careful attention needs to be given to the fundamental problem of low-cost tags which provide less space on the tags and reduced memory capability. The security attributes necessary to handle a cloning attack include the following:

- A tag identifier must always be encrypted (e.g., hashed) before transmission between tag-reader-server begins. This reduces skimming and eavesdrop attacks on RFID tags and the system.
- Immediately after a reader has been authenticated, the tag must refresh a secret key. As long as the tag output changes, the chances of a replay attack can be reduced and there are no opportunities to fake a tag. Without knowledge about the secret key, an adversary can never create a set of encryption values.
- Three-way mutual authentication should always take place in any system including encryption and hash on tags, readers, and the data entries in databases.
- Synchronisation between tags and databases should always be consistent to eliminate cloning and eavesdropping.
- The number of communication rounds and operation stages should be minimal without any redundant operations to maintain scalability and eliminate the chances of replay and DOS attacks.
- The server for coordinating the global item tracking should be designed with a timely tracking system to maintain the freshness of randomness of the keys used in inter-organisational item-tracking activities. This helps against DOS attacks and cloning. It ensures that even though a key is compromised, an adversary can only capture a single tag rather than a bulk of tags.
- The most appropriate supply chain prevention mechanism should consider efficiency with a low-cost and practical approach. The techniques employed will need to be performed within the limitation of tags and RFID constraints. Therefore, techniques such as the physical uncloneable function (PUF) (Devadas et al., 2008) and watermarking technology (Potdar & Chang, 2006) are out of the question. The first is too costly and the latter is not efficient and practical when utilised on low-cost RFID tags.

- EPC-PAS and EPC-TAS should be modelled into the current EPC global network (Lehtonen, 2007).
- Item-level tracking should be used to diminish counterfeiting especially for luxury products such as jewellery and wine.
- A novel trust solution with an associated prevention mechanism via authentication for tag readers and supply chain partners is required. The trust model should be designed with some human interaction and feedback capability to enhance trust even more.

We also propose a simple prevention mechanism which is able to prevent cloning and fraudulent tags in a supply chain management. Since RFID tags are the most vulnerable point for any security attack in an RFID system, the tags should not be embedded with any important or confidential information. They should always function as pointers in which essential information such as secret key information or random numbers is stored in the database. In this proposed model, we make use of the message authentication code (MAC) algorithm. The function of the MAC algorithm is similar to the hash function in which it authenticates a message using a key and produce an authenticated code (Menezes et al., 1996). Message authentication codes are useful in many situations. If we need to perform basic message authentication without resorting to encryption for efficiency reasons, MACs are the right tool for the job. In addition, we add the public key cryptosystem to provide an added security capability which is signature capability. The concepts of random numbers and timestamps are used to track the liveness of the tags and to eliminate replay attacks. We make use of the Certificate Authority (Menezes et al., 1996) a third party trusted entity to maintain a higher security level of authenticating the readers. The benefit of this approach is that it eliminates the risk of compromised readers.

At this point it is important to articulate the assumptions for the cloning prevention system. These assumptions are:

- Channel between reader and database is secured.
- Trusted party, CA authenticates readers upfront.
- A Key Distribution Centre (KDC) is required to distribute and manage the secret key used by the tags and database.
- Tags used here are passive and compliant to Class 1 generation 2 (CIG2) tag with security function such as 16 bit pseudorandom generator.
- Timestamp values will be used to prove the authenticity of the tags based on the timeline starting from the movement information. For example, at location 1, the duration between the lifetime will be recorded according to the tags. The database on the trusted server will update the range of timeframe for any particular location and add the duration of the time. Finally, both timestamps will be similar or the difference of the timeline will be derived by a value of + 0.5 seconds or less.
- The random number will be generated from the CIG2 capability to produce the sequences from a 16 bit generator.

Figure 4 below provides a graphical representation of how the IPS framework will function, and shows the framework of how the required algorithms and security requirements will function.

The cloning technique that can be applied in the RFID-enabled supply chain functions through a number of steps. The readers in an RFID system should always be authenticated to ensure authenticity and eliminating the replay attack scenario from arising. First, the readers will read and send a query to the RFID tag. We assume that RFID tags only function as identifiers without any sensitive and important information on the tag. The only information on the tags will be the ID, random number and the timestamp. Next, the reader will send the information from the tag to the database. Here, the MAC algorithm will be used to distinguish whether the tag ID and the random number between the tags and the one stored in the database is similar. The KDC server will be used to generate the secret key each time a tag is checked for its authenticity. The benefit of the MAC value is that it protects both the data integrity of the message as well as its authenticity, by allowing the verifier (which possesses the secret key and which in our example is the KDC server) to detect any changes to the message content. Based on the calculation of the timestamp to ensure the authenticity of the tag ID, the response will then be sent to the tag by the reader.

<table>
<tr><td colspan="3">Pseudorandom generator - PNRG</td></tr>
<tr><td>CA</td><td colspan="2">Message Authentication Codes</td></tr>
<tr><td rowspan="2">Reader</td><td colspan="2">Timestamp</td></tr>
<tr><td>Database</td><td>Tags</td></tr>
<tr><td colspan="3">The Notation of the system are :</td></tr>
<tr><td>CA</td><td colspan="2">Trusted server</td></tr>
<tr><td>ID</td><td colspan="2">Tag ID</td></tr>
<tr><td>R(0,1,...n)</td><td colspan="2">Reader's ID</td></tr>
<tr><td>D</td><td colspan="2">Database</td></tr>
<tr><td>x</td><td colspan="2">Secret key distributed by Key Distribution Center</td></tr>
<tr><td>TS</td><td colspan="2">Timestamp</td></tr>
<tr><td>MAC[m]</td><td colspan="2">A MAC computed by applying secret key x to message m</td></tr>
<tr><td>r</td><td colspan="2">Random number</td></tr>
<tr><td>→</td><td colspan="2">Information movement (Send/Receive)</td></tr>
</table>

Based on method illustrated in Figure 4, we are able to provide the below system analysis on how the proposed prevention approach is able to reduce the chances of counterfeiting in a supply chain plant:

The use of the CA – the CA will have the list of authorised readers upfront and will only authenticate the trusted reader. This eliminates the possibility of a compromised reader.

The use of MAC with a secret key which is hashed and encrypted will protect the integrity of the message and eliminate the eavesdropping attack and skimming attack from occurring. The security of the communication channel between the database and tags is guaranteed because of this.

The use of KDC – the Key Distribution Centre function provides a secret key to both tags and database. The use of a trusted dedicated server will reduce the chances of the key being compromised by an adversary. In addition, the key in the KDC will be generated randomly. The number of bits used to generate the keys will impact on the security level. Using higher numbers of bits will guarantee a stronger key. If a particular key is being compromised, the adversary is only able to clone the particular tag and not the entire batch.

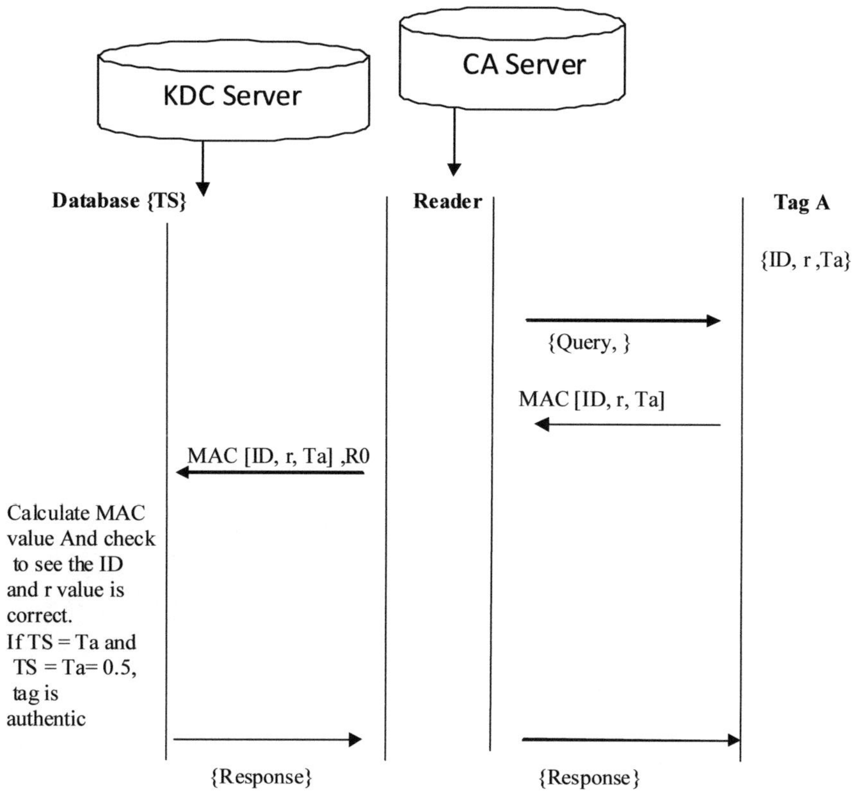

Fig. 4. Cloning Prevention Method

The use of timestamps will reduce the chances of the replay attacks that allow cloning to take place. The duration of time from each location will show the authenticity of a tag. The duration will be added and a rounded-up value for the TTL will be stored in the database.
The use of random numbers will increase the difficulty for an adversary to guess the key value of the tag.
It is worth mentioning that we have shown how three different attacks which are skimming, eavesdropping and active attacks through replay attack are able to be removed by utilising the above algorithm. However, physical attacks will only be addressed by using a higher level of key values. In addition, reverse engineering attacks could only be addressed by using a secure hardware implementation such as PUF (Devadass, 2008). Hence, we do not discuss these two attacks in our chapter. As supply chain management uses passive tags with low capabilities, we are not able to protect the RFID tags by using high-end security properties. However, by employing the trust framework, we are able to use third party solutions such as the CA server and KDC server. All the calculations of the MAC algorithm keys will be done at the database end. RFID tag information will store only minimal ID information. With minimal information, the probability of being skimmed and

eavesdropped upon will reduce. This model could be used for any RFID application such as the wine supply chain in our context.

4.2 Detection of cloning and fraud wine bottles in RFID system

This section explains RFID supply chain, RFID data structure and how TTL will be used in our proposed system. There are four different attacks in an RFID system (Mahinderjit-Singh & Li, 2009; Mahinderjit-Singh & Li 2010). Skimming attack occurs when RFID tag are read directly without anyone knowledge. Eavesdropping attack happens when an attacker sniffs the transmission between the tag and reader to capture tags data. On the other hand, man-in-the-middle attack occurs when a fake reader is used to trick the genuine tags and readers during data transmission. RFID tag data could also be altered using this technique and as a result, fraud tags could be generated too. Physical attack which requires expertise and expensive equipment takes places in laboratory on expensive RFID tags and security embedded tags.

The strength of any RFID application is fully capitalised when the temporal and location information are correctly utilised in eliminating data security issue in RFID. Real time monitoring of events such as fraud and cloning attacks in RFID application are still rare.

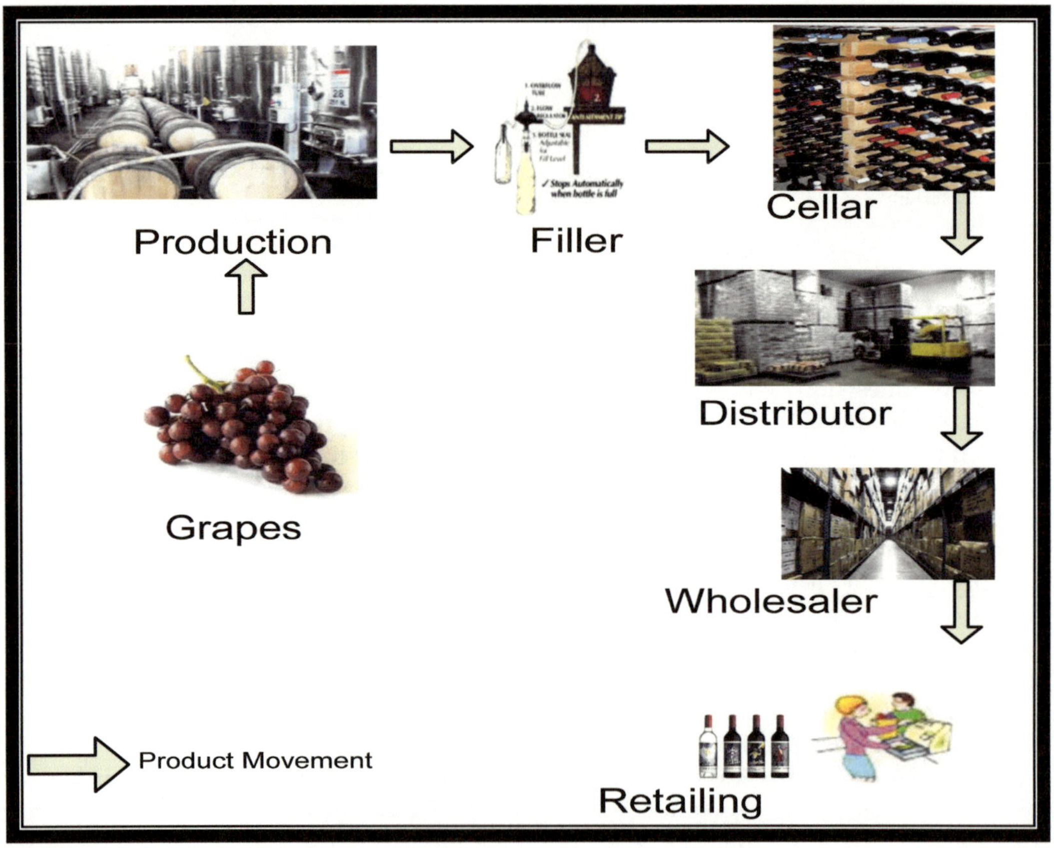

Fig. 5. Wine Supply Chain

Figure 5 shows a wine SCM environment with four different sites (Manufacturer, Distributor ,Wholesaler and Retailer). RFID tags are attached to the products for instance wine bottles. RFID based supply chain system involves the movement and flows of millions of data. The data generated consists of RFID tuples of the form of (EPC, location, time), where *EPC* is the unique identifier read by an RFID reader, *location* is the place where the RFID reader scanned the item, and *time* is the time when the reading took place. Tuples are usually stored according to a time sequence.

Each sites will have their own database system and this distributed manner database system are combined with a centralized EPC global server; EPC- Information Server. (EPC-IS). Our trust framework will resides in centralized server with ONS and EPC-IS (Verisign,2004) . The trust framework, fifth layer, mainly the detection module will consists predefined rules of real time monitoring and tracking system. The tracking and monitoring system can even play role as an intrusion detection system by using events rules and triggers function in database. Among the rules are as below:

- If, for instance, a product was identified at specific read points, e.g., 'shelf' (R3) and then 'exit' (R6),without having first been identified at the read point 'checkout' (R4 or R5), then it could be a matter of cloned or fraud.
- If a pallet P, which is containing the objects O1, O2, and O3 when leaving the production facility (M2 or M3) was identified as having only the objects O1 and O3 at the distributors receiving dock (D1 or D2), then the object O2 could have been replaced with O4 during transportation. These mean counterfeit products are injected.

i) Data structure time to live(TTL)

TTL indicates the time restriction that targets events should satisfy. Since most RFID application has a restriction time, we believe if carefully defined, we can use the notion of TTL to detect clones and fraud tags in a typical SCM. Based on TTL taxonomy (Li.X et.al , 2009), there are 4 different notions of TTL given based on the event types, both primitives and complex categorised based on events as Absolute TTL (TTL*a*), Relative TTL (TTL*r*), Periodic TTL (TTL*p*) and Sequential TTL (TTL*sE*). The detection process of cloned and fraud tags are able to manipulate all the above TTL notions. However, based on RFID applications, we determine that three relevant TTL notion for a SCM transactions and monitoring process is mainly *TTLa, TTLr* and *TTLs.* We also argue that the absolute *TTL (TTLa)* notion can be further categorised based on RFID applications. Some applications such as drugs and fast moving products for e.g. diary and foodstuff requires restriction in expiry date as the *TTLa* compare to product such as wine and jewellery. These expensive products emphasize more on manufacturing time. We will introduce the new notion of *TTL* called *Initial TTL (TTL i).*

TTLi specifies the period of time a RFID tag is tagged on the product. By tracking, monitoring and storing the *TTLi* in the system; we are able to classify cloned RFID tags from genuine tags. Below are some examples to show the practicality of the usage of *TTL.*

i. *Example 1 - Initial TTL (TTLi):* Suppose 1000 new RFID tags have been purchased from its manufacturer. Each tag is then scanned by the reader denoting the birth time of the tags. Once the tag is tagged to a product such as wine, the expire time of tag is also stored. The period between this birth time and expire time areconcluded as *Initial TTL.* For products such as wine, *TTLi* is extremely important. Since the *TTLi* is an event happening at the manufacturer site, any fraud injection of fake wine bottles after the manufacturer site can be detected.

ii. *ii) Example 2 – Relative TTL (TTLr)* – In a wine based SCM, when the wines bottles are transported from the manufacturer site to the distributor site, the transportation period need to be carefully tracked. If the time to reach a destination is more than its relative *TTL*, an alarm will be raised as the state of bottles are suspicious. *Relative TTL* also indicates the period time the bottles are scanned by multiple readers at the front door of the distributor up to the time period the bottles leaves the site. Thus the *TTLr* can be categorised as *transfer TTL (TTLt)* and *site TTL (TTLs). TTLt* is the restriction time for all the movement time from one point to the other. Meanwhile *TTLs* is the whole site location e.g. Manufacturer, Distributor and Retailer period from the time it enter a site where it will be processed for unpack or repack up to the time it leaves the site.

iii. *iii) Example 3 – Sequential TTL (TTLsE)* – The products movement from the manufacturer site upto the retailer site is denoted by the *TTLsE. TTLsE* is the sum of all the *TTLr* in a supply chain. If the time from the manufacturer site and till the retailer site exceed or lesser than the *TTLsE*, the event could be suspicious.

SiteTTL (TTLs) = Time of RFID within a site such as manufacturer, Distributor and Retailer
TTLs = tend (Distributor site) – tstart (Distributor site)
TransferTTL (TTLt) = Time taken when moving products from site A to site B
Sequential TTL (TTLsE) = Overall accumulated time from Manufacturer site up to Retailer site
The audit data for a single RFID is given below:

Audit tag, for a single RFID tag ,

T = < Po, Pm, Psd, Pt, Pr > where:

Po= operation match rate,

Pm =mean of TTL, where TTL = { TTLs,TTLt, TTLsE}

Psd =standard deviation of TTL, where TTL = { TTLs,TTLt, TTLsE}

Pt = rate of tag responses, and

Pr = R/W (mean and standard deviation) rate.

ii) Cost- Sensitive learning

Cost-Sensitive Learning is a type of learning in data mining that takes the misclassification costs (and possibly other types of cost) into consideration. The goal of this type of learning is to minimize the total cost (Turney, 2000). Many works for dealing with different misclassification costs have been done, and they can be categorized into two groups. One is to design cost sensitive learning algorithms directly (Turney,1995; Domingos,1999). The other is to design a wrapper that converts existing cost-insensitive base learning algorithms into cost-sensitive ones. The wrapper method is also called cost-sensitive meta-learning (Witten and Frank , 2005., Domingos,1999) sampling (Zadrozny,2003), and weighting (Ting,1998). Cost-sensitive meta-learning converts existing cost insensitive base learning algorithms into cost-sensitive ones without modifying them. Cost-sensitive meta-learning techniques can be classified into two main categories, *sampling* and *nonsampling*,in terms of whether the distribution of training data is modified or not according to the misclassification costs. This paper focuses on the nonsampling cost-sensitive meta-learning approaches. The non-sampling approaches can be further classified into three subcategories: relabeling, weighting, and threshold adjusting, described below. The first is *relabeling* the classes of instances, by applying the minimum expected cost criterion (Witten and Frank , 2005). *Relabeling* can be further divided into two branches: relabeling the training instances (Witten and Frank , 2005) and relabeling the test instances (Domingos, P. 1999) .

In Relabeling approach such as Metacost (Domingos, P. 1999)and Cost Sensitive Classifier (Witten and Frank , 2005), cost C is known at the learning time. The technique to modify the inputs to the learning algorithm to reflect cost C includes :

i. If there are 2 classes and the cost of a false positive is λ times larger than the cost of a false negative, put a weight of
   λ on each negative training example
   $\lambda = C(1,0) / C(0,1)$
ii. Then apply the learning algorithm as before
iii. Setting λ by class frequency (less frequent class has higher cost)
   $\lambda \sim 1/nk$, nk - number of training examples from class k
iv. Setting λ by cross-validation

WEKA (http://www.cs.waikato.ac.nz/~ml/weka.), an open source Java package which contains machine learning algorithms and Metacost algorithm are used for solving the RFID cloning issue in SCM.

iii) Cost -based Counterfeiting Detection Architecture and Result

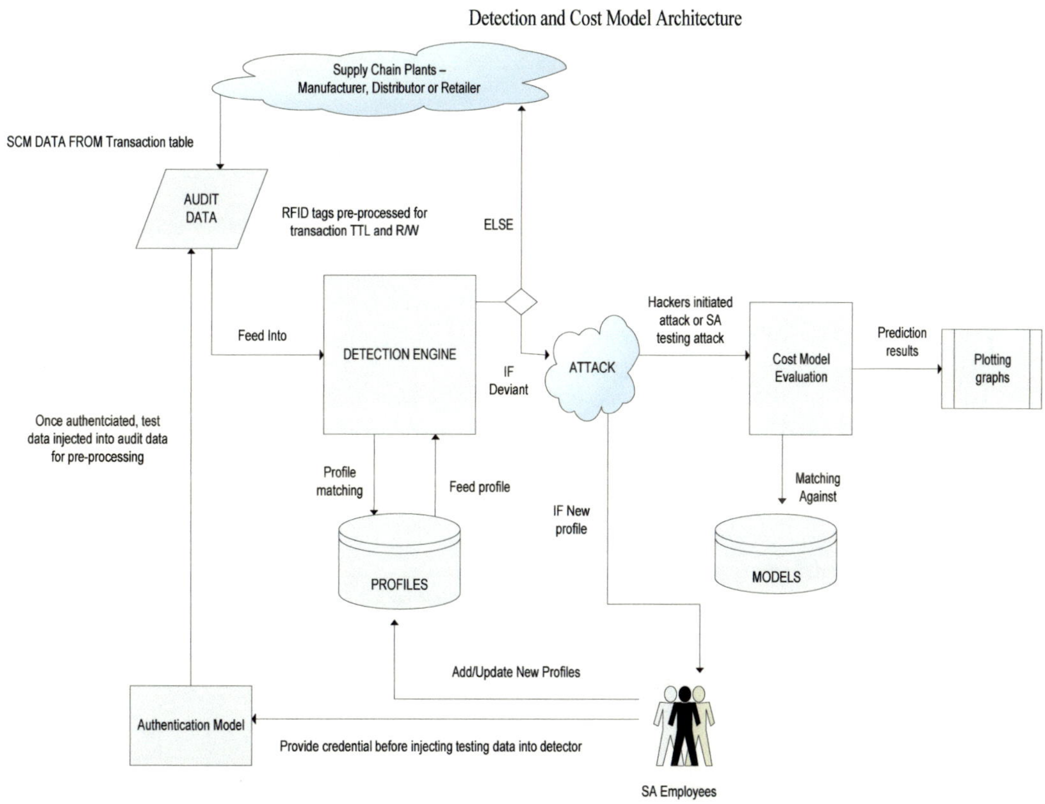

Fig. 6. Detection and Cost Model Architecture

```
Input: Training data: T= {t1,... tm} where each example Ti has attributes { Po, Pm, Psd, Pt, Pr} and a
class ci
        : Classifier C with learning algorithm L
        : Misclassification cost, Cij

Output: W: the predicted test class, alarm log, response
For ∀ T ∈ {ti,tm}
 C        L(T) ←
 Create a Root node for the tree
Initialize all the weights in T, Wi=1/N, where N is the total number of the examples.
                                                                                    n
Calculate the prior probabilities P(Cj) for each class Cj in T. P (Cj) = ∑Wi  /  ∑Wi
                                                                         Ci        i=1
Calculate the conditional probabilities P (Aij | Cj) for each attribute values in T. P (Aij | Cj) = P (A)
/ ∑Wi
  Ci
Calculate the posterior probabilities for each example in D.P(ei | Cj) = P(Cj) Π P(Aij | Cj)
Update the weights of examples in D with Maximum Likelihood (ML) of posterior probability
P(Cj|ei);  Wi= PML (Cj|ei)
If (all the examples in T are in the same class ci)
{
  Return (the single node tree Root with label ci)
}
Else
{
   Let a be the Best attribute (T)
   For (each possible value v of a) do
      {
        Add a new tree branch below Root, which correspond to the test a = v
        If (Dv is empty)
{
   Below this branch add a new leaf node with label equal to the common class
   Value in D.
}
    Else
     {
       Below this branch add the subtree (Dv,A-a)
      }
}
}
Return Root
End learning phase
C = {Ti , Tx}
 For ∀ Tx ∈ {Cloned,Fraud}
A (k x k) misclassification cost matrix L,
L = a classification algorithm
Output: W
Estimate the class probabilities P(yi|xi)
Relabel
W= L (x,y)
Return W
```

Fig. 7. Pseudo code for Decision Tree (J48 algorithm) with Metacost

In this section we discuss how RFID tag cloning and fraud detection as well as cost modelling are supported seven layer trust framework (Mahinderjit-Singh & Li, 2009; Mahinderjit-Singh & Li 2010). Our RFID detection system has three main components:pre-processing; detection; and response and decision module as shown in Figure 6.Pre-processing is the component that collects a RFID event set *E* that is supplied by different supply chain partners. RFID event sets are then sent to the detection component where the information sources are analysed. Several detection functions are performed in this component, such as pattern matching; traffic or protocol analysis; finite state transition; etc. The response and decision component notifies the system administrator where and when an intrusion takes place and calculate the total cost of any attack.

Applying the dataset from the simulated RFID supply chain, 3000 example of RFID traces are generated from manufacturer site up to retailer site. RFID traces is then pre-processed into audit dataset which includes attributes such as Tags ID, location ID, TTLs (mean), TTLt (mean) , TTLsE(mean and standard deviation) and Read/write (mean and standard deviation. The datasets are then feed into Weka engine by applying Metacost algorithm shown in Figure 6. The audit data will then be feed into a filtering system upfront for normalization purposes. CfsSubsetEval with Best First technique are used to determine the evaluation of attributes and search methods.

The base classifiers used were Naive Bayes, Random Forest and Weka's implementation of a Support Vector Machine (SMO), JRIP and C4.5 (J48) decision tree. Default Weka options were used for the Naive Bayes , Random Forest and JRIP but for the SMO "build logistic models" was set to true and for the J48 tree "Pruning" was disabled. Receiver Operating Characteristic (ROC) curve is a plot of the probability of true positive (recall) as a function of the probability of false alarm across all threshold settings. An ROC curve provides an intuitive way to evaluate the classification performance of RFID detection system. Recall represents the probability of detection of cloned tags and precision is the proportion of the correctly predicted genuine tags in each prediction class. In this study, we will utilize ROC for models evaluation.

The engine is trained with a training dataset. Cloning attacks such as skimming, eavesdropping and man-in the middle are simulated. To train the models cross-validation was employed. Cross-validation is a standard statistical technique where the training and validation data set is split into several parts of equal size, for example 10% of the compounds for a 10 fold cross validation. An independent test dataset is simulated as well. However, for the differing classifiers they have used across-the-board costs of 20, 40, 60, 80,100, 200, 500 etc. Weka normalises (reweights) the cost matrix to ensure that the sum of the costs equals the total amount of instances. Next we will illustrate one of the algorithms, J48 used with Metacost in WEKA tool. The pseudo code for Decision Tree (J48 algorithm) with Metacost is shown in figure 7. The ROC curve plotted in figure 8 takes in to account a few classifiers in WEKA. Based on this ROC curve, we could conclude that various classifier provide different performance based on the setting and nature of the classifier itself. For instance, Naïve bayes provide the larger area of ROC curve which indicate, it has the best performance. In addition, the true positive is almost 98% with only less than 2% of false alarm.

In a cloned detection RFID enabled supply chain, misclassifying cloned tag as genuine is undesirable. Result shows that when we increase *cost-ratio* from 20 to 10,000, recall rate increases, although the rate of increase depends on the algorithm. However, although not unexpected, is the decrease of *precision* which implies needless analysis of large number false positives (shown in fig.9) SMO, JRIP and J48 algorithms consistently reach *Recall* rates close to 1 at high cost ratios, with precision slightly above 0.1.

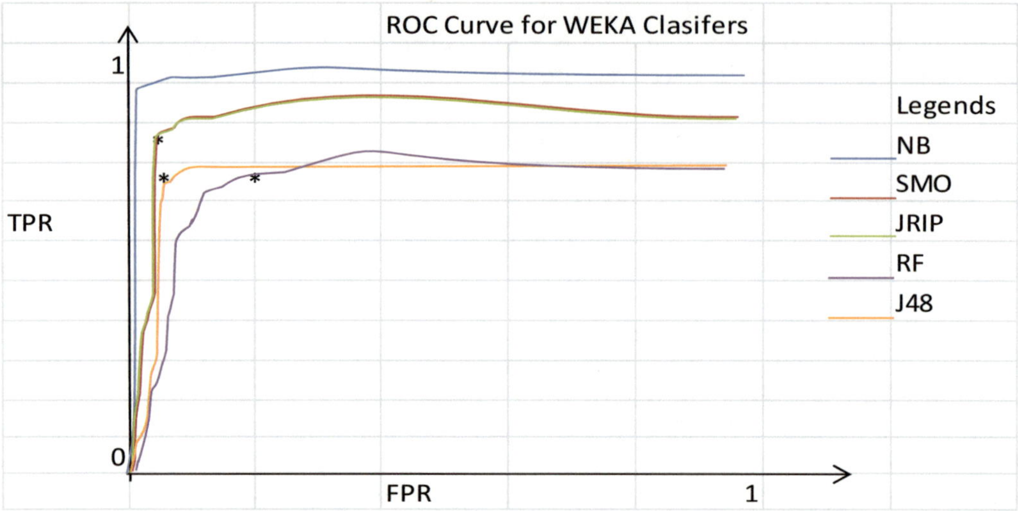

Fig. 8. ROC Curve plot for WEKA Classifiers

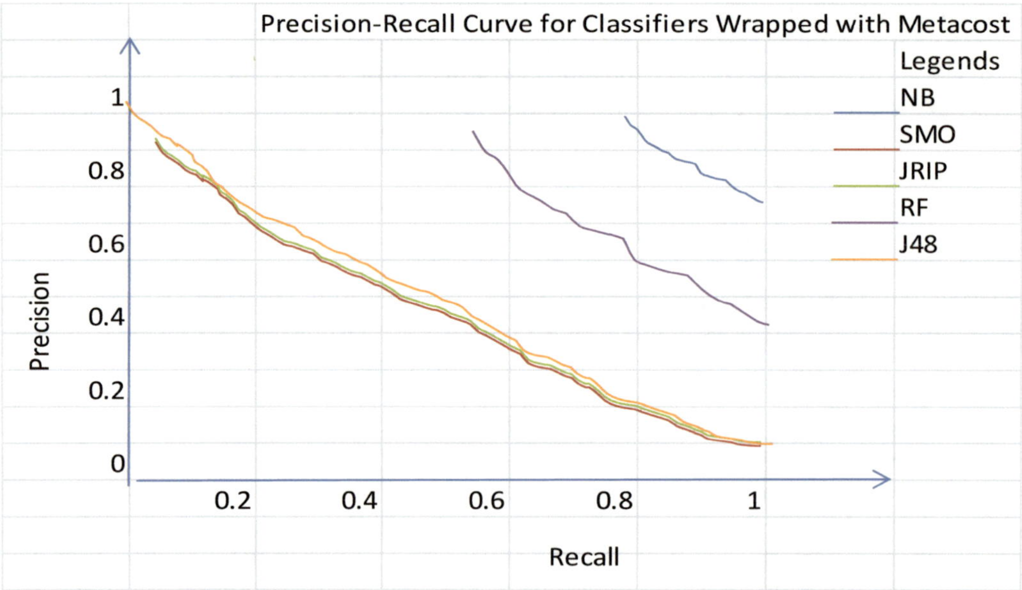

Fig. 9. Precision-Recall Curve for Various Classifiers in WEKA

Figure 10 indicates the accuracy of various classifiers against misclassification costs. We could conclude that as cost ratio increases, the accuracy of classifier decreases as well. An important implication from this study is that we can use cost to choose suitable operational threshold (based on different *cost-ratio*) to control a classifier's performance.

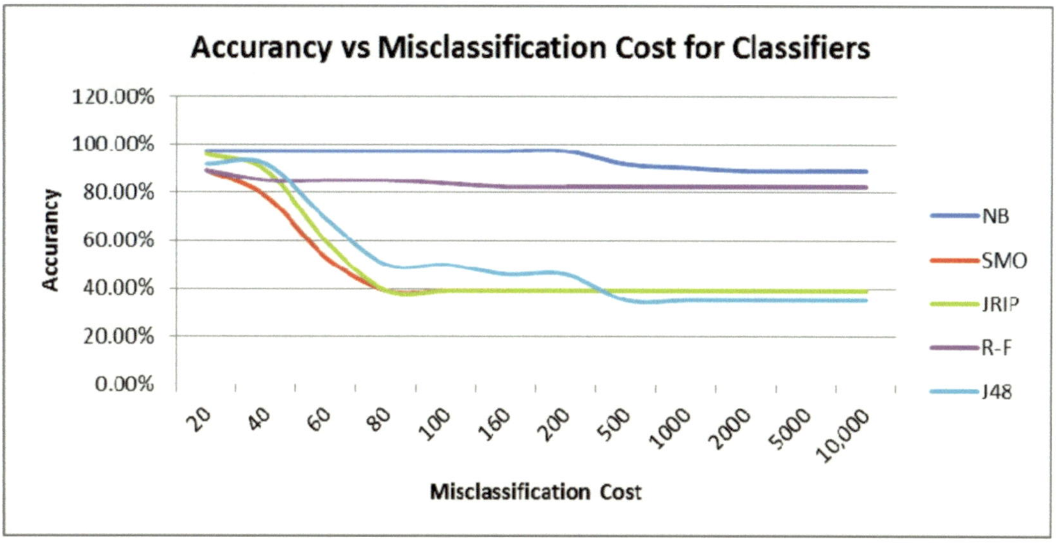

Fig. 10. Accuracy vs. Misclassification Cost for Classifiers

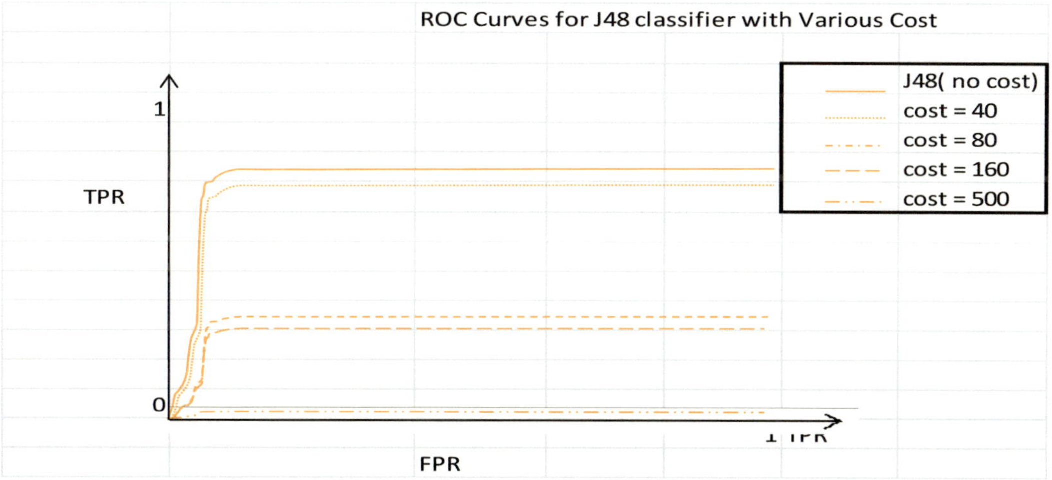

Fig. 11. ROC Curve for J48 classifier with various Costs

In practice, exact costs are rarely known and could change as we learn more about system requirements, its design, operational environment, etc. When considering a wide range of cost ratios the resulting models differ significantly. For instance from Fig 11, J48 classifier is made cost sensitive when the cost ratio was set to be 500 with accuracy of 35.1%. This means that FN needs to be 500 times more expensive than FP for J48 to transform to cost sensitive. Overall, J48 provides the most robust and versatile classifier for imbalanced RFID dataset compared to other classifiers.

With respect to construct validity, cost ratios in our experiments, which vary from 20 to 10,000 might not include all meaningful cost differentials. Different intrusion detection systems may have their own cost ranges of interests. The selection of classifiers is another possible source of bias. We cannot exclude the possibility that a classifier not studied here could show significantly better performance. Nevertheless, based, we believe that the chance of such a classification algorithm being in existence is rather low. The results above could be implicated by the small datasets used in the training models. When small dataset are used, classifier cannot accurately estimate the class membership probabilities and the imbalanced in class distribution of the dataset.

Any RFID cloned detection classifiers used must be correlated with cost since lower cost properties projects to lower or zero cloned tags in the system. This also impact positively in reducing the counterfeit attack which risks billions of dollars losses yearly in the market. Overall, we could conclude that by using WEKA tool, we are able to detect cloned and fraud tags in a supply chain plant. In addition, when various cost files are utilised we are able to reduce the misclassification cost of testing dataset. The important of the above experiment are to show the relationship between false positive rate and false negative rate. The trade-off shows that by increasing cost values, the false negative or the misclassification cost can be reduced. As a result, the false positive rate increase and this reduced the classifier accuracy overall. We also conclude that among the various supervised learners used, J48 is more sensitive to cost effects and outperform other classifiers when used together with Metacost.

In an RFID based wine supply chain, our main concern will be to eliminate the possibility of any counterfeit wine bottle passing through any detection classifier without generating any alarm. We believe the risk of counterfeit wines bottles passing through our detection system is greater than any genuine wines bottles detected as counterfeit one. Thus, even though the overall accuracy of classifiers decreases under the cost effects, we are able to reduce the losses in term of money and trust in RFID technology when used in supply chain. By minimising the counterfeit rate flowing in the market, human trust in this technology increases dramatically.

Next section provides a comprehensive privacy guideline in handling counterfeiting in a supply chain environment.

4.3 Privacy - countermeasures in preventing privacy violations

In the clone detector, some ways to prevent privacy violations in a Wine based RFID-enabled supply chain include:

1. The EPCglobal Discovery Service (DS) is equipped with key management mechanisms using ElGamal or RSA encryption algorithms. The clone detector is installed on the DS. The partners that need to access the clone detector will have to go through the DS for authenticity, and only permitted personnel are given permission to access information.

Before using the clone detector, all players obtain the necessary information to establish a connection to each other through the DS, which knows who owns an event on a certain ID and can the bootstrap the network upon a partner request for detecting clones of ID.

2. Distributed network architecture is employed. The distributed network architecture eliminates the problem of information overload and makes it easier to exchange information (VeriSign, 2008). Manufacturers as well as all trading partners create and store their own serialised information about each and every product. The manufacturer will manage and host a database that stores information about the generation of products, while trading partners host and manage similar databases storing information about product movement through the supply chain. Each involved partner will make this information available to authorised parties over the internet. This will ensure minimal sharing of local tracking data (times and places) with the EPC network.
3. The ONS could be used to point to an address on the EPCglobal network where information about the product being questioned is stored. The information stored here should be in minimal granularity that has limited timestamp information. By limiting timestamp accessible data, the effect of data leakage and data privacy can be minimised.
4. Default killing of RFID tags at store exits or password protection of RFID tag content could be set up. This means that the production tag which is used for tagging on the product within the supply chain will be deactivated at the POS exit. This will reduce the possibility of tracking and inventorying for the purposes for profiling done by the supply chain partners especially the manufacturer in learning the behavior of the consumer. In addition, a new tag can be placed on the tag after the purchase of the product that comes along with warranties. This information should be accessible only by the manufacturer and consumer.
5. Partial or no saving of the full EPC serial number should always be applied on RFID tags in an RFID-enabled supply chain environment.
6. There can be rigorous controls and transparency of EPC network access rights. A role-based access control (RBAC) policy should always be implemented together with item-level tagging (Illic et al., 2007). The main purpose of the RBAC policy is allowing only certain individuals to access certain levels of information. By applying this policy, we are able to limit accessible information by different role of personnel in an organisation.
7. Deletion of all product data after a certain period of time. After a while, the entire product data linked by the tag ID and the database should be deleted. This requirement reduces any form of tracking violation and curbs fraud situations from occurring. However, this will stamp out the advantages of an RFID system in a supply chain such as providing visibility and traceability.
8. Any supply chain partner could exercise control over personal information on sold products available on the EPC network. This will limit any misuse of product information by the consumer and competitors in learning about the supply chain partner"s financial gain in forecasting sales information. In addition, a competitor could also use this information in creating cloned tags with similar product information on fake products for future transactions.
9. All RFID transactions and information transmissions in the RFID supply chain require consent from both parties, namely, the business owner and consumer. By complying

with Garfinkel et al' s proposed policy (2005), RFID organisations in a supply chain environment need to be aware of their full rights especially to know when, where and why an RFID tag is being read. To comply with it, organisations could post a sign wherever RFID readers operate. Embedding this policy with a detection system is possible when a tag equipped with memory could count the number of times it has been read.

In preservation of RFID privacy, besides employing user policies in accessing the information in system, ownership transfer between partners can also be supported. By using one of the ownership transfer protocols discussed in Section 2.2, the security of the protocols can be maintained if the communication channel is protected. Another way to ensure a secure transfer of information will be to allow access to information to all the partners in the local EPC-IS without handing out any sensitive information such as sales and forecasting information. The conclusion we could draw here is that by following one or more of the privacy guidelines are able to protect the whole supply chain running on an EPCglobal network platform.

5. Conclusion

In this paper, three layers - Layer 1 - Security, Layer 2-Privacy and Layer 5-Detection - from our seven-layer trust framework are investigated for tackling counterfeit problem in a wine industry RFID-enabled supply chain. We have directed the security (prevention and detection of counterfeiting) and privacy preservation by using the RFID-enabled wine supply chain application. In an RFID-enabled supply chain system, privacy concerns require urgent attention especially to control the counterfeit issue. Security principles such as authorisation, authentication and encryption need to be combined with privacy procedures to maintain data integrity and privacy. Protection of privacy is essential for both consumers and business owners in order for a trustworthy relationship to be maintained between them. We have demonstrates that by applying MAC technique and third party services such as CA and KDC service, we are able to protect the low cost tags from being counterfeit.

In addition, we argue that RFID clone detection classifiers must always be correlated with cost since lower cost properties project to lower or zero cloned tags in the system. This also impacts positively in reducing the counterfeit attack which risks billions of dollars in losses every year in the market. We have shown that when the relabelling approach is used, we are able to reduce the misclassification cost and eliminate the scenario of having cloned and fraudulent tags in the system.

Nevertheless, RFID tag cloning and fraud can be detected in a supply chain at an initial stage if there is proper transfer of ownership with secure and authorised information exchange. This is made possible by integrating the monitoring, detection, and security and privacy functions from the seven-layer trust framework model which focuses on reducing risks and increasing benefits such as eliminating counterfeiting tags in SCM systems and boosting supply chain players" confidence. In future work, we aim to extend our RFID cloning and fraud detection work by using an outlier detection technique to identify illegitimate RFID tags and designing an improved cost decision model to calculate the damage, response and operational cost for a typical RFID clone detector system in a supply chain application. In addition, we would like to enhance RFID supply chain privacy and security in terms of context-awareness.

6. Acknowledgements

This work is partially sponsored by University Sains Malaysia (USM)

7. References

Ayoade, J (2007). Privacy and RFID Systems: Roadmap to Solving Security and Privacy Concerns in RFID Systems. *Computer Law and Security Report*, 23(6):555–561, 2007.

A.J. Menezes, P.C. van Oorschot, S. Vanstone. Handbook of Applied Cryptography, CRC Press , Florida , USA (1996), 780 pages, ISBN 0-8493-8523-7.

Domingos, P. (1999). MetaCost: A general method for making classifiers cost-sensitive. In Proceedings of the Fifth International Conference on Knowledge Discovery and Data Mining, pp. 155-164, ACM Press.

Drummond, C. and Holte, R. (2000). Exploiting the cost (in)sensitivity of decision tree splitting criteria. In Proceedings of the 17th International Conference on Machine Learning, pp.239- 246.

Devadas, S., Suh, E., Paral, S., Sowell, R., Ziola, T., & Khandelwal, V. (2008). Design and implementation of PUF based "Unclonable" RFID ICs for anti-counterfeiting and security applications. In Proceedings of the 2008 IEEE International conference on RFID, 2008 (pp. 58- 64).

Garfinkel, S., Juels, A., and Pappu, R. (2005). RFID Privacy: An Overview of Problems and Proposed Solutions.*IEEE Security and Privacy*, 3(3):pp 34–43, May–June 2005.

Gao, X., Wang, H., Shen, J., Huang, J., Song, B. (2004). "An Approach to Security and Privacy of RFID System FOR Supply Chain," Proceedings of the *IEEE International Conference on E-Commerce Technology for Dynamic E-Business (CEC-East'04), pp. 164-168, 2004.*

G.Johnston, "An anticounterfeiting strategy using numeric tokens. International journal of pharmaceutical medicine", pp 163-171 2007

Hargraves, K., Shafer, S. (2004). Radio Frequency Identification (RFID) Privacy The MicrosoftPerspective [Online]: http://www.microsoft.com/twc (2004)

Ilic, A., Michahelles, F., Fleisch, E. (2007). Pervasive Computing and Communications Workshops, 2007. PerCom Workshops '07. Fifth Annual IEEE International Conference on pp. 337-341.

Juels, A. (2006). „RFID security and privacy: a research survey" IEEE Journal on Selected Areas in Communications, vol. 24, no. 2, February 2006, pp. 381-394.

Juels. A. (2005). „Strengthening EPC tags against cloning", in Proc of the 4th ACM workshop on wireless security. 2005, Cologne, Germany, pp. 67-76.

Kutvonen, S. (2005). Trust management survey, Proceedings of iTrust 2005, number 3477 in *LNCS, pp. 77--92 , Springer-Verlag.*

Koh, R., *et al.* (2003). White Paper: Securing the pharmaceutical supply chain, Auto-ID Center, Massachusetts Institute of Technology, 2003, http://www.autoidlabs.org/uploads/media/MIT-AUTOID-WH021.pdf (accessed 5 Nov 2009).

Lehtonen, M., Michahelles, F. and Fleisch, E. (2007). „Probabilistic approach for location-based authentication", Auto-ID Labs White Paper WP-SWNET-020, Auto-ID Labs ETH Zurich. pp. 3-17.

Lehtonen.M (2007) , "Trust and Security in RFID-Based Product Authentication Systems" Systems Journal, IEEE, pp 129 - 144

Lehtonen.M et.al (2006). "From Identification to Authentication – A Review of RFID Product Authentication Techniques." Workshop on RFID Security—RFIDSec,pp 169-181 2006 - Springer

Li, X., Liu, J., Sheng, Q.Z., Zeadally, S., and Zhong, W. (2009), TMS-RFID: Temporal Management of Large-Scale RFID Applications, International Journal of Information Systems Frontiers, Springer, July. 2009 pp.1-20.

Mahinderjit-Singh, M. and Li, X. (2009). "Trust Framework for RFID Tracking in Supply Chain Management," Proc of *The 3rd International Workshop on RFID Technology – Concepts, Applications, Challenges (IWRT 2009), Milan, Italy, pp 17-26, 6-7 May 2009.*

Mahinderjit-Singh, M. and Li, X. (2010). Trust in RFID-Enabled Supply-Chain Management, in International *Journal of Security and Networks (IJSN), 5, 2/3 (Mar. 2010), pp 96-105. DOI= http://dx.doi.org/10.1504/IJSN.2010.032208*

Nochta, Z., T. Staake, and E. Fleisch. "Product specific security features based on RFID technology."in Applications and the Internet Workshops, 2006. SAINT Workshops 2006. International Symposium on, pp 23-27 2006.

Osaka, K., Takagi, T., Yamazaki, K. and Takahashi,O. (2006). "An Efficient and Secure RFID Security Method with Ownership Transfer" Computational Intelligence and Security, 2006, vol. 2, pp. 1090-1095.

Pedro, P.L et al. (2010).Vulnerability analysis of RFID protocols for tag ownership transfer, Comput. Netw. (2010), doi:10.1016/j.comnet.2009.11.007

Potdar.V and Chang.E, "Tamper detection in RFID tags using fragile watermarking," 10th IEEE International Conference onIndustrial Technology (ICIT2006), Mumbai, INDIA, Dec. 15–17,2006

R. Derakhshan, M. E. Orlowska, and X. Li. (2007). RFID data management: Challenges and opportunities. In: D. W. Engels, *IEEE International Conference on RFID 2007. IEEE International Conference on RFID 2007, Grapevine, Texas, USA, (pp 175-182). 26-28 March, 2008*

Ranasinghe. D. C and Cole , P.H, "EPC Network Architecture," In: Cole, P.H. and anasinghe, D.C., (eds.) Networked RFID Systems and Lightweight Cryptography: Raising Barriers to Product Counterfeiting. Springer; 1 edition . ISBN 9783540716402, 2007.

Staake, T. Thiesse, F., and Fleisch, E. (2005). „Extending the EPC network: the potential of RFID in anti-counterfeiting", Proc of ACM symposium on Applied computing, Santa Fe, New Mexico, 2005, pp. 1607-1612.

Sarma, S., Ashton, K., Brock, D. (1999). The Networked Physical World, Technical Report IT-AUTOID -WH-001, 1999. http://www.autoidcenter.org/research/MITAUTOID-WH- 001.pdf.

Seong. D et. al , "Access Control and Authorization for Security of RFID Multi-Domain Using SAML and XACML," presented at Computational Intelligence and Security, 2006 International Conference on, pp 1587 - 1590 2006.

Saito, J., Imamoto, K., Sakurai, K.: Reassignment scheme of an RFID tag's key for owner transfer. In: Enokido, T., Yan, L.,Xiao, B., Kim, D.Y., Dai, Y.-S., Yang, L.T. (eds.) EUC-WS 2005. LNCS, vol. 3823, pp. 1303-1312. Springer, Heidelberg (2005)

Song, B.(2008). RFID tag ownership transfer, in Proceedings of Workshop on RFID Security, Budapest, Hungary, July 2008.

Turney, P.D. 1995. Cost-Sensitive Classification: Empirical Evaluation of a Hybrid Genetic Decision Tree Induction Algorithm. Journal of Artificial Intelligence Research 2: pp. 369- 409.

Turney, P.D. 2000. Types of cost in inductive concept learning. In Proceedings of the Workshopon Cost-Sensitive Learning at the Seventeenth International Conference on Machine Learning, Stanford University, California pp. 15-21.

Ting, K.M. (1998). Inducing Cost-Sensitive Trees via Instance Weighting. In Proceedings of the Second European Symposium on Principles of Data Mining and Knowledge Discovery, pp. 23-26. Springer-Verlag.

Turney, P.D. (1995). Cost-Sensitive Classification: Empirical Evaluation of a Hybrid Genetic Decision Tree Induction Algorithm. Journal of Artificial Intelligence Research 2: pp.369- 409.

Verisign - Expanding value of Supply Chain, (2008) http://www.verisign.com/static/DEV044098.pdf

Verisign Inc : “EPC Network Architecture” (2004) http://www.verisign.com/static/DEV044097.pdf

Witten, I.H., and Frank, E. (2005). Data Mining - Practical Machine Learning Tools and Techniques with Java Implementations. Morgan Kaufmann Publishers.

Hall.M and Frank.E et.al (2009); The WEKA Data Mining Software: An Update; SIGKDD Explorations, Volume 11, Issue 1. Mark Frey: "EPCglobal Certificate Profile [online]," Available http://www.epcglobalinc.org/standards/cert/cert_1_0_1-standard-20080514.pdf.

Frey.M, 2008 "EPCglobal Certificate Profile [online],"Available http://www.epcglobalinc.org/standards/cert/cert_1_0_1-standard-20080514.pdf.

Zadrozny, B., Langford, J., and Abe, N. (2003). Cost-sensitive learning by Cost-Proportionate instance Weighting. In Proceedings of the 3rd International Conference on Data Mining pp. 435-445.(2005) GS1 :

Wine Supply Chain Traceability” [Online] Available:http://www.gs1.org/docs/traceability/GS1_wine_traceability.pdf(2006, Sep)

Vivian Yeo : Bedding, wine get a taste of RFID[Online]. Available: http://www.zdnetasia.com/news/communications/0,39044192,61953022,00.htm(2007 Mar.).

Australian IT, 2009: "RFID to fight wine fraud" [Online]. Available: http://www.australianit.news.com.au/story/0.24897,21355653-I5841,00.html.

Domenitz.L and Kravitz.J (2011); e-Provenance [Online] : Available: http://eprovenance.com/WNYUV76B/index.htm?
Jared Sagoff : New bottle cap thwarts wine counterfeiters [Online]. Available: http://www.anl.gov/Media_Center/News/2008/NE0

15

An RFID-Based Anti-Counterfeiting Track and Trace Solution

Ioan Ungurean, Cornel Turcu, Vasile Gaitan and Valentin Popa
Stefan cel Mare University of Suceava
Romania

1. Introduction

As markets become more global and competition intensifies, firms are beginning to realize that competition is not exclusively a firm versus firm domain, but a supply chain against supply chain phenomenon (***a, 2008). Under these circumstances, an increasing strategic importance to any organization independent of size or of sector is to deliver information, goods and services in full, on time and error-free to customers.

Radio Frequency Identification (RFID) technology represents one of a number of possible solutions to enhance supply chain. RFID technology permits the unique identification of each container, pallet, case and item to be manufactured, shipped and sold, thus allowing an increased visibility throughout the supply chain. Also, an RFID anti-counterfeiting mechanism could be implemented.

This chapter focuses on how RFID technology can be used to solve problems faced by supply chain, such as track and traceability, anti-counterfeiting. It proposes a track-and-trace anti-counterfeiting system using RFID technology. The submitted system (hereinafter referred to as ATPROD system) is aimed at relatively high-end consumer products, and it helps protect genuine products by maintaining the product pedigree and the supply chain integrity. Our system integrates mobile systems to extend corporate data outwards to mobile devices for viewing and querying. Also, users can use any mobile device endowed with an RFID reader for data collection. In this way, manual entry data has been eliminated. Moreover, users can read the tags wherever the items are placed, which enables a more flexible storage environment and an efficiency increase of supply chains and anti-counterfeiting.

We developed an RFID embedded system based on an eBox with an RFID reader attached. This system, named MICC (Interfacing, Command and Control Module), enables many applications to run at the same time as concurrent processes.

Each entry or/and exit gate of the warehouse in a supply chain could be managed by a MICC module. If there are multiple gates the installed MICC modules (from warehouse or company) could be linked together into a network.

From a functional perspective, the MICC module must meet the following requirements: to read/write data on RFID tags attached to items passing through a gate, to manage a large number of RFID tags passing through a gate at the same time, to provide data transmission via the network to a central server, to process local data and to provide the possibility of online and offline operation, as well as a set of commands in order to adapt it to a range of applications by software configurations.

From a hardware perspective, the MICC module is an embedded device, built around Vortex86SX SoC (System on Chip) device. This device integrates an x86 processor, different input/output interfaces (RS-232, parallel, USB, GPIO), BIOS, power management, MTBF counter, LoC (LAN on Chip), JTAG on the same chip. Two versions of MICC module have been designed and developed: MICC01 - without VGA output and MICC02 with VGA output. The RFID reader could be directly connected to the eBox using the USB or serial port. As for the operating system employed, each module runs Windows CE 6.0, which can perform real-time operations.

An OPC (OLE for process control) data server will run on a MICC module. This server is designed according to OPC specifications and can be a possible support for RFID middleware. The OPC data server ensures communication with the RFID reader/writer and sends information to the central database.

Thus, the application from the MICC module will be developed as an OPC data server that will communicate via RS232 or USB with RFID reader/writer. The communication between data server and RFID reader is based on a communication model that uses real-time capabilities of the operating system. This communication model allows the management of a large number of tags at the same time.

The developed system offers a good price-performance ratio. Also, it will satisfy a large number of customer requirements for fields such as industry, retail, supply chain, logistics etc.

2. Functional description of the authentication system

The MICC module is designed as a component of an RFID-based authentication and track & trace system for supply chains. This subchapter displays a short description of this system.

The main role of the system is to authenticate well-known brand products. Such authentication is carried out at various points within the supply chain (starting with the manufacturer, up to the end user). The secondary function of the system is to track and trace products.

Each product has an RFID tag (also named transponder) attached to it. An RFID tag has a factory-programmed identification code stored in a non-volatile memory. This tag also provides a limited capacity memory that is used to store required information. Thus, an RFID tag could store data concerning the trace of the product to which the tag is attached (for every point of the trace; such information will also be sent to the manufacturer's server).

Operationally, product authentication and track & trace can be structured on three levels (see Fig. 1): manufacturer, distributor and retailer. At the manufacturer's level, there is a server by which product authentication is performed.

At the manufacturer's level, each manufactured product will have an attached tag that identifies the product (at the encasement phase). This tag may display initial information (manufacturer, product code, manufacturing date, warranty period, server's address where authentication of products can be performed, or any other information). If a product needs special transport and storage conditions, the RFID tag may hold a temperature sensor and memory, in order to carry out temperature logger (automatic data recorder), which will include data read from the temperature sensor. Also, at this level, products can be grouped into packages or pallets.

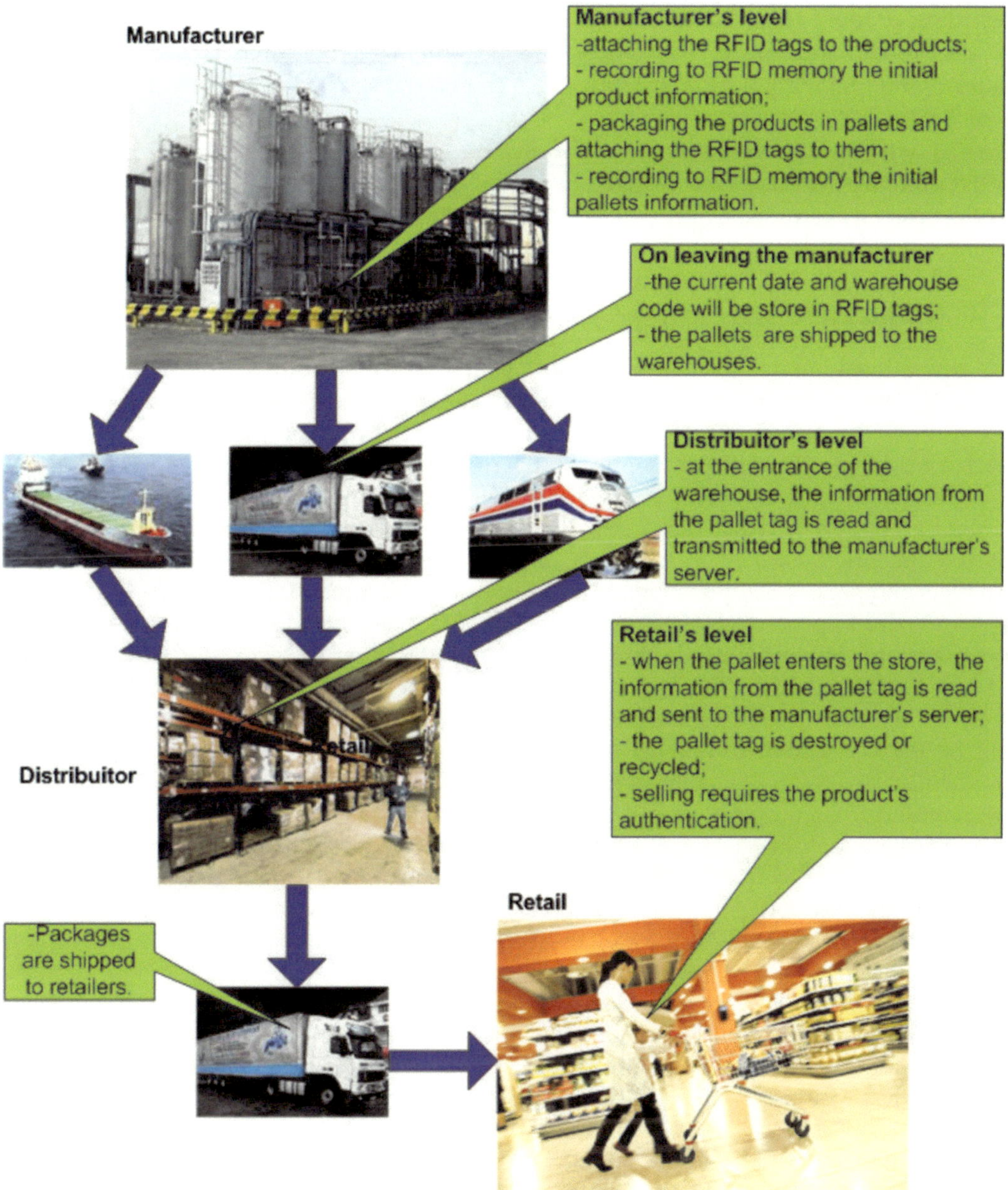

Fig. 1. Authentication and track and trace of products from manufacturer to end user customer

At the distributor's level, when a product is received into the warehouse, the information saved on the attached RFID tag will be read and sent to the manufacturer (if authentication is required). To accomplish the authentication process, a comparison is carried out between the information received from the distributor and the information from the manufacturer (stored in the manufacturer's database server). The distributor receives the results of this comparison. If the authentication process confirms the product's origin, then the data

regarding product reception into the distributor's warehouse will be automatically written to the product tag. Since distributors can be organized on three levels (international, national and regional), products or packages can be transported to any other distributor or retail dealer in the supply chain.
In retail, when products are received into the warehouse, authentication can be performed in the same manner as at the distributor's level. Afterwards, the RFID tag attached to the product can be destroyed or kept attached to the product in order to preserve product ID for future maintenance.

3. Hardware architecture of the ATPROD system

The operational model proposed in this chapter aims to authenticate, track and trace products, starting from their manufacturing phase up to their selling to end users (Fig. 1). The ATPROD system may also be developed in order to track and trace products, until they reach a recycling center. Block diagram and hardware elements are illustrated as follows:
Fig. 2 emphasizes the general hardware architecture of the operational model, at one of points in the supply chain where a pallet/product passes: manufacturing, distribution or retail. Elements illustrated in the block diagram are the following: RFID tags attached to products and pallets, RFID readers, MICCs (Interface, Command and Control Module), PCs used in order to process information read from RFID tags and to authenticate products in accordance to this information; firewall and/or router used to secure Internet connection, manufacturer's server - installed at the manufacturer's level for each manufacturer, PCs provided in order to read information associated to each product, MICC. The MICC module will be used in order to process information read from RFID tags, to authenticate the products according to this data, as well as to write the product tag and send information to the manufacturer's database server.
For each manufacturer, a central server is employed to store data specific to each product. This server will be connected to the Internet (and protected by a firewall), and used to perform authentication. All PCs connected to the Internet can connect to this server, through a password and security certificate, to access information about products of a specific manufacturer.
Pallets and/or products are provided with RFID tags. Information stored on these tags is read/written/updated by RFID readers. Readers are connected by a serial port or USB port to PC or MICC module, which controls and processes information read from RFID tags. By means of these connections, the MICC modules should receive information read from RFID tags; the modules should also transmit new information to be written on the RFID tags. The ATPROD system should operate on 13.56 MHz and should allow multiple tag readability (reading of tags attached to products included in a pallet) enabled by its anti-collision function.
MICC modules are similar, in function, to dedicated computing systems (Barr, 2007), being connected to the Internet (and protected by means of firewalls), in order to carry out the authentication of products on manufacturer's server. Dedicated computing systems are used more often, because they center round systems designed and optimized to carry out specific tasks. In contradistinction to the general use of computing systems (personal computers), a dedicated system includes a hardware subsystem specialized and optimized to carry out the tasks for which the system was designed. Since the system is dedicated to reach specific tasks, the designers can optimize its architecture, in order to reduce its dimension and final cost.

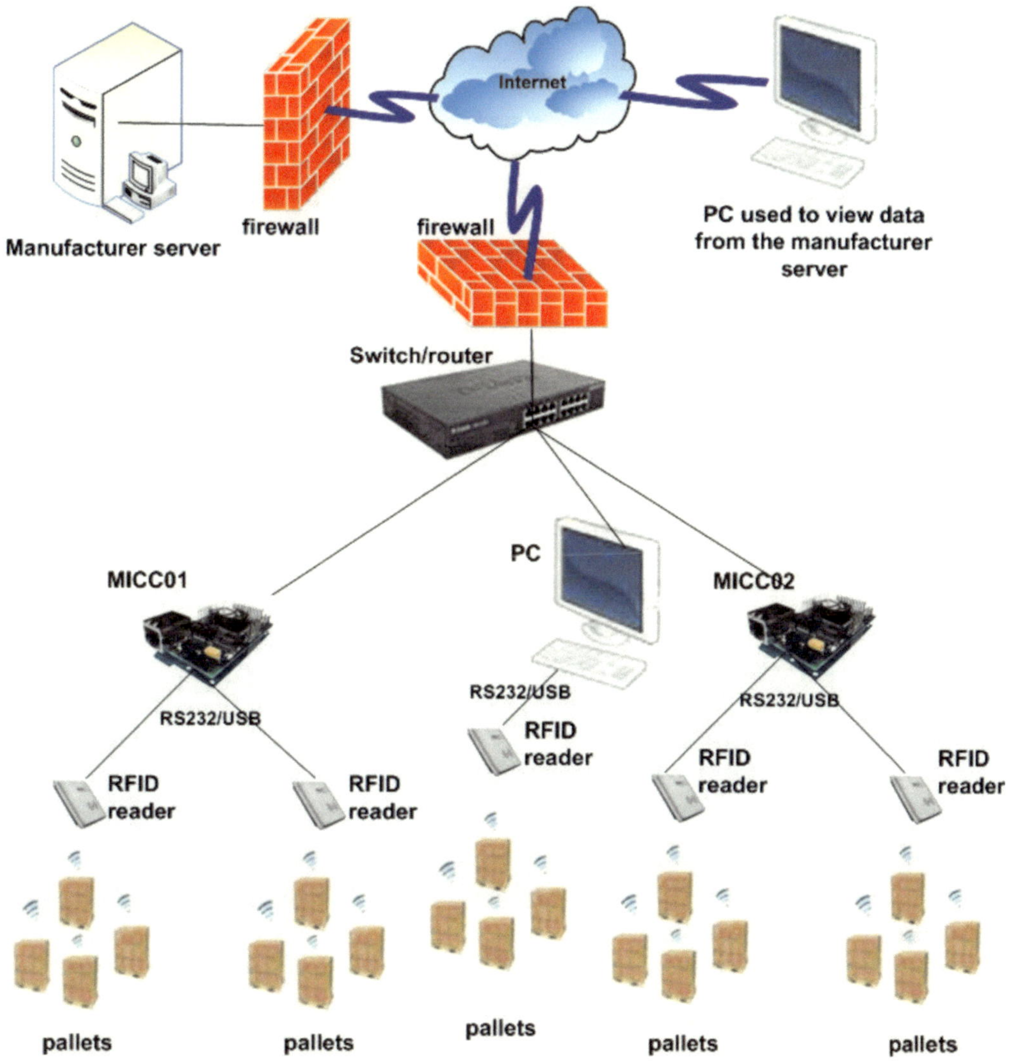

Fig. 2. Hardware architecture of the ATPROD system

The architecture illustrated in Fig. 2 is part of the class of dedicated architectures; these types of architectures face intensive network traffic better, the latter being specific to the infrastructures of a high number of RFID tags. A router of firewall equipment will isolate the network, and ensure data security. In this way, bandwidth is saved, which would have been otherwise busy with various passwords, authentications and other security information. Computing time is also saved, which otherwise would have been spent with different encryption/decryption methods, key generators etc.

In defining hardware architecture at unit level (manufacturer, reseller or retailer), hardware resources should be taken into consideration (storing capacity, communication interfaces, computational ability), required by the software packages used on implementing the system.

In what concerns the software, the use of OPC specifications helps improve the system (Lange et al., 2010) (Gaitan et al., 2010), as a potential support to RFID middleware. These specifications are used by many manufacturers to implement many applications of their work field. The current OPC specifications reached the maturity phase (Mahnke et al., 2009). OPC specifications allow the connection of any OPC server to any other OPC client. Several clients can be connected to the same server and a client can connect to more than one server. Thus, a server can become another server's client, or servers can connect directly with one another. A special versatility will therefore be achieved for application configuration. These characteristics may generate a configuration of servers hierarchically connected in a tree type structure.

Clients at the level of MICC modules can also be created, in the view of configuring and local tracking, for packaging of cases, pallets or charging/discharging at docks level. If servers from other levels are clients of other servers, then servers from the last level will represent the clients of readers. The diversity and complexity of readers should be hidden by these servers. As result, they will implement a software level, often named HAL (Hardware Abstraction Level), especially to operating systems, which signifies a level of hardware abstracting. Each unit is connected to the Internet, and all units are grouped in a VPN network.

4. Block diagram of MICC module

After an analysis of hardware requirements for the MICC module, its designing in accordance with SoC Vortex86SX device was proposed. Depending on functions provided by the Vortex86SX processor (***a, 2010), carrying out a MICC module is proposed. Its architecture can be seen in Fig. 3. The Vortex86SX processor (***a, 2010) is compatible with x86 family and is of a SoC type. By using this solution, Windows XP or Linux operating systems can be set up on a MICC module, where the system operates as a desktop system of limited resources. If the aim is the executing of tasks in real time, operating systems as Linux Embedded or Windows CE can be used. As illustrated in the figure, the MICC module provides 1 port of Ethernet, 3 USB ports, 2 RS232 ports, 32 programmable digital I/O (0-5V), 1 IDE port for HDD connection, an interface for connecting a CF flash card, one PS2 port for connecting mouse and keyboard, as well as one VGA output used for monitor connection.

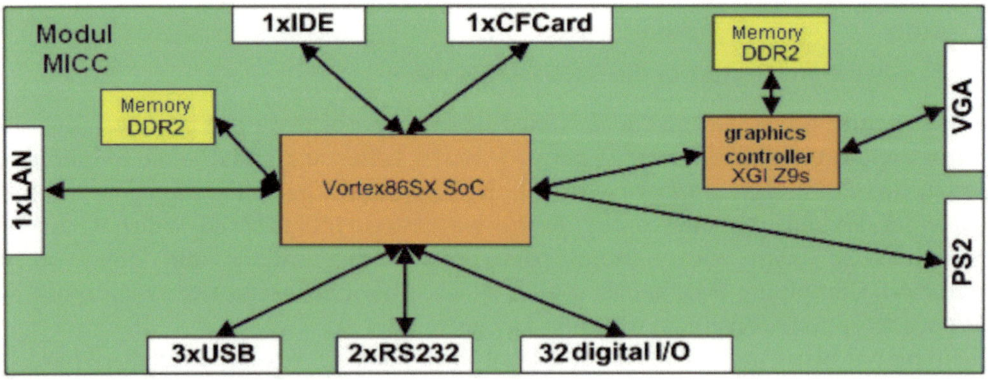

Fig. 3. Block diagram of MICC module

Most of the facilities provided by the MICC module are integrated on a SoC Vortex86SX chip (***a, 2010). Besides, this chip includes a DDR2 memory of 128 MB connected to

Vortex86SC by a DDR2 interface, and a graphical controller XGI Volari Z9s connected to Vortex86SX by a PCI interface. After defining these requirements, printed circuit board (PCB) design can be started. Electrical circuitries can be performed for MICC02 version (with VGA port), and the MICC01 (without VGA) can be obtained by using the same PCB, on which the graphical controller and DDR2 memory used will not be mounted.

The MICC module is designed according to Vortex86SX (***a, 2010) produced by DMP company (***b, 2010). Vortex86SX is a SoC x86, manufactured by using 0.13 microns technology and a model of very low power consumption (less than 1 Watt). This intelligent SoC displays important features, such as: various interfaces of input/output (RS-232, parallel, USB or GPIO), BIOS, WatchDog type timer, management of power consumption, MTFB counter, LoC (LAN on chip), JTAG, etc., features that are not integrated on a single chip of 27x27mm (BGA-581). Vortex86SX is compatible with Windows CE, Linux and DOS operating systems. It integrates, on the same chip SoC 32KB a cache memory L1, ISA bus on 16bits, PCI bus Rev. 2.1 of 33MHz on 32 bits, SDRAM, DDR2, ROM controller, IPC (peripheral internal controllers with DMA and timer/counter of interruption included), SPI (serial peripheral interface), Fast Ethernet MAC, FIFO UART, USB 2.0 main and IDE controller.

5. Designing PCB circuit board for MICC module

The next step in designing the MICC module is the PCB circuit board design. We aimed to obtain a board of 11x11cm, resulting in an ergonomic MICC module of low sizes. The PCB circuit board is structured on 3 layers (Top, Middle and Bottom) of minimal width of a running wire of 10 mil (use of three layers is preferred, due to the high number of pins for Vortex86SX SoC chip). Fig. 4 illustrates the PCB circuit board for designing the MICC device.

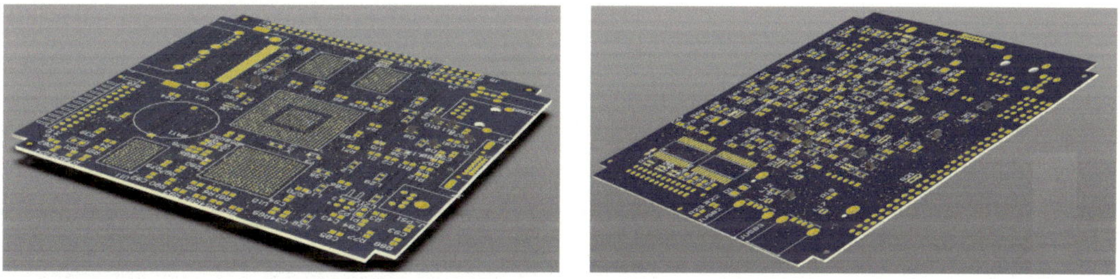

Fig. 4. Front and back images of the PCB

6. Software architecture of the ATPROD system

The general architecture of the ATPROD system is illustrated in Fig. 5. It is obvious that several manufacturers may co-exist in this architecture. Each of them could have more production lines, geographically distributed in more locations. The warehouses and retailers are also geographically distributed in some more locations, provided with Internet connection, in order to have access to the manufacturers' servers and to be able to authenticate products and send information related to their traceability.

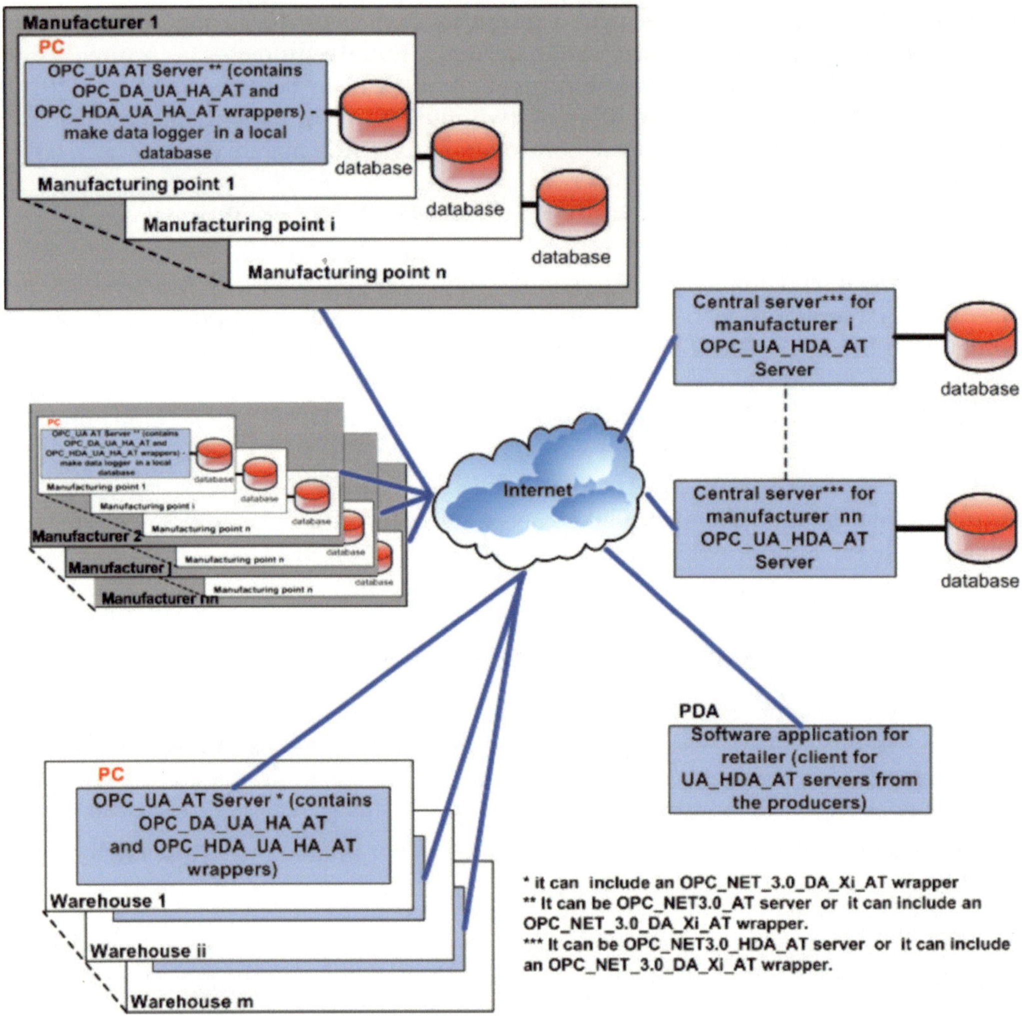

Fig. 5. Software architecture of the ATPROD system

Each manufacturer has a central server, on which an OPC_UA_HDA_AT server runs. This server will store information regarding the products that exit the production line and are sent to the resellers' warehouses. At the level of each distribution point, there is a PC that runs an OPC_UA_AT server. This server is able to send requests for information about the manufacturing process, as well as to save relevant information within a local database historian.

It is important to remember that this server should not be in the same location as the manufacturing points, where the only condition to be met is the existence of an Internet connection. In this way, OPC_UA_AT servers from the manufacturing points are in fact clients of the OPC_UA_HDA_AT server. Fig. 5 also illustrates the way in which the shared database is developed: for each reseller, the database is shared to all distribution points and central server.

Each warehouse is provided with a server, on which an OPC_UA_AT server runs; this server also represents a client of OPC_UA_HDA_AT associated to each producer, in order to require necessary data for authentication. The manufacturer's server will also receive information about input of products, storing conditions or exit of products from warehouses. In what concerns the retail dealers, a PDA with an RFID reader can be used, on which a client of OPC_UA_HDA_AT servers runs. These servers are clients for OPC_UA_AT manufacturers' servers and can be used in order to authenticate products.

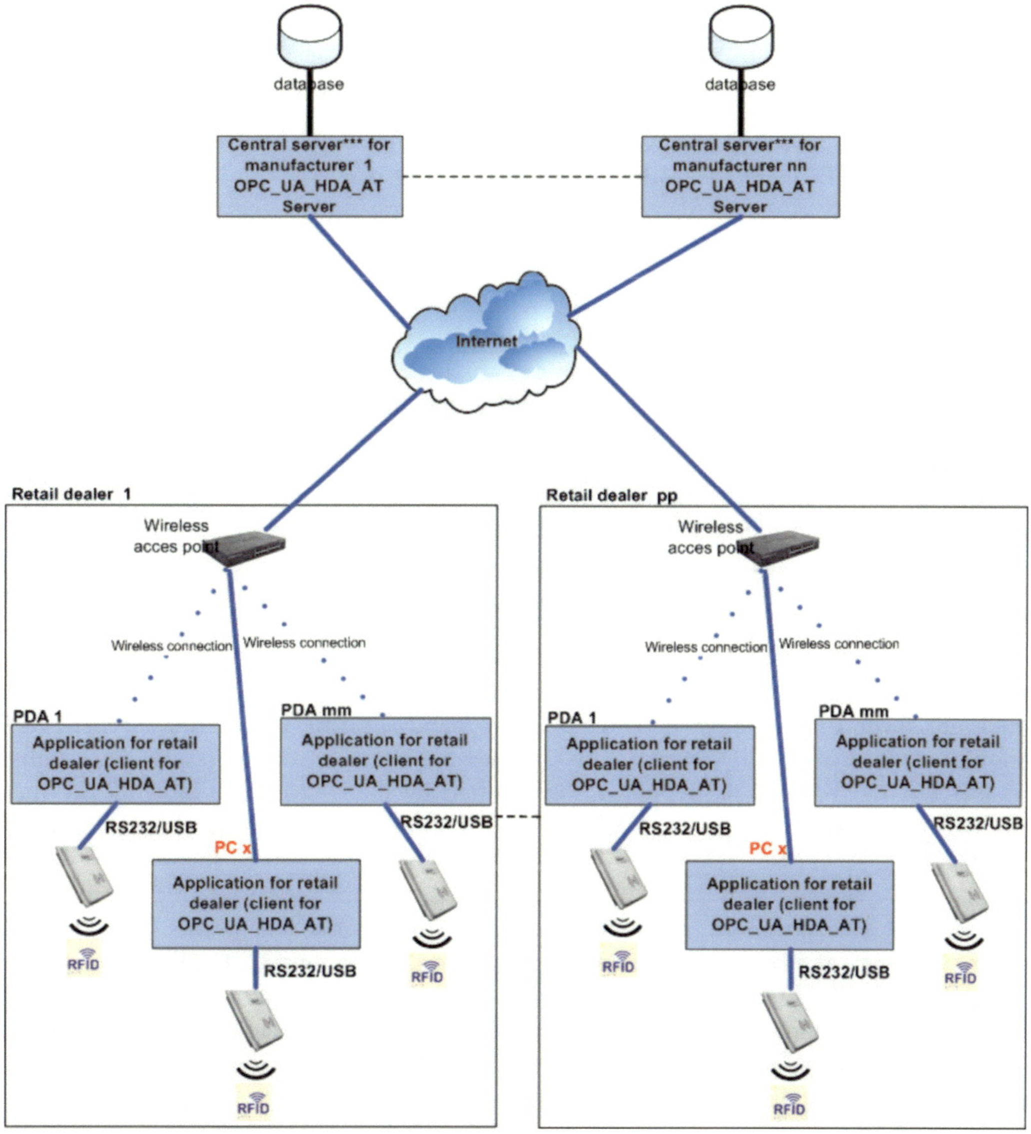

Fig. 6. ATPROD system seen from the perspective of a retail dealer

A client from a warehouse or retail dealer can send an authentication request to the manufacturer's servers (central or local). If the needed information cannot be found within the central database server associated to the manufacturer, this will require data from a server existing in the manufacturing point. After information is achieved, it is sent to the client which has required it, in order for the client to identify products. Using such mechanism, information existing within the shared database related to the tagged products can be freely accessed.

As can be seen in Fig. 5, the shared database between manufacturers' local servers and the central servers is illustrated in red color. For each point, a SQL database server is set up. OPC_UA_AT and OPC_UA_HDA_AT servers can access the local database by means of SQL commands.

Fig. 6 emphasizes the way a shared database can be accessed by dealers. Therefore, one or several PDA or tag-reading PCs can be provided for each dealer. It is very important that PDA devices and PCs used for authentication should be connected to the Internet, so as to make possible the access to the manufacturers' servers. If the Internet connection of a manufacturer associated server does not work properly, the authentication of products will not be accomplished. On such computing systems, an OPC client application runs for OPC_UA_HDA_AT servers associated to each manufacturer. When a product is sold, the client application requires information from the manufacturer's server in relation to the authentication of that product.

The requested information is sent to client application, and the authentication of product is carried out. After authentication, information concerning the selling of product is sent back to the manufacturer's server.

Information concerning the selling of products is not erased from the database. By means of client application, the database administrator will be able to carry out erasing operations related to products sold or to create an archive with this information (for each manufacturer) that users can use at a later time.

Operationally, servers within ATPROD system are placed in two different locations: at manufacturers and in warehouses. Fig. 7 illustrates the structure of servers, as seen from the manufacturer's perspective.

There are several RFID tagging points at each manufacturing point, and several MICC modules connected to RFID readers. The OPC servers running at these points are assigned as OPC_DA_CE_AT and their presence is justified by the necessity to label all products that exit the production line. Since MICC module should work both online and offline, a historian OPC server can also run on it; therefore, a historian will be carried out for data that should have been sent on network, but cannot be transmitted when the connection with the server is no longer active at the level of manufacturing points. OPC_UA_AT server existing at the manufacturing point level includes OPC_DA_UA_AT and OPC_HDA_UA_AT wrapper, so as to connect to OPC_DA_AT and OPC_HDA_AT servers on MICC modules. This server stores a historian with all operations performed at the manufacturing point. The central server OPC_UA_HDA_AT from the manufacturer's level should also be a client of the OPC_UA_AT servers, at the level of manufacturing points, so as to require the necessary data. However, such data is not stored within the central server database, but within the database of manufacturing points. OPC_UA_AT servers can be replaced with OPC_NET3.0_AT servers that use WCF (Windows Communication Foundation) technology.

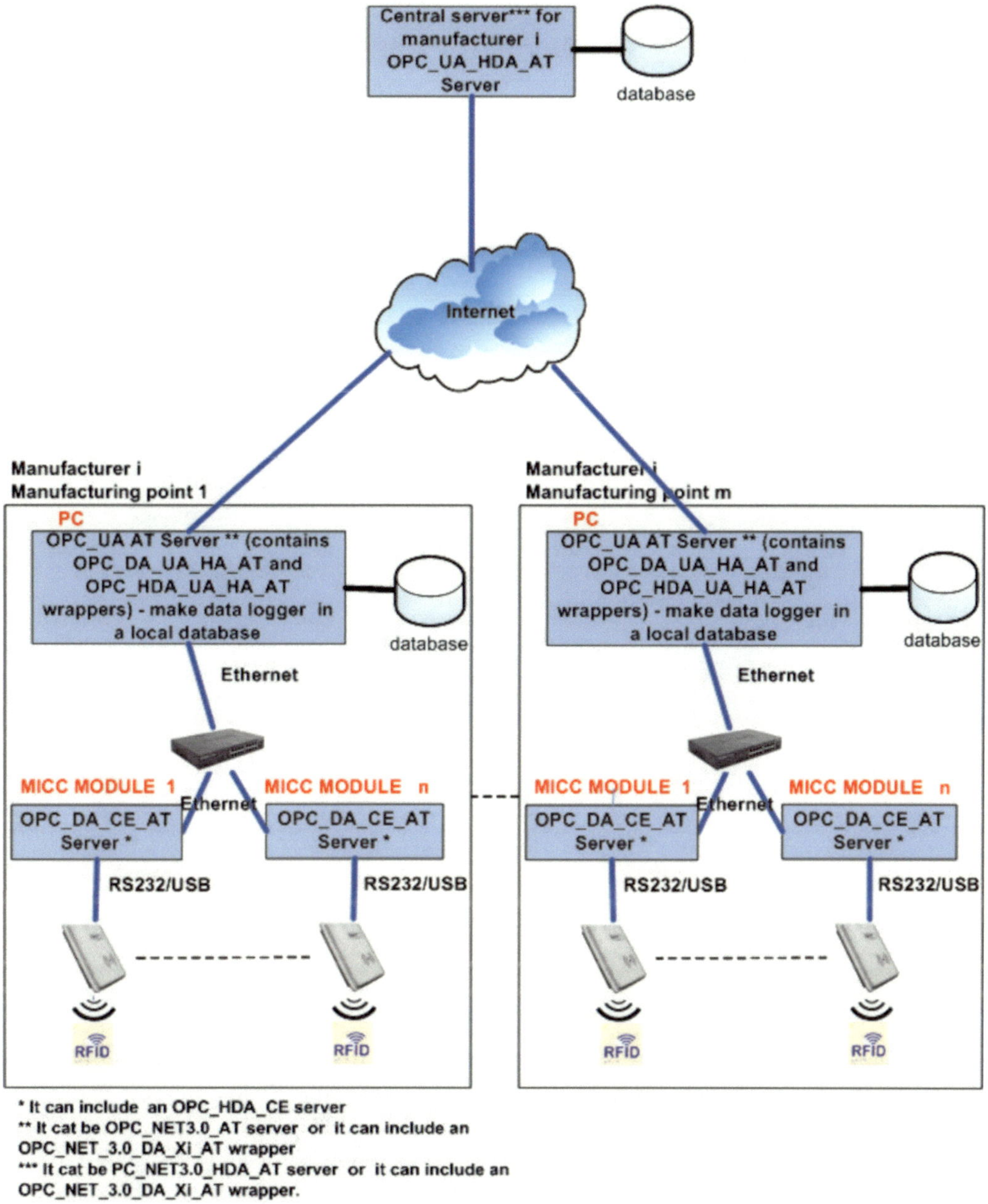

Fig. 7. Placing of servers to the manufacturer site

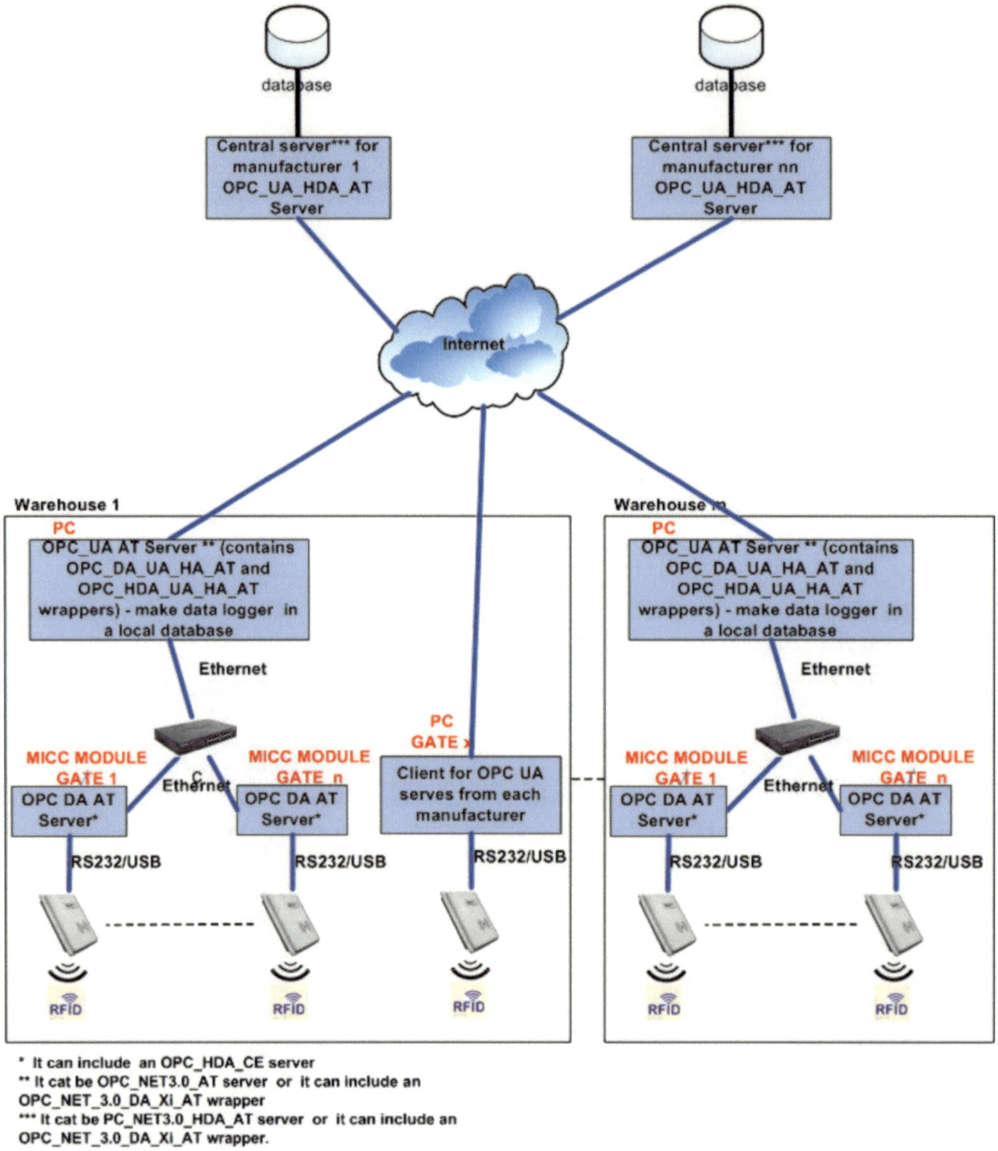

Fig. 8. Placing of servers to warehouses' site

Fig. 8 emphasizes the architecture of servers as regards the warehouses. One might see that here, at the level of each system's gate, there is a MICC module. Such a module is provided with an RFID reader, connected by RS232 serial port or USB port, in order to read/write information from RFID labels. This module runs OPC_DA_CE_AT and OPC_HDA_CE_AT servers. An OPC_UA_AT server runs on a server at the warehouse level. This server is a client of OPC_DA_CE_AT and OPC_HDA_CE_AT from MICC modules. This fact is accomplished by including the OPC_DA_UA_AT and OPC_HDA_UA_AT wrappers. In order to achieve the information necessary to product authentication, this server signifies a client of the servers existing at manufacturers' level. Any input or output of products from the warehouse will be sent to manufacturers' servers. In case of small distribution chains, OPC_UA servers can be replaced with OPC_NET3.0_AT servers that use the WCF (Windows Communication Foundation) technology.

Up to the present, six types of servers have been identified: OPC_DA_CE_AT, OPC_HDA_CE_AT, OPC_UA_AT, OPC_DA_UA_AT, OPC_UA_HDA_AT, OPC_NET3.0_AT.

OPC_DA_CE_AT runs on MICC modules, specific to manufactures and warehouses. This server includes a driver for the communication with RFID reader. Such a server should run on WINDOWS CE 6.0 operating system. OPC_HDA_CE_AT runs on MICC modules, specific to manufacturers and warehouses. This server is a client of OPC_DA_CE_AT server and will carry out a history if the local connection is interrupted. The server should run on WINDOWS CE 6.0 operating system. OPC_UA_AT runs at the level of manufacturing points and warehouses. It is a client of OPC_DA_CE_AT and OPC_HDA_CE_AT servers, including the OPC_DA_UA_AT and OPC_DA_UA_AT wrappers. This is also a client of OPC_UA_HDA_AT servers, at the level of manufacturers. OPC_UA_HDA_AT runs at the level of manufacturing points and centralizes all data corresponding to products circulating within the distribution chain. OPC_NET3.0_AT- emphasizes an alternative of OPC_UA_AT and OPC_UA_HDA_AT servers.

7. Software architecture of the MICC module

MICC module is connected by means of RS232 or USB ports to a RFID reader/writer, using 13.56 MHz frequency band and ISO 15693 standard. This module should allow the reading or writing of information on RFID tags.

One should mention that products can use two types of RFID tags: standard self-adhesive RFID tags, 13.56 frequency band, ISO 15693, variable memory (I code SLI - 896 bits, Tag-it TM HF-I - 2048 bits) and RFID tags of types: flexible card, self-adhesive, active, 13.56 MHz frequency band, ISO 15693, provided with integrated temperature sensor and temperature values history, Variosens model made by KSW Microtec, 8kbits EEPROM, which can store 1 720 values of 10 bits temperature values. All products have tags attached to them, mostly of them of first class mentioned above; those products that need special conditions of warehousing and transport can also have attached RFID tags of the second type.

Operationally, the MICC device should meet the following requirements: reading of information from RFID tags; setting up the tags so as to establish the sampling rate; if necessary, reading the temperature and storing its value into RFID tag's memory; storing of information read from RFID tag into its own memory; sending data to central server by means of Ethernet; possibility of both on-line and off-line operations.

Fig. 9 illustrates the position of MICC module within ATPROD system. Therefore, it is placed at the input or exit of a warehouse. In what concerns the input, the MICC device reads the RFID tags attached to products that enter warehouses, sends the information read to the central server, accomplishes the authentication of products (by comparing information from tags to information existing within manufacturer's server), and writes the information related to the input on RFID tag. When products have attached RFID tags provided with temperature sensors, the MICC module reads the history of temperatures and deletes this history from RFID tag.

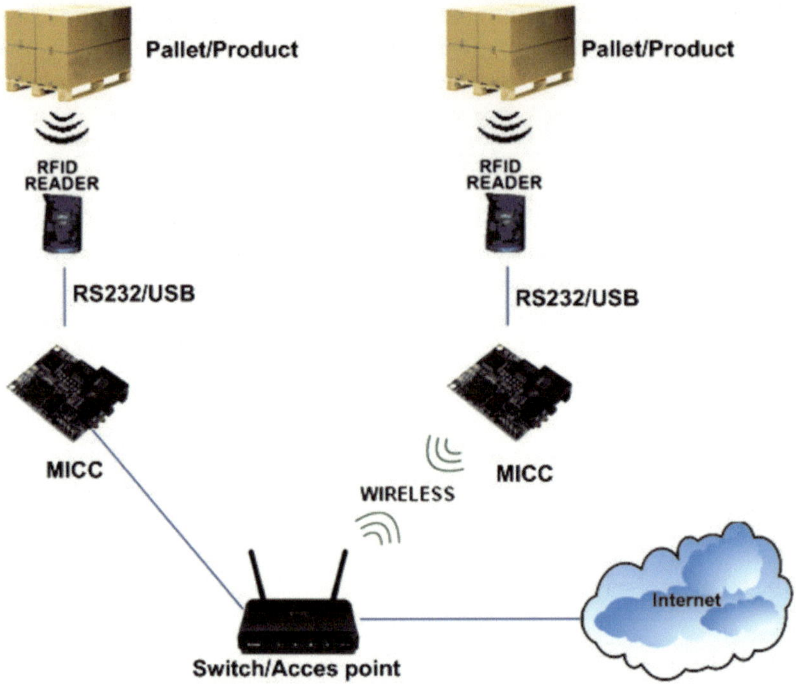

Fig. 9. MICC module operating mode

Fig. 10 also shows the UML diagram with a view to using the MICC module. From this diagram, the main two operations carried out in this module can be identified, as follows: reading of information from RFID tag, as well as writing of information on this RFID tag.

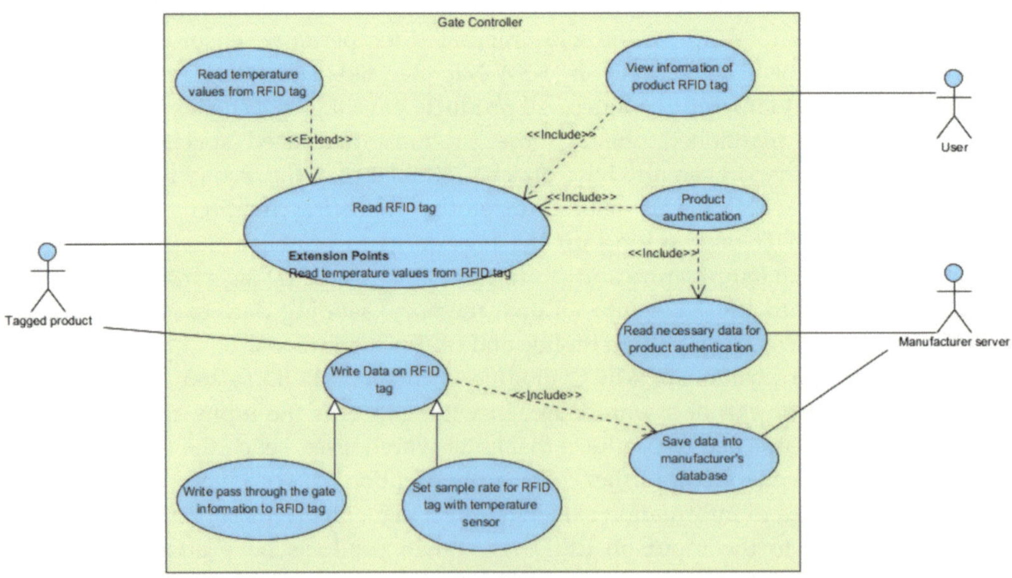

Fig. 10. UML diagram of using MICC module

Reading of information from RFID tag is done depending upon RFID tag's type (with or without temperature sensor). The operation of product authentication is carried out after reading information from RFID tag, and includes the connection to the manufacturer's server, the reading of information from this server about products, and finally a comparison between this information and that provided on RFID tags. Due to this procedure, authentication of products can be performed only if the module is provided with direct connection to manufacturers' servers.

As previously stated, data writing is performed depending upon the RFID tag under use. If an RFID tag provided with temperature sensor is used, then the sampling rate can be set up in order to save temperature values, and to empty the memory after reading the information included on RFID tag. If a classical tag is used, that is, without temperature sensor, then the tag will be written with information concerning the input or exit in or from warehouse, input/exit date, warehouse code, etc.

The application running on MICC device will be further developed in C++, under an OPC server type, able to communicate with the RFID by means of RS232 or USB. The application will be developed by using SDK package, performed after creating an image on Windows CE 6.0 (Samuel, 2008).

8. Conclusion

RFID will have significant impacts on the economy as well as on the operational and financial performance of companies in the focus areas: productivity, employment, markets, goods and services, and innovations and new products. RFID will especially generate significant impacts in applications with a unique selling proposition, e.g. anti-counterfeiting, secure supply chains, and cold chain and quality monitoring, as well as better information for decision makers.

This chapter proposes an RFID-based system to track, at the item level, material flows among partners until they reach the consumer, while maintaining data accuracy. The presented system helps small, medium companies and enterprise organizations to improve productivity and provide better service to their customers. Thus, our system has the potential of helping retailers provide the right product at the right place at the right time, allowing maximizing sales and profits.

The system is still under construction and in the near future some security aspects will also be taken into consideration.

9. Acknowledgment

This work was supported by the project "Computer system for controlling and checking the authenticity of products - ATPROD" - Contract no. 12082/2008, project co-funded by 2007-2013 PNCDI Program.

10. References

Barr, M., (2007). *Embedded Systems Glossary*, Available at http://www.netrino.com/Embedded-Systems/Glossary

Chalasani, S.; Boppana, R.V.; Sounderpandian, J. (2005). RFID Tag Reader Designs for Retail Store Applications, *AMCIS 2005 Proceedings*, Paper 149, Available at http://aisel.aisnet.org/amcis2005/149

Gaitan, N. C.; Gaitan, V. G.; Pentiuc, S. G.; Ungurean, I.; Dodiu, E. (2010). Middleware based model of heterogeneous systems for scada distributed applications, Advances in Electrical and Computer Engineering. Vol.10, No.2, pp. 121-124, ISSN: 1582-7445

Lange, J.; Iwanitz F.;, Burke, T.J. (2010). *OPC - From Data Access to Unified Architecture,* fourth edition, revised and extended, 431 pages, ISBN 978-3-8007-3242-5

Mahnke, W.; Leitner, S.H.; Damm, M. (2009). *OPC Unified Architecture,* Springer; 1 edition (May 4, 2009), ISBN: 978-3-540-68898-3

Preradovic, S.; Karmakar, N.C.; Balbin I. (2008). RFID Transponders, *IEEE MICROWAVE MAGAZINE,* Vol. 9, No. 5, pp. 90-103, ISSN: 1527-3342

Samuel P.(2008). *Professional Microsoft Windows Embedded CE 6.0,* Wrox; New edition (November 3, 2008), ISBN: 978-0470377338 ***a (2008). *A Summary of RFID Standards, RFID Journal,* Available at: http://www.rfidjournal.com/article/view/1335/1/129 ***a (2010). *Vortex86 System-On-Chip (SoC),* Available at http://www.vortex86sx.com/ ***b (2010). *DMP Electronics INC.,* Available at http://www.dmp.com.tw/

16

A Knowledge-Based Approach for Detecting Misuses in RFID Systems

Gennaro Della Vecchia[1] and Massimo Esposito[1,2]
[1]*Institute for High Performance Computing and Networking of the National Research Council of Italy*
[2]*University Parthenope*
Italy

1. Introduction

In the last few years, the Radio Frequency IDentification (RFID) technology has gained increasing attention as an emerging solution for automatically identifying remote objects, people, animals. Early successful applications in asset tracking and supply-chain management and the falling cost of RFID tags have fostered a broadening of the application domain, with new pervasive, RFID-based solutions supporting more user-oriented services. As a result, RFID technology is going to contribute to the massive deployment of sensors in an ever more networked society -a coming Internet of Things where everything is alive, that is, where common objects (including those that are inanimate and abstract) can have individual identities, memory, processing capabilities, along with the ability to communicate and sense, monitor and control their own behavior (Thompson, 2004). In previous works we have explored this technology's potential to facilitate everyday life by seamlessly integrating virtual and physical worlds, varying from personnel tracking and localization to healthcare monitoring (Ciampi et al., 2006; Coronato et al., 2009; Della Vecchia & Esposito, 2010; Esposito et al., 2009; Coronato et al., 2006).

As known, typical RFID systems use a combination of tags, readers and middleware as sketched in Fig. 1. Basically, a reader broadcasts a radio frequency signal to get the data stored on the nearby tags. Data can be a static identification number, user written data or data computed by the tag itself. Having obtained tag data, the reader informs via a wired or wireless network the middleware that in turn stores both tag and reader data in a back-end database.

RFID systems deal with information which very often, if not always, may be critical. Such systems are intrinsically insecure and vulnerable, being prone to threats that can affect tag, reader and middleware as well .

In particular, *tag cloning* is one of the most serious threats to the security of RFID systems. Tag cloning simply consists in catching a tag's unique identifier with the aim of making an exact copy (*clone*) of the cloned tag, so that the clone can pose as the genuine tag, being indistinguishable from the original. Once legitimate tag data are obtained, attackers can reproduce their clone tags on a wide scale and gain access to secured facilities, make fraudulent purchases, alter or even disrupt supply chains, etc.

One conventional approach to secure RFID systems against tag cloning might use cryptographic tags that enable strong tag authentication and make tag cloning a rather daunting task, but this would skyrocket the cost of the single tag. As a result, a viable solution to defend against tag cloning in RFID systems seems yet to be developed due to the RFID industry's desire to manufacture commercially affordable tags.

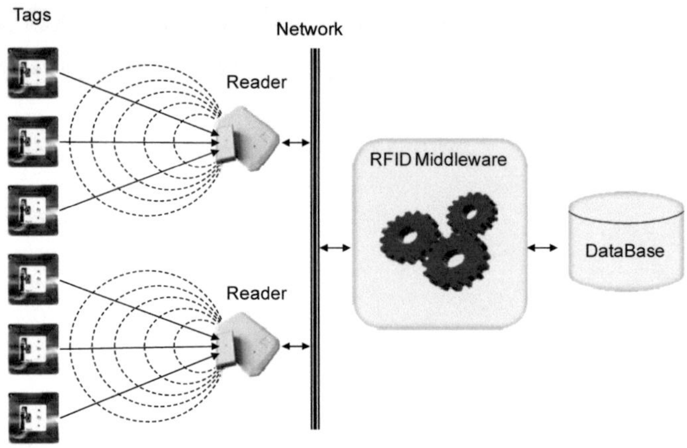

Fig. 1. A typical RFID System Architecture

A less conventional approach to address the tag cloning issue may exploit the well-known security paradigm of "intrusion detection". Generally speaking, an intrusion detection system monitors a given environment and implements a detection method to reveal suspicious activities and respond accordingly. One diffused intrusion detection method is "misuse detection", which utilizes a knowledge base that explicitly models the concept of what is deemed to be "suspicious". Everything that does not match the expected behavior formalized in the knowledge base is considered to be "normal".

In this book's chapter we propose a methodology in which misuse detection uses a knowledge base built upon a "*track & trace*" model relying on the notion of "tag location" to gather all the information required to identify an attack of tag cloning. The knowledge base embedding the track & trace model is formalized in an ontology by means of semantic web languages in order to achieve an unambiguous, well-defined and machine-readable knowledge representation.

The methodology here described stands on three key points: i) the definition of an ontology model formalizing expected and actual profiles, each of them being based on location and tracking information about RFID tagged objects; ii) the application of an inferential engine that, exploiting inference patterns proper to the logic underlying the ontology formalism, checks the consistency between expected and actual profiles of tagged objects; iii) the detection and identification of anomalous conditions in presence of inconsistencies. This methodology led to the design of a misuse detection system aimed at detecting and characterizing tag cloning in RFID applications. Such a system exhibits a reactive and event-dependent behavior in response to new tracking information coming from a network of RFID systems and shows an architecture structured in a set of components operating at middleware layer that can be transparently integrated into existing RFID applications.

The rest of the chapter is organized as follows. Section 2 introduces some preliminary notions, Section 3 discusses motivations and related work, Section 4 describes the proposed methodology and Section 5 illustrates the MDS architecture along with a proof of concept of the knowledge-based approach. Finally, some concluding remarks are reported in Section 6.

2. Preliminaries

2.1 Intrusion detection taxonomy

Intrusion Detection Systems (IDS) can be classified in several ways. It is common to classify an IDS according to the detection method, the audit source, the usage frequency and the response mechanism (Debar et al., 1999).
Classification by the detection method is the most diffused. Mainly, two kinds of detection methods are considered: misuse detection and anomaly detection.
Misuse detection systems utilize a knowledge base that explicitly models what is not allowed. Everything that does not match the knowledge base is allowed.
Anomaly-based systems, on the contrary, use a model of normal activity and anything that does not match the model of normality is considered an attack. An anomaly detector assumes that all anomalous events are signs of an attack and that all attacks produce anomalous events. Since an anomaly-based system does not model attacks specifically, it can detect previously unknown attacks.
Misuse-based systems, on the other hand, can detect attacks of which they have prior knowledge, being unable to detect new forms of attack but some mutation of those already in the rule base. However, a misuse detection approach paves the way to a clear understanding of the application domain, where users need to be aware of and formalize their knowledge about specific misuse scenarios. This leads to low false positive rates and permits a simple and efficient processing of the audit data.
A different method of classification of IDSs takes into account the type of audit data processed. Three different categories of audit data are common, namely network-based audit data, host-based audit data and application-based audit data.
Network-based sensors collect packets from the protected network in order to perform detection. Some network-based systems use firewall logs as input. These firewall logs contain the headers of the network packets that have been blocked by the firewall. Host-based sensors process audit data generated by a host's operating system. It is very common for this type of sensor to perform detection on the log of system calls that have been executed (Hofmeyr et al., 1998; Kruegel & Robertson, 2004). Application-based sensors process logs created by a user-space application. This kind of sensor is usually used to protect network demons, as several systems exist that process web logs.
IDSs systems can also be classified according to their usage frequency. Online systems operate in real-time and consume audit data as they are generated. This is the most common mode of operation. Other systems are run in offline mode, where the system is activated periodically to look for signs of attack.
Finally, it is also possible to classify Intrusion Detection Systems according to the type of response the system yields when an attack is detected. The most common is passive response, where an attack occurrence is logged or the administrator is alerted by other means (e.g., SMS or email). Active response systems block an incoming attack so that it cannot succeed. They are usually referred to as Intrusion Prevention Systems. Depending on the implementation, an active system could, for instance, send a reset packet to tear down the attacker's connection or update the firewall rules so that the attacker is blocked.

2.2 Ontology modeling

Historically, *Ontology* is the philosophical study of the nature of being, existence or reality as such, as well as the basic categories of being and their relations. Traditionally listed as a part of the major branch of philosophy known as metaphysics, Ontology deals with questions concerning what entities exist or can be said to exist, and how such entities can be grouped, related within a hierarchy, and subdivided according to similarities and differences. By extension, the core meaning of "ontology" within Computer Science is a model for describing the world that consists of a set of types, properties, and relationship types. What ontology has in common in both computer science and philosophy is the representation of entities, ideas, and events, along with their properties and relations, according to a system of categories.

The term "ontology" is currently used to mean "a formal, explicit specification of a shared conceptualization" (Gruber, 1995). In this perspective, "conceptualization" relates to an abstract model that identifies the relevant concepts of a certain domain. "Explicit" means that the type of concepts used and the constraints on their use are explicitly defined. "Formal" means that the ontology should be formalized to be machine understandable and enable intelligent agents to infer new statements from existing ones based on a set of rules. "Shared" means that an ontology captures consensual knowledge of a community. Simply put, ontology refers to a formalization of knowledge in a given domain.

An ontology can be used to explicitly represent the meaning of terms in vocabularies and the relationships between those terms. In other words, ontology is the concept which is separately identified by domain users, and used in a self-contained way to communicate information. In particular, some of the reasons why someone wants to develop an ontology are to share common understanding of the structure of information among people or software agents, to analyze domain knowledge and enable its reuse, to make domain assumptions explicit, to separate domain knowledge from the operational knowledge.

An ontology structure holds definitions of concepts, binary relationships between concepts and attributes. Three types of relationship may be used between concepts: generalization, association, and aggregation. Relationships may be symmetric, transitive and have an inverse. Concepts, relationship types and attributes and rules, put together, enable the description of a schema in terms of abstraction. On the other hand, concrete objects populate the concepts, concrete values instantiate the attributes of these objects and concrete relationships instantiate relationships.

The semantic web languages used to formalize an ontology are defined with a model-theoretic semantics. In particular, for the language OWL (Web Ontology Language) (Patel-Schneider et al. 2004), a semantics was defined so that very large fragments of the language can be directly expressed using so-called description logics (Baader et al., 2005). Description logics are a decidable subset of first order predicate logic. Namely, OWL DL (where DL stands for "Description Logic") was designed to support the existing description logic business segment and provide a language subset that has desirable computational properties for reasoning systems.

DL-based knowledge bases are built using concept language expressions, and they are usually divided in two distinct parts: intensional and extensional. The intensional part takes the name of T-Box and describes the general conceptual domain model made of concepts and relationships between concepts and attributes, whereas the extensional part is named A-Box and constitutes a (partial) instantiation of the model, since it contains assertions about a set of individuals.

2.3 Ontology reasoning

The effective use of ontology modeling in real applications is critically dependent on the provision of efficient reasoning services to support both ontology design and deployment. In such a direction, DL-based ontologies have stretched the capabilities of DL inference engines, offering a collection of reasoning services implemented through automated reasoning techniques.

The reasoning services provided by a DL inference engine can be classified as basic services, which involve the checking of the truth value for an assertion, and complex services (Donini et al., 1996). Generally speaking, basic services are, for instance, the verification of the subsumption between two concepts or the satisfiability of a concept, whereas complex services implement tasks such as finding all the individuals being in a concept expression, or organizing in form of taxonomy the concept names appearing in a terminology. In more detail, basic services can be classified in services of terminological reasoning and hybrid reasoning, respectively. Terminological reasoning involves only the terminology (i.e. without considering A-Box assertions), whereas hybrid reasoning takes account of both the parts of a knowledge base, i.e., T-Box and A-Box.

On one hand, terminological reasoning services are intended to verify both *Concept Satisfiability* and *Subsumption*, i.e., to check whether a newly defined concept makes sense or is contradictory with respect to the existing T-Box, and to check if a concept C is more general than another concept D, respectively. On the other hand, hybrid reasoning services are aimed at verifying *A-Box Consistency* (with respect to the T-Box) and executing *Instance Checking*. Specifically, A-Box Consistency checks whether a new assertion in the A-Box generates an inconsistency with reference to the T-Box, whereas Instance Checking allow to decide whether an individual is an instance of a concept or not.

The provided complex reasoning tasks vary from system to system, and are defined on top of the basic services above described. The most common are *Classification* and *Retrieval*, which are terminological and hybrid reasoning services, respectively. Classification consists of explicitly representing the concept taxonomy entailed by the knowledge base, being this taxonomy a graph whose nodes are the concept names appearing in the knowledge base, and the edges represent the subsumption relation between them. This graph can be built by checking the subsumption between every pair of concept names. Retrieval (or *Query Answering*) consists in collecting all the individuals in the knowledge base that are instance of a given concept in every model of the knowledge base.

3. Motivations and related work

3.1 Motivations

The most challenging security threat in RFID applications is tag cloning. The conventional approach to secure RFID systems against tag cloning is to use cryptographic tags that enable tag authentication and make tag cloning considerable harder. The fundamental difficulties of such an approach revolve around the trade-off between tag cost, level of security, and hardware functionalities. As a matter of fact, RFID tags are typically deployed in great amount and the end-user companies have a strong financial incentive to minimize the tag cost and, thus, the features the tags provide (Lehtonen et al., 2009).

As a result, it is extremely difficult to use cryptography for protecting low cost tags from cloning, due to their limited power, storage and processing resources. Moreover, in order to supply cryptographic components with sufficient power, tags would need to be read from a shorter distance, which would degrade the read-rate of readers (Ranasinghe et al., 2005).

A less conventional approach makes use of location information to detect RFID clone tags. Location-based product authentication is an anti-counterfeiting measure that brand-owners may use in many situations to fight against product forgery. For instance, in the last years the pharmaceutical industry has been planning to track and trace the history of each single medicine using RFID information as an effective and proactive measure against cloning, instead of using expensive cryptographic tags. Obviously, the goal of this approach is not to make tag cloning harder. Rather, by properly detecting the presence of clone tags and acting accordingly, it aims at nullifying their effects: a tagged product that lacks valid track and trace history can be easily singled out and labeled as not genuine, thus posing a substantial barrier against counterfeit players.

Keeping in mind the aforementioned trade-off between cost and effectiveness, there are at least a couple of good reasons to assume that the location-based approach can be suitably followed when it comes to fight tag cloning. First, efforts focused on the tag itself are intrinsically insecure, because in an RFID system tags constitute the weakest link in the whole chain due to their limited functional capabilities: an attacker, even with poor resources, can violate their security quite easily; on the other hand, it is rather questionable whether it will be ever possible to produce a truly secure RFID tag, able to address all known vulnerabilities without increasing the overall cost. As a second but not secondary consideration, even if some solution may prove itself effective on preventing tag cloning, a really secure RFID system should go beyond prevention measures and provide detection capabilities *when tag cloning has already occurred* (Mirowski & Hartnett, 2007).

All these considerations constitute the rationale which led us to adopt the location-based approach in developing the methodology proposed in this chapter.

3.2 Related work

RFID technology raises a number of security and privacy concerns, which may substantially limit its deployment and reduce potential benefits. Among the great deal of papers addressing these concerns, an interesting survey can be found in (Rotter, 2009), with the focus put on the technical aspects of security and privacy.

Most specific literature covers the topic of tag cloning and the efforts made by the research community to tackle this threat through a wide range of solutions.

In costly RFID tags, where resources are less subject to strict constraints, several countermeasures have been devised to combat tag cloning, such as deactivation of tags, encryption, authentication and hash codes (Karygiannis et al., 2007). In (Juels, 2005), some techniques are illustrated for strengthening the resistance of EPC tags against cloning attacks, using PIN-based access to achieve challenge response authentication. In (Weis et al., 2004), the authors proposed a cryptographic approach to lock the tag without storing the access key, but only a hash of the key on the tag instead. The key is stored in a back-end server and can be found using the tag's meta-ID.

Duc et al. (Duc et al., 2006) proposed a communication scheme to protect user privacy in RFID system which is based on a synchronous session key between tags and back-end database server to authenticate each other. A further development of Duc's scheme which overcomes some vulnerabilities has been proposed in (Cheng et al., 2009).

Avoine and Oechslin (Avoine & Oechslin, 2005) proposed another hash-based RFID protocol providing modified identifiers for improving privacy that can be applied for authentication. In addition, hash-based RFID protocols for mutual authentication have been proposed in (Choi et al., 2005; Lee et al., 2006). All these protocols rely on synchronized secrets residing on the tag and back-end server and require a one-way hash function from the tag.

In contrast, in low cost RFID passive tags, due to their small size and strictly constrained resources, complex cryptographic solutions like hash functions cannot be implemented. In (Sarma et al., 2003) the authors mention scarcity of tag resources in low-cost RFID systems as a primary challenge in providing security and privacy mechanisms, and in combating cloning as well. In this perspective, few lightweight authentication protocols that do not require cryptographic hash/keys in the tag have been proposed (Karthikeyan & Nesterenko, 2005; Chien, 2007). Yet another approach to tackle tag cloning uses a Physical Unclonable Function (PUF) (Devadas et al., 2008). PUFs significantly increase physical security by generating volatile secrets that only exist in a digital form when a chip is powered on and running. Its main property is that it is easy to generate but hard to characterize.

As it clearly appears, all these efforts aim at *preventing* tag cloning. However, *detecting* fake tags plays a not lesser role. Once genuine tags have been cloned, the success of a potential attack greatly depends on the capabilities of the whole system to timely recognize it and react accordingly.

Clone detection can be achieved through the gathering of information at the middleware layer. For instance, a very interesting approach has been proposed in (Mirowski & Hartnett, 2007, ib.). The authors have proposed an intrusion detection system for RFID systems, called Deckard, based on a statistical classifier and oriented to the detection of change of tag ownership. This is one of the early researches devoted to the need of intrusion detection systems in RFID. Another remarkable approach, similar to Deckard's intrusion detection architecture, has been proposed in (Thamilarasu & Sridhar, 2008). Its concern goes beyond the change in tag ownership and provides a more generic security framework to detect a variety of RFID malicious attacks.

The methodology we are going to describe in this chapter shares with these latter works the basic concept of applying intrusion detection techniques to identify tag cloning. But, differently, our research efforts have been primarily focused on integrating the principles of ontology modeling and reasoning in the intrusion detection paradigm.

Until recently, few literature can be found about the adoption of the ontology-based approach in developing IDSs. In particular, Raskin et al. (Raskin et al., 2001) advocated the use of ontology modeling in the field of information security in order to provide a common ontology that lets IDS sensors agree on what they observe. Undercoffer et al. (Undercoffer et al., 2003) proposed a target centric ontology for intrusion detection that models properties that are observable and measurable by the target of an attack. Li et al. (Li et al, 2008) described a hierarchical knowledge model to support alert correlation, formalized in an ontology and a set of rules built on top of it. However, to the best of our knowledge, the RFID security literature has not yet addressed applications of inferential engines and ontology modeling to implement intrusion detection techniques in RFID systems, neither system-oriented researches appear to have been developed in that direction.

4. Misuse detection system methodology

4.1 Track and trace model

The Misuse Detection System (MDS) described in this chapter belongs to the class of IDS characterized by misuse detection as detection method, application-based sensors as type of audit data processed (i.e., the RFID readers), real-time as usage frequency (i.e., audit data are processed as they are generated) and passive response (i.e., an attack occurrence is logged and the administrator is notified of it) (see Sect. 2.1).

The methodology devised to design the MDS relies on a knowledge base containing a "track & trace" model that formalizes all the information required to identify an attack of tag cloning. Such a model essentially relies on the reasoning that *when you know where the genuine tagged object is, the fake/clone ones can be detected.*

More in detail, the model includes static and dynamic profiles to be associated to RFID tagged objects. A static profile is a path composed of points of interest (POI) which a specific tag must visit during its life-cycle under normal working conditions, like for instance those disseminated along a supply chain. A dynamic profile, instead, is a path composed of a tag's actually visited POI, dynamically detected through the supply chain. Dynamic profiles can be built exploiting the set of location events retrieved from a tracing system, such as the EPC network (EPCglobal Inc., 2009). Moreover, the dynamic profile also stores time and type of access (i.e., read/write) of a tagged object visiting a POI.

Both static and dynamic profiles make use of the concepts of *physical location perspective* as opposed to *semantic location perspective*, detailed described in (Coronato et al., 2009, ib.). In principle, a physical location is a precise region identified in terms of proximity to well referenced spots, whereas a semantic location represents the significance of a location within a specific context and may cover more physical locations. In our case, physical locations are the areas covered by RFID readers, while semantic locations can be a country, a city, a building, a room inside a building, a railway station, a watched gate, and so on. A simple typical layout is shown in Fig. 2, where the relationships between physical and semantic perspectives are put into evidence.

The static profile of a tagged object is thus a fixed sequence of semantic locations to be visited through the supply chain, whereas its dynamic profile is the sequence of physical locations actually visited.

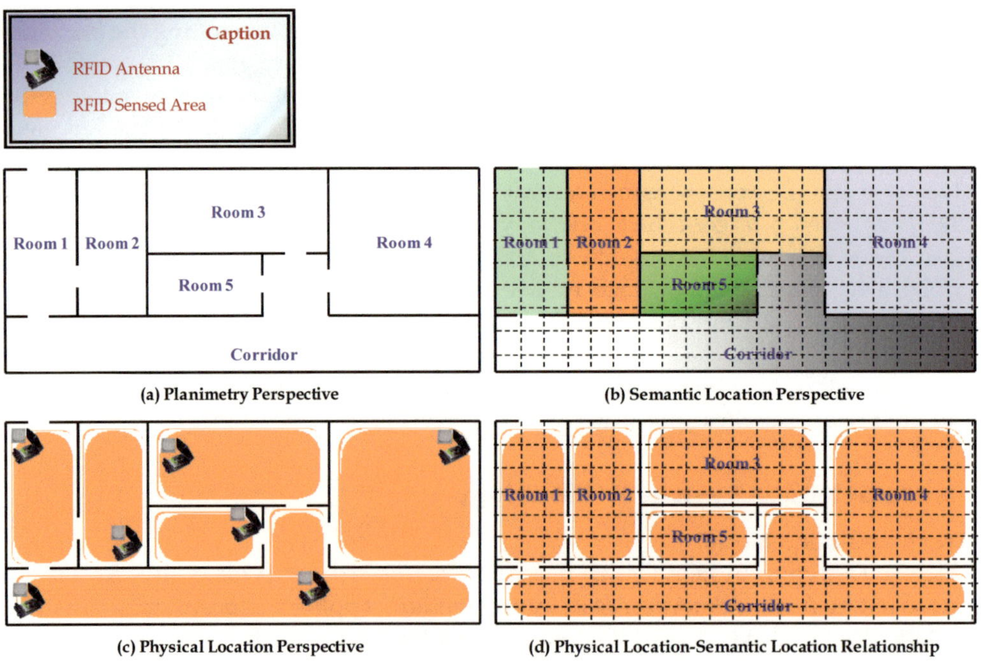

Fig. 2. Representation of the physical and semantic location perspectives

The association between physical location and semantic location is essential to the track & trace model which indeed is based on location-awareness. This means that high-level location information is required, while positioning systems like RFID readers are only able to collect raw location information. In order to fill such a semantic gap, a suitable mapping scheme based on the target supply-chain layout can be established to map physical locations onto semantic ones. This scheme is used in conjunction with the data stored in a tagged object's dynamic profile to identify the semantic locations it in fact passed through.

Physical locations information can be gathered from the audit records generated by RFID read/write operations. As a matter of fact, a typical audit record can be logically structured in <*tagID, readerID, RFIDoperation, timestamp*>, meaning that the *tagID* has been read/written by the *readerID* at the time *timestamp*. The physical location in which *tagID* has been detected is the one associated to *readerID* in the physical location perspective. Then, the mapping scheme established for the target semantic perspective is used to identify the corresponding semantic location. When a tag visits a semantic location, that location becomes a POI for that tag.

As said before, the static profile defined for a given tagged object models its expected behavior within the supply chain, while the dynamic profile stores the ongoing behavior of that object. By comparing static vs. dynamic profiles, the absence of inconsistencies would indicate a "validated" behavior. In contrast, if for any reason the dynamic profile does not satisfy the specifications modeled in the static profile, this fact implies an "abnormal" behavior due to a misuse and -potentially- a threat. In order to effectively detect a real attack, an *a priori* knowledge of possible tag cloning scenarios must be defined (recall that misuse detection techniques rely on a knowledge base that must explicitly model abnormal situations - see Sect. 2.1).

For this reason, in addition to the "normal" static profile associated to each RFID tagged object, a set of "abnormal" static profiles has been formalized with the aim of characterizing known types of clone attack. Once an inconsistency between normal static and dynamic profiles has been revealed for a certain tag, this tag's dynamic profile is checked against the set of abnormal profiles to detect if and what type of attack is going on. Some instances of abnormal static profiles indicating a clone attack are here reported:

- an abnormal profile which includes multiple and non-bordering semantic locations visited at the same time, meaning that a tagged object expected to be at a particular POI has been simultaneously detected in a different semantic location of the supply chain;
- an abnormal profile which includes non-bordering semantic locations visited in a time interval that is inconsistent with the distance between those locations, meaning that a tagged object is moving along the supply chain too fast to be genuine;
- an abnormal profile which includes one or more semantic locations visited by a tagged object not in accordance with the normal static profile, meaning that the object has been detected at one or more POI where it is not expected to be;
- an abnormal profile which includes multiple detections of the same tagged object in a given semantic location, indicating the presence of multiple copies of the same tag. This situation must not be confused with the well-known issue of *collision detections* which affects raw RFID interactions and is usually handled by RFID systems at a lower architectural level;
- an abnormal profile which includes operations performed on a tagged object not allowed under normal working conditions, like for instance too many access to the tag occurring in a fixed time interval at the same semantic location.

More abnormal profiles can be formalized to exhaustively cover all the scenarios and adaptively respond to new kind of attacks as well.

4.2 The ontology formalization of the model

The knowledge base at the core of the "track & trace" model has been formalized in an ontology by means of the semantic web languages in order to achieve an unambiguous, well-defined and machine-readable knowledge representation.
In particular, this ontology -called "Track and Trace"- has been devised and implemented in terms of relevant concepts and properties. It is depicted in Fig. 3 as a graph whose nodes represent concepts and sub-concepts while the edges represent properties.
A property can be used to model either binary relationships between concepts or simple attributes. Concepts and properties have been formalized in OWL DL. The choice of this language is due to i) the high degree of expressiveness and modeling power, that enable to formalize complex models in an accurate and sound way; ii) its model-theoretic semantics, that allows to automatically apply reasoning techniques, as discussed in the next section.
For the sake of clarity, we subdivided the ontology in four logical sections. Each property is defined in terms of domain (the set of possible subject concepts) and range (the set of possible object concepts or data types). Besides, the inverse property is reported, if applicable, and the transitiveness is specified, if existing. It is worth noting that each sub-concept inherits super-concept properties and adds new specialized ones.
The first section of the ontology is devoted to model the static profile of a tagged object. In Fig. 3 (a), main concepts and properties are outlined , while the complete list of properties for each concept and sub-concept, not shown in figure for conciseness, is reported in Tab. 1.

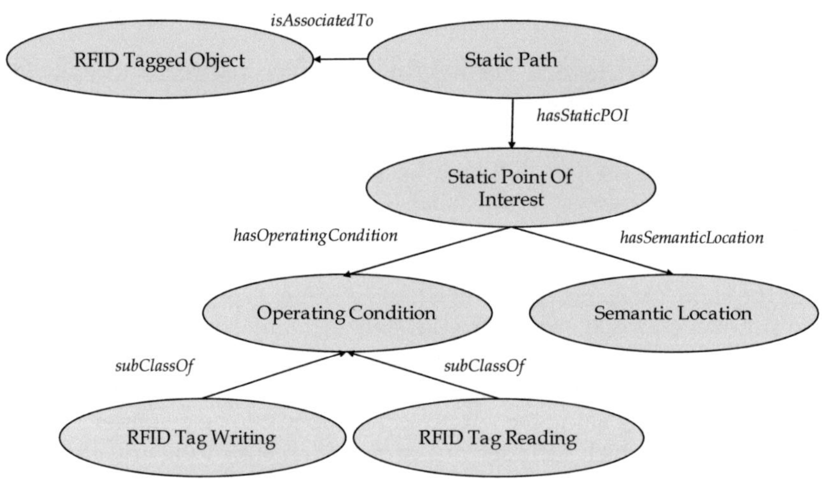

Fig. 3. a) "Track & Trace" Ontology: Static Profile

Property	Domain	Range	Inverse	Trans.
hasStaticPOI	StaticPath	StaticPointOfInterest	isStaticPOIOf	No
hasStaticPathID	StaticPath	*Datatype*: String	-	-
isAssociatedTo	StaticPath	RFIDTaggedObject	hasStaticPath	No
hasOperatingCondition	StaticPointOfInterest	OperatingCondition	isOperatingConditionOf	No
hasSemanticLocation	StaticPointOfInterest	SemanticLocation	isSemanticLocationOf	No
hasMaxOperationNumber	OperatingModality	*Datatype*: Integer	-	-
hasMaxTimestamp	OperatingModality	*Datatype*: Time	-	-

Table 1. Properties defined for a static profile

The second section of the ontology models a dynamic profile. Main concepts and properties are outlined in Fig. 3 (b), and a detailed list of properties is reported in Tab. 2.

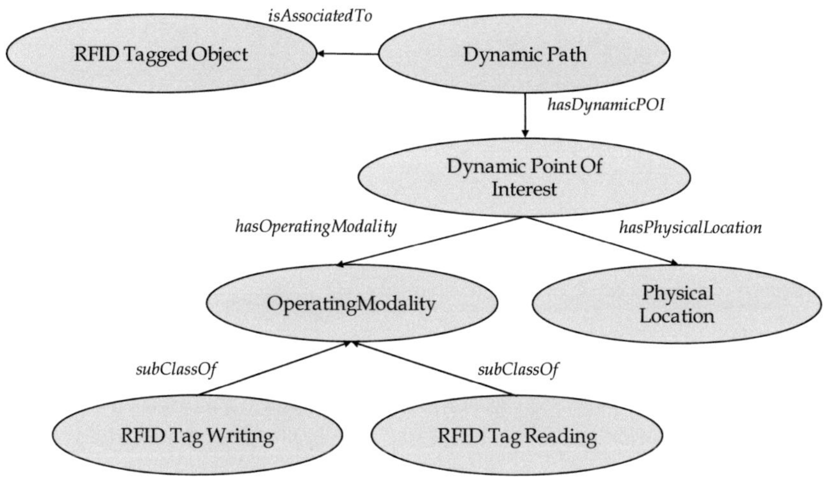

Fig. 3. b) "Track & Trace" Ontology: Dynamic Profile

Property	Domain	Range	Inverse	Trans.
hasDynamicPOI	DynamicPath	DynamicPointOfInterest	isDynamicPOIOf	No
hasDynamicPathID	DynamicPath	*Datatype*: String	-	-
isAssociatedTo	DynamicPath	RFIDTaggedObject	has DynamicPath	No
hasOperatingModality	DynamicPointOfInterest	OperatingModality	isOperatingModalityOf	No
hasPhysicalLocation	DynamicPointOfInterest	PhysicalLocation	isPhysicalLocationOf	No
hasSemanticLocation	DynamicPointOfInterest	SemanticLocation	isSemanticLocationOf	No
hasOperationCounter	OperatingModality	*Datatype*: Integer	-	-
hasTimestamp	OperatingModality	*Datatype*: Time	-	-

Table 2. Properties defined for a dynamic profile

The third section of the ontology is devised to specify location information as outlined in Fig. 3 (c) and detailed in Tab. 3.

Property	Domain	Range	Inverse	Trans.
isPartOf	SemanticLocation	SemanticLocation	hasPart	Yes
isBorderingOn	SemanticLocation	SemanticLocation	-	No
hasSemanticLocationName	SemanticLocation	*Datatype*: String	-	-
hasPhysical LocationID	PhysicalLocation	*Datatype*: String	-	-
maps	PhysicalLocation	SemanticLocation	isMappedWith	No
covers	RFIDReader	PhysicalLocation	isCoveredBy	No
hasReaderID	RFIDReader	*Datatype*: String	-	-

Table 3. Properties defined for location information

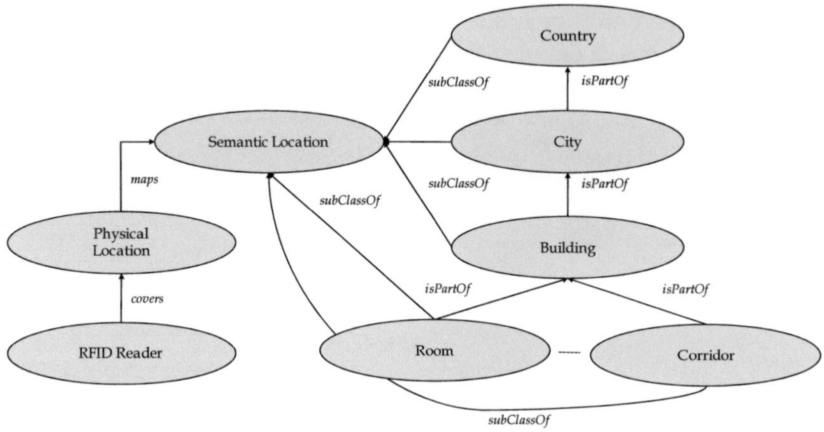

Fig. 3. c) "Track & Trace" Ontology: Location Information

Finally, the last part of the ontology specifies information contained in an audit record in terms of concepts and properties, as outlined in Fig. 3 (d) and reported in Tab. 4.

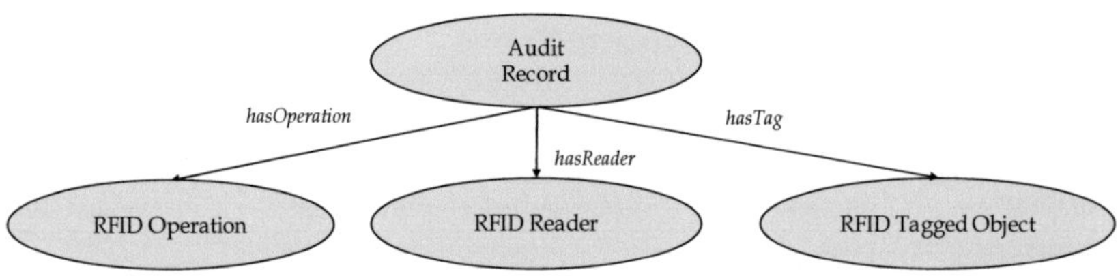

Fig. 3. d) "Track & Trace" Ontology: Audit Record

Property	Domain	Range	Inverse	Trans.
hasOperation	AuditRecord	RFIDOperation	isOperationOf	No
hasTag	AuditRecord	RFIDTaggedObject	isTagOf	No
hasReader	AuditRecord	RFIDReader	isReaderOf	No
hasTimestamp	AuditRecord	*Datatype*: Time	-	-
hasTagID	RFIDTaggedObject	*Datatype*: String	-	-
hasTagData	RFIDTaggedObject	*Datatype*: String	-	-
hasValidID	RFIDTaggedObject	*Datatype*: Boolean	-	-
isClone	RFIDTaggedObject	*Datatype*: Boolean	-	-

Table 4. Properties defined for an audit record

4.3 The detection procedure

The application of DL reasoning in the context of intrusion detection requires that a class of attacks be modeled and DL formalism be utilized in such a way that the formulas of the logic can be directly and automatically evaluated. This is where the integration between ontology and DL reasoning plays its role.

In particular, the detection procedure defined for the MDS described in this chapter is a specific application of the description logic on the "Track and Trace" ontology, written in

OWL DL, as previously stated. According to the OWL DL model-theoretic semantics, the detection procedure relies on the application of the hybrid reasoning service "A-Box Consistency" (see Sect. 2.3) to verify whether a dynamic profile of a tagged object is consistent with the corresponding static profile.

In particular, starting from the ontology, T-Box and A-Box have been arranged as follows:

- the T-Box contains closed-form definitions of the static profiles associated to the tagged objects, formulated in terms of concepts, properties and axioms on properties.
- the A-Box contains the individuals (instances of concepts) and the instances of properties. It is populated with the actual information pertaining to dynamic profiles of tagged objects, built in terms of audit records, physical and semantic locations.

The A-Box Consistency check allows to verify whether the terminology of the ontology (i.e., the T-Box) admits the existence of an interpretation that satisfies all the assertions and axioms contained in the A-Box. In the MDS perspective, this functionality has been implemented on the basis of the *tableaux reasoning* (Baader and Sattler, 2001) as provided by the DL inference engine proposed in (Esposito, 2007). This tableaux reasoning searches for an interpretation through a process of completion which starts by constructing an initial completion graph from the A-Box. The nodes in the completion graph intuitively stand for individuals and are associated to their corresponding types. Property-value assertions are represented as directed edges connecting nodes. The reasoning iteratively applies the tableaux expansion rules until a clash (i.e., a contradiction) is detected in the label of a node, or until a clash-free graph is found to which no more rules are applicable. Tableaux reasoning has many advantages: not only it eases the design of provably sound, complete and decidable algorithms, but it is also usually quite efficient at solving many problems that commonly affect real applications.

The detection procedure has been devised and developed on top of the reasoning service for A-Box Consistency, as outlined in Fig. 4 and described in detail as follows.

Through the ontology, the user provides both a high level representation of the terminology and the assertive part to be checked for each tagged object under evaluation. The terminology to be checked consists in a normal static profile stored in the T-Box, whereas the assertive part is represented by the dynamic profile, built on demand with the information coming from the audit records and stored in the A-Box.

The DL inference engine executes the task of verifying the A-Box Consistency and may terminate with the answer "true", indicating that the dynamic profile satisfies the normal static profile, i.e., there exists an interpretation of the assertive part that satisfies the terminology specified in the ontology. In other words, this means that the tagged object under investigation is recognized as genuine. On the contrary, if the inference engine terminates with the answer "false", thus indicating that the dynamic profile is not consistent with the normal profile, this implies that the tagged object is not genuine and is expected to be a clone. In order to determine the reason why the terminology is not satisfied by the assertive part, that is to say why a coherent match between dynamic and normal static profiles has not been found, a set of additional executions are launched by the inference engine.

First, such additional executions require a change of the terminology loaded in the T-Box, i.e., the normal static profile is replaced by one of the abnormal static profiles previously described in section 4.1. Then, the A-Box Consistency task is launched again in order to verify whether the dynamic profile satisfies the abnormal static one loaded in the T-Box.

Additional executions continue until the inference engine returns the answer "true" and, finally, a report is generated describing the kind of abnormal static profile in the T-Box that has been just satisfied by the dynamic profile currently in the A-Box.

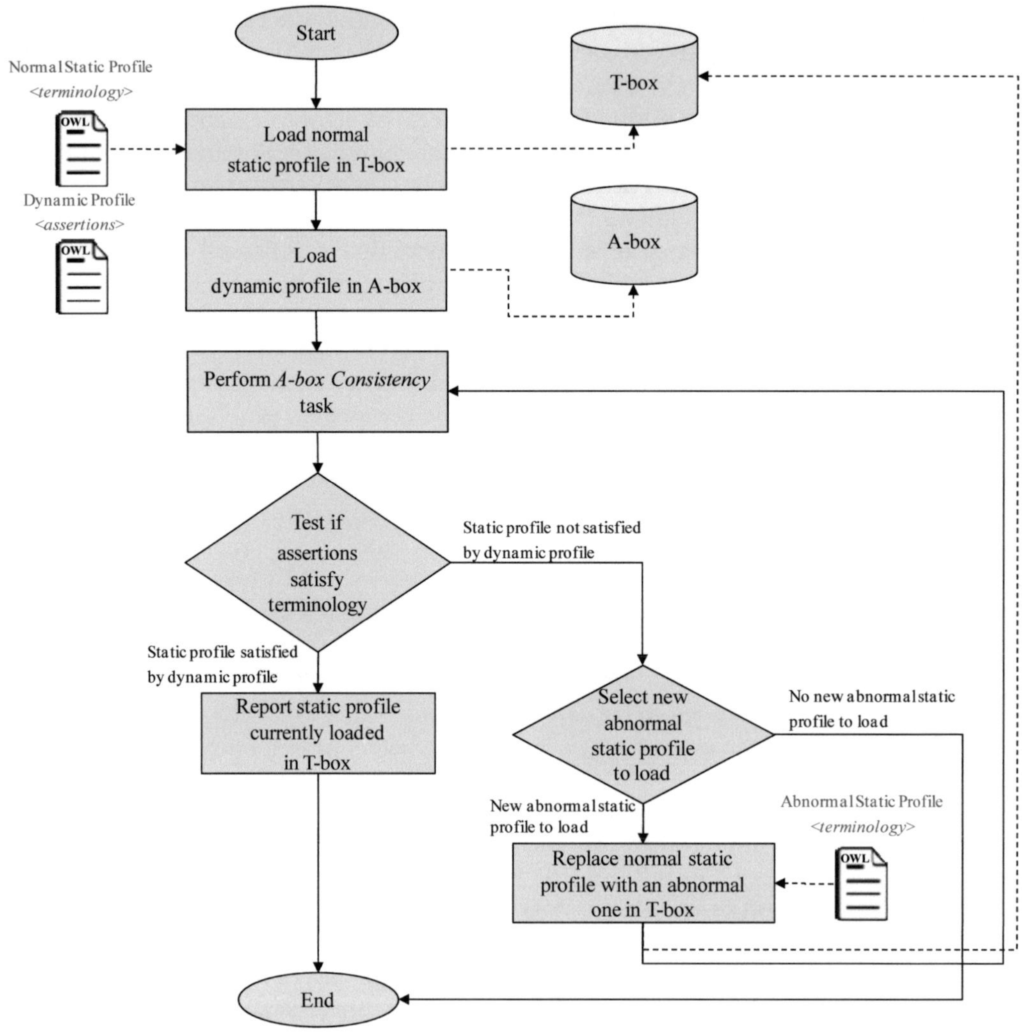

Fig. 4. The detection procedure

In summary, not only can this detection procedure determine the kind of attack, but it also reports a detailed description of how the attack has manifested itself, using the rich and very expressive formalism guaranteed by ontology languages.

5. The ontology-based Misuse Detection System

5.1 The Misuse Detection System architecture

We designed the architecture of the Misuse Detection System described in this Section on the basis of the methodology illustrated in Sect. 4.

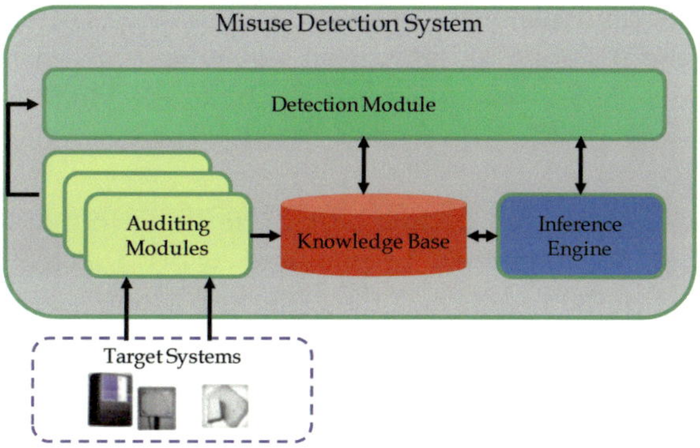

Fig. 5. The MDS architecture

This architecture takes its place at the middleware abstraction layer, on top of the target system layer. A target system is a typical RFID system capable of gathering data that summarize the activity of tags detected in its coverage area. Essentially, a target system produces an audit record for each performed RFID read/write operation.

The MDS architecture is composed of the set of components shown in Fig. 5 and here described:

- *Auditing Modules*. These are the component responsible for collecting the audit records generated by the target systems and storing them into the Knowledge Base. There are as many Auditing Modules as physical locations, each module being associated to a specific RFID reader.
- *Knowledge Base*. It is constituted by the domain knowledge formalized in the ontology described in Sect. 4.2.
- *Inference Engine*. This is the "smart" component with advanced capabilities for performing reasoning services on the ontology stored into the Knowledge Base.
- *Detection Module*. This is the component which executes the core business of the whole MDS. By interacting with the Inference Engine, it is in charge of deciding if and what kind of attack has occurred. The Detection Module exhibits a reactive and event-dependent behavior in response to new generated audit records.

With such components in mind, the MDS operates in the following manner. When a tagged object is sensed by a reader in a Target System, a new audit record is generated with the details of the performed RFID operation (see Sect. 4.1). The corresponding Auditing Module collects this record, stores it in the Knowledge Base and notifies the Detection Module of this event. The Detection Module, in turn, invokes the Inference Engine for processing the data contained in the new audit record. First, the dynamic profile associated to the sensed tagged object is identified. Then, on the basis of the location information stored in the Knowledge Base, the Detection Module looks for the physical location where the tag has been detected, and successively determines the corresponding semantic location.

After that, the dynamic profile is updated in the Knowledge Base with the information pertaining to the visited POI (type of access and timestamp reported in the audit record) or, if the semantic location was never visited before, a new POI and its related information is added to the dynamic profile.

This behavior has been formalized by means of a set of rules conforming to the Event-Control-Action (ECA) architectural pattern. Generally, ECA rules have the form "on<event>if<condition>then<action>", where <condition> specifies the circumstances that must be verified for the <action> to be carried out whenever an <event> occur. In our specific case, the <event> is the one generated by an Audit Module while <condition> is constituted by the dynamic profile current state. Finally, <action> consists in the updating of the dynamic profile.

Once the Knowledge Base has been updated, the Detection Module launches the detection procedure by invoking the Inference Engine and waits for the answer. Finally, the administrator is notified of the results and the Detection Module is ready to process a new event generated by one of the Auditing Modules.

5.2 Proof of concept

In order to give a proof of concept of the knowledge-based approach for detecting misuses in RFID systems presented in this book's chapter, an experimental prototype has been developed at the Institute for High Performance Computing and Networking (ICAR) of the National Research Council of Italy (CNR). This prototype implemented the MDS architectural components as fully portable Java entities on a platform having the following characteristics:

- 2.80GHz Intel® Core™ i7 CPU;
- 8GB of RAM; 240GB of hard disk;
- Windows 7 Professional OS;
- Java SDK 1.6; 1GB of max heap size.

Tests have been performed on this prototype for assessing the response to cloning attacks and the detection accuracy rate. For this purpose, a network of target systems has been simulated through an additional software component in charge of feeding the MDS with suitable devised sets of audit records (see Sect. 4.1) able to reproduce various working conditions.

The simulated scenario consisted of a layout of 50 semantic locations relying on an RFID network made of 50 readers handling up to 20000 tags. For each tag, a static profile has been built with a randomly chosen number of POI between 15 and 50. The minimum time required by a tagged object to move from a POI to another one was set in accordance with the simulated distance among the semantic locations. The maximum number of read/write operations for a single tag allowed at each semantic location was set to 4.

As many as 10 test cases have been designed with the aim of reproducing some combinations of one, more or all the abnormal profiles described in Sect. 4.1 under an increasing traffic of both genuine and clone tags. For each test case, a testing session has been executed on the prototype by injecting the Auditing Modules with the appropriate set of audit records, simulating the traffic of tagged objects in the RFID network. In details, depending on the particular test case, a single attack of clones has been simulated by injecting n copies of a genuine tag with n randomly chosen between 1 and 1000, while multiple attacks have been simulated by injecting k distinct single attacks with k randomly chosen between 2 and 10, for a maximum of 10000 clone tags. Upon completion of a testing session, the resulting data set has been stored for further analysis.

5.2.1 Data analysis

With the aim of assessing the capability of the MDS to respond to cloning attack, a detailed analysis of the data set resulted from the testing activities above has been carried out.

In the following discussion, the term *positive* indicates a clone tag as well as the term *negative* indicates a genuine tag, so that *true positive* (TP) means a clone tag correctly detected whereas *true negative* (TN) means a genuine tag correctly recognized as such. Likewise, *false positive* (FP) stands for a clone tag wrongly treated as a genuine as opposed to *false negative* (FN), i.e., a genuine tag wrongly detected as a clone.

The occurrences of true positives and true negatives have been counted along with the number of false positives and false negatives. Tab. 5 shows the detection performance in terms of accuracy, true positive rate (TPR) and false positive rate (FPR), calculated as follows:

$$ACCURACY = \frac{TP + TN}{TP + TN + FP + FN} \quad (1)$$

$$TPR = \frac{TP}{TP + FN} \quad (2)$$

$$FPR = \frac{FP}{FP + TN} \quad (3)$$

Furthermore, response times have also been collected, as reported in the last column of Tab. 5. Looking at the outcome of the data analysis, it can be observed that the experimental Misuse Detection System achieves both high accuracy levels and elevated true positive rates in detecting cloning attacks. Even when these ratings lower as the number and complexity of attacks increase, they still remain quite good, with values of 92.5% of accuracy and 93% of true positive rate for the worst attack scenario considered in the simulation.

The variability of the accuracy rate shown by these experimental results is motivated by the threshold values set respectively for the minimum time required by a tag to move from a POI to another one and for the maximum number of read/write operations for a single tag allowed at each semantic location. Stricter thresholds might yield lower false positive rate but may also let many actual attacks go unnoticed, whereas relaxing the thresholds may increase the true positive rate but also lead to too many false alarms. The actual threshold values were chosen in order to strike a right balance between true positive and false positive rates, with the relief that, in the secure RFID application domain, it is anyhow better to detect false attacks rather than neglect truly dangerous treats.

Clone tags	Total tags	Accuracy %	TPR %	FPR %	Response Time (min)
1000	11000	100	100	0	< 8.0
2000	12000	98.33	95	1	< 9.0
3000	13000	98.46	96.66	1	< 10.5
4000	14000	97.14	97.5	3	< 9.5
5000	15000	96.66	94	2	< 11.5
6000	16000	96.25	96.66	4	< 13.0
7000	17000	93.52	94.28	7	< 14.5
8000	18000	94.44	93.75	5	< 9.5
9000	19000	93.68	93.33	6	< 15.0
10000	20000	92.5	93	8	< 16.5

Table 5. Experimental results for the simulated scenario

Moreover, it should be noted that the achieved response times, while affordable, show some non-monotonic trends as the number of tags increases, contrary to what one could expect. This is due to the composite nature of the various test cases devised to stimulate as exhaustively as possible the detection capability of the MDS to respond to cloning attack. In fact, different abnormal profiles require different computational time for reasoning, and it may happen that a more time-consuming profile is checked against a lower traffic of tags in a time shorter than the time required to check a less time-consuming profile against a great deal of tags, or vice versa. However, the overall linearity exhibited by response times –a nearly twofold increase in traffic calls for nearly doubled execution times– indicated a good scalability of the MDS.

6. Conclusions

Detection may be seen as the first step in defending against tag cloning and preventing RFID-enabled crime.

As an effort in this direction, the research presented in this book's chapter was intended to investigate whether it is feasible to integrate the principles of ontology modeling and reasoning in the intrusion detection paradigm with the final aim of identifying clone RFID tags. A suitable methodology has been devised for designing a Misuse Detection System which relies on an ontology that explicitly models both normal and anomalous working conditions and uses an inferential engine to dynamically reveal possible inconsistencies in the traffic of RFID tagged objects.

The MDS has an architecture made of a set of components placed at the middleware abstraction layer and exhibits a reactive and event-dependent behavior in response to new tracking information coming from the underlying network of RFID systems. By invoking an inference engine to apply reasoning technique on the formalized knowledge base, the MDS is able to detect cloning attacks and characterize their nature.

An experimental prototype of the MDS has been developed with the purpose of attaining a proof of concept of the knowledge-based approach at the core of this research. Several tests have been carried out on this prototype for assessing its response to various cloning attacks and detection accuracy rate in a simulated scenario. A detailed analysis performed on the experimental data derived from the testing activity showed both high accuracy levels and elevated true positive rates in detecting cloning attacks. Also, the overall linearity exhibited by response times indicated a good scalability of the MDS.

Some concluding remarks can then be drawn.

First, the knowledge-based approach here proposed makes use of semantic-rich models formalized in an ontology which paves the way to a clear understanding of a possible scenario of tag cloning, thus achieving high reliability in the detection process. In addition, the ontology simplicity and intuitiveness can significantly facilitate the tasks of specifying novel attacks of tag cloning, thus keeping the MDS knowledge base up to date.

Second, the basic approach we adopted goes beyond the mere prevention of RFID tag cloning since it allows for the actual detection of clone tags. This is a key point, because when prevention –often very costly to implement– fails, the only way to reduce damages is to react as soon as possible to the presence of clones. Without an automatic mechanism able to timely signal an undergoing cloning attack, it can go unnoticed for days or even weeks. The MDS here proposed constitutes a highly automatic and computationally affordable solution for processing dynamic RFID audit data and identifying attacks within quite acceptable response times. Moreover, by launching several executions where the terminology to be checked is in turn replaced in accordance with a specific abnormal static profile, it is possible to generate sound reports of how attacks have manifested themselves, each of them being terminologically described with the rich and very expressive formalism proper to the ontology languages .

Third, the MSD has been thought to be transparently and seamlessly integrated into existing RFID systems at the middleware level, with no specific requirements concerning the kind of RFID technology used (passive/active tags, HF/UHF, etc.).

In summary, the outcome of this research seems to be encouraging, suggesting that an actual, scaled-up deployment of the Misuse Detection System here proposed could effectively and proficiently support the detection of clone tags in RFID systems.

Finally, the experimental tests gave a proof of the feasibility of the methodology devised for designing the MDS, which was our main goal. However, an issue remains open as far as MDS's response times are concerned. Improving the performance of the MDS was beyond the initial scope of our research, nonetheless we are confident that the overall execution times might be significantly reduced through a more performance-oriented implementation of the architecture of the MDS, for instance by proceeding with a proper parallelization of its components.

7. References

Avoine, G. & Oechslin, P. (2005). A scalable and provably secure hash based RFID protocol, *Proceedings of the Third IEEE International Conference on Pervasive Computing and Communications Workshops*, pp. 110-114, ISBN 0-7695-2300-5, Kauai Island, Hawaii, March 8-12, 2005.

Baader, F.; Horrocks, I. & Sattler, U. (2005). Description logics as ontology languages for the semantic web, In: *Mechanizing Mathematical Reasoning: Essays in Honor of Siekmann on the Occasion of His 60th Birthday, Lecture Notes in Artificial Intelligence 2605,* D. Hutter & W. Stephan, (Eds.), Springer-Verlag, pp. 228-248, , ISBN 3-540-25051-4, New York, NJ, USA.

Baader, F. & Sattler, U. (2001). An overview of tableau algorithms for description logics. *Studia Logica*, Vol. 69, No. 1, pp. 5–40, ISSN 0039-3215.

Cheng, L. M.; So, C.W. & Cheng, L.L. (2009). An Improved Forward Secrecy Protocol for Next Generation EPCGlobal Tag. *Development and Implementation of RFID Technology*, ISBN 978-3-902613-54-7, I-Tech Education and Publishing, Available from:
http://www.intechopen.com/articles/show/title/an_improved_forward_secrecy_protocol_for_next_generation_epcglogal_tag.

Chien, H. Y. (2007). SASI: A new ultralightweight RFID authentication protocol providing strong authentication and strong integrity, *IEEE Transactions on Dependable and Secure Computing*, Vol. 4, No. 4, pp. 337-340, ISSN 1545-5971.

Choi, E.Y., Lee & S.M., Lee, D.H. (2005). Efficient RFID authentication protocol for ubiquitous computing environment, *Embedded and Ubiquitous Computing, Lecture Notes in Computer Science 3823*, T. Enokido, L. Yan, B. Xiao, D. Kim, Y.S. Dai & L.T. Yang (Eds.), Springer, pp. 945-954, ISBN 3-540-30803-2.

Ciampi, M.; Coronato, A.; De Pietro, G. & Esposito, M. (2006). A Location Service for Pervasive Grids, In: *Advances in Systems, Computing Sciences and Software Engineering*, T. Sobh & K. Elleithy (Eds.), Springer, pp. 119-123, ISBN 978-1-4020-5263-7.

Coronato, A.; Della Vecchia, G. & De Pietro, G. (2006). An RFID-Based Access andLocation Service for Pervasive Grids, In: *Emerging Directions In Embedded And Ubiquitous Computing, Lecture Notes in Computer Science 4097*, X. Zhou, O. Sokolsky, L. Yan, E.S. Jung, Z. Shao, Y. Mu, D.C. Lee, D. Kim, Y.S. Jeong & C.Z. Xu (Eds.), Springer, pp. 601-608. ISBN 3-540-36850-7.

Coronato, A.; Esposito, M. & De Pietro, G. (2009). A Multimodal Semantic Location Service for Intelligent Environments: An Application for Smart Hospitals. *Journal of Personal and Ubiquitous Computing*, October 2009, Vol. 13, No. 7, pp. 527-538, ISSN 1617-4909.

Debar, H.; Dacier, M. & Wespi, A. (1999). Towards a taxonomy of intrusion detection systems, *Computer Networks*, April 1999, Vol. 31, No. 8, pp. 805-822, ISSN 1389-1286.

Della Vecchia, G. & Esposito, M. (2010). A Pervasive System for Nuclear Medicine Departments, *Journal of Wireless Personal Communications*, September 2010, Vol. 55, No. 1, pp. 105-120, ISSN 0929-6212.

Devadas, S.; Suh, E.; Paral, S.; Sowell, R; Ziola, T. & Khandelwal, V. (2008). Design and implementation of PUFbased "unclonable" RFID ICs for anti-counterfeiting and security applications, *Proceedings of the 2008 IEEE international conference on RFID*, pp. 58–64, ISBN 978-1-4244-1711-7, Las Vegas, Nevada, USA, April 16-17, 2008.

Donini, F. M.; Lenzerini, M.; Nardi, D. & Schaerf, A. (1996). Reasoning in description logics. In: *Foundation of Knowledge Representation*, G. Brewka, (Ed.), pp. 191-236, CSLI-Publications, ISBN 1-57586-056-2, Stanford, CA, USA.

Duc, D.N.; Park, J.; Lee, H. & Kim, K. (2006). Enhancing Security of EPCglobal Gen-2 RFID Tag against Traceability and Cloning, *Proceedings of the 2006 Symposium on Cryptography and Information Security*, Abstracts pp. 97, Hiroshima, Japan, January 17-20, 2006.

EPCglobal Inc. (2010). EPCglobal Architecture Framework Version 1.4, In: GS1 - The global language of business, 15-12-2010, Available from www.epcglobalinc.org/standards/architecture/.

Esposito, M. (2007). An Ontological and Non-monotonic Rule-based Approach to Label Medical Images, *Proceedings of the 2007 Third International IEEE Conference on Signal-Image Technologies and Internet-Based System*, pp. 603-611, ISBN 978-0-7695-3122-9, Shanghai, China, December 16-18, 2007.

Esposito, M.; Gallo, L.; Coronato, A. & Della Vecchia, G. (2009). An Infrastructure for Pervasive Access to Clinical Data in eHospitals, In: *New Directions in Intelligent Interactive Multimedia Systems and Services, vol. 226/2009 of Studies in Computational Intelligence*, E. Damiani, J. Jeong, R.J. Howlett & L.C. Jain, (Eds.), pp. 431-442, Springer, Berlin/Heidelberg.

Gruber, T. (1995). Towards Principles for the Design of Ontologies Used for Knowledge Sharing. *International Journal of Human-Computer Studies*, December 1995, Vol. 43, No. 5-6, pp. 907–928, ISSN 1071-5819.

Hofmeyr, S. A.; Forrest, S. & Somayaji, A. (1998). Intrusion detection using sequences of system calls, *Journal of Computer Security,* August 1998, Vol. 6, No. 3, pp.151-180, ISSN 0926-227X.

Juels, A. (2005). Strengthening EPC tags against cloning, *Proceedings of the 4th ACM workshop on Wireless security*, pp. 67-76, ISBN 1-59593-142-2, Cologne, Germany September 2.

Karthikeyan, S. & Nesterenko, M. (2005). RFID security without extensive cryptography, *Proceedings of the 3rd ACM workshop on Security of ad hoc and sensor networks,* pp. 63-67, ISBN 1-59593-227-5, Alexandria, VA, USA, November 07-10,2005.

Karygiannis, T.; Eydt, B.; Barber, G.; Bunn, L. & Phillips, T. (2007). Guidelines for securing radio frequency identification (RFID) systems, *NIST Special Publication 800-98*, April 2007, available from http://www.nist.gov/manuscript-publication-search.cfm?pub_id=51156.

Kruegel, C. & Robertson, W. (2004). Alert Verification Determining the Success of Intrusion Attempts, *Proceedings of the First Workshop on the Detection of Intrusions and Malware and Vulnerability Assessment*, pp. 1-14, ISBN 3-88579-375-X, Dortmund, Germany, July 6-7, 2004.

Lee, S.; Asano, T. & Kim, K. (2006). RFID Mutual Authentication Scheme based on Synchronized Secret Information, *Proceedings of the 2006 Symposium on Cryptography and Information Security,* Abstracts pp. 98, Hiroshima, Japan, January 17-20, 2006.

Lehtonen, M.; Ostojic, D.; Ilic, A. & Michahelles, F. (2009). Securing RFID Systems by Detecting Tag Cloning, *Proceedings of the 7th International Conference on Pervasive Computing*, pp. 291-308, ISBN 978-3-642-01515-1, Nara, Japan, May 11-14, 2009.

Li, W. & Tian, S. (2009). Preprocessor of Intrusion Alerts Correlation Based on Ontology, *Proceedings of the 2009 WRI International Conference on Communications and Mobile Computing - Volume 03*, pp. 460-464, ISBN 978-0-7695-3501-2, Kunming, Yunnan, China, January 6-8, 2009.

Mirowski, L. & Hartnett, J. (2007). Deckard: A system to detect change of RFID tag ownership, *International Journal of Computer Science and Network Security,* July 2007, Vol. 7, No. 7, pp. 89-98, ISSN 1738-7906.

Patel-Schneider, P.F.; Hayes, P. & Horrocks, I. (2004). OWL Web Ontology Language Semantics and Abstract Syntax, In *W3C Recommendation,* Available from www.w3.org/TR/owl-semantics/.

Ranasinghe, D.C.; Engels, D.W. & Cole, P.H. (2005). Low-Cost RFID Systems: Confronting Security and Privacy, in *Auto-ID Labs Research Workshop,* available from http://www.autoidlabs.org/single-view/dir/article/6/80/page.html.

Raskin, V.; Hempelmann, C.F.; Triezenberg, K.E. & Nirenburg, S. (2001). Ontology in information security: a useful theoretical foundation and methodological tool, *Proceedings of the 2001 workshop on New security paradigms,* pp. 53-59, ISBN 1-58113-457-6, Cloudcroft, New Mexico, USA, September 10-13, 2001.

Rotter, P. (2009). Security and Privacy in RFID Applications, *Development and Implementation of RFID Technology,* ISBN 978-3-902613-54-7, I-Tech Education and Publishing, Available from: http://www.intechopen.com/articles/show/title/security_and_privacy_in_rfid_applications

Sarma, S.; Weis, S. & Engels, D. (2003). Radio-frequency identification: Security risks and challenges, *RSA Laboratories Cryptobytes*, Vol. 6, No. 1, (Spring 2003), pp. 2-9.

Thamilarasu, G. & Sridhar, R. (2008). Intrusion detection in RFID systems, *Proceedings of IEEE Military Communications Conference*, pp. 1-7, ISBN 978-4244-2677-5, San Diego, CA, USA, November 17-19, 2008.

Thompson, C. (2004). Everything is Alive. *IEEE Internet Computing*, Vol. 8, No. 1, (Jan-Feb 2004), pp. 83-86, ISSN 1089-7801.

Undercoffer, J. L.; Joshi, A.; Finin, T. & Pinkston, J. (2003). A target-centric ontology for intrusion detection, *Proceedings of the 18th International Joint Conference on Artificial Intelligence*, pp. 47-58, Acapulco, Mexico, August 9-15, 2003.

Weis, S.; Sarma, S.; Rivest, R. & Engels, D. (2004). Security and Privacy Aspects of Low-cost Radio Frequency Identification Systems, In: *Security in Pervasive Computing, Lecture Notes in Computer Science 2802*, G. Goos, J. Hartmanis & J. van Leeuwen (Eds.), Springer, pp. 50-59.

17

A Study on Implementation and Service of Digital Watermark Technology Architecture for Distribution Management

Manabu Hirakawa
Department of Industrial Engineering and Management, Tokyo Institute of Technology
Japan

1. Introduction

Recently, fake brand-name products and other problems concerning the manufacture of counterfeit goods, as well as the abundance of pirated music and movies, and the misuse of personal information, have been the subject of extensive news coverage. Numerous problems related to security have also been reported, in addition to the falsification of expiration dates and production location information labels on food products. As can be seen from these examples, consumers' trust, especially in regard to food safety, is at risk of being damaged. Information technology has advanced, and traceability has become technologically possible. However, I believe that a societal system for preventing such problems is lacking, and a fundamental reason for this may be a gap in perspective at the business management level with respect to the low-cost requirements of the market place. In society, individuals managing information are assumed to have good intentions. I believe that this system itself is beginning to fail. These problems are seen in the lower reliability of distribution systems, for example, the falsification of information about freshness dates and product origin. At present, the social system for preventing these problems is mostly defective, despite the technologically availability of traceability by means of information technology. Accordingly, new measures for dealing with these problems are urgently needed. It is important to note that ensuring safe distribution, improving security, and managing cost are not independent of each other, but are actually interconnected.

In this research, I analyze these problems in the context of the societal system and propose a solution that uses digital watermarking technology. My focus is on different types of information carriers including multimedia content such as movies and music, distributed manufactured products. At present, the security rules for information carriers are complex and are not uniformly applied to specific objectives and applications. Applying digital watermarking technology to information carriers will allow uniform information to be given to various media regardless of the application and environment; further, mobile services can be provided that do not depend on specific hardware and software. These mobile services be affected neither by structural disparities in applicable systems and codecs, nor by differences of copyright management policy.

2. Background of this research

Distribution is what connects the producer with the consumer. Physical products that we can see with our eyes, and those that we cannot, such as data, are both distributed. Images, videos, and audio that can be used on a computer are all forms of digital media. Therefore, recently pressing DVDs and CDs and creating packaging and other types of physical processing have become unnecessary. There are now cases where the data alone can be sent and received thus providing the service. Traditional distribution systems run the risk of increasing production costs due to media creation and packaging, and due to the need to hold unnecessary inventory. On the other hand, digital content services use the infrastructure of the Internet to transmit the digital data directly; therefore, these services have the following advantages: 1) no need to hold unnecessary inventory, 2) reduced distribution time, 3) reduced overhead costs, and 4) the ability to have customers around the world, without borders. Such distribution systems resolve the problems with existing systems, and are still expanding today. Digital content businesses that handle the distribution of images, videos, and audio are able to use the Internet to disclose, transmit, and distribute copyrighted work directly to the consumer. Many Web sites already use this service. On the other hand, there are many illegal Web sites that infringe on copyrights and negatively impact legitimate digital content businesses.

Relevant actors in the upstream process where content is created include the producer, the copyright holder, the secondary copyright creator, and the license manager. At this stage copyright comes into play, so we know that it is necessary to add copyright information to the content. Next, looking at the downstream processes of distribution and disclosure, the irrelevant actors include the distributor, the network company, the broadcasting company, e-commerce site managers, and administrators. Other content is distributed for offline use via Internet downloads, magazines, or DVDs and CDs. In these cases it is necessary to add information on the use of the content. There is an urgent need for a framework to be constructed that can take this copyright information and detect the illegal use of content as copyright infringement, and that can legally enforce copyright.

To ensure the solid growth of the promising digital content industry, content protection technology is necessary, which will be the used as the mechanism to protect copyright holders and their content. Content protection technology is an all-inclusive concept that involves the prevention and deterrence of unauthorized copying of content, as well as copyright protection technologies. Digital watermarking is an example of an effective content protection technology. Digital watermarking technology development began around 1995, and its full-fledged application began around 1998. Digital watermarking places an imperceptible mark that identifies the copyright holder into the digital content itself. In the event that the content is copied, the watermark can be used as evidence for tracking. Digital watermarking does not prevent unauthorized copying. However, it can be applied broadly, and it is effective in enforcing copyright.

3. Problems

Digital content businesses that deliver images or music make it possible to release, transmit, and sell copyrighted data directly to users via the Internet. Numerous Web sites already provide this service. On the other hand, there are many illegal Web sites that infringe on copyrights and negatively affect digital content businesses [1]. Music, images, and video that can be used on a computer are digital data, so the full service can be provided by simply sending and receiving the data. This eliminates the need for pressing CDs and DVDs, packaging, and other physical processing. This is the concept behind the digital content business. Conventional distribution systems have problems of increased production costs due to CD or DVD manufacturing and packaging, and the risk of carrying unnecessary inventory. With these conventional general distribution systems, there is a fear that it is difficult to commercialize content that has a low sales outlook.

On the other hand, because digital content services use the Internet as the infrastructure to send digital data, it has developed into a distribution system that resolves the problems of conventional systems as follows.

- No need to carry unnecessary inventory
- People around the world can be customers in a "borderless" manner
- Short distribution time
- Reduced costs

Moreover, with the development of infrastructure, the range of customer categories has expanded from the conventional range. Reaching target customer audiences and diversifying categories has become a recent remarkable trend.

As previously noted, the handling of digital content is highly anticipated in the future business scene, but the news is not all good. Because digital data can easily be copied, the user can sell it to a third party without permission, and there is also the possibility that the content will be illegally copied while en route over the Internet. Because there are no markings on the content itself that shows who holds the copyright, who sold it, or who purchased it, it is difficult to determine the route if the content is redistributed. If there is no evidence, then it is impossible for the copyright holder to prove a copyright claim when the content is illegal copied. Because of this, if illegal copies of content are made on a regular basis then the distributor cannot collect income appropriate for the content provided, and the business model will collapse. From the perspective of digital content businesses that use the Internet and construct their business models based on the ability to protect their digital content, the anticipation for success is high [2]. On the other hand, they also bear the risk of loss due to illicit copying of their content [3].

In recent years there have been major changes in the environment surrounding digital music [4]. There have been many reports of illegal MP3 Web sites [5]. These Web sites illegally copy music data from commercially available CDs, or from regular broadcasts, and then convert the data into MP3 files. They then publish the MP3 files on Web sites that they run and answer the requests of their users by making the files freely available for download. Commercially available CDs and other distribution media have a legally recognized specific copyright that makes it illegal for users without rights to copy and distribute the content without permission. Because this type of use ignores the legally recognized rights of copyright holders, the Web sites are considered to be illegal MP3 Web sites. The number of Web sites similar to illegal MP3 sites has increased. When digital content is distributed for free, it negatively affects the state of CD distribution, harming its commercial viability. One technology that will form a pillar of the solution is digital watermarking.

4. Comparison with existing technology

In recent years, there have been many cases where RFID[1](Fig. 1) and QR codes[2](Fig. 2) technology have been introduced as new technologies for distribution management. In this

[1] Radio-Frequency IDentification. RFID is an automatic identification method.
[2] Quick Response Code. A matrix code (or two-dimensional bar code).

research, as shown in Fig. 3, I apply digital watermarking to a variety of information media, I examine objectives and applications such as copyright protection

Fig. 1. Example of RFID Card

Fig. 2. Example of QR code

Fig. 3. Example of digital watermark to various information carriers

I compared RFID and QR codes and digital watermarking technology.
The following items were compared.

- Raw material processing and age degradation: heat resistance, waterproofing properties
- Degrees of freedom of the markings: Minimum required area, degrees of freedom for shape
- Security: Confidentiality, protection against duplication, protection against alteration
- Reading: Ease of reading, reading rate, compatibility with reading devices
- Cost

The results of the technical comparisons of the above items are shown in Table 1.
In terms of the raw materials during the processing stage and in the environment of practical use, RFID has inferior heat resistance. RFID uses RF tags to perform wireless communication. RFID can be constructed from multiple elements on a circuit board, or can be implemented on a single chip, both of which are prone to destruction by heat.
Although the impact of heat is reduced in QR codes as compared to RFID, preserving the print condition of the markings becomes a challenge.

In digital watermarking, a laser directly burns the markings into the raw materials, and therefore it has heat resistance and waterproof properties that are superior to those of the conventional technologies.

	RFID	Watermark	QR Code
Heat resistance	△	○	△
Waterproof	△	○	△
Minimun required area	○	○	×
Degrees of freedom	○	○	×
Confidentiality	△	○	×
Protection against duplication	○	○	×
Protection against alteration	○	○	×
Ease of reading	○	△	△
Cost	△	○	○
Reading device	○	×	○

Table 1. Comparison of copyright protection technologies
(○: Applicable △: Partially applicable ×: Not applicable)

Next, in regard to the degrees of freedom of the markings and the reading environment, QE codes have more restrictions. QR codes have between 21 × 21 cells in version 1 and 177 × 177 cells in version 40. The required minimum area is determined by the amount of embedded data and the resolution of the reader. If the area of the managed materials is greater than the minimum area of the QR code, there is no problem. However, if the available area is less that this minimum, it is not possible to mount the marking. Also, reading might not be possible if the managed material is curved, such as a sphere or cylinder (error correction can improve the reading rate). RFID is strong in regard to this point: if it is possible to mount the RF tag, then recognition is certain. Marking for digital watermarking is performed in accordance with the shape; thus, the markings have a high degree of freedom, and reading can be performed easily regardless of the shape.

There are many security concerns with QR codes. QR codes are compatible with reading devices such as specialized readers and mobile phone terminals, and are the most common of these three technologies. However, they are weak in terms of confidentiality and protection against alteration.

The benefit of RFID is that it can ensure non-contact recognition by using wireless communication. However, there is the problem that RFID reader eavesdropping can be performed from an unintended location. In terms of cost, RFID requires that RF tags be installed in all of the target objects. Although the cost is currently lower than 10 yen per RF tag, when the number of target objects is great, this amounts to a cost that cannot be ignored. In QR codes and digital watermarking, the cost can be controlled relatively well since the markings are constructed by printing or burning.

One challenge for digital watermarking is its compatibility with reading devices. Although specialized terminals are used as readers in the current stage of development, the range of usability should be increased in the future by using readers for conventional PCs and mobile terminals.

5. Media types and their objectives

Numerous types of information media surround us. In this section, I will discuss the types of media in which digital watermarking technology can be used, and the objectives and applications of its use. Copyright in this digital and networked environment has been debated from a variety of perspectives [6-7]. However, in regard to technology, the advance of digital technology has led to proposals of new copyright protection technologies. In recent years, digital watermarking has been gathering attention as one technology for copyright protection [8-9]. Digital watermarking is technology that directly embeds additional information into content at a level that cannot be detected by the human sense of hearing or sight. Including copyright protection information into these digital watermarks makes it becomes possible to protect the copyright of the author. A variety of engineering methods have been researched regarding digital watermarking technology that can be embedded into a variety of data formats, such as static images, videos, and audio [10-12].

Generally speaking, "multimedia" data comes in three forms: static images, videos, and audio. Here, I have included documents such as public documents and research papers as a type of image medium. From the background to this research, the following five points regarding the objectives and applications for digital watermark use can be noted: 1) copyright protection, 2) distribution traceability, 3) proof of authenticity, 4) security advantages, and 5) sales promotion. Table 2 summarize the objectives and applications of digital watermarking for physical media and static images, and for video and audio, respectively.

Digital information has the characteristic that even if it is processed or edited, the quality will hardly deteriorate at all. Therefore, copyright protection, an item listed in the table 2, is a critical issue. In the past, the © mark has been displayed to indicate the copyright holder, but a common problem is that this mark can be removed through illegal processing or editing [13]. In response to cases like this, digital watermarking can be used on video, image, and voice media to implement a mechanism to prevent the alteration of the copyright owner information, thus protecting the copyright.

In relation to this, the distribution traceability of information media is discussed. Recently, with the spread of digitization and the Internet, the situation is such that content distribution is done over networks, sharing the information with the world [14]. Such an environment makes thorough compliance extremely important. The improvement of people's morals in regard to information must be maintained in tandem with defense mechanisms built into the system; however, the reality is that weak security can cause people's moral sense to decline. By using digital watermarks to embed distribution route information into image, video, and music media, in the case that the information is leaked, it will be possible to clearly determine what route the information followed. Similar to copyright information protection, information traceability can also be achieved, which should already exist, and can prevent a malicious user from intentionally altering the information during the distribution process.

It is not easy for the user to determine whether public documents, research papers, or other purchased products are actually legitimate, which is referred to as proof of authenticity in Table 2. From the fact that counterfeit goods of famous brands are being sold extremely cheaply, it can be inferred that a large quantity of these counterfeit goods are detected [15]. Digital watermarks can be used as one method to differentiate between authentic and counterfeit products. Until now, digital watermarks have almost exclusively been used in digital data such as images, videos, and audio. However, current research has shown that it is possible to use digital watermarks to embed information into physical media such as metals, printed-circuit boards, acrylic boards, and cloth [16].

During the manufacturing process, invisible digital watermark information is embedded into the patterns or logos of legitimate products. In the distribution process and at the purchase stage, if the digital watermark is detected, the product can be determined to be legitimate. If the digital watermark is not detected, then the product can be determined to be a counterfeit. There is also a method to determine authenticity from another perspective. If the strength of the digital watermarks is purposely reduced, the digital watermarking information in the areas that are altered or processed will be lost. This makes it possible to determine what areas have been tampered with. In this way, depending on the application, two different models can be selected. In one, the strength of the digital watermarks can be increased to improve its evidential capacity, and in the other, the strength can be reduced to enable identification of areas that have been altered.

In regard to the previously mentioned copyright protection, traceability, and proof of authenticity, I believe that adding information that cannot be seen by the human eye to the medium can be effective. On the other hand, displaying a visible mark on the medium could have the effect of deterring illegal use; I refer to this as the deterrent effect. Explicitly displaying visible logos or names on products has the potential to have a deterrent effect, thus providing defense against illegal copying. Credits are often displayed on the edge of images or videos. However, the major difference between credits and visible digital watermarks is that by purposely using a release key afterwards, the visible portion can be removed, allowing the original content to be extracted without leaving any excess. In other words, a service model can be created in which content is first released having the deterrent effect, and users can then be provided the original content upon completing official procedures.

Until now, I have focused on means of protecting media. Next, I will discuss the application of digital watermarking for sales promotion. Digital watermarking is a technology that was originally designed considering strong security elements. However, the use of digital watermarking for advertising purposes can be easily considered [17-19]. By embedding URL information into image, video, or audio content using digital watermarking, a mobile phone camera can be used to read the watermarks and guide the user to Web pages that contain information related to the media content. In the case of video and audio media, unlike images and physical media, the content changes with the passage of time. Therefore, this method has a significant advantage in that users can acquire and view information related to the content of interest, unconstrained by time. As an example of practical use for music content, a model can be devised in which the user can easily be guided to an artist's Web site while listening to music content of interest.

		Digital watermarking for physical media	Digital watermarking for static images	Digital watermarking for video	Digital watermarking for audio
Medium characteristics	Visible/ Invisible	Visible	Visible	Visible	Invisible
	Physical/ Electronic	Physical media	Electronic data	Electronic data	Electronic data
	Changes over time	Does not change over time	Does not change over time	Changes over time	Changes over time
Copyright protection		Embedding copyright information	Embedding copyright owner information	Embedding copyright owner information	Embedding copyright owner information
		Copyright protection for physical media (metals, cloths, plastics, etc.)	Copyright protection for images and pictures	Copyright protection for DVDs, broadcasts, and movies	Copyright protection for CDs, broadcasts, and music

		Copyright protection for public documents and research papers		
Distribution traceability	Add the producer's information, and the tracking information for the distribution route to the products and goods	Add the tracking information to the contents for first-time use (use by the copyright holder), and second-time use (reselling)	Add the tracking information to the contents for first-time use (use by the copyright holder), and second-time use (reselling)	Add tracking information to contents for first-time use (use by the copyright holder), and second-time use (reselling)
	Detect a product's unique information from the physical medium (Apply to products that cannot support QR codes and RFID) Adding traceability to products and goods in high-temperature or high-humidity environments	Understand the distribution process by adding copyright holder and buyer information to movie and photo download sales	Tracking illegal movie recording	Understand the distribution process by adding copyright holder and buyer information to music download sales
Proof of authenticity	Detect unique information about products from physical media Detection of counterfeit brand name goods, or discovery of the illegal export using the vehicle identification number	Detection of illegal copying and alteration of public documents and research papers	Detection of pirated DVDs and illegal video distribution websites Tracking illegal movie recording	Detection of pirated CDs and illegal music distribution websites
Can be used on printed materials		Detection of illegal copying for magazines, gravure, and CD jacket images	-	-
		Detection of illegal copying of public documents and research papers	-	-
Deterrent effect (Deterrent effect from using visible digital watermarks)	Intimidation against illegal copying of products and goods	Protection for copyright and portrait rights for images and pictures	Protection for copyright and portrait rights for images	-
		Removed depending on the situation (e.g., legitimate sales) Intimidation against illegal copying	Removed depending on the situation Intimidation against illegal copying	- -
Sales promotion	Guidance to related websites and websites with product details	Guidance to related websites and websites with product details	Guidance to related websites and websites with product details	Guidance to related websites and websites with product details
	Detect unique information about the product from the physical media (Services available to the purchaser only, introduction campaigns for new products, etc.)	Online provision of the latest magazines and books (Services available to the purchaser only, introduction campaigns for new products, etc.)	Acquire detailed information from TV broadcasts and videos	Download ring tones or entire song for karaoke Provide visual information from the audio for individuals with hearing disabilities

Table 2. Types of digital watermarking and their objectives and applications

6. Solution by integrated framework

Digital watermarking technology embeds information that cannot be detected by the human eye into the content. By using the redundancy in digital contents to slightly change the values of the pixels across the image, data that the user cannot normally see can be stored in the image in addition to the usual image data. Fig. 4 shows the use case chart of the digital watermark.

As shown in Fig. 4, this data is directly embedded into the image, and therefore it has the characteristic feature that it cannot be removed even if the image is compressed, formatted, modified, cropped, or printed. This enables the automatic discovery of unauthorized copies of an image file among the enormous number of images on the Internet by embedding the image ID, the copyright holder's name, or other conditions into the image. In addition, by proactively communicating to users that this feature is in use, it will deter unauthorized use.

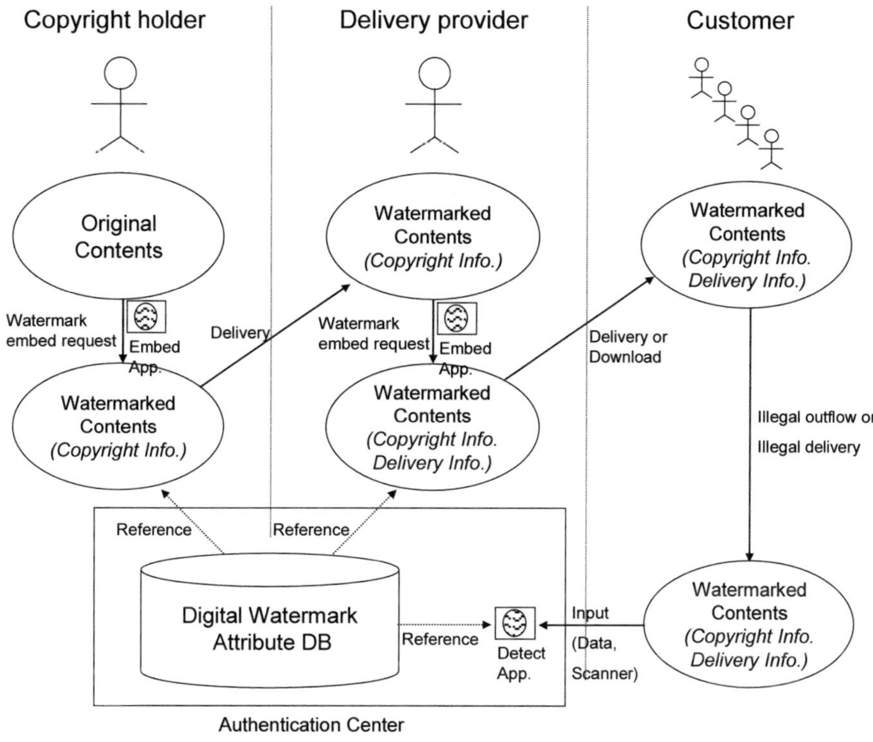

Fig. 4. Use case chart of the digital watermark

I have discussed a variety of ways in which digital watermarking technology can be used to secure information media, and also how it can be used for sales marketing. The framework that integrates these is shown in Fig 5.

The platform's general utility is a critical element in implementing this technology on mobile devices such as mobile phones [20]. There are two methods for detecting watermarking on mobile devices. The first is a client-server method where the file itself is sent to a server and the server detects the watermark. The second is a method where the mobile device itself detects the digital watermark. Performance is rarely an issue with the client-server method as detection is performed on servers with high processing power. However, the disadvantage of this method is that the file must be sent to the server, a process that is time consuming and entails communication costs. In cases where a digital watermark cannot be detected, this result can only be known after sending the file to the server. By performing the entire process of detecting digital watermarks on the mobile device, users can be guided to a variety of network services such as Web services and e-mail to obtain information without incurring communication costs. For this reason, digital watermarking processes in a mobile environment are preferably based on a method where the detection of digital watermarks is performed on the mobile device, preferably using a digital watermark detection application that supports mobile OS middleware such as Java and BREW.

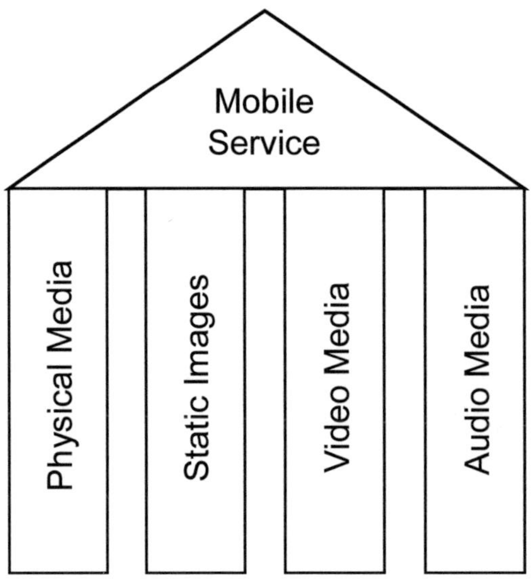

Fig. 5. Integrated framework

As shown in Fig. 6, since middleware neutralize the various differences in the bottom-most hardware layer, generic utility of the platform and good development efficiency are achieved. Device manufacturers can outsource the development of embedded software and concentrate on hardware development. Software developers can develop applications without worrying about differences between platforms. Also, developers do not have to develop different applications for different device vendors. This will eventually lead to applications for not only mobile phones but also smart phones and PDAs.
Next, I explain a mobile solution of the digital watermark to each media.

6.1 Physical media

Data management is performed by using a laser to burn a digital watermark into raw materials such as iron, aluminium, stainless steel, and plastic, and then reading the marking with a reader. The read data is linked to a database, after which management, acceptance examinations, sorting, and distribution can be monitored. An encryption algorithm protects the marking itself, and user authentication and alteration prevention are considered.

As shown in Fig. 7, DPM[3] is a method where markings are made directly onto a product itself. As part of this research project, digital watermarking is implemented using DPM. One characteristic of DPM is that because the markings are made directly onto the material, there is no need to worry that the markings might peel off like an adhesive label. Markings made using DPM can be used in harsh environments and deterioration of the markings is slow, so they can be used over a long period of time [21].

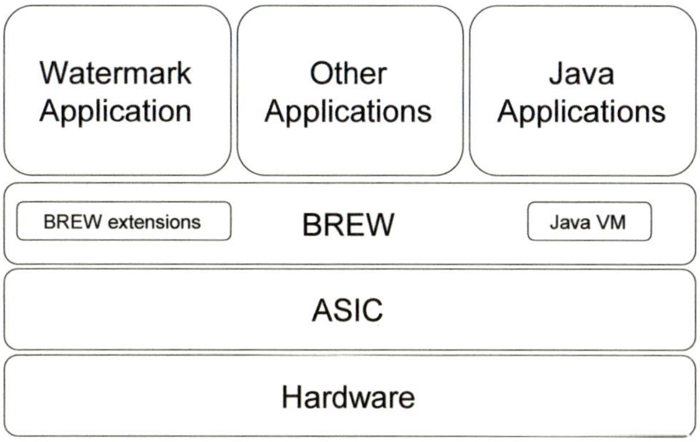

Fig. 6. Software stack for mobile equipment

Fig. 8 is an example of the mobile application for physical media. In one example, the application of DRM technology to the design of brand name products allows for legitimate product authentication and makes it possible to identify the distribution route. Also, directions to the Web site of the relevant brands can also be added. An example applies DRM technology to the coating and parts of an automobile. By recording the automobile data, which cannot be altered by the user, DRM can be used as a theft detection and crime-prevention measure. In addition, services can be provided, such as promotions with information about new car models from the car manufacturer, or the current market price for used cars. These markings can be used in the manufacturing industry to embed a unique ID or lot number into metallic parts for the purpose of automobile detection and accident prevention.

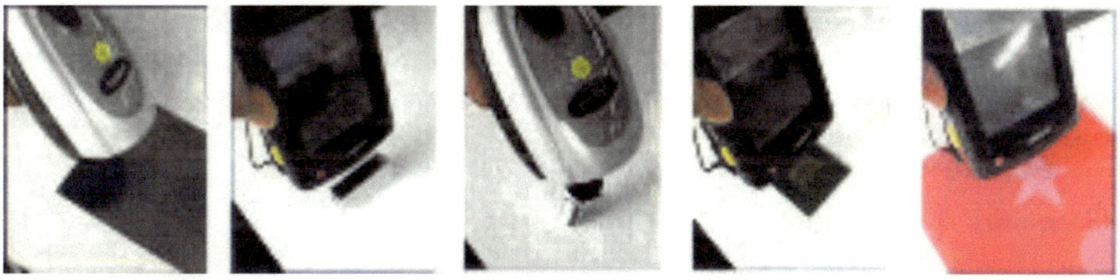

Fig. 7. Example of information detection from physical media

New data management and business schemes can be constructed by using the previously mentioned digital watermarking. Digital watermarking technology was originally used to enforce copyright and for certifying authorship. However, new business models have been established on the basis of this technology.

[3] Direct Parts Marking. Direct Parts Marking is a process to permanently mark parts.

Fig. 8. Example of mobile application for physical media

Additionally, a system that confirms distribution routes with absolute certainty can be designed and applied in other fields, for example, by embedding digital watermarks into medical equipment (e.g., scalpels or scissors) and managing the equipment with a database. In this way, equipment that needs a lease renewal or quantity check can be managed.

6.2 Static images

Presently many posters, pamphlets, magazines, and company brochures have Internet URLs printed on them. Details that cannot be printed on paper such as the latest information or information about related services are published on the Internet. However, there are various problems. Inputting a URL while looking at the printed material is troublesome, input mistakes can occur, and even if there are multiple information items introduced on the printed material, usually only the URL to the top page of the Web site is noted, making it difficult to locate the desired information. Also, the use of many recognition technologies has been hampered by problems with the layout and design of existing media.

A characteristic of using digital watermarking technology in a mobile environment is that mobile content is accessed directly from printed material, providing cross-media marketing. Digital watermarking facilitates the normally difficult task of measuring the effectiveness of promotions using printed medium. When the specialized application software is downloaded, the customer's data is collected, and a unique digital watermarking ID is embedded based on the type of printed medium, the distribution time, and region. Because the connection destination URL information is managed on the server, the user sends the ID information to the content management server and then receives the URL information. In this way, a detailed access log with information on when and where it was accessed, who accessed it, and what print medium was used can be collected, allowing the effectiveness of the printed material to be measured and analyzed.

6.3 Video media

In the introduction, I explained that the distribution of pirated media created from the illegal recording of motion pictures has become a major societal problem [22]. In recent years, motions pictures have been distributed not only on film, but also, in many cases, in digital formats as video data. Information about the time and location that the movie is shown can be embedded into the movie as digital watermarking information and then the movie is shown. The viewers will watch the movie without noticing that this information has been embedded. If the movie is pirated and shown illegally via DVD or over the Internet, when the movie was copied and from what movie theater can be determined by detecting the digital watermarking information. Fig. 9 is an example of detecting information from movie media.

Fig. 9. Example of information detection from movie media

Next, let us consider a service that links the movie database with mobile devices. In recent years, increasing numbers of TV stations and other companies that possess movie content have been managing their digital data as a media archive stored in a database. They can reuse the content by redistributing them on television or over the Internet, or create added value and use them as services for the mobile market. Information about related sites can be embedded as digital watermarks into the movie content that is managed using the database.

If digital watermarking is used to embed a URL into movie content, the URL can be extracted from the video by inputting it via the mobile phone's camera by simply holding the phone up to the display. On TV programs information is sometimes displayed temporary on the screen, such as sports scores, or recipes in the case of cooking shows. However the amount of time that this information is provided is extremely short and the necessary information is often missed. With paper media such as magazines it is possible to use a QR code or a URL to link to a mobile Web site, but it is difficult to acquire information from movie content because the screen is constantly changing. It is also undesirable to constantly display a QR code or URL on a screen. However, if digital watermarking for movies is used, the required information can be extracted from the movie itself and linked to the mobile phone. This eliminates the need to display information on the screen for long periods of time. The mobile phone only needs to be held up to the screen to access the information, so there is no need for the viewer to hurry to write down the information. The extremely convenient interface places little burden on the user. For example, the following information could be provided: shop information during a gourmet TV program, recipes during a cooking program, or lodging information during a travel program. It would also be possible to search for and acquire the necessary information from mobile phone sites after the TV program has ended. Also, by holding up a mobile phone to the screen when a favorite musician is playing during a music program, the user can easily access the music data and the artist's Web site. Finally, similar to the measures to prevent the illegal recording of movies, copyright information could be embedded into the watermark to use in measures to prevent illegal usage.

6.4 Audio media

Next, let us turn our attention to services that utilize digital watermarking on audio media. By embedding digital watermark information onto music or audio files, and reading these files on mobile phones and mobile devices equipped with microphones, such as PDAs, information and content can be displayed that is relevant to the music or audio the user is listening to, and users can also be guided to predefined sources of information such as Web sites. Depending on the specification of the digital watermark detection program and the content of embedded information, users, in addition to being guided to a particular Web site, will also be able to access phone numbers and e-mail information, as well as view relevant video.

In Fig. 10, this schematic shows an example where the mobile device reads digital watermark information embedded in broadcasts or karaoke tunes, and the user is guided to an advertisement site or presented with information such as coupon information or information on the site of a particular manufacturer. For example, the user can select a favorite karaoke tune, visit the artist's site, the URL for which is extracted from the audio being played back, and then download the original song.

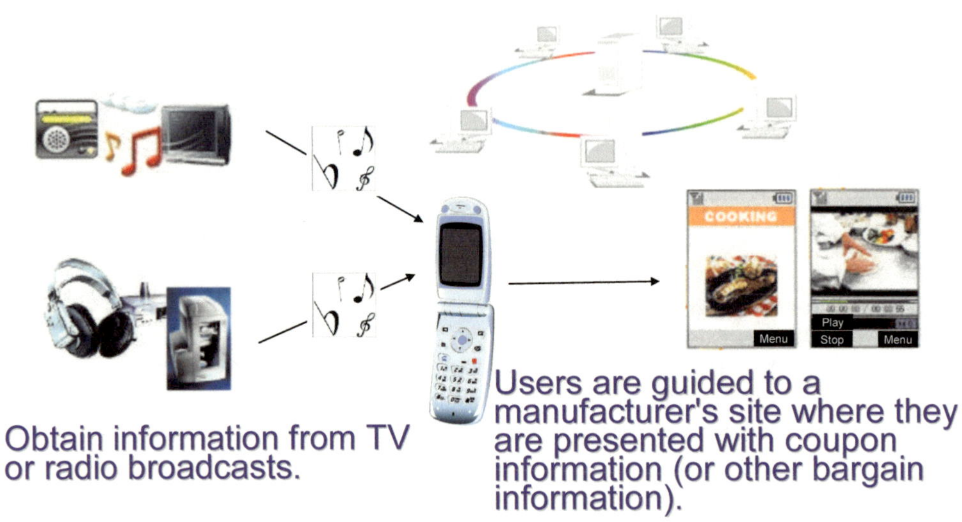

Fig. 10. Guiding users to mobile services from audio media

Digital watermarking can also be used as an information feature at public facilities such as zoos and museums. Since the input source is in audio format, information can be obtained easily over a broad area as opposed to QR codes, which require users to take close-up shots of a specific point. This will be an effective way to provide guidance information in large venues where visitors can often get lost.

Audio files do not require special equipment and can be played back on any consumer market speaker. This technology enables operators to provide interactive services in situations where only a mobile phone or PDA is used. More information can be provided if the mobile device is equipped with Internet connectivity. Compared to similar services based on RFID, digital watermarking, which extracts information from audio files without the use of IC tags, delivers the same effect at a lower cost. Since generic mobile phones can be used instead of special devices, development costs, and device purchase costs can be reduced, expanding the technology's range of applications.

7. Conclusion

In this way, watermarking technology can identify certain copyrighted content and the related rights during the distribution process or after distribution. If revolutionary and new content distribution and usage models can be realized, the confirmation of the copyright owner and rights management will become possible. However, there has been little research done on the implementation method. Especially in the case of embedded software, which is widely used in mobile devices, implementation is seen to be difficult due to the limited implementation resources. Also, in regard to the assimilation of different implementations and platforms of the wide variety of digital watermarking algorithms, the amount of development work is becoming huge and existing resources are only rarely being reused. Moreover, the increasing complexity of errors and memory management that has accompanied this increase in development scale has become a problem that can no longer be ignored. Uncertainty can be actively managed and exploited. The unexpected developments in an environment can be actively managed through flexible designs that can adjust to changed conditions. Moreover, these new states can provide opportunities. Thus, it is important to broaden one's perspective to consider not only risk but also opportunities. Architecture is important. It is useful and productive to consider explicitly how the parts of an engineering system interact with each other. It may often be essential to do so, to enable us to deal effectively with the need to reconfigure systems in response to new possibilities and new requirements.
Digital multimedia content requires a copyright protection system to be constructed due to the ease by which they can be copied and edited, and due to the ease of high-volume distribution through digital distribution channels. By embedding copy control information into digital watermarks, which are robust against digital disturbances, it is possible to construct a copyright protection system that can prevent illegal copying. The control information that is embedded into digital watermarks can be securely transmitted to the system through both the packaging for physical distribution, and through the network for digital distribution, thus strongly protecting the author's copyright. Many security rules for information media are complex, and they are not unified across regions and for every application. By using digital watermarking technology on information media, information can be embedded into different media without regard to the application or environment. Moreover, the service can be provided regardless of what hardware or software is used. In sum, digital watermarking is a technology that is not affected by various conditions such as business models or the structure of the system that is used, software differences, coding differences, or copyright management policies. I anticipate that using an economical copyright protection system that uses digital watermarking will promote the digitization of multimedia content, and protect the author's copyright. This will also foster more open and global multimedia content distribution.

8. References

[1] Kineo Matsui. (1998). *Base of Digital Watermark* (In Japanese), Morikita Publishing Co., Ltd.,
[2] Liquid Audio, Music on the Net, A Topographic Tour of the Online Music World, http://www.minidisc.org/music_internet.html (Accessed: Oct. 8, 2010)
[3] Kazuhiro Okamura. (2008). The one that electronic watermark brings -First part-(In Japanese), *Monthly Automatic Operation Recognition*, Feb. 2008 Vol.21 No.2, Japan Industrial Publishing Co. Ltd., pp.36–38

[4] Business Software Alliance, Web site, (In Japanese), http://www.bsa.or.jp/press/related/2010_Global_Piracy_Studyj.html (Accessed: Oct. 8, 2010)
[5] Manabu Hirakawa, Junichi Iijima. (2009). "Validating the effectiveness of using digital watermarking technology for e-commerce website protection," *Proceedings of the 9th Asian eBusiness Workshop*, No.21, pp.127-132
[6] Kotaro Nawa. (1996). *Copyright of Cyberspace* (In Japanese), Chuokoron-sha, p.194
[7] Kenji Naemura. (1997). *Copyright of Multimedia Society* (In Japanese), Keio University Press Inc., p.285
[8] Fumitada Takahashi. (1997). "The digital watermark keeps to the multimedia era (In Japanese)," *Nikkei Electronics*, Nikkei BP Marketing Inc., Vol.683, No.2-24, pp.99-124
[9] Satoshi Nanamatsu, Toshihiro Masumoto, Kazuyoshi Tanaka. (2000). "Multimedia digital contents and copyright protection (In Japanese)," *Information Management*, Vol.42 No.12 pp.1013-1021
[10] Kineo Matsui. (1998). *Base of Digital Watermark - New Protection Technologies for Multimedia* – (In Japanese), Morikita Publishing Co., Ltd.,
[11] Kineo Matsui. (1998). "Electronic watermark and the evaluation item," *Institute of Image Electronics Engineers of Japan Magazine*, Vol.27, No.5, pp.483-491
[12] Kiyoshi Yamanaka. (1998). "Problem in application to electronic watermark and copyright protection," *Information Management*, Vol.40, No.10, pp.933-940
[13] Naohisa Komatsu, Kenichi Tanaka. (2004). *Digital watermarking technology – Digital content security* (In Japanese)," Institute of Image Electronics Engineers of Japan Magazine
[14] Tsukasa Ono. (2001). *Digital Watermark and Contents Protection* (In Japanese), Ohm-sha Co., Ltd.,
[15] Manabu Hirakawa. (2008). "A digital watermark service model's effectiveness of verification in copyright protection (In Japanese)," *Proceedings of National Spring Research Conference 2008*, The Japan Society for Management Information, pp.G4-2
[16] Manabu Hirakawa, Junichi Iijima. (2008). "A study on usage of digital watermark in distribution management (In Japanese)," *Proceedings of National Autumn Research Conference 2008*, The Japan Society for Management Information, pp.B1-3
[17] Key Pousttchi, Dietmar G. Wiedemann. (2006). "A Contribution to theory building for mobile marketing: Categorizing mobile marketing campaigns through case study research," *Proceedings of ICMB '06. International Conference*, pp.1-1
[18] Andreas Albers, Christian Kahl. (2008). "Design and implementation of context-sensitive mobile marketing platforms," *Proceedings of E-Commerce Technology and the Fifth IEEE Conference on Enterprise Computing*, 10th IEEE Conference, pp.273-278
[19] Phyoung Jung Kim, Young Ju Noh. (2003). "Mobile agent system architecture for supporting mobile market application service in mobile computing environment," *Proceedings of Geometric Modeling and Graphics International Conference*, pp.149-153
[20] Hidemi Mizoguchi, Yukinori Miyakita, Yuta Tokoro. (2007). *EZ Applications Explained (BREW) Programming*, RIC Telecom Co., Ltd.,
[21] Kazuhiro Okamura. (2008). "The one that electronic watermark brings -Latter part-(In Japanese)," *Monthly Automatic Operation Recognition*, Mar. 2008 Vol.21 No.3, Japan Industrial Publishing Co. Ltd., pp.33–36
[22] Japan and International Motion Picture Copyright Association, Inc. Homepage, http://www.jimca.co.jp/index.html (Accessed: Oct. 8, 2010)

18

RFID Middleware Design and Architecture

Mehdia Ajana El Khaddar[1], Mohammed Boulmalf[3],
Hamid Harroud[2] and Mohammed Elkoutbi[1]
[1]*SI2M Lab, ENSIAS*
[2]*WML Lab, Alakhawayn University in Ifrane*
[3]*Canadian University of Dubai*
[1,2]*Morocco*
[3]*UAE*

1. Introduction

Radio Frequency Identification (RFID) is a form of *Automatic Identification and Data Capture* (AIDC) technique (Ishikawa et al., 2003). RFID is recently being used in a wide range of areas such as Supply Chain Management (SCM), health care, traffic monitoring, retail, and access control (Polniak, 2007). The ability to store large amounts of data and identify items which are not in the line of sight has given RFID technology an edge over other automatic identification approaches such as the barcode based systems (Ishikawa et al., 2003) and optical character recognition systems (OCR) (Phoenix Software International, 2006). As an example, RFID technology integration in SCM systems has resulted in the reduced losses and improved visibility in various stages of supply chaining (Sheng et al., 2008), reduced numbers of data entry errors, efficient inventory management, and lower human labor costs in distribution centers (Tutorial-Reports, 2007).

A binary code comprising a field of bars and gaps arranged in parallel configuration is used by the barcode based identification systems. The analysis of the reflected beam on the bar gaps, allows the numerical and alphanumerical interpretation of the barcode sequence made up of narrow and wide bars. The interpreted value obtained specifies a unique code that is used for object identification. The disadvantage of the barcode system is that the barcode needs to be aligned in order to be read by the laser scanner (Ishikawa et al., 2003). The OCR based systems consist of optical machine readers used to recognize alphanumeric codes which are placed on the objects to be uniquely identified. The drawbacks of this system consist of the cost of operation, and the complexity of the OCR readers (Phoenix Software International, 2006).

The RFID systems basically consist of three elements: a tag/transponder, a reader and a middleware deployed at a host computer. The *RFID tag* is a data carrier part of the RFID system which is placed on the objects to be uniquely identified. The *RFID reader* is a device that transmits and receives data through radio waves using the connected antennas. Its functions include powering the tag, and reading/writing data to the tag. As shown Fig. 1, the signals sent by the reader's antennas form an *interrogation zone* made up of an electromagnetic field. When a tag enters this zone, it gets activated to exchange data with the reader (Al-Mousawi, 2004). Later, the identification data read by the RFID reader is processed by the software system, known as the *RFID middleware*. The RFID middleware manages readers, as well as filters and formats the RFID raw tag data so that they can be

accessed by the various interested enterprise applications (Floerkemeier & Lampe, 2005). Hence, the middleware is a key component for managing the flow of information between tag readers and enterprise applications (Burnell, 2008).
Major advantages of using RFID as an auto-ID system are the following:

- RFID readers do not require a line of sight to access data from the RFID tags.
- RFID systems can read data over varied range from few centimeters to few hundred meters.
- RFID readers can interrogate, and make RFID tags readings much faster.
- RFID systems can read and write different sizes of data from / to the tag, based on the type of tag.
- RFID systems can read tags in harsh environments, without any human interference.

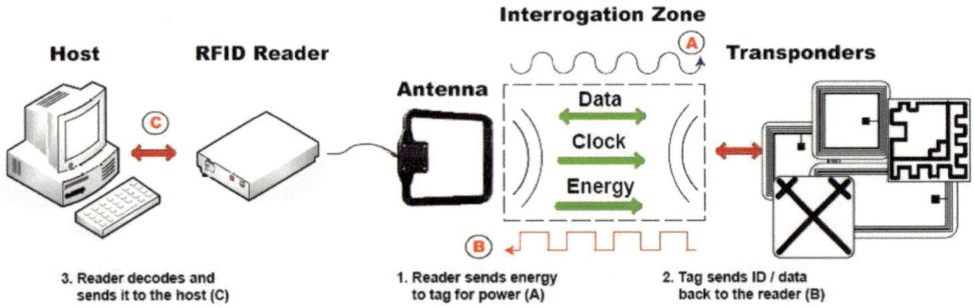

Fig. 1. RFID system components (Glasser et al., 2007)

RFID technology is becoming ubiquitous as RFID systems have recently undergone significant improvements. A variety of makes and models of RFID tags and readers, combined with decreasing RFID hardware prices, are making RFID deployment more attractive (Glasser et al., 2007). In the traditional applications of RFID such as access control, networking was not a concern and there was barely a need for a RFID middleware solution. However, in the novel application areas such as SCM, a number of RFID readers could be used to capture RFID data which need to be disseminated to a variety of enterprise applications. Hence, there is no longer a one-to-one relationship between reader and application (Floerkemeier et al., 2007).
The researchers in this area have reported a vast amount of research (e.g. (Burnell, 2008; Molnar & Wagner, 2004; Parliament Office of Science and Technology, 2004) about the benefits, possible misuses, ethical issues (e.g. privacy), and technical issues (Floerkemeier & Lampe, 2004) involved in the RFID technology. However, less significant attention has been paid to the issues involved in the RFID middleware that manages large deployments of readers producing high volumes of captured data, and encapsulates applications from the low level data by transforming them into more meaningful events (Burnell, 2008). Considering this void in the RFID middleware research, herewith, we discuss the design issues of RFID middleware, present our solution called FlexRFID which addresses the above aspects, and compare it to other middleware solutions. We analyze FlexRFID to the extent to which it addresses applications' needs, and allows an easy management of devices.

2. RFID system components

RFID systems are produced by many manufacturers and exist in countless variants. However, a RFID system consists mainly of three components; the transponder/tag, reader, and RFID middleware.

2.1 RFID transponder/tag

A RFID transponder, or tag, consists of a chip and an antenna. A chip can store a unique serial number or other information based on the tag's type of memory. The tag's type of memory can be read-only, read-write, or write-once and read-many (United States Government Accountability Office, 2005). *Read-only* tags are much cheaper to produce and are used in most current applications. *Read-write* tags are useful when information needs to be updated (Al-Mousawi, 2004). The antenna is used to transmit information from the chip to the reader, and the larger the antenna the longer the read range. The RFID tag can be either attached or embedded in an object to be identified, and can be scanned by mobile or stationary readers using radio waves (United States Government Accountability Office, 2005).
RFID tags exist in three different versions: passive tags, active tags, and semi-passive / semi-active tags.

2.1.1 Passive tags

Present the simplest version of RFID tags which do not contain their own power source, such as a battery, and cannot initiate communication with the reader. The passive tag derives its power from the energy waves transmitted by the reader and responds to the reader's radio frequency emissions, therefore the passive tag relies entirely on the reader as its power source. A passive tag should store, at a minimum, a unique identifier for the item tagged, and can be read from a range of about 10 to 20 feet under perfect conditions (United States Government Accountability Office, 2005). Passive tags have lower production costs, meaning that they can be applied to less expensive disposable goods (e.g. a bottle of shampoo).
The cost of passive tags varies based on the radio frequency used, amount of memory, and design of the antenna, and other tag requirements. Passive tags can operate at low, high, ultrahigh, or microwave frequency. The development of passive RFID tags has made wide scale use of them in many organizations. Examples of passive tag applications include mass transit passes, building access badges, and consumer products in the supply chain (United States Government Accountability Office, 2005).

2.1.2 Active tags

Unlike passive tags, active tags contain a power source and a transmitter, in addition to the antenna and chip, and send a continuous signal. These tags typically have read/write capabilities; tag data can be rewritten and/or modified. Active tags can initiate communication and communicate over longer distances up to 750 feet, depending on the battery power. Because these tags contain more hardware than passive RFID tags, they are more expensive and are reserved for costly items that are read over greater distances (United States Government Accountability Office, 2005). RFID manufacturers typically do not quote prices for active tags without first determining their storage type and quantity, and range.

2.1.3 Semi-passive tags

This type of tags is called also semi-active tags. Semi-passive tags do not initiate communication with the reader but contain batteries that allow the tag to perform other functions, such as monitoring environmental conditions and powering the tag's internal electronics. In order to conserve battery life, some semi-passive tags do not actively transmit a signal to the reader. Instead, they remain dormant until they receive a signal from the reader. Semi-passive tags can be connected to sensors to store information for container security devices (United States Government Accountability Office, 2005). Semi-passive tags have the middle transmission range and cost (Vacca, 2009).

As a summary, passive tags are consequently much lighter than active tags, less expensive, and offer a virtually unlimited operational lifetime. The trade off is that they have shorter read ranges than active tags and require a higher-powered reader (Association for Automatic Identification and Mobility, n.d.). Table 1 shows a comparison among passive, semi-passive, and active tags.

	Passive Tags	Semi-Passive Tags	Active Tags
On board power supply	No (From Reader)	Yes (Internal Battery)	Yes (Internal Battery)
Transmission range	Short (up to 6.096 meters)	Medium (up to 30.48 meters)	Long (up to 228.6 m)
Communication pattern	Passive	Passive	Proactive
Cost	Cheap	Medium	Expensive
Type of memory	Mostly Read-Only	Read-Write	Read-Write
Life of tag	Up to 20 years	2 to 7 years	5 to 10 years

Table 1. Characteristics of passive, semi passive and active RFID tags (United States Government Accountability Office, 2005; Vacca, 2009)

2.1.4 RFID tags by type of memory

RFID Tags have various types of memory (United States Government Accountability Office, 2005):

Read-Only tags: have minimal storage capacity (typically less than 64 bits) and contain permanently programmed data that cannot be altered. These tags primarily contain item identification information and have been used in libraries and video rental stores.

Read-Write tags: in addition to storing data, they can allow the data to be updated when necessary. Consequently, they have larger memory capacity and are more expensive than read-only tags. These tags are typically used where data may need to be altered throughout a product's life cycle, such as in manufacturing or in supply chain management. Read-Write tags have three main procedures for managing and storing data:

- *EEPROM* (Electrically Erasable Programmable Read-Only Memory): is a type of non-volatile memory used to store small amounts of data that must be saved when power is removed. It is the most dominant procedure in many RFID systems, but has the disadvantages of high power consumption during the writing operation and a limited number of write cycles (Al-Mousawi, 2004).
- *FRAM* (Ferromagnetic Random Access Memory): its read power consumption is lower than the EEPROM by a factor of 100 and the writing time is 1000 times lower. Because of manufacturing problems, its widespread introduction onto the market was affected (Al-Mousawi, 2004).
- *SRAM* (Static Random Access Memory): SRAM are used for data storage in microwave system which facilitate very fast write cycles. The disadvantage of this procedure is that the data requires an uninterruptible power supply from an auxiliary battery (active transponder) (Al-Mousawi, 2004).

Write-Once, Read-Many tags: allow information to be stored once, but does not allow subsequent updates to the data. This tag provides the security features of a Read-Only tag while adding the additional functionality of Read-Write tags.

2.1.5 RFID tags operation frequencies

RFID tags operate in several frequency bands. Most of the used frequencies are those that are in the Industrial, Scientific or Medical (ISM) frequency ranges. RFID frequencies are divided into the following three basic ranges:

- *Low Frequency (LF)*: this range operates between 30 and 500 KHz. However 125-134 KHz is the most ordinary range used in animal tracking, car immobilizers, security access, asset tracking etc. LF tags are commonly used where there are liquids, electrical noise, or metals present and when a fast read rate is not required. Most of low frequency systems operate without the need of integrated battery in their tags, have short reading ranges, and are lower system costs (Al-Mousawi, 2004).
- *High Frequency (HF)*: this range operates between 10-15 MHz, but 13.56 MHz HF tags are the most commonly used, due mainly to the relatively wide adoption of smart cards based on RFID technology. The cost of the high frequency systems is inexpensive, but higher than the low frequency systems, they have longer read ranges and higher reading speeds than the LF systems. The HF systems are used in access control and smart cards (Al-Mousawi, 2004).
- *Ultra High Frequency (UHF) and Microwave Frequency*: Ultra High Frequency Systems operate between 400 and 1000 MHz and microwave frequencies between 2.4 and 2.5 GHz. These systems are the most expensive compared to the others. UHF tags are considered as being the most practical for item-level tracking as they offer a good balance between range (typically less than a few meters), a high reading speed, and the ability to read multiple tags. Unlike the other systems, line of sight is required for the communication between RFID reader and tags. UHF systems have a very long read range, and are used for such applications as railroad car tracking and automated toll collection. Microwave frequency band is also used by many other systems e.g. Bluetooth and Wi-Fi systems (Al-Mousawi, 2004).

2.2 RFID reader

A RFID Reader is a scanning device that reliably reads the tags and communicates the results to the middleware. A reader uses its own antennae to communicate with the tag by broadcasting radio waves to which all tags within range will respond. Readers can process multiple items at once, allowing for increased read processing times. They can be either mobile or stationary, and they are differentiated by their storage capacity, processing capability, and the frequency they can read (United States Government Accountability Office, 2005).

RFID reader consists of the following functional blocks:

2.2.1 HF interface

The master part of the reader which has these functions (Al-Mousawi, 2004):

- Supplying RFID transponders with power by generating high frequency power;
- Modulation of the signal to the transponder;
- Reception and demodulation of signals from the transponders.

2.2.2 Control unit

The slave part of the reader that performs the following functionalities (Al-Mousawi, 2004):

- Communication and execution of the application software's commands;
- Signal coding and decoding;
- Communication control with a transponder.

Some RFID readers have additional functionalities like *anti-collision algorithm*, *encryption* and *decryption* of transferred data, and *transponder-reader authentication* (Al-Mousawi, 2004).

Different designs of readers exist, because different applications have different requirements from each other. RFID readers are classified into three types (Al-Mousawi, 2004):

- *OEM readers*: Original Equipment Manufacturers readers are mostly used for data capture systems, access control systems, and robots.
- *Industrial use readers*: used in assembly and manufacturing plant.
- *Portable readers*: These readers are more mobile than the other readers, and supported with a LCD display and keypad. This kind of readers is used in animal identification, device control and asset management applications.

2.3 RFID middleware

The middleware refers broadly to software or devices that connect RFID readers and the data they collect, to enterprise information systems. RFID middleware helps making sense of RFID tag reads, applies filtering, formatting and logic to tag data captured by a reader, and provides this processed data to back-end applications (Burnell, 2008). RFID middleware serves in managing the flow of data between tag readers and enterprise applications, and is responsible for the quality, and therefore usability of the information. It provides readers connectivity, context-based filtering and routing, and enterprise / B2B integration. RFID middleware design and components will be discussed further in the next sections.

When designing a RFID middleware solution, the following issues need to be considered:

- **Multiple hardware support**: The middleware must provide a common interface to access different kinds of hardware offering different features.
- **Synchronization and scheduling**: There should be intelligent scheduling and synchronization among all the processes of the middleware. This minimizes the latency and improves the efficiency of the middleware.
- **Real-time handling of incoming data from the RFID readers**: The middleware should handle the huge amount of data captured by the connected readers in real time without read misses.
- **Interfacing with multiple applications**: The middleware should be capable of interacting with multiple applications simultaneously, by catering to all the requirements of the applications with minimal latency.
- **Device neutral interface to the applications**: The application developer should only use the generic set of interfaces provided by the middleware independently of the type of hardware connected to the system.
- **Scalability**: The middleware design must allow easy integration of new hardware and data processing features.

3. RFID middleware components

A RFID middleware is the interface that sits between the RFID hardware and RFID applications. It provides the following advantages:

- It hides the RFID hardware details from the applications;
- It handles and processes the raw RFID data before passing it as aggregated events to the applications;
- It provides an application level interface for managing RFID readers and querying the RFID data.

A layer of the RFID middleware incorporates all the device drivers of different hardware and exposes to the application standard interfaces to access this hardware. If the application was provided with all the device drivers of all connected readers, it will be a hard job to manage and interface each of the devices. The application developer will then need to understand all the hardware specific internals and operations. Also, the application, if provided with the huge amount of raw tag data reported by the readers, will find it very difficult to process the data in real time. A RFID middleware provides a standardized way of dealing with this flood of information, which processes the raw data and provides the application with clean and filtered data.

As shown in Fig. 2 a RFID middleware is generally composed of four major layers:

- Reader Interface
- Data Processor and Storage
- Application Interface
- Middleware Management

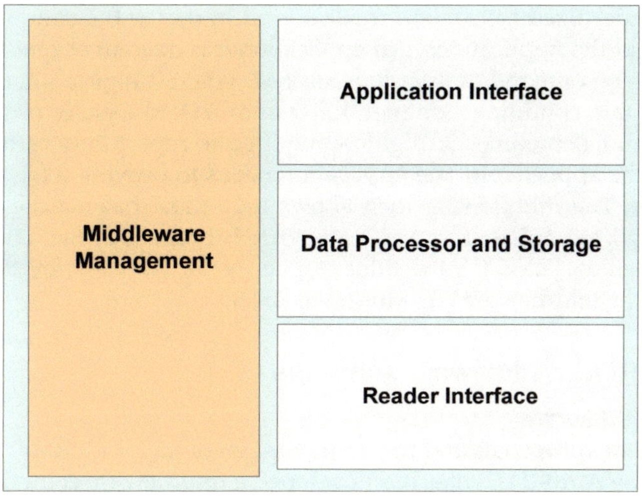

Fig. 2. RFID middleware components

3.1 Reader interface

The reader interface is the lowest layer of the RFID middleware which handles the interaction with the RFID hardware. It maintains the device drivers of all the devices supported by the system, and manages all the hardware related parameters like reader protocol, air interface, and host-side communication.

3.2 Data processor and storage

The data processor and storage layer is responsible for processing and storing the raw data coming from the readers. Examples of processing logic carried by this layer are data filtering, aggregation, and transformation. This layer also processes the data level events associated with a specific application.

3.3 Application interface

The application interface provides the application with an API to access, communicate, and configure the RFID middleware. It integrates the enterprise applications with the RFID middleware by translating the applications' requests to low level middleware commands.

3.4 Middleware management

The middleware management layer helps managing the configuration of the RFID middleware, and provides the following capabilities:

- Add, configure, and modify connected RFID readers;
- Modify application level parameters such as filters, and duplicate removal timing window;
- Add and remove services supported by the RFID middleware.

RFID readers are typically abstracted as a logical reader which is either a collection of several readers or a part of the reader. This grouping mechanism is used where there is a need to have a set of readers capturing data from a particular area such as a warehouse with many loading docks. The advantage of this is that the application can query a small number of logical readers rather than having to aggregate events from each of the individual readers.

There are two standardized interaction models used to define the communication between the middleware and the applications. An application can operate at *synchronous mode* when requesting services on demand or *asynchronous mode* when it registers for information to be sent to it when certain conditions are met. RFID middleware usually provide some kind of data filtering, because sometimes it might be required to report only certain type and value of the tag data to the application. The application needs to provide a set of defined patterns to the middleware. The middleware then allows only data that matches the pattern to be reported to the application. E.g. if an application needs to see only tag data that starts with a specific pattern such as "XYZ20", the filter can be set to this value by the application and communicated to the middleware (Al-Mousawi, 2004).

4. Examples of RFID middleware solutions

4.1 The savant middleware

There have been some proposals and research work involving middleware design and RFID data processing. The Auto-ID Center has developed a middleware component called Savant (Clark et al., 2003) that collects, accumulates, and processes Electronic Product Code (EPC) data obtained from several RF readers. It adjusts multiple readings of a tag, and performs tasks such as archiving data, and inventory control (Ishikawa et al., 2003).

The Savant has a set of *Processing Modules* or *Services* which may be combined to meet the user's application's needs. This modular structure allows innovation to be promoted by independent groups of people, which helps avoiding the creation of a single monolithic specification that attempts to satisfy all needs for everybody (Clark et al., 2003). Fig. 3 shows the three key elements of the Savant middleware architecture: *Event Management System* (EMS), *Real-Time in-Memory Data Structure/ Real-Time in-Memory Event Database* (RIED) and *Task Management System* (TMS) (Ishikawa et al., 2003).

The EMS provides a JAVA API for different types of RF readers and it serves to collect tag read events. The EMS allows adapters to be written for various types of readers, collecting

EPC data from readers in a standard format, allowing filters to be written to smooth or clean EPC data, allowing various loggers to be written, and buffering events to enable loggers, filters and adapters to operate without blocking each other.

The EMS is composed of the following elements (Auto-ID Center, n.d.):

- *Reader Interface*: Allows readers and adapters to communicate events detected by the Auto-ID readers
- *Reader Adapters*: Communicate with readers to capture EPC events
- *Event Loggers (Event Consumers):* Allow for varied processing of events; store the information in the database, store events in a memory data structure, and broadcast the events to remote servers
- *Event Queues (Event Forwarders):* Handle multiple reader event loggers with synchronous implementations

The RIED is an in-memory database that can be used to store event information by Edge Savants. It provides the same interface as a database, but offers much better performance. The RIED should be a high-performance in-memory and a multi-versioned database.

The TMS manages tasks, just as the operating system manages processes, and provides an interface for task management. Task examples include data gathering, remote task scheduling, personnel alerts, and remote upload. The TMS should be a platform-independent system requiring little memory processing power, should automatically upgrade the tasks it executes, and should present a well-defined, interoperable external interface to schedule, monitor, and remove tasks. Tasks should also be written in a platform-independent language using a simple well-defined SDK.

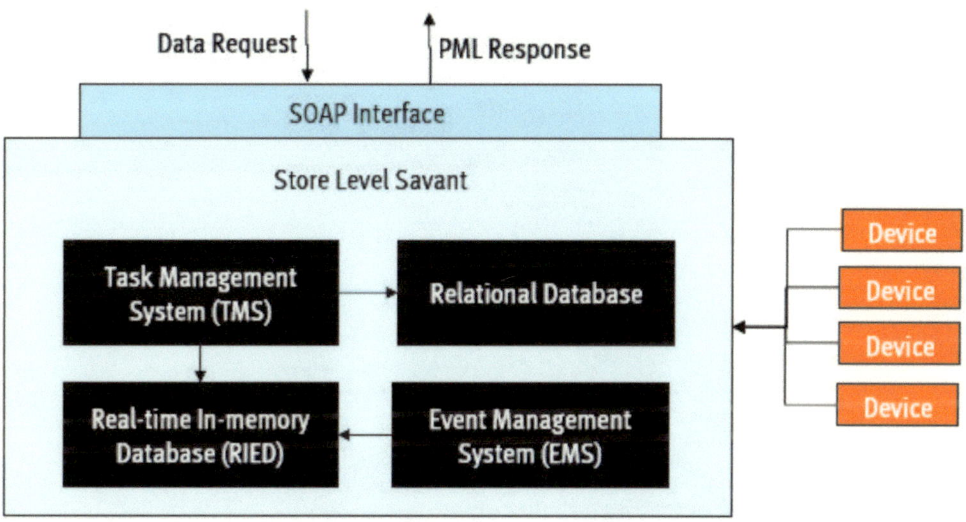

Fig. 3. Savant middleware key components

4.2 WinRFID middleware

WinRFID (Prabhu et al., 2005a) developed at the University of California Los Angeles (UCLA), is another middleware architecture that uses web services and enables rapid RFID applications development. It is a multi-layered middleware that consists of five main layers shown in Fig. 4.

The *physical layer* deals with the hardware consisting of readers, tags and other sensors. This layer abstracts the hardware elements; readers, tags and host I/O interfaces. This abstraction allows extending the middleware capabilities in the advent of introduction of new RFID technology (Prabhu et al., 2005).

The *protocol layer:* The ability to support multiple tag protocols and add new ones is becoming imperative in middleware designs. The protocol layer of the WinRFID middleware allows abstracting the reader-tag protocols. It wraps the command syntax and semantics of a variety of published protocols such as ISO 15693, ISO 14443, ISO 18000-6 A/B, ICode, EPC Class 0 and EPC Class 1. It also deals with protocol specifics such as byte-based, block or even page reading and writing, structure and length of the command frames, partitioning of the tag memory space, checksums, etc (Prabhu et al., 2005 b).

The *data processing layer* deals with processing data streams generated by the network of readers. It includes processing rules that deal with problems due to tag density, read/write distance, orientation of tags and material of item that introduce inconsistencies in reading or writing such as multiple reads of the same tag, some tags not being read, erroneous reads, etc. All of these discrepancies are processed as exceptions and a variety of altering systems are available for resolution such as emails, messages, and user defined triggers (Prabhu et al., 2005 b).

The *XML framework layer* formats the cleaned tag data in a variety of ways to a higher level XML based representation. The purpose of this layer is to provide data in a suitable format to the application layer for decision making (Prabhu et al., 2005 b).

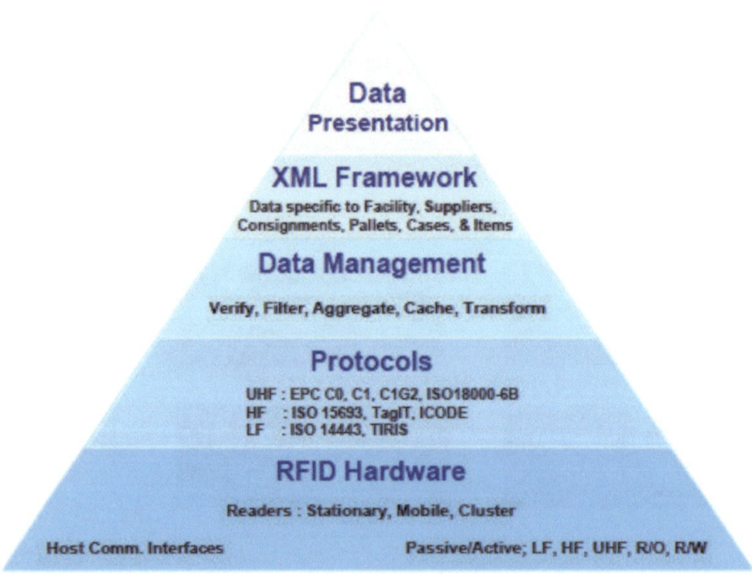

Fig. 4. WinRFID middleware multi-layered architecture (Prabhu et al., 2005 b)

The *data presentation layer* presents the data as per the requirements of end-users or different applications requirements. It facilitates data visualization for decision making. This layer supports two components the portal and the database connector. The portal provides the users with an interface to subscribe to the information of interest. For the database connector, currently the middleware can populate SQL Server and Oracle

RDBMS. The databases get populated in an asynchronous fashion in a trickle mode - a process with least priority so as to avoid the edge hosts getting locked up (Prabhu et al., 2005 b).
WinRFID exploits the .Net framework's runtime plug-in feature to support the addition of new readers, protocols, and data transformation rules with minimum disruption of the existing infrastructure (Prabhu et al., 2005 a).

4.3 The WebSphere RFID middleware

The WebSphere RFID middleware solution, designed by IBM, consists of three main components as shown in Fig. 5: *RFID devices*, *WebSphere Premises Server*, and *Websphere Business Integration Server* (IBM Corporation, 2009).
The IBM WebSphere is a sensor enabled product that allows sensor data aggregation and analysis, deriving insights from sensor data and integrating those insights with the SOA business processes. The software provides the use of intelligent business rules that manage complex event identification and processing (IBM Corporation, 2009).
This solution expands device services allowing a single platform to support multiple sensor types, and supports workflow tooling for sensor data integration with business processes (IBM Corporation, 2009). Therefore, it delivers new and enhanced capabilities to create a robust, flexible, and scalable platform for capturing new business value from sensor data.

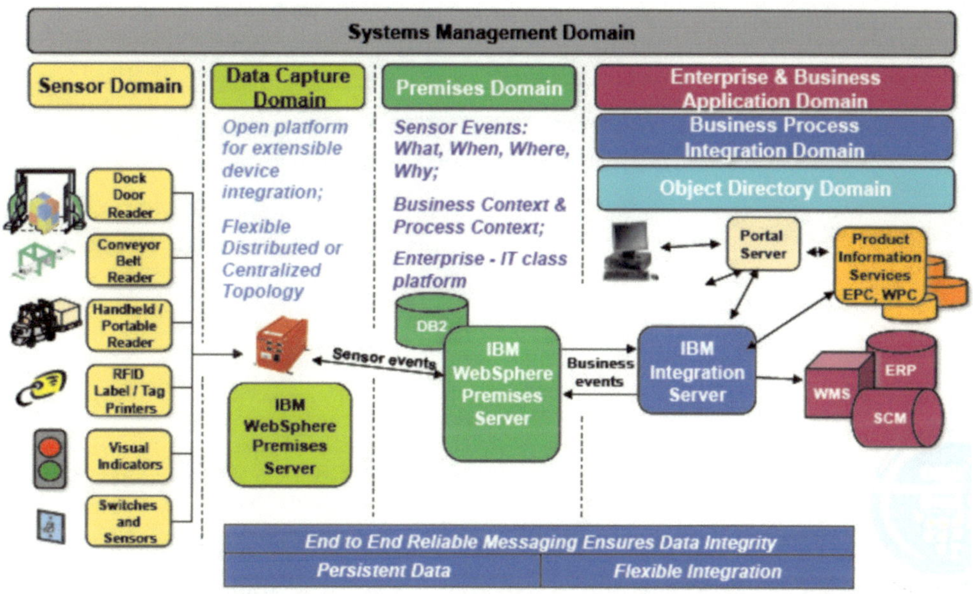

Fig. 5. The IBM sensor and actuator solutions framework (Eisma, 2008)

4.4 The Sun JAVA RFID system

Sun Java System RFID software is a Java based commercial middleware platform provided by Sun. It is a critical RFID infrastructure component that allows a safe, secure, and efficient data and device integration from the edge of the enterprise into enterprise application systems. It has a dynamic, service provisioning architecture that enables scaling from small pilots to large deployments with high data volume (Sun Microsystems, 2006 b).

The Java System RFID Software supports a variety of new and existing standards, such as EPC, ISO, Gen 2, passive and active tags and devices, read/write tags, and commercial and government standards. It is a part of the Java Enterprise System (JES) and has four components as shown in Fig. 6: the *RFID Event Manager*, the *RFID Management Console*, the *RFID Information Server*, and a *Software Development Kit* (SDK). The RFID Event Manager is a Jini-based event management system that facilitates the capture, filtering, and eventual storage of events generated by RFID readers. The RFID Management Console provides a browser based management interface, which allows configuration of various attributes and parameters of the middleware. The RFID Information Server is responsible for storing and querying the EPC related data, it also manages inter Enterprise handling of the data. The SDK provides a development platform to build custom applications (Sun Microsystems, 2006 a).
The Sun middleware exposes to the application, the hardware as logical readers. These logical readers may be a collection of one or more physical readers that the application can select and apply the various processing parameters to the group (Sun Microsystems, 2006 a).

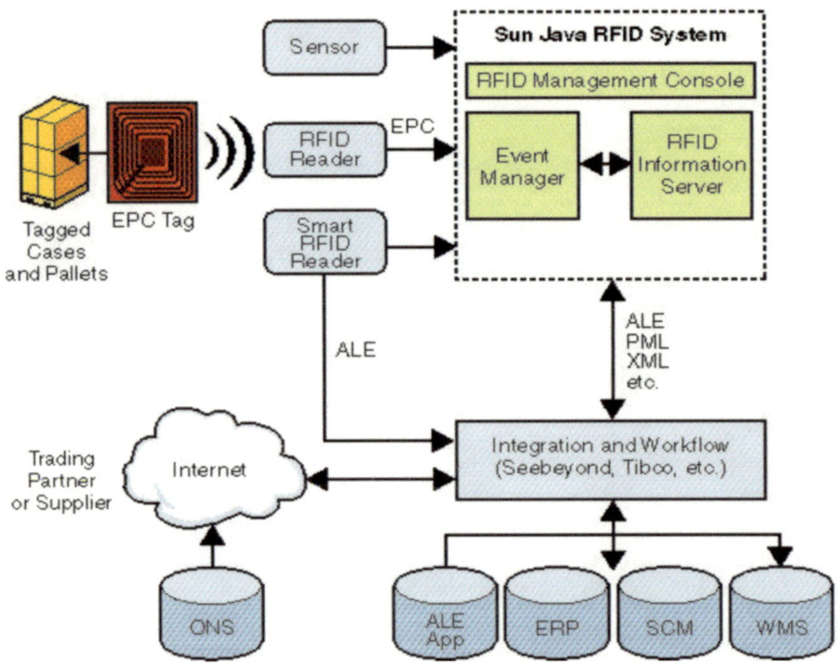

Fig. 6. The Sun Java RFID system function in the EPC network (Sun Microsystems, 2006 c)

All of these middleware designs aim at providing a scalable solution for gathering, filtering, and providing clean RFID data to the end-user. However, there are still many open issues. The reliability of RFID data needs to be improved since inaccurate data could misguide the application users. The accumulation of RFID data generated in high volumes, may lead to slower queries and updates, therefore efficient RFID data management solutions such as data transformation, aggregation, and dissemination should be investigated. Raw RFID data is not of significant value until it is aggregated with other data to obtain appropriate inferences, and transformed into a suitable form for application level interaction. Also, the applications with high security requirements are increasingly using RFID; therefore support for data security and confidentiality is needed. However, such support should maintain a desirable system performance. RFID also raises the privacy concerns because of its potential to leak proprietary information and ability to track private information such as the spending history of a consumer. Technical solutions must be implemented to ensure that private data is not compromised with (Sheng et al., 2008).

While the Savant middleware architecture provides features for cleaning the data and interfacing with different kinds of RF readers, it has limited built-in functionality for addressing business rules management, dealing with all types of sensor devices and providing data dissemination, filtering, and aggregation. Also, none of WinRFID and IBM WebSphere considers the business rules policies implementation, especially the ones concerned with security and privacy.

As compared to the related work described herewith, the distinguishing aspects of our FlexRFID middleware solution are as follows: the FlexRFID design aims to provide the applications with a device neutral interface to communicate simultaneously with many different hardware devices, creating an intelligent RFID network. It also provides an interface to access the hardware for the management and monitoring purposes. The FlexRFID provides all data processing capabilities along with the security and privacy features included in the data processing layer and enforced by a policy based management module for the business events, referred to as the Business Rules layer. The modular and layered design of FlexRFID allows integration of new features with little effort. The design also permits seamless integration of different types of enterprise applications. More detail about the FlexRFID middleware architecture is presented in the next section.

5. FlexRFID: a flexible middleware for RFID applications development

The FlexRFID middleware architecture takes into account the design issues discussed above. As shown in Fig. 7, FlexRFID is part of a three-tier architecture consisting of: the backend applications layer, FlexRFID middleware layer, and hardware layer consisting of diverse types of sensors and devices.

The *Diverse Types of Sensors and Devices layer* comprises RFID readers, sensors and other industrial automation devices. Such approach allows incredible flexibility in the selection of devices, lets companies build their enterprise solutions without handling low-level programming, and allows creating an intelligent sensor network, where RFID readers are choreographed with other devices. There are diverse makes and models of devices, which require a middleware layer that monitors, manages, coordinates, and obtains data from the different devices. In FlexRFID, these functions are taken care of before processing the raw data and applying business logic to them. Our approach is to use a *Device Abstraction Layer* (DAL) that abstracts the interaction with the physical network of devices. The FlexRFID middleware incorporates three other layers which are: *Business Event and Data Processing Layer* (BEDPL), *Business Rules Layer* (BRL), and *Application Abstraction Layer* (AAL) (Ajana et al., 2009).

5.1 Device abstraction layer (DAL)

The Device Abstraction Layer of the FlexRFID middleware is responsible for interaction with various devices and data sources independently of their characteristics. The *Data Source Abstraction Module* (DSAM) of the DAL provides a standard view of data regardless of the data source protocol (e.g. EPC Gen2, ISO 15693, and ISO14443A), air interface (e.g. UHF, HF), power supply, type, and memory size of a device. The *Device Abstraction Module* (DAM) of the DAL provides a common interface to access hardware devices with different characteristics such as protocols, air interface, and host-side communication interface (e.g. USB, Serial Port, Ethernet port). The DAM exposes simple functions like open, close, read, write, etc. that trigger the complex operations of the devices. Both, the DSAM and the DAM allow the FlexRFID middleware to be extendable to support various data sources and devices. The *Device Management and Monitoring Module* (DMMM) of the DAL is responsible for dynamic loading and unloading of the driver libraries or device adaptors. This allows the FlexRFID middleware to be light weight as libraries are loaded based upon request. The DMMM configures the devices as specified by the upper layers, and also monitors and reports their status (Ajana et al., 2009).

5.2 Business event and data processing layer (BEDPL)

The BEDPL acts as a mediator between the DAL and the AAL. The services accepted by the BEDPL are first authorized by the *Business Rules Layer* (BRL) and then allowed to issue commands to the DAL in order to get the raw data and process them accordingly. Similarly the raw data are carried from the DAL, processed, and passed on to the AAL by this layer. Services provided by the BEDPL are described as follows (Ajana et al., 2009).

5.2.1 Data dissemination

A diverse set of applications across an organization are interested in the captured information. The captured data are therefore broadcasted by the data dissemination service to all the interested entities. In addition, different applications require different latencies. For example, low latency for the notifications is desired by the applications that need to respond immediately to objects' events. In contrast, some legacy applications need to receive batched updates on a daily schedule (Floerkemeier et al. 2007).

5.2.2 Data aggregation

The fine-grained data has implicit meanings and associated relationships with other data, and need to be aggregated into summaries and/or proper inferences for applications that can not deal with the increased granularity. For example, it is common that an application is only interested in an event when an object enters or leaves a certain area. Other applications may only need a total count of objects belonging to a specific category rather than a serial number of each object detected. The data aggregation service provides such kind of functionality (Floerkemeier et al. 2007).

5.2.3 Data transformation

Raw data present little value until they are transformed into a form suitable for application-level interactions. So, from an application perspective, it is desirable to provide a mechanism that turns the low-level captured data into the corresponding business event. For example, a detection of a number of tagged books at the exit door of a library can be automatically translated into a books checked out event. This requirement is taken care by the data transformation service (Floerkemeier et al. 2007).

5.2.4 Data filtering

The volumes of data generated by the different devices require significant data filtering to extract the most important information. Also, different applications are interested in different subsets of data captured. There are filtering policies available in the FlexRFID middleware policy repository of the BRL, therefore the data filtering service filters data depending on the filter characteristics provided by the application. This offers flexibility in handling multiple filtering formats (Floerkemeier et al. 2007).

5.2.5 Duplicate removal

Multiple devices may generate duplicate readings of the data, for example tags in the vicinity of a RFID reader are read continuously. This results in a large amount of repeated data, and therefore duplicate removal service prevents the reporting of these duplicate data. The application specifies a time window, so that the same data read within it are only reported once (Ajana et al., 2009).

5.2.6 Data replacement

Usually the rate at which the devices insert data in the channel buffer is slower than the read rate of the applications. However, in case the application is not responsive enough or not executing, the channel buffer gets full, and leads to buffer overflow problem. The data replacement service allows the application to specify the action to be taken in case of channel buffer overflow. The application specifies the data replacement policy stored in the BRL policies repository, which will be executed by the data replacement service (Ajana et al., 2009).

5.2.7 Data writing

Certain special data sources like RFID tags provision additional memory space for both ID and additional data. The FlexRFID middleware handles both the reading and writing of data to this additional memory (Floerkemeier et al. 2007).

5.2.8 Privacy

RFID based tracking solutions could trigger RFID tags attached to the personal belongings to reply with their ID and other private information, therefore increasing the potential of unauthorized surveillance mechanism that would pervade large parts of our lives. FlexRFID design supports dedicated privacy enhancing feature through the privacy module. The business rules of this module are stated in the privacy policy of the BRL (Ajana et al., 2009).

5.3 Business rules layer (BRL)

The BRL is a policy-based management engine that defines the rules that grant or deny access to resources and services of the FlexRFID middleware, and enforces different types of policies for filtering, aggregation, duplicate removal, privacy, and different other services. This is achieved by determining the policies to apply when an application requests the use of a service in the BEDPL. The *Middleware Policy Editor* (MPE) allows storing, retrieving, and removing policies from the *Middleware Policy Repository Database* (MPRD). When an application needs to access a service that is protected by the Business Rules Layer, the request passes through the *Middleware Policy Enforcement Point* (MPEP) which asks the *Middleware Policy Decision Point* (MPDP) whether to permit or deny access to the service by applying the privacy rule, and how the service will be processed depending on its type. The MPEP gives the MPDP the authority of decision making; whether or not to grant the application access to the service based on the description of the application attributes, and

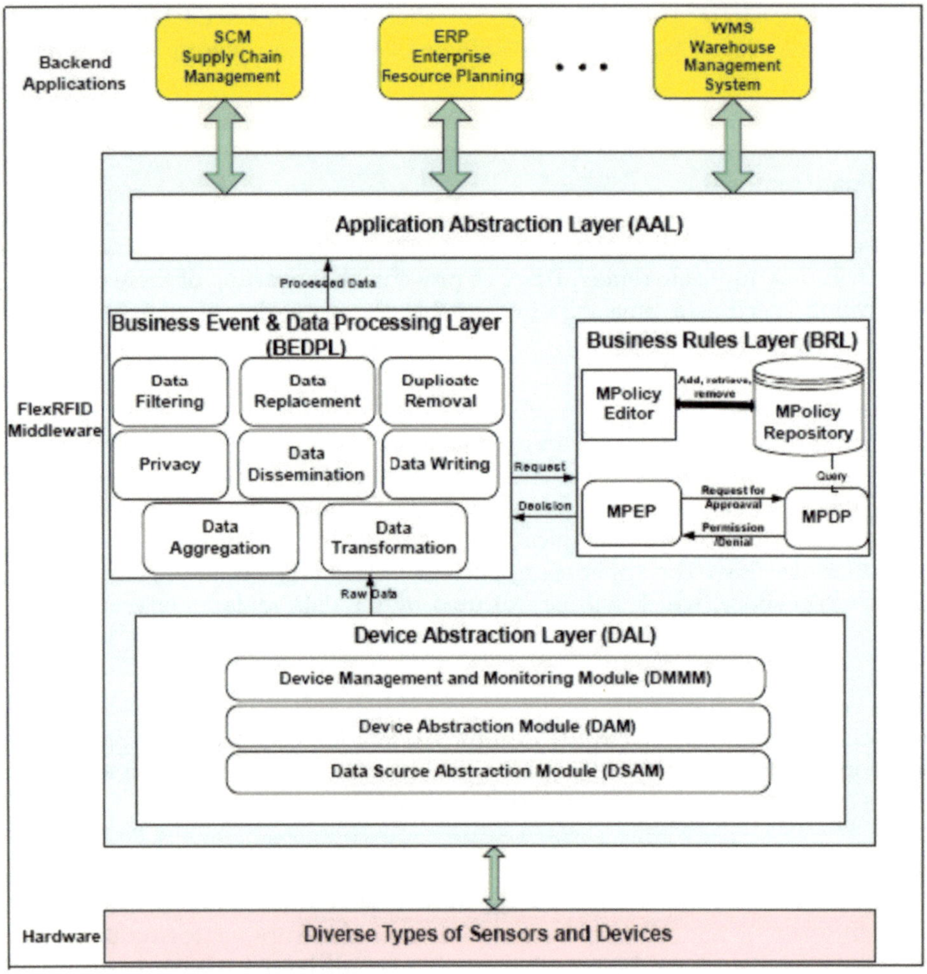

Fig. 7. FlexRFID middleware architecture (Ajana et al., 2009)

which policies will be applied to the services used by this application. The MPDP makes its decision based on the applicable policies stored on the system. The returned decision is Permit, Deny, Indeterminate or Not Applicable. Indeterminate is returned when there is an error in processing the request and Not Applicable when no policy that applies to the request could be found (Ajana et al., 2009). Policies are operating rules used to maintain order, security, consistency, or other ways of successfully achieving a task. Examples of policies that should be available in the Business Rules Layer are: Access policy, data replacement policy, quality of service policy, and privacy policy.

Different types of applications using the FlexRFID middleware may define rules to detect events and process them using the services provided by the middleware. *Primitive events* such as observations from readers may lead to actions such as change of location. *Sequence events* consist of a sequence of primitive events of the same type, defined by the order and closeness of intervals. *Composite events* are a combination of primitive events and sequence events, and may lead to actions such as aggregation of data. Here we present some examples of rules enforced by their corresponding policies (Ajana et al., 2009):

- The *filtering rule* filters data according to predefined policies by the applications. For example, multiple readers may generate duplicate readings. To filter this, the filtering policy will scan data within a sliding window to find if there are duplicate RFID tag readings from multiple readers, and delete the duplicate if it exists. A policy for duplicate removal could specify that if readings from reader Rx and Ry have the same tag ID value within time T, then one of them is dropped.
- The *location transformation rule* serves to transform RFID readers' observations into location changes. For example, Reader R1 is mounted at a warehouse departure zone and will scan objects before their departure. A policy for this transformation could state that any observation generated from reader R1 will change the object's location to a value different from its current location.
- The *data aggregation rule* is used to detect a sequence of ordered events and generate an aggregation relationship. For instance when pallets are loaded into a truck to depart, a sequence of readings on the pallets are done, followed by (with a distinctive distance) a separate reading of the truck's EPC. This sequence of events will aggregate as a containment relationship between the pallets and the truck.
- Privacy threats in an RFID application can include covert reading, tracking over time, and individual profiling. The *privacy rule* specifies whether an application has the right to access RFID tag data, can track them over time, and use them to generate events. Applications can load into the FlexRFID middleware's Business Rules Layer privacy policies specifying how to use and configure the RFID technology to maintain the privacy of data and prevent data from tracking and hotlisting.

5.4 Application abstraction layer (AAL)

The AAL provides various applications with an interface to the hardware devices, through which the applications request the set of services provided by the FlexRFID middleware with hidden complexity (Ajana et al., 2009).

6. FlexRFID applications

6.1 Smart library application

In the late 1990s, libraries began using RFID systems to replace their electro-magnetic and barcode systems. In North America approximately 130 libraries are using RFID systems, and hundreds more are considering it. The RFID self-check systems are increasingly becoming popular since they allow patrons to check-in or check-out many items, rather than one at a time. This reduces the number of library staff needed at the circulation desk. Inventory related tasks could also be done in a fraction of the time, as a portable reader can read a whole shelf of books, and then report which are missing or misplaced. Moreover, as books are dropped in the book return station, the reader reads the tag and uses the automatic sorting system to return the book back to the shelves. A RFID tag can be used for both identifying items and securing them, and there is no need to purchase additional tags for security or use security strips separately. As patrons leave the library, the tags are read to ensure that the items have been checked out. If the item is not checked-out, the RFID readers placed near the exit detect the presence of the tag and trigger an alarm (Ayre, 2004).

A significant impediment to library use of RFID is privacy concerns associated with an item-level tagging. The tag contains static information that can be easily accessed by unauthorized readers. The privacy issues are generally described as tracking and hotlisting.

Tracking refers to the ability to track the item movement or the person carrying the item by correlating multiple observations of the item's RFID tag. Hotlisting allows building a database listing the items and their corresponding tag numbers and then using an unauthorized reader to get who is checking out items on the list. Therefore, libraries implementing RFID should use and configure the technology to maintain the privacy of patrons (Ayre, 2004).

Smart library management applications require data to be automatically read, analyzed and written back. Every patron is issued a RFID tagged library card that stores both personal information and information of the library items borrowed. Upon borrowing an item, the patron card is checked if he/she is permitted to borrow. Then, depending on the permissions, the application updates the borrowing status of the patron and the internal library database or rejects the request.

We developed a smart library RFID prototype using FlexRFID, which provides services to borrowers without having to go through an employee at the library. This prototype aims also at helping library staff to track items placed at the wrong places, and identifying most read documents in the library. This allows the visualization of important events and alerts in real time. The most important events are: item check-in, item check-out, shelf management, and item theft.

In order to illustrate the value and maturity of the FlexRFID middleware, the smart library prototype makes use of its services such as filtering, duplicate removal, transformation, aggregation, and is tested with different devices such as bar code readers, RFID readers, and sensors. A solution to the security and privacy concerns is also provided by the FlexRFID's security and privacy modules managed by policies. The smart library prototype is developed using Microsoft Visual Studio .Net. The prototype is coded using C# as a language and uses the Data Writing, Data Replacement, and Duplicate Removal services of the FlexRFID BEDPL module. The hardware used in testing the prototype consists of Intermec IF4 fixed RFID reader, Intermec 915 MHz ID Card, Intermec passive tags, and sensors used to initiate and stop the reading of tags at the entry/exit points of the library.

6.2 Supply chain management application

RFID technology has gained greater prominence and a higher level of adoption due to its recent advancements and decreasing costs across the years. The applications of RFID in the SCM have vast potential in improving effectiveness and efficiency in solving supply chain problems. RFID tags are placed on objects so that they can be uniquely identified. These objects in motion are traced throughout the supply chain from manufacturer's shop floor, to warehouses, to retail stores. Such a visibility of accurate data brings opportunities for improvement and transformation in various processes of the supply chain, and allows a wide range of organizations to realize significant productivity gains and efficiencies (Ajana et al., 2010).

Some of the key questions to be answered when applying RFID to SCM are: (1) what would be the benefits of RFID integration in supply chain? (2) What are the risks, challenges, and recommendations in adopting and implementing RFID in supply chain? (3) What processes in supply chain will be affected by RFID, and where does this technology have the potential of creating the most business value? (Ajana et al., 2010)

RFID promises to revolutionize supply chains and usher in a new era of cost savings, efficiency and business intelligence. Some of the main benefits of integrating RFID in SCM are: Automatic non-line-of-sight scanning, labor reduction, enhanced visibility, asset tracking, item level tracking, traceable warranties and product recalls, quality control and regulation, and ability to withstand harsh environments (Ajana et al., 2010). Major issues that inhibited the adoption of RFID in SCM are: the cost of tags, tag readability, the need for new data structures for RFID data management, data ownership and sharing, standardization, business process changes, and privacy (Ajana et al., 2010).

RFID can provide major benefits in the following SCM processes (Ajana et al., 2010):

- **Demand Management:** The use of RFID allows eliminating inaccuracies in data due to human errors, and provides timely data both at the item level and in aggregate about the market demand of a particular product.
- **Order Fulfillment:** Order fulfillment is a key process in meeting customer requirements and improving the effectiveness of supply chain. RFID can reduce the cost of operations in order fulfillment, and enables suppliers to automatically and accurately determine the location of an item, to track its movement through the supply chain, and to make instantaneous business decisions.
- **Manufacturing Flow Management:** The use of RFID helps manufacturers with their Just-in-Time (JIT) assembly lines by tracking where every item is in the manufacturing process and supply chain.
- **Returns Management and RFID:** RFID facilitates return management by helping retailers know if they sold the item being returned. Through the use of the *ESM* (Electronic Security Marker), RFID can tie the relationship of a particular product to a given sale and then to the return.

SCM applications target many aspects depending on supply chaining processes. One of these major aspects is inventory control. We focused on the use of FlexRFID middleware to provide input to existing tools and applications of inventory control. FlexRFID middleware deals with RFID data streaming, reactivity, integration, and heterogeneity that represent a challenge for e-logistics and SCM systems (Ajana et al., 2010):

- **Streaming**: RFID devices are becoming cheaper and widely deployed and it is now increasingly important to perform continual intelligence analysis of data captured. To relieve the SCM applications from dealing with the streaming nature of data and the fact that the data might be redundant, even unreliable in certain cases, the FlexRFID middleware is able to process such unreliable real time sensing data before delivering it to the backend system.
- **Reactivity**: RFID has promised real time global information visibility for SCM participants. To benefit from such visibility, the SCM participants have to be able to identify the interested situations and react to such situations when they happen. The events associated with the triggers have to be reported in a timely manner and notification has to be sent to interested SCM participants. The FlexRFID middleware handles this through its Business Event and Data Processing Layer and policy based Business Rules Layer.
- **Integration**: The design of FlexRFID middleware allows it to scale and support different devices and data sources that may be used at numerous points of inventory control such as Point of Sale (PoS), and smart Shelves.

The advantages of using FlexRFID for inventory control can therefore be summarized as follows:

- Report RFID data about location and inventory level in real time so that the inventory control application could place an automatic order whenever the total inventory at a warehouse or distribution center drops below a certain level.

- Report and aggregate accurate data at the PoS that will be used by the SCM application to monitor demand trends or to build a probabilistic pattern of demand that could be useful for products exhibiting high levels of dynamism in trends.
- Reduction of the Bullwhip effect, which means an exaggeration of demand in upward direction in a supply chain network. FlexRFID will provide accurate and real time information on actual sales of items that can be used for decision making and that will diminish the magnitude of the bullwhip effect. Reducing bullwhip effect would benefit industries where instances of supply-demand imbalances have high costs attached to them.
- Capturing data that gives total visibility of product movement in the supply chain. This will help to make early decisions about inventory control in case there is any interruption in the supply. This results into reduction of total lead-time for arrival of an order. Pharmaceutical and perishable product industries could benefit from this to increase total useful shelf life of items.
- Reduced inventory shrinkage: FlexRFID can transform the capture of RFID data into inventory shrinkages events including thefts and misplacement of items.
- FlexRFID allows issuing policies by the inventory control applications for items as per the requirements. E. g.: first-in-first-out (FIFO) policy for items such as, vegetables, and bread.

7. Conclusion and future work

A number of enterprise applications using RFID technique introduce a need for an infrastructure that hides proprietary device interfaces, facilitates configuration and monitoring of the devices, and processes the captured data. This chapter introduces RFID middleware and its design issues, presents some existing middleware solutions, and details the FlexRFID middleware framework that we developed to address the application requirements stated above. FlexRFID has four important layers: the Device Abstraction Layer (DAL), the Business Event and Data Processing Layer (BEDPL), and the Application Abstraction Layer (AAL). FlexRFID enables the following: communication with different types of devices; implementation of functionalities by ensuring the business rules using policy-based management; and seamless integration of various enterprise applications. The smart library application has been developed to show the usefulness of the designed middleware solution. Also the scenarios of integrating FlexRFID with an inventory management application have been set.

With respect to the future work we intend to develop all the possible scenarios and specific events that could be triggered in an SCM application for inventory control, integrate the FlexRFID middleware with an open source system for inventory control (e.g. TechLogic Inventory Control System, Opentaps…), and show how the different layers of FlexRFID middleware will work to deliver enhanced visibility of inventory in various stages of supply chaining.

Next we are intending to integrate FlexRFID with a healthcare application, and in the context of Situational Awareness; being aware of what is happening around users and understand how information, events, and actions will impact their goals, both now and in the near future. This will allow us to evaluate the FlexRFID middleware with multiple hardware configurations and applications' requirements.

8. Acknowledgements

We would like to express our sincere appreciation to AlAkhawayn University and ENSIAS School in Morocco, for their support of this research work.

9. References

Ajana, M. E., Boulmalf, M., Harroud, H. & Elkoutbi, M. (2010). FlexRFID in the Supply Chain: Strategic Values and Challenges, *Proceedings of NGNS 2010 5th International Conference on Next Generation Networks and Services*, Marrakesh, Morocco, July 08-10, 2010

Ajana, M. E., Boulmalf, M., Harroud, H. & Hamam, H. (2009). A Policy Based Event Management Middleware for Implementing RFID Applications, *Proceedings of WiMOB 2009 5th International Conference on Wireless and Mobile Computing, Networking and Communications*, ISBN 978-0-7695-3841-9, Marrakesh, Morocco, October 12-14, 2009

Al-Mousawi, H. (2004). Performance and Reliability of Radio Frequency Identification (RFID), 01.01.2011, Available from: http://student.grm.hia.no/master/ikt04/ikt6400/g28/Document/Master_Thesis

Association for Automatic Identification and Mobility (n. d.). What is RFID?, In: *AIM*, 27.02.2011, Available from: http://www.aimglobal.org/technologies/RFID/what_is_rfid.asp

Auto-ID Center (n. d.). EMS Specification, 28.02.2011, Available from: http://www.quintessenz.org/rfid-docs/www.autoidcenter.org/media/feb03_board/oatsystems.pdf

Ayre, L. B. (2004). Position Paper: RFID and Libraries, In: *Galecia Group*, 05.03.2011, Available from: http://www.galecia.com/included/docs/position_rfid_permission.pdf

Burnell, J. (2008). What Is RFID Middleware and Where Is It Needed?, In: *RFID Update*, 27.02.2011, Available from: http://www.rfidupdate.com/articles/index.php?id=1176

Clark, S., Traub, K., Anarkat, D. & Osinski, T. (2003). Auto-ID Savant Specification 1.0, In: *Auto-ID Center*, 28.02.2011, Available from: http://www.amece.org.mx/amece/Documentos/estandares/epc/WD-savant-1_0-20030911.pdf

Eisma, A. (2008). Data Capture in IBM WebSphere Premises Server™, In: *OSGi Alliance*, 28.02.2011, available from: http://www.osgi.org/wiki/uploads/CommunityEvent2008/23_Eisma.pdf

Floerkemeier, C., Roduner, C. & Lampe, M. (2007). RFID Application Development with the Accada middleware Platform. *IEEE Systems Journal*, Vol.1 No.2, pp. 82-94, ISSN 1932- 8184

Floerkemeier, C. & Lampe, M. (2005). RFID Middleware Design: Addressing Application Requirements and RFID Constraints, *Proceedings of SOC'2005 Smart Objects Conference*, pp. 219-224, ISBN 1-59593-304-2, Grenoble, France, October, 2005

Floerkemeier, C., & Lampe, M. (2004). Issues with RFID Usage in Ubiquitous Computing Applications, 04.03.2011, Available from: http://www.vs.inf.ethz.ch/res/papers/RFIDIssues.pdf

Glasser, D. J., Goodman, K. W. & Einspruch, N. G. (2007). Chips, Tags and Scanners: Ethical Challenges for Radio Frequency Identification. *Ethics and Information Technology*, Vol.9, No.2, pp. 101-109, ISSN 1388-1957

IBM Corporation (2009). IBM WebSphere Sensor Events, 27.02.2011, Available from: http://www-01.ibm.com/software/integration/sensor-events/index.html

Ishikawa, T., Yumoto, Y., Kurata, M., Endo, M., Kinoshita, S., Hoshino, F., Yagi, S. & Nomachi, M. (2003). Applying Auto-ID to the Japanese Publication Business to Deliver Advanced Supply Chain Management, Innovative Retail Applications, and

Convenient and Safe Reader Services, In: *Auto-ID Center*, 27.02.2011, Available from: http://www.autoidlabs.org/uploads/media/KEI-AUTOID-WH004.pdf
Molnar, D. & Wagner, D. (2004). Privacy and Security in Library RFID: Issues, Practices, and Architectures, *Proceedings of ACM CCS 2004 11th Conference on Computer and Communication Security*, ISBN 1-58113-961-6, Washington, DC, USA, October, 2004
Parliament Office of Science and Technology (2004). Radio Frequency Identification (RFID), 01.03.2011, Available from: http://www.parliament.uk/documents/upload/postpn225.pdf
Phoenix Software International (2006). Optical Character Recognition (OCR): What You Need to Know, 27.02.2011, Available from: http://www.phoenixsoftware.com/pdf/ocrdataentry.pdf
Polniak, S. (2007). The RFID Case Study Book: RFID Application Stories from Around the Globe, In: *Abhisam Software*, 02.03.2011, Available from: http://www.bin95.com/case_studies/RFID_Technology_Applications.htm
Prabhu, B. S., Su, X., Ramamurthy, H., Chu, C. & Gadh, R. (2005 a). WinRFID: A Middleware for the Enablement of Radio Frequency Identification (RFID) Based Applications, In: *Wireless Internet for the Mobile Enterprise Consortium (WINMEC)*, 27.02.2011, Available from: http://www.techrepublic.com/whitepapers/winrfid-a-middleware-for-the-enablement-of-radio-frequency-identification/2349745
Prabhu, B. S., Su, X., Ramamurthy, H., Chu, P.,Qiu, C. & Gadh, R. (2005 b). WinRFID: Middleware for Distributed RFID Infrastructure, In: *Wireless Internet for the Mobile Enterprise Consortium (WINMEC)*, 27.02.2011, Available from: http://www.wireless.ucla.edu/techreports2/winrfid-middleware.pdf
Sheng, Q. Z., Li, X. & Zeadally, S. (2008). Enabling Next-Generation RFID Applications: Solutions and Challenges. *IEEE Computer*, Vol.41, No.9, pp. 21-28, ISSN 0018-9162
Sun Microsystems (2006 a). Sun Java™ System RFID Software 3.0 Developer's Guide, 27.02.2011, Available from: http://download.java.net/general/sun-rfid/Release30/Docs/Developers_Guide_819-4686.pdf
Sun Microsystems (2006 b). Sun Java™ System RFID Software 3.0, 28.02.2011, Available from: http://www.slgroup.com/Portals/0/docs/sample_docs/sun_rfid_datasheet.pdf
Sun Microsystems (2006 c). Introduction to the Sun Java System RFID Software, In: *Sun Microsystems*, Available from: http://download.oracle.com/docs/cd/E19486-01/819-4684/RFID-intro.html
United States Government Accountability Office (2005). Information Security Radio Frequency Identification Technology in the Federal Government, 27.02.2011, Available from: http://epic.org/privacy/surveillance/spotlight/0806/gao05551.pdf
Vacca, J. R. (2009). Computer and Information Security Handbook, In: *Morgan Kaufmann Publishers*, Available from: http://books.google.co.ma/books?id=TnE85sckwMAC&pg=PA206&lpg=PA206&dq=rfid+tags+power+supply+read+range+cost+type+of+memory&source=bl&ots=tWEVtCBId0&sig=1fy75A0dYwfKhIpN4sxwOzQz14Q&hl=fr&ei=UaBjTfzuI8H7lweKzIz7Cw&sa=X&oi=book_result&ct=result&resnum=1&ved=0CBcQ6AEwAA#v=onepage&q=rfid%20tags%20power%20supply%20read%20range%20cost%20type%20of%20memory&f=false
Wal-Mart and RFID: A Case Study RFID Tags Advantages and Limitations (2007). In: *Tutorial-Reports*, 27.02.2011, Available from: http://www.tutorial-reports.com/wireless/rfid/walmart/tag-advantages.php

19

A Study on the Influence of RFID Tagging on Circulation Services and Collection Management: a Case Study of the Taipei Public Library

Shu-hsien Tseng and Chien-ju Chou
National Central Library / National Yang-Ming University Library
Republic of China (Taiwan)

1. Introduction

In 1998, the National Library Board Singapore undertook a trial application of RFID (Radio Frequency Identification) on acquisitions, cataloguing, and circulation, and in 2002, it put out the first RFID library management system in the world. The range of applications includes: (Zhou, 2009)checking out and returning books by readers, sorting and delivering books, setting up an automated check-out machine outside library, managing library property, and taking inventories of materials and managing stacks. The management of library property of the National Library Board Singapore is now automated at a high rate of efficiency with its administration making effective use of RFID applications. Inventory work has been simplified resulting in significantly fewer mistakes and manual tasks such as shelving require fewer man-hours. The radio wave sensor also makes it easier to do book searching. Concrete results include: US$2.8 million can be saved every year; costs for up to 2,000 workers can be eliminate every year; and the number of borrowers can increase to an equivalent of more than 31 million, up from 10 million annually. In addition, many public libraries in the US have begun using RFID. By using RFID, for example, readers can check out and return books by themselves at the San Antonio Public Library. The circulation of library materials has been expedited and made more convenient, the management of stacks is more efficient, the efficiency of librarians has been enhanced, the range of services has been expanded, and the number of patrons has been increasing at a rate of 3% annually over the past few years. For librarians, reading RFID tags by means of hand-held inventory readers makes locating books and confirming the quantity of books much faster. (Zhou, 2009) The RFID automated book sorting system at the Seattle Public Library makes it possible for librarians to serve significantly more patrons in the same amount of time than was previously possible, and the range of services is not limited to clerical duties of checking out and handling returned materials. Even when the library is closed after operating hours, readers can return materials through the return slot outside the library. A conveyor belt then brings the materials to the sorting room, and after the circulation record of the book recorded on a chip is read by the RFID reader, the book will be ready to be re-shelved. If someone from another branch library has placed a reserve on the book, however, the book will automatically be sent to a box designated for that branch library. Such material is delivered on the following day ready for the patron who reserved the book to pick it up. Through RFID automated circulation system, the 1.4 million books in the Seattle Public Library System can be returned to shelves automatically, and the 28 branch libraries can obtain requested books in short order. Such an intelligent system has replaced the traditional time- and energy-consuming work of librarians. (Industrial Technology Research Institute, 2009)

Although there are many advantages in using RFID so that it should play a key role in managing library collections, there are some hidden problems that need to be addressed before it can reach the goals of high quality management and meet the demand of actual operations. One of the essential elements in the successful use of RFID in performing library services lies in the quality of RFID tags. Since it is a new service recently launched in the public library systems in Taiwan, potential problems in using it to provide library circulation services may not have surfaced as yet. Thus, one year after the intelligent library management system was set up in the Taipei Public Library, through observation of patrons using the automatic check-out system and gathering their opinions, this study tries to come to grips with related problems to understand the influence of the RFID management system on library patrons and library operation for the reference of all libraries in the effort to enhance service quality and maximize the usefulness of RFID for library services in the future.

2. Using RFID in the library

2.1 Introduction

RFID is a denoting radio detector that uses radio waves to deliver information to identify people or objects carrying encoded microchips. (Chen, 2006) It is comprised of three parts: (Zhuang, 2004)

2.1.1 The RFID tag

The RFID tag is formed by an antenna, RF Front End, a digital block, and a memory chip. There are usually two types, active and passive, according to whether or not batteries are used. The passive tag receives energy delivered by a reader and transfers the electric energy inside the tag, so no battery is needed. The advantages of a passive tag are its smaller size, cheaper price, and that it is longer lasting.

2.1.2 The reader

With the delivery of energy and signals by high frequency radio waves, the identification rate of the tag can reach 50 per second. The use of wire line or wireless communication can be combined with its application system.

2.1.3 The application system

Combined with techniques such as a database management system, the internet, and a firewall, the RFID can provide automatic, safe, and convenient instant surveillance functions.

Presently the RFID standards are commonly used ISO standards, including 1) ISO 14443, commonly used in tickets and cards for public transportation; 2) ISO 15693, used in most entry cards; and 3) ISO 18000, used in the circulation control of RFID. (RFID Technology Center, 2007)

RFID tags can be divided into three types according to different ranges of radio frequency: 1) 30-300kHz low frequency; 2) 3-30MHz high frequency; and 3) 300MHz-3GHz super-high frequency. Among them, 13.56 MHz is used in many fields, mainly for managing objects, and its advantages include wide-range deployment and imperviousness to moisture; its drawbacks are its limited reading range (within 1.5 meters) and susceptibility to interference by metal objects. (Yu, 2005)

2.2 Advantages of using RFID in library collection management over the traditional barcode

The main reason why a library chooses to replace barcodes with the new technique of RFID is that it drastically increases the efficiency of circulation services and inventory operations. Traditionally, the library clerk at a library's circulation desk would need to use a desktop or a handheld sensory barcode reader to read the information on the barcode of each borrowed item. But the RFID technique simplifies the operation of checking out and returning materials mainly because the information of related materials are encoded and stored on RFID tags, and the RF Wireless can transmit information on the tag instead of just reading the traditional barcode in "Line Sight". The system can identify information in a large quantity of built-in chips and the remote cursor can retrieve the information immediately. (Hong, 2005)

Table 1 is differences between barcode functions and RFID functions made by the Information Data Center of the Industrial Technology Research Institute, and it clearly shows the advantages of RFID.

Function	Barcode	RFID
Reading quantity	One barcode is read at a time	Many RFID tags can be read at the same time
Remote reading	Infrared rays are needed to read a barcode	RFID tags can be read or renewed without infrared rays
Information volume	Low volume of information saved	High volume of information saved
Reading and writing capacity	Barcode information cannot be replicated	Electronic information can be read and written repeatedly
Reading convenience	Only barcodes in good condition can be read.	RFID tags can be very thin and can be read even inside packaging.
Information accuracy	Barcodes need to be read by humans, so human errors are possible.	RFID tags can deliver information for tracking materials and for security purposes
Duration	A stained or damaged barcode cannot be read, and have low durability.	RFID tags can be read even when stained or dirty.
High-speed reading	Reading barcodes is more time-consuming.	High-speed reading is possible.

Table 1. Differences between barcode functions and RFID functions

Moreover, after summing up the advantages RFID has over the traditional barcode, RFID can be said to have the following characteristics: (Yiu, 2006; Cheng, 2006)

2.2.1 It can access saved information repetitively and it has high storage capacity

Information in RFID tags can be added, revised, deleted repetitively, and it has many megabytes' storage capacity.

2.2.2 It can read information of many individuals at the same time without having to read from a stable angle

The RFID Reader has a wide range reading capacity than can read many overlapping RFID tags simultaneously and saves time and energy.

2.2.3 It reads and identifies information easily and quickly

The RFID tag is read through a radio frequency which can transmit information even when the tag is not visible.

The above advantages in reading information in RFID tags indeed serve to increase the efficiency of managing materials in an intelligent library and simplify procedural operations of the services offered by librarians at the front line.

3. Using RFID in Taiwan's public libraries and related studies

In Taiwan, besides the Taipei Public Library, the National Taichung Library, the Kaohsiung Public Library, the Xinbei Public Library, and the Library of the Department of Cultural Affairs of Taichung also utilize RFID for their collection management and circulation services. However, limited by the reliability and high costs of RFID, applications of RFID in most of these libraries are limited. The Central Library of Taipei Public Library and its 2 Micro Self-service Libraries use RFID,(Taipei Public Library, 2009) but the other branch libraries continue to use the traditional system. The Xinbei Public Library set up an intelligent library with low-carbon emission at the Banchiao Train Station which uses RFID.(New Taipei City Library, 2010) In Kaohsiung, the Kaohsiung Public Library set up a Micro Self-service Library at an MRT Station. (Kaohsiung Public Library, 2009) All three Micro Self-service Libraries set up by National Taichung Public Library at Taichung Train Station, Taichung Hospital, and China Medical University Hospital each use RFID.(National Taichung Library, 2010)

As many public libraries have some experience using RFID, scholars and graduate students in Taiwan have begun doing research on uses of RFID in libraries and related topics. As for using RFID in libraries, students in the graduate program of Library and Information Science are interested in doing research on this topic; graduate students in technology management, information management, business administration, electronic and information engineering, information communication, applications of information technology, management science, and archival science are also engaged in similar types of research. The main topics include factors for introducing RFID and factors for its successful application, results of introducing RFID, satisfaction of users and acceptance by librarians, as well as the application of RFID in conducting library searches.

Liu Guang-ting explored the service quality, recognition value, and the relationship between using RFID and library patron satisfaction through a questionnaire survey, mainly to determine whether the source of patron satisfaction was influenced by the use of RFID or not. His research subjects were the patrons of a technical college library and a public library. The results show that using RFID has a positive impact on service quality, recognition value, and patron satisfaction. Liu Guang-ting contended that the results of his research could offer a clear direction for using RFID and can serve as a reference for libraries in their management and choice of system. (Liu, 2008) Moreover, Researcher Tsai Ji-jin chose to undertake a study on the acceptance of RFID by librarians in the libraries of Taiwan (Tsai, 2007)

Pan jing-mei chose to apply TOE (Technology-Organization-Environment) as the framework for her research and collected information on the application of RFID in Taiwan's libraries through a questionnaire to understand the key factors in a library's decision to use RFID. She found that financial readiness greatly influences a library's choice in setting up an RFID system. (Pan, 2008)

Fan Guo-ji explored the procedures required for introducing an RFID system and the results. His research shows that the benefits of using RFID technology in libraries are mainly limited to automated checking out and returning of materials and inventory work. It is especially useful in inventory work as libraries using the system tend to have a much clearer understanding of their collections and can offer accurate information to their patrons. With the automation of library procedures, manpower is replaced by machine-power, and patrons need to fully participate in the procedure of checking out and returning materials. As patrons express satisfaction in the automated procedures for checking out and returning materials, the introduction of RFID technology in libraries is a positive trend. (Fan, 2004)

Chen Xue-zhu compared the differences between RFID and the present identification management procedure through interviews with librarians and a field survey to understand the operational mode on the management of a featured collection. Another research subject of this study was the Archives of Chinese Information in the World at Shih Hsin University as the application of RFID may differ depending on specific factors of special collections because of differences in surroundings (temperature, moisture, metal shelves), space (controlled entrance, open shelves, closed stacks), and arrangement of materials (new books, categories, target readers). Finally, it examined the application of RFID in related fields and proposed a new management model suitable for special collections. Through in-depth interviews and understanding the advantages and disadvantages of using RFID in libraries, Chen analyzed, sorted, and set up the planning for introducing RFID in the special collections of a library to make it possible to open a special collection to the public, improve its automated management, and promote its service efficiency to library patrons. (Chen, 2007)

Xiung Ya-fei explored "The Key Factors of Successful Introduction of RFID to Library Video and Audio Materials: A Case Study of a Technical College" through the "analytical hierarchy process (AHP)" by coming up with a draft of the framework of the AHP levels in "factors determining the successful introduction of RFID in a library" and designing a questionnaire for experts. She conducted a survey investigation for her study which focused on libraries which successfully introduced RFID and observed the experience and procedures of such introduction. Through observation of individual cases, she obtained more thorough information for the revision of the draft on the framework of AHP levels. She then designed the second questionnaire for experts in her second survey investigation. She compared different aspects at different levels to measure their influence and used the software "Expert Choice 2000" to calculate the degree of influence of each factor before ranking the factors and determining the "key factors of successful introduction of RFID in the library." (Xiung, 2010)

This study focuses on individual understanding of the application of RFID in library management and the key factors of its successful introduction. The goal is to identify the most important key factors of the successful introduction of RFID in the library. The result of the study shows that of the five most influential factors of the Level 2 measurement index, "policies of the institution" is ranked number one, ahead of "skills and support system", "staff of the institution", "efficiency of the supplier", and "outside surroundings". Of the 25 influential factors of the Level 3 measurement index, the top five factors affecting the successful introduction of RFID are: "active participation and support of highly ranked superintendents" (ranked the first), "affordable cost for the institution", "stability of the system (reading percentage, interference)", "conformity with the prospects and strategy for development of the institution", and "effective interaction and communication among staff members of different ranks." (Xiung, 2010)

Hou Fu-yuan attempted to combine mobile devices and RFID to develop a GIS indoor navigation system as a searching guide for patrons. (Hou, 2008) Zhang Rong-hui combined RFID technology with Wireless LAN for library applications. He designed an information service system to help patrons look for and obtain materials. When a patron inputs the index number of the material he wants into the system, the system locates the shelf containing the material and suggests the most convenient access to the material, so that the patron can find what he or she wants in the least amount of time.

4. Intelligent collection management of the Taipei public library

4.1 Setting up intelligent collection management at the Taipei public library

The Taipei Public Library first evaluated the possibility of applying RFID technology to the management of the library and drew up plans for the direction and the method of application. In 2005, RFID was first applied to collection management under the project "Constructing an intelligent library for the new century." The first open-book intelligent library to adopt the self-service checkout system was established in Taiwan. In December 2005, the Central Library of the Taipei Public Library System also began using an automated checkout system and changed the original collection management system from barcodes and magnetic strips (See Figure 1) to RFID. Distinct from the traditional checkout system which required a clerical staff, the automated checkout system increases both the speed and the efficiency of checking out materials. In general, the following goals have been achieved through this project:

1. An intelligent library creates a new kind of library service and presents a new image of Taipei as a city of technology, one that promotes reading to its public.
2. The various sites of the intelligent library meet the public's demands for more libraries and longer service hours.
3. Patrons can enjoy convenient services from information technology and the internet, and they can participate actively to enhance their information literacy.
4. Librarians are able get an instantaneous grasp of the collection status which promotes the efficiency and control of collection management as well as the working efficiency of the library staff.

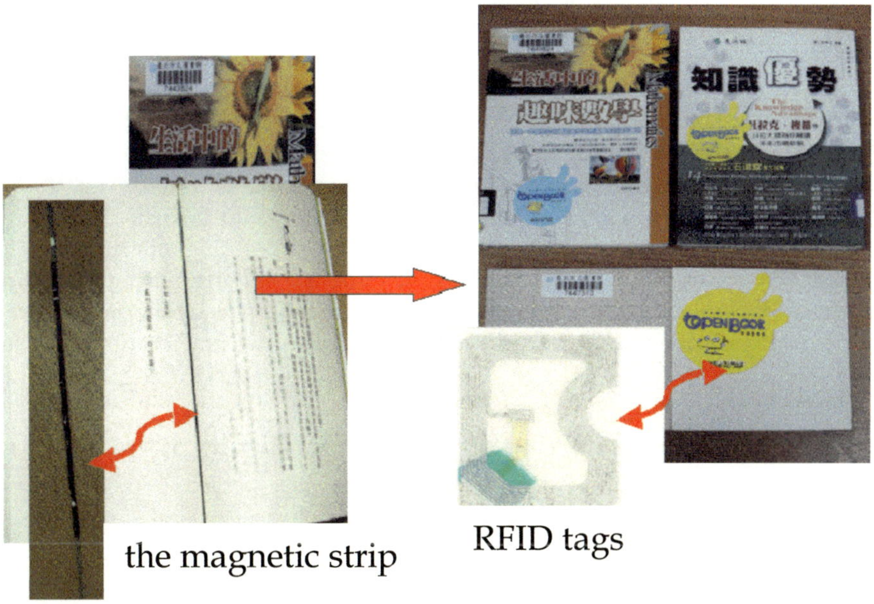

Fig. 1. The change to RFID tags from barcodes and magnetic strips as a control mechanism of the library collection

4.2 Problems stemming from the application of RFID Intelligent collection management

The Taipei Public Library was the first public library to install the RFID automated checkout service in Taiwan. To further understand the influence of this service on how patrons use the library and the functions of librarians, one year after the establishment of the management model of the intelligent library, the Taipei Public Library initiated a study to observe how librarians function at the service desk while eliciting the opinions of patrons and collecting related data for the purpose of improving library services.

After an analysis of the librarians' functions and the patrons' opinions, it was found that the three most popular of patrons' opinions of the intelligent library are:

1. After returning materials to the library through the automated system, records of material checked out are not erased by the system.
2. The sensory system at the entrance emits a signal for unchecked-out materials when the materials checked out through the automated system are carried out at the entrance.
3. Information about checked-out materials of other patrons appears in one's check out record when checking out materials with the RFID tag.

After careful analysis of the above problems, it was concluded that the problems occurred because using RFID for collection management was a new technique and library staff had insufficient experience in dealing with it. The readers' satisfaction affected the persistence of the new service. The following factors also contributed to the occurrence of problems.

The Circulation and Preservation Section of the Taipei Public Library believed that the frequency of the occurrence of these problems when checking out materials was related to the accuracy of wafer processing. When wafer processing was improperly functioned, patrons would not be able to check out materials smoothly, and they would voice complaints complicating librarians working at the checkout counter. In order to raise the service quality, decrease problems at the checkout counter, and provide patron satisfaction, a standard operating procedure was set up not only as a basis for librarians, but also as a reference for other libraries when changing to an RFID collection management in the future. The staff members who worked in the Circulation and Preservation Section in Taipei Public Library formed a quality-control circle to explore the relationship between the problems occurring during checking out through the automated system and the use of RFID tags under the title "Decreasing the frequency of improper wafer processing of the RFID collection management". They also applied the method of "problem solving" in their quality-control management approach to conduct the status investigation, draw up improvement policies, and practice the policy operation mechanism of the P-D-C-A pattern. Then they reviewed the outcome as a reference to improve the collection management quality of the intelligent library.

5. Analysis of problems occurring when using RFID tags at the Taipei public library

To make sure that the efficiency of RFID tags is the main factor affecting the quality of collection management of intelligent libraries, the staff of the Quality Control Circle of the Taipei Public Library analyzed the main factors of the above problems with a fishbone diagram (Fig. 2) and proposed several policies to address these problems:

Decrease the frequency rate of the security alarm at the entrance to 3%; the alarm is activated when patrons take out materials by mistake.

Improve the quality of operations and decrease the frequency rate of processing mistakes of RFID tags to 0.5%.

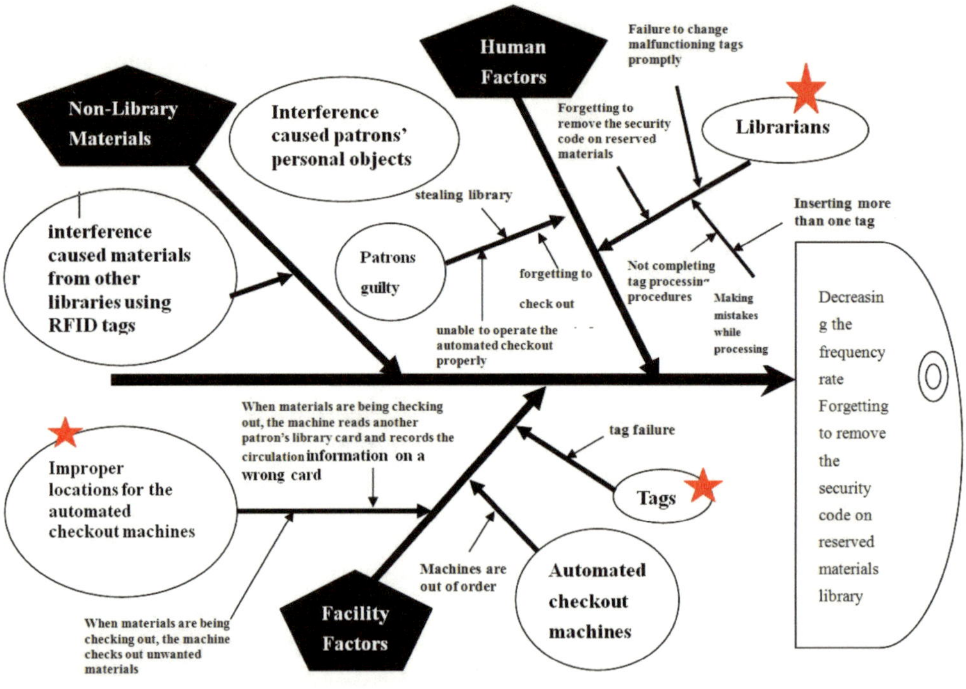

Fig. 2. Analysis of Main Factors (a fishbone diagram)

5.1 Policies for promoting the efficiency of RFID tags

In the analysis shown in the fishbone diagram in Figure 2, three possible factors causing the malfunction of an RFID tag are indicated--human factors, facility factors, and problems caused by outside library materials. The analysis explores possible problems in these factors. Human factors stem from both patrons and librarians; the facility factors include problematic tags and the automated checkout machines, and poor location the machines; the third set of factors include personal objects and materials with RFID tags carried into the library by patrons. The Taipei Public Library proposed improvement strategies to address these three factors: 1) the improper placement of automated checkout machines, 2) processing mistakes made by librarians, and 3) the improper operation of automated checkout machines. The staff of the Quality Control Circle of the Taipei Public Library investigated possible ways of improving the above three factors which are described below:

5.1.1 Improper location of automated checkout machines

5.1.1.1 Description of the problem

The three automated checkout machines are located at the service desk on the first floor of the Central Library of the Taipei Public Library. As the RFID's sensory zone of RFID can read material 30 cm. away, many mistakes occur when a crowd of patrons line up to check materials out.

5.1.1.2 Suggested improvement

After discussion, the staff of the Quality Control Circle of the Taipei Public Library found locations for the automated checkout machines to enlarge the service area for patrons checking out material, thereby decreasing reading errors made by the machine.

5.1.1.3 Results

After repositioning the automated checkout machines, the frequency rate of mistakes occurring when materials are checked out decreased to 1.9% from the previous 5.1 % (Fig. 3), much more efficient than the anticipated 3%.

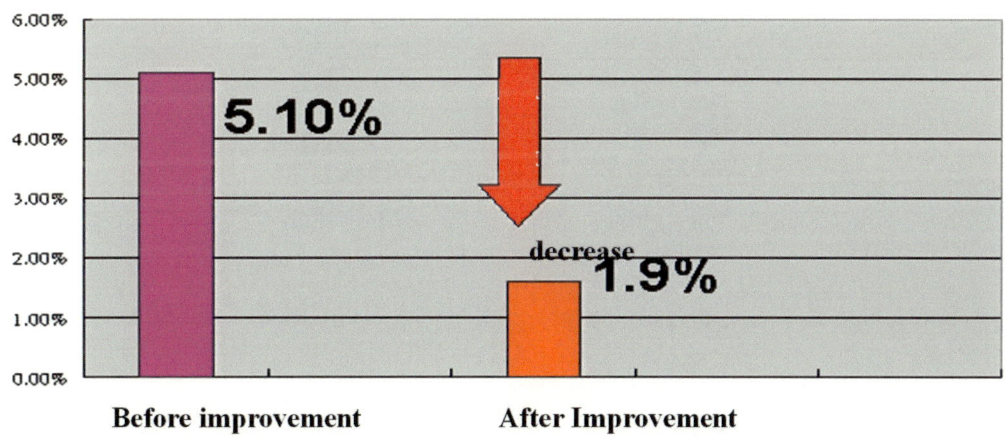

Fig. 3. Frequency Rate of the security alarm at the entrance activated due to patrons taking out materials by mistake

5.1.2 Processing mistakes made by librarians

5.1.2.1 Description of the problem

Librarians made mistakes when processing the RFID tags, so the RFID security code could not be removed.

5.1.2.2 Suggested improvement

After discussion, the staff of the Quality Control Circle of the Taipei Public Library proposed the following strategies:

1. Collect all the material with problematic RFID tags and identify the problems of these RFID tags.
2. Set up standard procedures for wafer processing and checking out requested materials
3. Revise the processing program of RFID tags in the Central Library and the intelligent libraries.
4. Review the regulations for the procurement of RFID tags.

5.1.2.3 Results

The staff of the Quality Control Department of the Taipei Public Library categorized the problems of RFID tags into three types: faulty tags, tags torn off, and 2 tags mistakenly put on one item by a librarian. After a standardized procedure was set up, problems concerning faulty RFID tags can now be tracked and controlled regularly. The situation in which patrons were unable to check out materials because of tag problems has been greatly improved. Fig. 4 and Fig. 5 show that the percentage of materials with problems has decreased to 0.62% from 4.02%. But it has not reached the expected 0.5 % because the RFID tags currently used come from the original procurement supply.

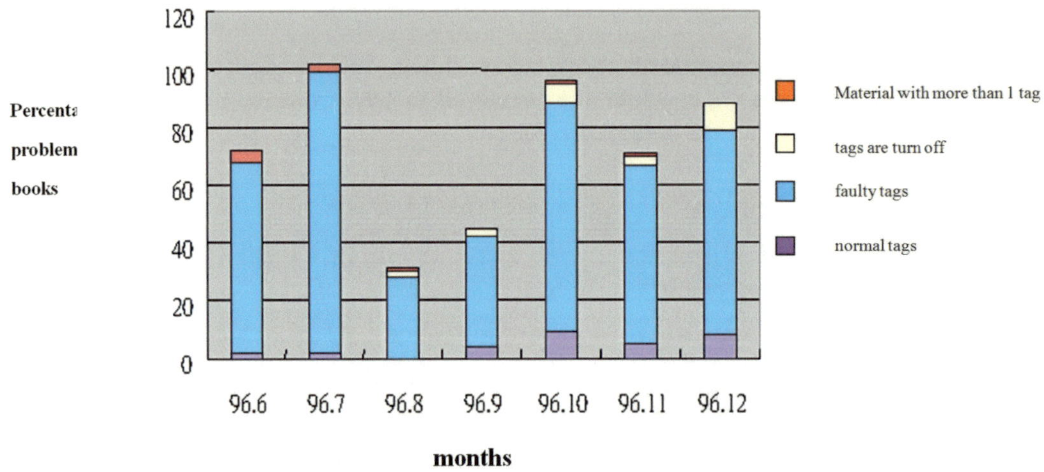

Fig. 4. Percentage of different types of problematic materials in different months

5.1.3 Improper operation of automated checkout machines

5.1.3.1 Description of the problem

Since patrons tend to be unfamiliar with the automated checkout machines, when they try to check out materials, they are unable to remove the security code of the RFID tags.

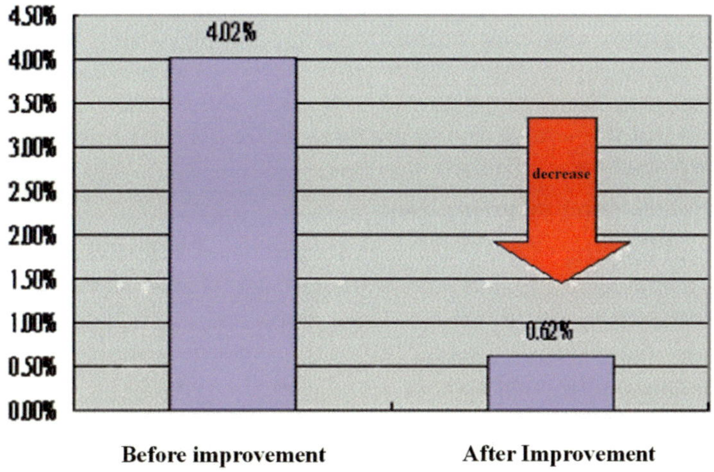

Fig. 5. Percentage of problems due to processing mistakes

5.1.3.2 Suggested improvement:

1. Clarify the instructions for operating the automated checkout machines. Place simple and clear illustrations of step-by-step procedures for patrons who have never used such machines to learn how to operate it in a short time.
2. Assign volunteer workers to help patrons operate the machines.

5.1.3.3 Results

The use of clear illustrations and the help of volunteer workers increased the efficiency of patrons' use of the automated check-out machines and diminished problems due to misreading.

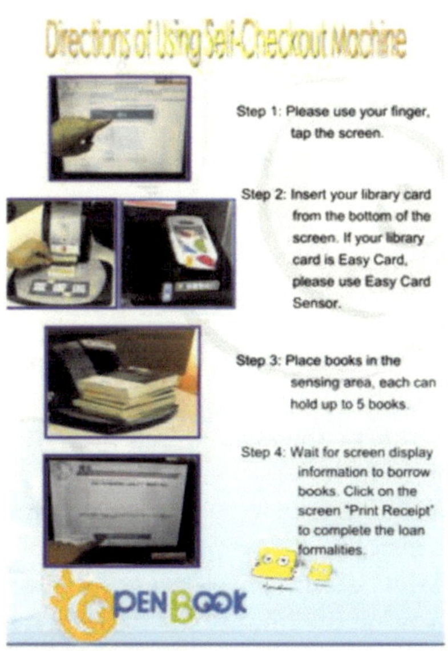

Fig. 6. Step-by-step illustrations for operating the automated check-out machine

5.1.4 Reasons of decreasing mistakes during process RFID

One policy proposed to eliminate mistakes during the processing of RFID tags was to enhance the standardized processing and revise the original wafer processing program. The statistics collected after the policy was put into effect show a decrease in the number of patrons unable to exit the library due to mistakes made during processing RFID tags. The reasons are analyzed below:

5.1.4.1 Standardizing operating procedures:

The Department of Reading of the Public Library produced a flow chart of the RFID wafer processing procedure for new staff and staff members liable to make mistakes during processing due to their unfamiliarity with the procedure. In addition, the supplier was requested to revise the wafer processing program so that librarian only had to choose the library for processing on the computer screen instead of having to revise the program code, thereby simplifying the processing procedure.

5.1.4.2 Reviewing and reproducing information on RFID tags:

Specific librarians were put in charge of the wafer processing procedure for placing RFID tags on problematic materials and new materials, so that faulty tags would be eliminated to ensure circulated materials could pass through similar security mechanisms.

5.1.4.3 Establishing procurement specifications of RFID tags for quality control:

A statistical analysis indicated that faulty tags made up the highest percentage of problematic tags. Therefore, starting from 2008, the procurement contract stipulated that suppliers are required to attach a certificate of inspection of RFID tags, and an increased number of tags will be tested upon delivery to decrease the number of defective products. Other terms listed on the contract include:

1. Tags need to meet the standard of ISO 15693 at the frequency rate of 13.56MHz.
2. The base of the tag and the antenna must be strengthened, and antenna is to be made of copper wire.
3. When bidding, the defective tags should not make up more than 5 % of the procured order.
4. Authorization by a certified notary public is required and the warranty period should be clearly listed.

5.2 Actual results

After going over the above policies, analyzing the important factors, setting up policies for improvement, confirming the results, and setting up standard procedures, the actual results are as follows:

1. The quality control circle staff proposed improvement policies leading to the following results
2. a decline in the frequency rate of mistakes occurring when patrons check out materials through automated machines to 1.9%.
3. a decrease in the rate of defective tags to 0.62%, leading to improvements in processing.
4. After the staff of quality control circle proposed and implemented new policies, patrons have voiced fewer complaints about problems when checking out materials using the automated checkout, improving the quality of circulation services.
5. The establishment of a standard operating procedure for processing RFID tags decreased the frequency rate of errors occurring during processing.

6. Quality testing of the RFID tags at the procurement stage and a quality control operation were set up.
7. Improvements made to the checking out system to facilitate patrons in checking out material led to increased efficiency and quality.

5.3 Future application

Managing the library collection with RFID technology shows the advantages of RFID. The results of further analysis and improvements made to related operations in the Taipei Public Library indicate that RFID technology can also be applied to the following:

1. As the experience of this project is shared and passed on, any branch library or reading room of the Taipei Public Library can apply RFID technology.
2. It can serve as a model for solving problems concerning RFID tags for other libraries.
3. Standardizing the specification of RFID tags and testing mechanisms can be used as references for other libraries at the time of procurement.
4. The Taipei Public Library can serve as a consultant for the application of RFID and related operations in other public libraries in Taiwan. The RFID management system of an intelligent library is a good model for others.

6. Future direction for the quality improvement of RFID tags

Even though RFID tags have many advantages and have become more broadly applied in managing library collections, problems still exit in the application of this technology. Only when these problems are resolved will this technology be successfully used in actual operations and eventually reach the goal of high quality management. The following is a summary of the problems that the Taipei Public Library encountered in practice and possible directions for improvement in the future:

6.1 The types of materials used for book covers and the edition shape of books affect the reading rate of RFID tags

Presently, publishers are striving for novel and diversified designs for publications, so that covers and edition types of books are well diversified. Metal or shiny book covers (Fig. 7 and 8) are not rare, and the shapes of books are often irregular. These affects the way libraries manage their collections with the application of RFID.

RFID transmits signals through electromagnetic waves, so it is extremely sensitive to liquids and metals. Book covers containing metal lower the success rate of RFID tag reading. Though some SMDs can counter the effect of metals, they are more expensive than the RFID tags themselves, so this solution is not cost-effective. (Zhang, 2006)

In the future, if the research and develop unit or the supplier of RFID tags can develop a less expensive product and solve the problem of metal interference in reading RFID tags, then metal-laced book covers can be read effectively, and ensure the implementation of a comprehensive automated checkout service.

6.2 The position of the RFID tag on a book affects the efficiency of inventory operations

Let us take the processing of the RFID tags in the Taipei Public Library as an example. Poor reading habits of certain patrons tend to ruin the wiring in the tag and the fixed position of RFID tags in books may affect the ability to read tags when books are stacked in piles. Most tags are pasted on book covers or on the inside of book covers.

Fig. 7. A book cover and text containing metal

Fig. 8. A book cover with reflective material

Usually the spine of the book is shelved facing outward; thus originally RFID tags were located in the inner part of shelves. The use of a racket-like portable reader, which is bigger and heavier, lowers the efficiency of librarians. Moreover, the reading range of a portable reader is unstable. Most libraries that have adopted RFID technology to manage their collections keep the bar code system and do not have the advantage of comprehensive RFID wireless reading. For example, the Taipei Public Library has not replaced all of its bar code label applicators with RFID label applicators, so the managing collections still has room for improvement.

If the function of RFID portable readers can be strengthened in the future and the hardness of the wiring and micro-circuitry in the RFID tag can be improved, the attrition rate can be effectively decreased. RFID tags suitable for publication can be developed and processing can be improved, making the application of RFID technology in the library more extensive and practical.

6.3 The size of RFID tags affects its widespread use

As mentioned earlier, with respect to the security mechanism of books, the relatively large size of RFID tags makes them more difficult to hide than the original bar codes. So the tags are more fragile. This increases not only the number of problems in checking out library materials by means of automated checkout machines, but also the frequency of replacing RFID tags.

Mr. Ogawa, the director of the Municipal Central Library in Yokohama, Japan, showed the RFID tags that his library used when he participated in the "International Conference on Operation Management and Service Trends of the Public Library" held by the Taipei City Library for its 55th anniversary. The shape and the size of the RFID tag is similar to the present bar code. Conforming to the management needs of the library, the library had RFID tags produced as a bar code and pasted the tag on the spine of a book, just like the existing magnetic strip, so that the tag would not be easily damaged and could be used for a longer period of time. However, the tag's super high frequency 300MHz-3GHz cannot be used in Taiwan which does not apply tag readers of such frequency.

In the future, if suppliers in Taiwan can develop a more compact RFID tag that conforms to the existing ISO standard and reading frequency and is easier to hide, this will solve the difficulties RFID tags encounter in library applications and make them more widely accepted and useful in a broader range of library services.

6.4 Data link of RFID tags and automated circulation system for shelved material

RFID technology can be applied in the circulation of materials. When materials are to be stored, staff members can decide where they should be put according to their suppliers, time of procurement, and user demand, so that materials procured first can be used first, and due dates can be controlled. When the materials are delivered to the processing unit, they are listed and catalogued by means of RFID. This assures accurate processing and avoids mistakes and misplacement of materials, making the material available to the public in a much shorter period of time. (RFID Flow and Supply Chain Resource Center, Chang Gang University, 2009)

The library may adopt the same mode of purchasing and processing to establish a complete management mechanism of books and shelving. When new material is placed on shelves in the library, librarians can determine their location in the library from a distance through the application of RFID, review how often an item has been checked out to determine how valuable the material is and if it is no longer being used. Patrons can easily learn about the circulation status of materials by checking on the computer adjacent to the shelving area, making it easy to find exactly what he or she wants. This is indeed the way intelligent collection management should work.

7. Conclusion

The rapid increase in the quantity and quality of information technology has transformed the fields of information science, mass communications, and broadcasting, as well as the way people obtain information and knowledge. The growth of information proceeds exponentially. The public needs to be able to acquire increasing amounts of information and develop information literacy. In order to meet the public's needs and demands, the library offers convenient access to information; consequently, it has to keep pace with developments in the field of high technology, and offer services that transcend traditional services. Applications of modern technology help libraries transcend the limitations of time and space and to improve the quality and efficiency of the services they provide.

Among new technologies, standard regulations have been gradually set up for RFID system beginning in 2001.(Zhuang, 2004) With its automated identification recognition, RFID has been ranked as one of the ten most important inventions of the 21st century and is an important tool in the future of industrial development throughout the world. (RFID Technology Center, 2009) RFID has drawn attention from scholars in a wide range of fields, and with its rapid development, it will soon be used in a manifold number of ways in widely different areas.

Presently, the application of RFID in libraries is still relatively new, so in most cases libraries have only made partial use of it. How to increase its stability, and control its quality, to increase the good will of library staff in adopting this new technology, and to lower its unit price are key factors in transforming circulation services and replacing the entire system of traditional bar codes and magnetic strips with RFID technology.

The staff of the Circulation and Preservation Section of the Taipei Public Library set up the Quality Control Circle for the purpose of identifying problems arising with the use of RFID. It analyzed the reasons by means of a fishbone diagram, and proposed several solutions to improve circulation operations and decrease the number of patron complaints. Their experience in dealing with problems stemming from the use of RFID and propose ways to improve its performance can serve as an important reference for other public libraries in Taiwan that have begun using RFID technology.

8. References

Chen, Xue-zhu. (2007). A Study on the Management of the Application of RFID for Library Special Collections. Unpublished Master's thesis, Graduate Program of Information Communication, Shi Hsin University, Taipei.

Chen, Zhi-huang. (2006). Application of RFID System on Sensory Systems, *Technique and Training*, Vol. 31, No.3, (September 2006), pp. 25-30 , ISSN 0254-5888

Fan Guo-ji. (2004). An Evaluation of the Procedures and Results of the Use of RFID in Libraries. Unpublished Master's thesis, Program of Business Administration, National Taipei University, Taipei

Hong, Guang-yi. (2005). An Analysis of the Application of RFID in Libraries. *Interdisciplinary Journal of Taiwan Library Management* Vol. 1, No. 3, (September 2006), pp. 20, ISSN 1813-6109

Hou, Fu-yuan. (2008). A Study of the Application of RFID on Book-Research Service in the Library. Unpublished Master's thesis, Graduate Program of Information Management, Providence University, Taichung.

Industrial Technology Research Institute. (December 2009). A case of innovative application of intelligent living space: intelligent learning environment, 10.05.2009, Available form http://www.ils.org.tw

Kaohsiung Public Library. (January 2010). The Intelligent Library System at the MRT of Kaohsiung Public Library, 30.01.2011, Available form http://211.20.224.211/MSLweb_ksml/w01.aspx

Liu, Guang-ting. (September 2008). A Study of the Interrelationship of Service Quality, Recognition Value, and Application of RFID Technology and Satisfaction in Libraries. Unpublished Master's thesis, Graduate Program of Technological Management, Leader University, Tainan.

National Taichung Library. (January 2011), The Search System for Micro Automated Library Collections, 30.01.2011, Available form http://microlib.ntl.gov.tw

New Taipei City Library. (January 2011). An Introduction to the Intelligent Library at Banchiao Train Station, 30.01.2011, Available form http://www.tphcc.gov.tw/library/lib_serv01.asp?id=355

Pan, Jing-mei. (September 2008). A Study of the Key Factors Influencing the Adoption of RFID for Use in Libraries, Unpublished Master's thesis, Program of Information Management, Chi-Nan International University, Nantou.

RFID Flow and Supply Chain Resource Center, Chang Gang University. (May 2009) . Application of RFID, 30.05.2009, Available form http://rfid.cgu.edu.tw/xoops/modules/tinyd3/index.php?id=3

RFID Technology Center. (April 2007). RFID Technology, 20.04.2009, Available form http://www.rtc.itri.org.tw/research/frid.htm

Taipei Public Library. (2009) . The Intelligent Library, 30.01.2011, Available form http://www.tpml.edu.tw/ct.asp?mp=104021&xItem=1139787&CtNode=33629

Tsai, Ji-jin.(2007). A Study of the Acceptance of RFID by Librarians in Taiwan, (Master's thesis, Graduate Program of Electronic and Information Engineering, National Kaohsiung University of Practical Technology)

Xiung, Ya-fei. (2010). Key Factors for Successful Application of RFID for Audio/Visual Materials in the Library—A Case Study of a Technology College. (Unpublished Master's thesis, Graduate Program of Information Management, National Changhua Normal University, Changhua)

Yiu, Zhang-song. Hong, Shu-fen. (2006). Practice of the Application of RFID Technology and the Establishment of E-commerce through RFID in the Library and on Campus. *College Library*, Vol. 10, No.1, (March 2006), pp. 71, ISSN 1682-2889

Yu, Xien-qiang.(2005). A Study of the Introduction and Application of RFID in Libraries, *Educational Data and Library Science*, Vol. 42, No. 4, (June 2005), pp. 510-11, ISSN 1013-090X

Zhang, Shou-jie. (2006). Application of Super High Frequency RFID and Label Design" *SoC Technical Journal* , No. 5, (October 2006), pp. 76.

Zhou, Wen-hao. (2009). The Digital Library and the Application of RFID in Libraries at Home and Abroad, In: *Commercial Status of RFID* , 15.05.2009, Available form http://www.rfidsalon.com/v.asp?id=583

Zhuang, Yi-zhang. (2004). Introduction and Application of RFID Technology, In: *Electronic Inspection and Quality Control*, No. 59, (July 2004), pp.28.

20

The Right UHF RFID Tags for Libraries – Criteria, Concern and Issues

Steve H Ching, Alice Tai, Henry Ip, Lau Lap Fai and Michael Cheng
City University of Hong Kong
Hong Kong

1. Introduction

In the logistics industry, UHF RFID has been widely deployed. Information and guidelines on the selection of the right tags for the different logistics applications are abundant. However, when it comes to the library arena where the use of UHF RFID is just in its infancy stage, such information is relatively scarce and sporadic. While the tag selection criteria for the logistics industry may also be applied to the library field, there are indeed fundamental differences between library business and logistics operation that libraries must take the initiative to formulate their own selection checklist.

In this chapter, the Run Run Shaw Library at the City University of Hong Kong *(hereafter the CityU HK Library, or the Library)*, attempts to share its experience on tag selections based on the research and studies that it has performed in the past few years. The pilot test on the use of UHF RFID in a selected small collection in the CityU HK Library has proved to be successful and the Library is now planning for large-scale implementation of the technology for its entire collection to improve overall service efficiency. As the collection contains over one million items, selecting the right tags that are reliable and sustainable in the long term is crucial. Research and studies in this regard therefore continue through (i) naturalistic evaluation and tests of different tags purchased or borrowed from the suppliers, and (ii) peer discussion with other libraries and UHF RFID practitioners in different occasions such as seminars and conferences and (iii) review of related literature.

By providing an overview of the different criteria, concerns and issues that the CityU HK Library has come across when evaluating different UHF RFID tags, the authors intend to create a momentum for more discussions to go on which for sure will benefit all who are also interested in deploying UHF RFID in their libraries.

2. Lessons learnt from the stories of barcodes and tattle tapes

For libraries, the basic issues behind any material check-in and check-out transactions are (i) item identification and (ii) security. There must be a means good enough for each book to be accurately identified so that the circulation status specific to each book can be recorded, mapped with the patron record concerned, and then reflected in the Integrated Library System. There must also be a mechanism good enough to safeguard the book from being illegally removed out of the library without going through the proper check-out procedures. For decades, libraries have been using barcodes and tattle tapes to handle the two issues.

RFID technology, however, has emerged recently with its capability to handle both item identification and security in an all-in-one manner. What more is, the contactless nature of the technology enables multiple book identification and thus can greatly improve loan transaction efficiency. The memory of RFID tags makes RFID more than just an identification technology but a data carrier that can keep track of the circulation status of the item concerned to ensure security and to store other necessary item information that facilitates collection management.

Since HF RFID was first deployed in libraries a decade ago, RFID technology has drawn the interests of many libraries. While the number of libraries adopting RFID technology is on the rise, nonetheless, many libraries are still stationing in the electro-magnetic domain of tattle tapes and barcodes observing the trend. At least this is the case in Mainland China and Hong Kong.

For many libraries, the decision to migrate to a new technology hinges not just to the capability of the technology concerned, but cost-benefit analysis and return on investment. A study called *"RFID Implementations in California Libraries: Costs and Benefits"* conducted in 2006 (Engel, 2006) surveyed different California public and academic libraries and provided an outline with different categories of possible costs and benefits for libraries to consider when planning RFID adoption. The report mentions that a large cost associated with the adoption of RFID is the expense on tags. Apart from the tag costs, tagging itself is also very labour intensive. New equipment, although costly, is comparable to the electro-magnetic systems currently in use in many libraries (Engel, 2006).

Since the survey was conducted in 2006, in fact the prices of tags have dropped, but migrating to a new technology still involves huge investment. To many libraries, barcodes and tattle tapes are still better choices because they have been in place for many years and have proved to be reliable, functional and stable. Adopting a new technology, however, involves risks and a lot of sunk costs in terms of technology research, product evaluation and testing, staff training, as well as new service development and planning.

After all, barcodes are widely recognized means of identifying items *(though they do need lines of sight and support single item identification for each scan only)*. Standards for barcodes have been well established and barcode scanners of almost any brand name can be tuned to read any barcode schema and thus there is no interoperability issue for resource sharing among libraries. *(The different RFID data models used in different systems however create obstacles for interoperability in the case of RFID tags as explained later in this chapter.)* Thus, without ensured interoperability and cost savings, many libraries are hesitant to change to any new technology. Thus, even though the 2-D barcodes nowadays have presented another possible choice for libraries to upgrade their barcode system so as to store more item information for books, very few libraries, if any, have chosen to do so. *(Some libraries, however, have adopted 2-D barcodes in other contexts, such as enriching holding information on library catalogues, providing web links to users on promotional brochures and alike.)* Although 2-D barcodes are comparatively newer technology, they are not backward compatible. Scanners for the traditional 1-D barcodes cannot read 2-D barcodes either. To change, it means hardware replacement and thus capital investment. So, if the traditional 1-D barcodes are good enough to store the basic information *(in most cases, the unique accession number of the book concerned)* to support any circulation transactions, why do the libraries need to bother about changing their 1-D barcodes to 2-D barcodes, not to mention about RFID tags?

For the case of tattle tapes, the common product specifications and requirements as well as the aggregate demand across different libraries have created the possibility for group purchase to drive the costs down. For example, among all the university libraries in Hong Kong, there are regular joint tendering exercises for bulk purchase of tattle tapes. The discount secured can be up to 30%.

The lessons learnt from the stories of barcodes and tattle tapes is that for UHF RFID technology *(despite its strengths and potentials to enhance library operations)* to be widely adopted among libraries, standardization, interoperability, aggregate demand supported by common product requirements are the necessary conditions to drive the costs down to ensure value for money. Moreover, efforts and sunk costs involved in technology research, evaluation and testing should be minimized through experience sharing among libraries.

At the CityU HK Library, studies and experiments on the use of UHF RFID in the library context have been carried out for years since 2007. After all, choosing the right tags is important as the tags are the souls that RFID-enable the books to make them identifiable throughout the RFID process. The performance of the RFID system hinges very much on the performance of the tags. For the UHF RFID hardware, the investment is basically one-off, though replacement may be necessary in a few years' time because of maintenance and system upgrade purposes. The expenses for tags with good performance, however, can be more significant when compared to the costs associated with the first-time one-off investment in hardware, especially when the collection to be converted is big. Moreover, the demand for tags is recurrent and ongoing throughout the years as the collection expands annually and so do the expenses. Similar to the case of tattle tapes, should libraries be able to agree on common requirements to ensure standardization, interoperability and compatibility, the aggregate demand will guarantee quality and sustainable supply of UHF RFID tags as well as the opportunity for consortial purchase to bargain for bigger discounts and long-term cost savings.

3. UHF RFID pilot test and long term implementation for the CityU library

When the National Library Board of Singapore first introduced RFID into their libraries in 1998, HF RFID was the only available technology and thus has naturally become the *de facto* choice by the system developers at that time. But then as technology emerges, UHF RFID has presented to libraries a possibly much cheaper and more powerful choice. The CityU HK Library thus joined hands with the Wireless Communication Research Centre *(RCW)* of the University in 2007 to form a project team *(hereafter, the CityU HK Library Project Team or the Project Team)* to look into the use of UHF RFID in the library environment. To compare the performance of HF RFID and UHF RFID, however, is not the intention of the authors (Ching & Tai 2009). Instead, based on the UHF RFID pilot test carried out at the CityU HK Library, this chapter discusses the criteria, concerns and issues behind the selection of the right tags for libraries that would also like to use UHF RFID.

In April 2008, the CityU HK Library selected its Semi-Closed Collection[1] as the site to carry out a pilot test that involved real users. UHF RFID applications of the beta version developed by the Library and RCW were put in the Collection for users to check-out and check-in books by themselves. The applications, being called the "EasyCheck System", include the EasyCheck Units (self-check machines), the EasyReturn Units (self-return machines) and the EasyDetect Gates (security detection gates). They have been well-received by the users and thus are still in

---

1 The Semi-Closed Collection consists of some 7,000 course-related library books that are for short loans (5 hours or 1 day) to students.

operation inside the Semi-Closed Collection till now. Studies and experiments, however, carry on as the Library is planning for large-scale implementation in the whole Library. In particular, the Project Team strongly felt that choosing the right UHF RFID tags is important if the utility and performance of the UHF RFID System is to optimize. Thus, tests have been performed with many different brands of UHF RFID tags.

4. Criteria, concern and issues behind UHF RFID tag selection

In the logistics industry, tags are for one-off use only. When the pallets/cases/items reach the end of the logistics chain, leaving the retailing line and settle in the hands of the customers, in most cases, the tags will be discarded together with the packaging. Nonetheless, for libraries, the tags have ever-lasting roles in the book circulation transactions, perhaps until the books concerned are withdrawn from the collection. Tags in libraries need to go through repeated check-in and check-out processes throughout the years and its anti-theft capability must last as long as the books concerned are still part of the library collection. Moreover, tags in libraries serve at the item level. Almost every book bears a tag and that constitutes to a dense tag environment. What complicates the case is the production life cycle of tags. With the rapid development in the UHF RFID technology, not just the readers are evolving, tags are also kept upgrading. Libraries cannot guarantee that they can use the same brands or the same models of tags throughout the years because of tag evolution. Thus, the dense tag environment will be one with a mixture of tags. Compatibility of tags of different generations to the same machines acquired years ago is a concern.

Other well known issues that libraries may consider also include compliance with regulatory standards, data model, interoperability among libraries, shapes of tags, read range and distance, physical mounting issues such as adhesive, position, orientation, suitability of the selected tags for efficient reading by foreseeable new applications (e.g. smart shelves) and so on. All these different considerations have something to do with the business nature of libraries and also the unique local situation and environments, or even loan rules of different individual libraries. Moreover, unlike the logistics industry where the major concern is smooth flow and tracking of pallets/cases/items throughout the supply chain, libraries' concern extends to customers' perception and transaction experience. Thus user behavior and expectation are determining factors too.

Since 2007, the CityU HK Library Project Team has been testing with different UHF RFID tags from different vendors. All the tags concerned are passive tags. Table 1 provides a snapshot of the tags that have been tested so far. To protect the interest of the tag suppliers and companies concerned, the brand names of the tags concerned are represented by the English alphabets only. The most distinctive features of the tags are listed in the table. The country of origin and also the EPC memory size of the tags are also provided.

Results and observations from the tests have provided valuable information to the Project Team for long term implementation of UHF RFID in the whole CityU HK Library. It is hoped that by sharing the findings, the other libraries that are also interested in adopting UHF RFID can benefit too or at least reduce their sunk costs in product testing and evaluation. To choose the right tags, the Project Group recommends that libraries concerned should pay attention to the following areas:

1. Standard Compliance
2. Data Models and Interoperability
3. Tag Memories

4. Form Factor, Orientation and Position of Tags
5. Interferences
6. Product Life Cycle and Compatibility

Tag	Description	Country	Memory size (EPC)
A	A general-purposed inlay intended for use by a wide variety of applications	US	96
B	Strong read range and provides a durable antenna that can withstand more physical abuse than a traditional dipole antenna due to its increased antenna surface	US	240
C	Comes with both EPC memory and user memory	China	EPC: 96 bits User: 224 bits
D	Powerful read performance with best in class reading capabilities for RF friendly contents at FCC frequencies. 240 bits EPC memory with an option for additional 512 bits of user memory	US	240
E	An Item-level inlay designed for best edge on performance, especially in close proximity to other tagged items	US	96
F	Offers far-field performance on RF-friendly materials & metals in a compact form factor	US	96
G	A general-purposed inlay intended for use by a wide variety of applications	US	96
H	Orientation sensitive to minimize cross-talk in dense reader environments	US	96
I	With a breakthrough antenna design that enables more reliable read/write functions in item level applications where tags may be stacked with millimeters of each other	US	240
J	Orientation insensitive inlay coupled with powerful read range performance. Ideal for reading randomly orientated tags like baggage tagging and pallet tracking	US	96
K	With better performance on items with metal	US	96
L	Orientation-insensitive, with high performance for pallet- and case-level applications.	US	96
M	Long and thin antenna which are long enough to prevent shielding of signals by human hands	Korea	96
N	Tailored, high-performance product for item level use. Reliable reads/writes when tags are in close proximity to each other.	US	240
O	Cost-efficient, high-performance product for a wide range of supply chain management and apparel applications.	US	96
P	Near field tag which is able to be detected by far field antenna. However, the tag cannot be read when it is too closed to the far field antenna, some distance is required.	US	96
Q	Designed for item level tracking and can be read in both near and far fields. Orientation insensitive with superb performance in dense tag environments.	US	96

Table 1. Tags that have been tested and tried out by the Project Team of the CityU HK Library

4.1 Standard compliance

4.1.1 Technical standards

The very basic consideration is compliance to standards. It is important that the selected UHF RFID tags should comply with existing and emerging standards so that they can be formatted and are readable by any RFID readers that have also incorporated the ISO standards. ISO18000-6 (UHF Generation 2 Standard) has been developed for UHF RFID. According to the EPC Global specifications (EPC Global, 2008), UHF RFID uses "EPC Gen2" standard as the air interface, standard protocol to communicate with readers and tags. It defines the frequency range, commands, memory bank and protocols for tags and it has been approved and included in the international standard organization (ISO 18000-6C).

4.1.2 Frequency band

As RFID makes use of radio waves, the technology is subject to governance by the radio telecommunication ordinance of each individual country. The UHF RFID bandwidths stipulated by different countries, however, are slightly different and sometimes incompatible. The following are some examples:

- The European Union defines 865 - 868MHz as the UHF RFID bandwidth in Europe.
- The Federal Communication Commission (FCC) of the US stipulates 902 - 928 MHz for their country.
- For Singapore, only frequencies between 923-925 MHz are allowed for UHF RFID applications.
- For China, the State Radio Regulation Committee (SRRC) under the Ministry of Information Industry (MII) has approved bandwidths in the 840.25 to 844.75 MHz and 920.25 to 924.75 MHz ranges to be used by UHF RFID tags and interrogators. Each band is divided into 20 channels, each consisting of 250 kHz of spectrum.
- For Hong Kong, the RFID restriction is less tight. The Office of the Telecommunications Authority (OFTA) has stated that for UHF RFID, the bandwidths are 865 – 868 MHz and/or 920 – 925 MHz. The Telecommunications Ordinance (Cap 106) has set out the technical requirements for RFID equipment operating in these frequencies.

The tags that the Working Group has tested so far (see Table 1) can support frequency range from 865MHz – 925MHZ and thus should have no frequency compatibility issue. However, caution should still be taken by libraries to ensure that their selected tags support the UHF RFID frequency bandwidth of the country or region where they belong to.

4.2 Data models and interoperability

Data models define the requirements for data elements and structure on the RFID tags and are somehow related to the standardization issue too. To ensure interoperability which is essential for interlibrary loan and resource sharing among libraries, data stored in the tags must be readable and usable by all libraries concerned irrespective of the UHF RFID system that they are using, whether the system comes from company A or company B. Therefore, data model standards are the keys to interoperability.

However, this has not been the case for HF RFID ever since it was first adopted by the libraries in Singapore a decade ago. Standard data models for HF RFID emerged only recently[2] when libraries started to realize that proprietary ways of formatting the tags have

---

[2] Different HF RFID library data model standards at the national level have emerged recently. They include data models from Denmark, the Netherlands, the UK and Finland that are examples of fixed

deprived them of the flexibility to use the equipment from any vendor they want. For libraries that have been using proprietary systems for years, changing the vendor or adopting the new data model standard means re-formatting all the old tags. *(This is possible only if old tags are compatible to the system of the new vendor, or otherwise, all items concerned will need to be re-tagged)*. This is contrary to the case of barcodes mentioned earlier. For barcodes, standards have been so well established and observed that basically libraries can buy any scanner from any supplier and be able to read barcodes of any schema such as Code 39, codabar, U.P.C. and so on.

Therefore, while UHF RFID is making its way into the library arena, libraries should take the opportunity to first compromise on the data model standards. So far, there is no ISO standard stipulating the UHF RFID library data models. However, instead of accepting whatever proprietary data models that the vendors may propose, libraries should present their own specifications to ensure vendor independence. Such specifications should at least be a consortial consensus among libraries that will have interlibrary loans among themselves, or preferably, a regional or national data model standard. Specifications as such are critical to the choice of tags as sufficient tag memory to support the data model standard concerned is a must.

Therefore, libraries should first make up their mind on the data model that they will adopt before making their tag selection or starting their tag conversion exercise. Once a certain data model is formatted in the tags, it cannot be easily transformed and rewritten. The following is what the CityU HK Library has experienced during its pilot test.

4.2.1 Data model used in the pilot test

When the CityU HK Library launched its EasyCheck System in its Semi-Closed Collection in 2008, the prevailing EPC memory size of UHF RFID tags available in the market was 96 bits only. Moreover, no other reference cases were available in Hong Kong as the CityU HK Library was the only pioneering library trying out UHF RFID in Hong Kong at that time[3]. No regional or international UHF RFID library data model standards could be identified either. Therefore, the Library has come up to its own proprietary data model which is a fixed length one of 12 bytes (96 bits).

The table below shows the structure of this 12-byte data model. The fixed length structure ensures that each data element is given its designed memory address to enable speedy identification of data location even without a precursor. However, the "fixed" approach also means lack of flexibility and the limited tag memory of 96 bits leaves no room for the Project Team to reserve space for additional data elements that other libraries may find necessary if they are to adopt the same data model. Thus, the 12-byte data model as outlined below is tailor made for the CityU HK Library only to suit its local circumstances and may not be suitable for other libraries. This in the long run can be an obstacle to interoperability if every other UHF RFID library devises its own proprietary data model.

---

memory models. Other examples are data models from Australia and the US which are examples of flexible memory models (ISO/IED 15962 encoding). ISO 28560 as an international standard which consists of 3 parts to provide general guidelines on the data elements and incorporate both the fixed memory approach and flexible memory approach came into place only in 2010.

[3] The Library of the Chinese University of Hong Kong, later on, also conducted a pilot test on UHF RFID during January to May 2010.

Offset	Length	Field
0	1 byte	Institution / Organization
1	1 byte	Library Branch / Location
2	1 byte	Classification
3	7 bytes	Barcode
10	2 bytes	CRC16

Table 2. The 12-byte proprietary data model used by CityU HK Library during its pilot test in 2008

4.2.2 Data model for library-wide implementation – the recommended standard

What added light to the situation, however, is the fact that the Moore's Law, (Moore, 2011) coined by the Intel co-founder Gordon Moore in 1965, also applies to UHF RFID tags. Within the few years since the CityU HK Library started its pilot test, the memory sizes of UHF RFID tags have been increasing yet with lower and lower costs. This has provided the Project Team the opportunity to re-plan the data model for future long term implementation of UHF RFID in the whole CityU HK Library. As tags of 240 bits or even larger EPC memory sizes are now available in the market, the Project Team can re-consider adopting a more flexible data model that can cater for more scenarios and possibly fits all UHF RFID libraries. Nonetheless, so far there is still no regional or international data model standard for UHF RFID. Therefore, modeled on ISO28560, the recently announced international data model for HF RFID, and with reference to the recommendations from the National Library of China on the adoption of ISO28560 by Chinese libraries, the Project Team has attempted to devise a data model standard specific to UHF RFID. Based on the ISO28560 data element table, the Project Team proposes that the starting block of the UHF RFID data model be as follows:

Offset	Length	Field
0	2 byte	Overhead
2	4 byte	Primary ID
6	3 byte	Owner Library
9	X bytes	Reserved (Title)
.....	......	Reserved (set information)
.....	......	Reserved (Type of usage)
.....	......	Reserved (call no.)
.....	......	Reserved (barcode)
28	2 bytes	CRC16

Table 3. Starting block of the proposed UHF RFID data model based on ISO28560

The lessons learnt from the stories of barcodes and tattle tapes as well as the evolution history of the data model standards for HF RFID have enlightened the CityU HK Library Management on the importance of standardization and interoperability. The data model so proposed by the Project Team should also be a regional consensus if not international. Thus, discussion and exchange of ideas with different stakeholders are the essential next steps.

In March 2010, the CityU HK Library, together with the Shanghai Jiao Tong University and the Tsinghua University, formed the *Higher Education Libraries "UHF RFID Application" Working Group (hereafter, the Working Group)*. In a meeting held in August 2010 organized by the Working Group, representatives from different libraries in Mainland and Hong Kong gathered together in Shenzhen, PRC, to discuss UHF RFID data model standardization. Then in March 2011, a conference called *The Development and Best Practices of UHF RFID Technology Applications* [4] co-organized by the Working Group and GS1 Hong Kong [5] involved not just participants from the library arena, but UHF RFID practitioners and organizations with expertise in standards to discuss and share ideas on standardization and best practices. The conference has created the nurturing ground for a regional UHF RFID data model standard to gradually emerge for libraries in Mainland China and Hong Kong. It has also provided a platform for libraries to collectively convey their needs and requirements to the UHF RFID practitioners.

The Project Team has a high hope that not long there will be a consensus on the UHF RFID data model, at least among the JULAC *(Joint University Libraries Advisory Committee)*[6] university libraries in Hong Kong. In fact, collaboration among the JULAC libraries in Hong Kong has already had a long history and for the adoption of UHF RFID, a few meetings have been held among the JULAC library directors in late-2010 to discuss the possibility of seeking external funding for collaborative implementation. This has naturally paved the way for adopting a common data model standard among the JULAC libraries.

4.3 Tag memories

As mentioned earlier, the UHF RFID data model so proposed by the Project Team was modeled on ISO28560 for which the 96-bit EPC memory size of the first generation UHF RFID tags is not sufficient. However, the development of UHF RFID Gen2 tags has been fast paced. Tags of 240 bits EPC memory are now available and some brand names even claim to have 496 bits. Moreover, apart from EPC memory, some suppliers can also provide an extendable memory that reaches 512 bits in their tags. Therefore, storage capacity is no longer an issue. What important rather is the choice of data elements.

Among the dozens of data elements outlined in ISO28560-1, libraries are to choose their own sets of data. The Project Team recommends that "primary item identifier" *(the unique identification of an item inside the Library and this usually is the accession number)* and "owner library (ISIL)" be the mandatory elements. Based on the description in ISO28560-2, libraries

---

[4] The conference called The Development and Best Practices of UHF RFID Technology Applications co-organized by the Higher Education Libraries "UHF RFID Applications" Working Group was held in Shenzhen, PRC, on 18 March 2011. The conference has attracted a total of 134 participants from Mainland China and Hong Kong. Participants include 88 librarians from 29 academic libraries, 29 UHF RFID practitioners and users from 17 companies, and 7 representatives from the National Library of China, the Hong Kong Public Library and GS1 Hong Kong. Details about the conference are available at: http://www.cityu.edu.hk/lib/about/event/rfid_conf2011/index_e.html

[5] Background about GS1 Hong Kong is available at:
http://www.gs1hk.org/en/hkana/gs1/aboutgs1/profile.html

[6] Details about JULAC is available at http://www.julac.org/. JULAC members include libraries of the following universities:

-The Chinese University of Hong Kong
-City University of Hong Kong
-Hong Kong Baptist University
-The Hong Kong Institute of Education
-The Hong Kong Polytechnic University
-Hong Kong University of Science and Technology
-Lingnan University
-The University of Hong Kong

have the flexibility to choose any other data elements that suit their local operations and circumstances. However, libraries should be cautious that the amount of data elements that they choose to include into the tags will affect the memory size and thus, the storage capacity of the tags they will need. The natural logic is that the more data a library would like to store in the tags, the larger the tag memory it will require. Moreover, between EPC memory and user memory, libraries will also need to decide what data elements are to be housed in the EPC memory and what data are to be housed in the user memory. In this regard, the reading speeds of different memory banks in the tags should also be taken into consideration.

4.3.1 Tests on reading speed

In terms of storage capacity, the different brand names of tags (see Table 2) that the CityU HK Library has tested so far are mainly of two types. The first type comes purely with EPC memory only and the second type comes with both the EPC memory bank and the user memory bank in a single tag.

The intention of the tests performed by the Project Team was to find out how different the reading speed can be for tags with different memory sizes. Test 1 compared the reading speed for tags with different EPC memory sizes *(96 bits versus 240 bits)* from a selected brand name (Brand I). Comparing tags from the same brand name ensured that all other possible deviations due to the difference in suppliers could be minimized. Test 2 compared the reading speed of tags with different memory combinations *(EPC memory versus EPC memory plus user memory)*, again from the same brand name only, though this brand name (Brand II) is different from the brand name used in Test 1.

For Test 1 and Test 2, both the 1-tag scenario and the multi-tag scenario *(10 tags have been involved)* have been examined. For both scenarios, the one tag or the ten tags concerned were read 100 times and the reading speed of each time was recorded. Table 4 and Table 5 show the results.

For Test 1 (see Table 4), when there was only one tag involved, the average time required for the reader to successfully read the data in the 96-bit EPC memory tags and the 240-bit EPC memory tags were 0.123 second and 0.126 second respectively. The difference has been insignificant. When ten tags were being read together, the average time required then became 0.193 second and 0.227 second for the two types of tags, meaning that when more tags were involved, the reading speed for the 240-bit EPC memory tag dropped, in this case, by 0.034 second. However, this 0.034 second was indeed minimal and even not noticeable by human beings during the transactions.

Reading Times	1-Tag scenario		10-Tag scenario	
	96 bits	240 bits	96 bits	240 bits
1	0.121	0.120	0.193	0.221
...	0.123	0.130	0.188	0.226
100	0.128	0.126	0.201	0.231
Average (Seconds)	**0.123**	**0.126**	**0.193**	**0.227**

Table 4. Reading speed for tags with different EPC memory sizes (96 bits versus 240 bits) from a selected brand name.*

*Comparing tags of the same brand name ensures that all other possible deviations due to the difference in suppliers could be minimized.)

For Test 2 (see Table 5), under the 1-tag scenario, the average time required for the reader to successfully read the data from the tags that provide EPC memory (96 bits) only was 0.103 second while that for the 10-tag scenario was 0.199 second. The difference is still less than one second. However, for the tags with both EPC memory (96 bits) and user memory (224 bits), the average reading speed for one tag was 0.492 second while that for reading ten tags together was 5.227 second which was more than ten times that of the 1-tag scenario. This in fact is an expected result by the Project Team as the reader used for the test provides only simple commands that support programming and reading of either the EPC memory alone or both the EPC memory and user memory together because the user memory cannot be separately read without mapping to the EPC memory to ensure correct association to the corresponding tags. Therefore, whenever the user memory is to be read, the reader must first read the EPC memory and thus requires longer reading time, though it is still a matter of a few seconds. The Project Team has tried out two other readers of the popular brand names and the same reading behavior was observed.

Reading Times	**1-Tag scenario**		**10-Tag scenario**	
	EPC 96 bits	**EPC 96 bits + User memory 224 bits**	**EPC 96 bits**	**EPC 96 bits + User memory 224 bits**
1	0.108s	0.519s	0.331s	5.279s
...	0.106s	0.500s	0.202s	5.039s
100	0.100s	0.480s	0.180s	5.389s
Average (Seconds)	**0.103s**	**0.492s**	**0.199s**	**5.227s**

Table 5. Reading speed for tags with different memory combinations (EPC memory alone versus EPC memory plus user memory)* #

* Comparing tags of the same brand name ensures that all other possible deviations due to the difference in suppliers could be minimized

\# The reader used during the test only provide simple commands to support programming and reading of either the EPC memory alone or both the EPC memory and the user memory together because the user memory cannot be separately read without mapping to the EPC memory to ensure correct association to the corresponding tags. With readers that support programming and reading of the EPC memory and the user memory separately, hopefully the reading speed for the user memory can be improved.

The tests involved tags from two different brand names only *(Tags from Brand I for Test 1 and tags from Brand II for Test 2)* and thus the sampling size may not be big enough for any authoritative conclusion. Moreover, when more and newer readers are involved as the technology evolves, the read rates can be different too. The tests therefore simply serve as preliminary references for libraries to select memory sizes for their tags and to decide on which memory is to be used for different data elements.

For the case of CityU HK Library and for the adoption of the UHF RFID data model standard recommended earlier (modeled on ISO28560), the Project Team will put data elements that are more transaction critical into the EPC memory. With the primary item identifier *(mandatory and for the CityU HK Library, it is the accession number)*, all other bibliographic information of the library item concerned will be readily retrievable from the Integrated Library System (ILS). Thus the primary item identifier must be read instantly in the first place for any check-in or check-out transaction to take place. It is therefore transaction critical and should be written in the EPC memory for speedy identification. As for owner library (ISIL), the Project Team strongly feels the need to have it mandatory too in view of the interlibrary loan and HKALL [7] transaction activities among the JULAC libraries. This data element enables libraries to quickly identify the ownership of the items concerned during the resource sharing processes and is therefore recommended to be written in the EPC memory too.

4.4 Form factor, orientation and position of tags

RFID tags consist of three components, namely, the integrated circuit (IC), antenna and substrate. The IC is connected to the antenna that is deposited or printed on the substrate. Even with an identical IC, tags with different antenna geometry will display completely different properties and behaviors. Tag antenna designs determine the frequency at which the tags concerned operate. They affect tag performance in terms of read range and orientation sensitivity. Also, as antenna is the largest component of a tag, its geometry impacts the form factor of the tags in terms of size and shape. However, just as much as how the antenna geometry requirements affect the form factor of the tags, form factor requirements appropriate to different applications also impact on antenna designs (Imprinj, 2005). For different purposes, the selected tags should exhibit a size and shape appropriate to the items to be tagged. Therefore, tags come in different sizes, shapes and forms. Generally speaking, larger tags with larger antennas support operations that require a long read range and are less orientation sensitive. On the contrary, for situations where only smaller tags can be used, the antenna geometry that conforms to the smaller form factor of the tags must also be compact and small, thus sacrificing the read range and orientation insensitivity. Of course, the extent of the shortfall in the tag performance also depends very much on the abilities and skills of the tag antenna designers.

For libraries, the size of the tags to choose depends very much on the types of materials to be tagged, how the tags are to be mounted on the library materials and the read range required in the real operational environment. This will have something to do with the relative distance between the tags to be read and the reader antennas that reside in the self-check machines and the detection gates. While choosing tags of a larger form factor seems to be advisable given its longer read range and less orientation insensitivity, libraries still need to practically consider if the tags would be too sensitive that a very large buffer area will be required to keep users with non-checked-out books in hands distant from the gates in order not to cause any false alarms. Of course, the power of the readers at the detection gates can be tuned down, but this will sacrifice security.

Moreover, libraries must also note that the tag masking phenomenon may occur when tags overlay each other in a stack of thin books. When tags mask each other, either one or both of the tags may become unreadable (Butters, 2008). Large tags may stand a higher chance of overlaying with each other when tagged at book covers (either front or back). Moreover, large tags may be too visible and easily subject to mutilation when noticed by naughty users.

---

[7] Based on a common on-line catalogue running on a server hosted in one of the JULAC libraries, HKALL seamlessly connects the library automation systems of all university libraries in Hong Kong and allows staff and students to request and borrow materials of the other local university libraries directly.

For the case of the CityU HK Library, the UHF RFID tags that the Project Team has used in its pilot test were of an optimal size with a dimension of 72mm x 30mm. The Semi-Closed Collection where the pilot test was conducted is a very small room of about 75sq.m. only. The self-check machines, the self-return machines and the security detection gates are all in proximity. For instance, the security detection gates are only 4.5 meters away from the nearest bookshelves. Therefore, the Project Team must be even more cautious when choosing the tags. Different tests have been performed to ensure that the size of the buffer zone required around the detection gates is kept to a minimum and at the same time the self-check machines can read the most number of tags when books are placed on top of them so as to maximize the benefit of multiple item identification at the check-out process.

In the Semi-Closed Collection, the UHF RFID tags have been placed at the back covers of books with a book plate of 150mm x 100mm on top to act as a camouflage to hide the tags from the scene (Photos 1 and 2). To reduce the tag masking probability, during the tagging process, tags have been randomly placed in four different positions behind the book plates. However, the additional book plates mean additional costs and labour.

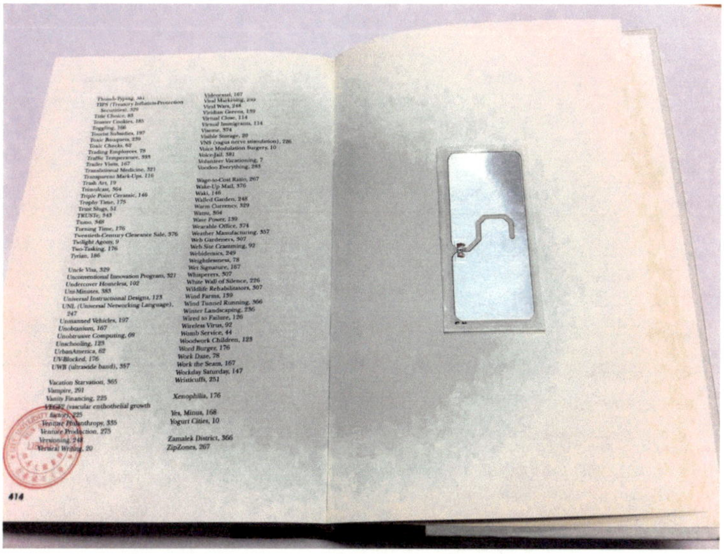

Photo 1. In the pilot test conducted by the CityU HK Library, UHF RFID tags were randomly placed in four different positions at the back covers of the books. The photo shows one of the selected positions.

In fact, some suppliers do provide long and narrow UHF RFID tags that resemble the shape of the traditional tattle tapes. These elongated tags can be put along the book spines, thus reducing the tag masking probability and also making the tags less visible to the users. These certainly are advantages. However, users usually hold the books on the spines and libraries must therefore be cautious enough to test beforehand to ensure that shielding of the spines by human hands will not affect the readability of the tags, especially in the case of the detection gates.

The Project Team chose to place the tags at the back cover of the books because, to maximize the read range, the tag orientation needs to match that of the reader antennas in the self-check machines. In the Semi-Closed Collection, the reader antennas lay flat horizontally inside the machines and they are circularly polarized ones that are supposed to be less orientation sensitive. However, tests conducted by the Project Team reveal that, even with circularly polarized reader antennas, when tags are placed on the book spines and exhibit a perpendicular orientation to the reader antennas, the read rate is far from satisfactory. When books with tags on the book covers are laid flat on the self-check machines and thus sharing the same orientation as the reader antennas, the read rate is much better and the read distance is long. The long read distance enables users to check out more books in any one go [8] and thus capturing the benefits of UHF RFID. *(Long read range, however, may generate a concern on misread if the detection area goes beyond the expected one. While the power of the reader antennas can be tuned to adjust the read distance when needed, the Project Team has also put in place a shielding mechanism in the self-check machines to guide the radio waves to go upright instead of sideway so as to safeguard against misread.)*

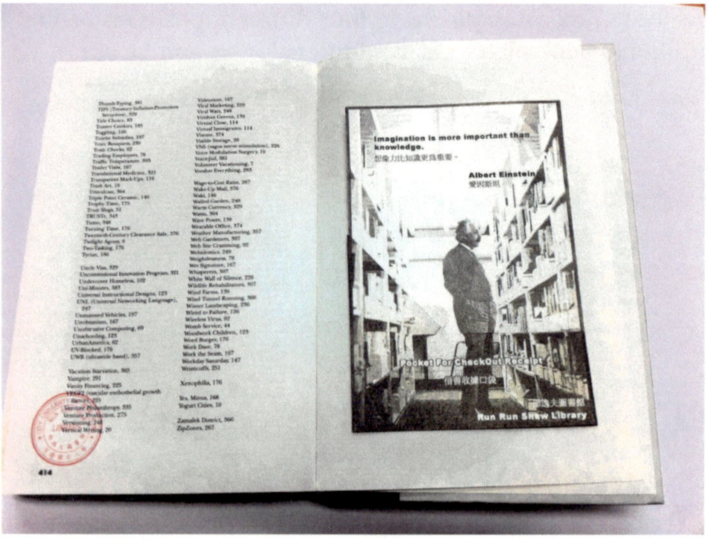

Photo 2. The tags were then covered by a book plate.

Nonetheless, on the book shelves, when books are read by hand-held readers, the test results present a different story (Photo 3). Books with tags on the spines are more readily identifiable by the handheld readers when compared to books with tags on the covers. This is because when books are vertically placed on the shelves, tags on the spines share the same orientation as the reader antennas in the handheld scanners while tags on the covers exhibit a perpendicular orientation. Handheld scanners are tools for stock taking and locating missing items on shelves. Some UHF RFID practitioners are developing smart shelves to provide similar functions in an automatic way. The performance of the smart shelves may possibly be related to the relative position and orientation between the reader antennas on

[8] In the Semi-Closed Collection, however, users are allowed to borrow up to 5 volumes of books at any one time. This is because materials inside the Collection are course-related and thus of very high demand. The restrictive loan rule ensures that every student gets a fair chance of using the books. However, the longer read range provides the CityU HK Library the flexibility to allow users to borrow more in one go when the general circulation collection is involved in the long term library-wide implementation of UHF RFID in the whole library.

the shelves and the tags on the books too. Smart trolleys and automatic book dispensers are examples of other foreseeable applications that many people have been talking about. Libraries forward-looking enough that will consider adopting these innovative applications in the future may also need to take into consideration the possible requirements of these end-use applications when selecting their tags at the present moment.

Photo 3. When books are vertically placed on the shelves, tags on the spines share the same orientation as the reader antennas in the handheld scanners while tags on the covers exhibit a perpendicular orientation.

4.5 Interferences

It is a fact of physics that metal reflects radio waves and water absorbs them. This makes tracking metal products and those with water content difficult in the logistics field. In the library environment, with the exception to the media collection which usually constitutes only a small part of the entire collection, the main subjects to be handled are mainly books. Therefore, to many people, interferences caused by metal and water seem not to be the problems for libraries. However, this is not the case.

To add elegance and a sense of luxury to the books, many publishers put metallic gold or silver printing on the book covers. To add varieties to the contents, some books contain CDs as the accompanying materials. All these are metallic elements that will cause interferences to the readability of the tags. For cases as such, libraries will need workarounds such as placing the tags sideway at positions that do not overlap with the metallic prints or, in the extreme case, sending the books to the binders to have the metallic covers replaced. For the CDs accompanying materials, they can be detached from the books concerned for separate handling.

As for water, it is rare that books or library materials will contain water, but humans do. As mentioned earlier, for books with spine tags that are not long enough, there are chances that human hands may shield the tags making them less sensitive to radio waves.

However, while metal and water have a detrimental effect on radio waves, the two factors are not necessarily negative with regard to the application of UHF RFID in libraries. In the library setup, given the long read range of UHF RFID, it is important that radio waves are confined only to the designated area and distance appropriate for the purposes intended. For example, as discussed earlier, for the self-check machines in the pilot test of the CityU HK Library, the radio waves are expected to go only upright in a distance long enough to allow the most number of tags/books to be identified in any one go. To ensure that the waves go far but not wide, the Project Team has made use of metal to provide shielding around the reader antenna (Photo 4).

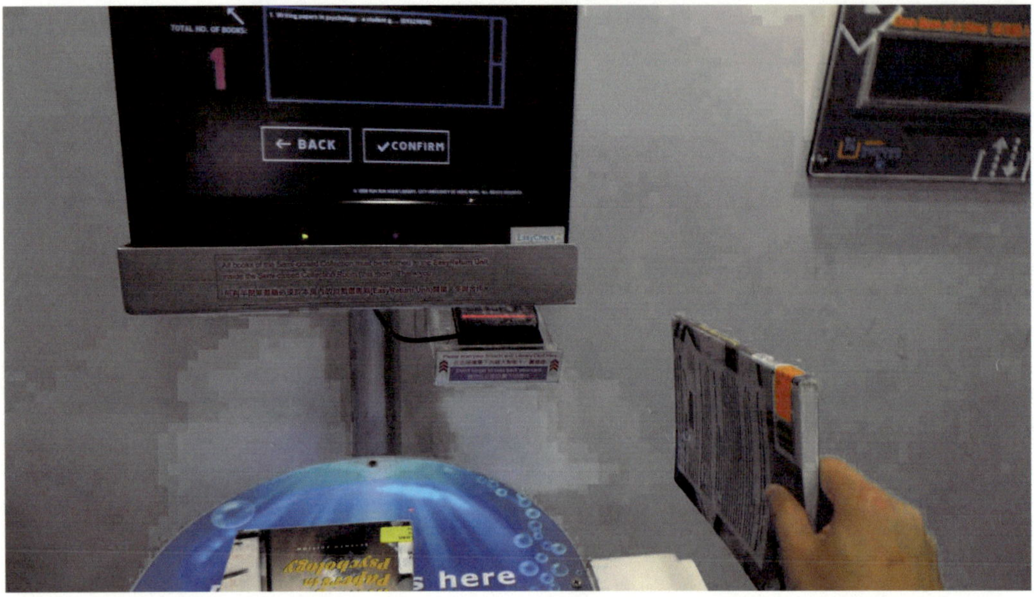

Photo 4. In the pilot test conducted in the CityU HK Library, the UHF RFID reader antenna is laid flat horizontally inside the self-check machines. The radio waves must go far upright to ensure the reading of the maximum number of books. However, the waves are not supposed to go wide to scan the books sideway as well. Any cases like that are misreads and must be rectified. The Project Team thus has made use of metal to provide shielding around the reader antenna and the result is good.

4.6 Tag production life cycle and compatibility

Usually, the biggest investment on tags takes place during the first-time conversion of the entire collection; the subsequent annual requirements will depend on the expected growth in the collection every year. To buy additional tags for the growing collection, it is natural that libraries will tend to buy the same tags as what they have used during the first-time implementation *(unless the tags have proved to be a wrong choice)*. The following is the experience of the CityU Library during its pilot test.

In the Semi-Closed Collection where the CityU Library carried out its pilot test on UHF RFID at the operational environment involving real users, all the 7,000 volumes of books were tagged with UHF RFID tags of Model A from Producer X acquired through vendor Y. *(To protect the interest of the parties concerned, the authors prefer calling them with English alphabets.)* The model was the final choice after a series of tests and careful consideration and has proved to be a correct choice. In the first purchase, the Project Team acquired 10,000 tags so that a stock of some 3,000 tags could be reserved for future use at least for the first two subsequent years before any further requisition was required. Materials in the Semi-Closed Collection are course-reserved materials. The Collection is subject to reviews and changes every semester according to changes in the curricula and the teachers' requirements. Books considered no longer relevant will be removed and returned to the general circulation collection while new items will be added in. Therefore new UHF RFID tags must be ready for new members in the Collection every semester. Everything has been so far so good until early-2010 when there were just several hundred tags left and the Project Team found it necessary to order more before the stock ran out.

The natural response was then approaching Vendor Y to buy more UHF RFID tags of Model A from Producer X again. However, the Project Team was told by Vendor Y that Producer X has ceased the production for tags of Model A. Only limited stocks were available and when they ran out, the Project Team must find substitutes. This has not been anticipated by the Project Team. Choosing tags of another model will mean creating a "mixed-tags environment" in the Semi-Closed Collection and also a series of tests to ensure that the reader antennas in the self-check machines, the self-return machines and the detection gates are compatible to the newly selected tags and at the same time do not upset the performance of the old tags. Consequently, UHF RFID tags of Model B from the same producer were selected as the substitutes. However, to start with, occasional misreads (though not too many) at the self-check machines and the detection gates were reported. That was rare when there were just tags of Model A in the Collection. The "Mixed-tags environment" did cause some concerns. Given that the relative distances between the different components in the RFID processes have been fixed in the Collection, what the Project Team could do was to adjust the power of the reader antennas. Through trial and error, the reader antennas were finally tuned to become just optimal for both the old and new tags.

The lesson learnt from the experience is that tags do have their production life cycle. While the tag specifications from the vendors claim that their tags can be used up to 10 years or more or the read/write times being 100000 times, they are talking about the life spans of the tag ICs and tag antennas. No matter how long lasting the tag ICs and antennas can be, the fact is that the tag model itself may not have a very long production life span depending on the producer's different manufacturing considerations. Libraries should not expect that they can stick to the tags of the same model forever for the collections despite the "claimed" life span of the tags. Libraries must be prepared to face a "mixed-tags environment" in the long run and be cautious to ask for compatibility guarantee from the tag providers as well as the RFID system providers. In fact, the same will apply to the reader antennas too. Because of wear and tear and system upgrade, machines will be upgraded or changed. Backward compatibility is therefore a must.

5. Conclusion

The compatibility issue discussed above has highlighted the fact that libraries are now playing a rather passive role in terms of UHF RFID product development. Apart from compromising on the data model to be used *(if libraries are collaborative enough)*, libraries do not have much influence on what the UHF RFID practitioners are offering as the demand generated from each individual library is indeed too small when compared to the transaction volumes in the logistics industry. In fact, while the operational environment that each library is facing can be quite unique, the nature of the transactions to be enhanced by UHF RFID is by and large similar. This provides a very good necessary condition for common specifications and requirements to be identified and thus aggregating the demand to make it large enough for libraries to influence the decisions of the suppliers as a consortial entity.

Moreover, it is important that experiences and test results are shared so that libraries learnt from each other to reduce the sunk costs and reserve more investigation time for newer findings. Libraries should play a more proactive role to work with the UHF RFID practitioners so that the latter know well what libraries are expecting and find it less risky to develop more innovative UHF RFID solutions for the library arena. When UHF RFID starts to transform library services, libraries together should act early enough to ensure that they can get the best out of it.

6. References

Butters, A (2008), "*New RFID Technologies & Standards – What Does it Mean for Your Library?" Paper Presented in VALA Conference 2008*. Available from: http://www.valaconf.org.au/vala2008/papers2008/66_Butters_Final.pdf

Ching, S.H. and Tai, A. (2009), "*HF RFID versus UHF RFID – technology for library service transformation at City University of Hong Kong*", Journal of Academic Librarianship, Vol. 35 No. 4, pp. 347-359.

Engel, E. (2006), "*RFID implementations in California libraries: costs and benefits*", available at: http://www.kcoyle.net/RFIDCostsBenefits.pdf.

EPC Global (2008). *Specification for RFID air interface*. Available from: http://www.gs1.org/sites/default/files/docs/uhfc1g2/uhfc1g2_1_2_0-standard-20080511.pdf

Imprinj (2005). *The RFID Tag Antenna: Form Factor*. Impinj RFID Technology Series. Available from: http://www.impinj.com/WorkArea/linkit.aspx?LinkIdentifier=id&ItemID=2553

Mc Carthy U, Ayalew G, Butler F, McDonnell K, Ward S. (2009) "*The effects of item composition, tag inlay design, reader antenna polarization, power and transponder orientation on the dynamic coupling efficiency of backscatter ultra-high frequency radio frequency identification*". *Packaging Technology and Science* 2009; 22(4): 241–248.

Moore, G (2011). *What is Moore's Law?* Available from: http://www.intel.com/about/companyinfo/museum/exhibits/moore.htm

R. H. Clarke, D. Twede, J. R. Tazelaar, and K. K. Boyer (2005), "*Radio frequency identification (RFID) performance:The effect of tag orientation and package contents*", Packaging Technology and Science, vol. 19, no. 1, pp. 45-54, 2005.

21

RFID- Application in Info-Documentary Systems

Angela Repanovici and Luciana Cristea
Transilvania University of Brasov
Romania

1. Introduction

The automatization process in all industrial and social fields requires large amounts of data processing. Data Acquisition and Control Solutions can be improved by collecting and processing data in real-time without human involvement through Automatic Identification or Auto ID.

Auto-ID technology provides the means to track any object, anytime, anywhere by using low-cost smart tags, readers, and unique object-identification schemes.

These technologies include:

- Electronic Product Code (EPC);
- Barcode (uniform product codes- UPC);
- Optical character recognition (OCR);
- Magnetic ink character recognition (MICR);
- Magnetic strip;
- Biometrics (such as retinal scans, fingerprints, etc);
- Voice recognition.

Modern libraries must provide quality services quickly and efficiently. This requires automation and computerization of libraries specific activities. Auto ID allows automated identification, recording and management books, magazines, CD's, tapes, videos and DVDs.

Until recently, bar code type indicators have great use in libraries, but lately have started to become inadequate in a number of increasingly large applications. The advantage is that bar codes can be purchased at extremely low prices, but their drawback is the limited capacity to store information, data having rescheduled.

Radio Frequency Identification (RFID - Radio Frequency Identification) or proximity is the latest and most advanced method automatic data collection technology, gaining a wide acceptance as people understand and use this technology.

With the advent of RFID technology, RFID has been introduced in the library. The free and efficient use of the newest resources of the information technology is a big step toward to the public free and rapid access to information and to the global documentation with high quality.

2. The RFID technology

RFID is a no touch technology, which identifies an object or person automatically by using radio waves through a serial number or an Electronic Product Code (EPC). RFID can be used in authentication, detection of tracking, checking, warehousing, inventory management, surveillance, security, library store, document management, transportation management, cashless payments and computation for objects in various fields of industry such as manufacturing, construction, library and health care.

The simplest applications of RFID can be compared with barcode systems, but the most sophisticated RFID products can be interface with external sensors to measure specific parameters, or even GPS (Global Positioning Satellite system) for tracking the position of objects via satellites.

RFID technology was invented in 1948 by Harry Stockman. Until 1960 RFID was experimented in laboratory and after that the theory was funded. After 1970, tests of RFID were accelerated and began the implementation and the development of RFID. From 1990 commercial applications and Standards are developed. Today, RFID becomes a part of everyday life.

The fundamental components of an RFID system are primarily a transponder (tag), an interrogator (reader), communication networks and host computers (fig. 1)

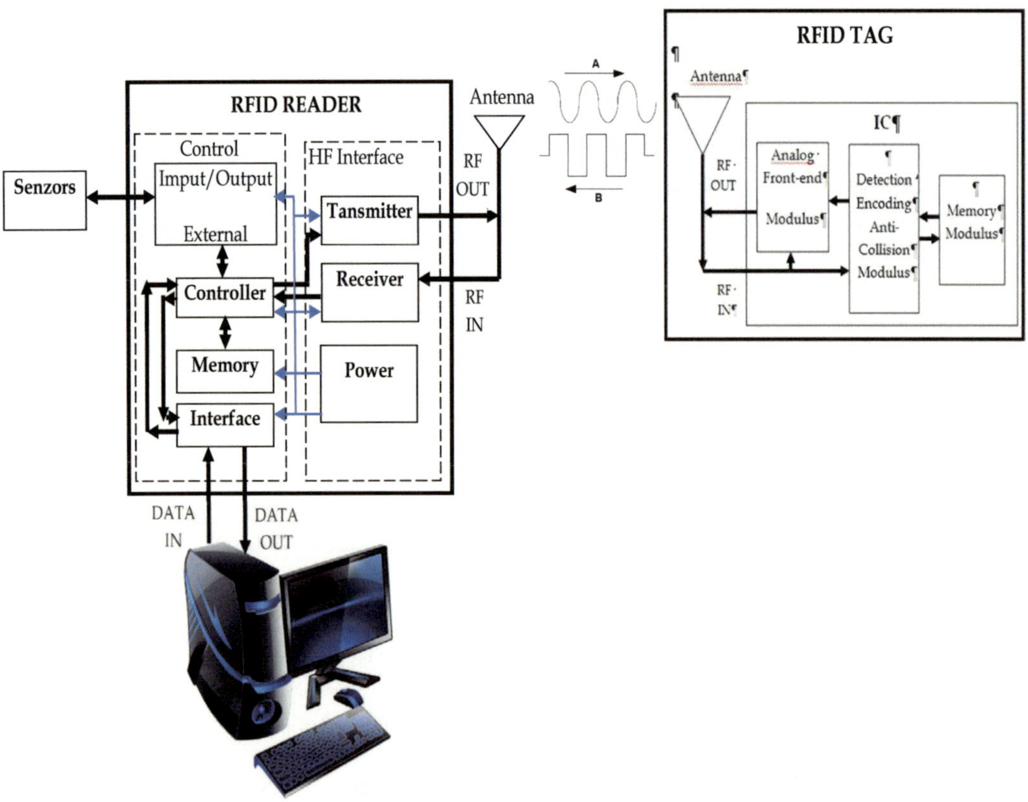

Fig. 1. RFID configuration - Block Chart.

In an RFID system there are two types of antennas: one is in the tag while the other is connected to the reader. The information flow during the RFID system (from simple tag to the host application) begins with host manages reader and issues commands. The reader and tag communicate using a radio-frequency (RF) signal. Reader generate carrier signal on request from the host application and send it out from reader antenna. This signal, hits the tag which receives and modifies it and reflects back the modulated signal. The reader antenna receives the modulated signal and sent them to the reader which decodes the signal into digital data. The digital data is sent to the host application.

2.1 RFID tags

The tag is a device that stores certain unique information. Tags are attached to objects or people and then communicate with a reader when the reader receives radio waves. It consists of an electronic circuit (ASIC) and an antenna integrated into one piece. "RFID tags are used in many applications, depending on the application"(3M, 2011) the purchaser will have different expectations for tag cost, read range and durability.

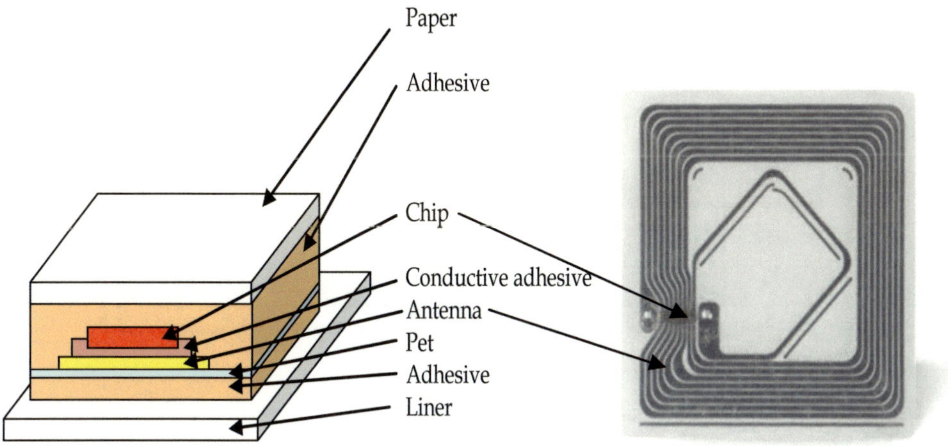

Fig. 2. Tag construction

A common HF RFID tag is a lamination of multiple categories of materials that can interact with each other (Create a new library, 2010)

The first layer is made usually of paper or polypropylene and is a protective layer. Under this layer is a layer of adhesive which can be hot melt or pressure sensitive. The integrated circuit or chip (IC) is linked with the antenna through a conductive adhesive which can be an epoxy, a tape or a paste.

The antenna is made of aluminium or copper and she is attached to a substrate of plastic, usually PET. The last layer is the liner which is a silicone-coated paper and this layer is attached to the others by an adhesive layer. The materials used in the tag construction can have a large impact on long-term reliability. In tag design, materials are chosen for each application. Tag designers select the best materials that assure the optimal configuration of cost, performance and durability.

The antenna receives and reflects radio-frequency (RF) waves coming from the reader antenna. The design of the antenna is according with the particular frequency of the application and it determines the size of the tag.

The chip assures the operational functionality of the tag. The main parts of the integrated circuit (IC; chip) are: RF front-end, (Course Hero) some basic signal processing blocks, logic circuitry (algorithm implementation), and memory for storage (Figure 3).

The RF front-end is the core interface between the antenna and signal processing unit. It is responsible of implementing modulators, voltage regulators, resets and connections to the external antenna. (Halayci, 2009)

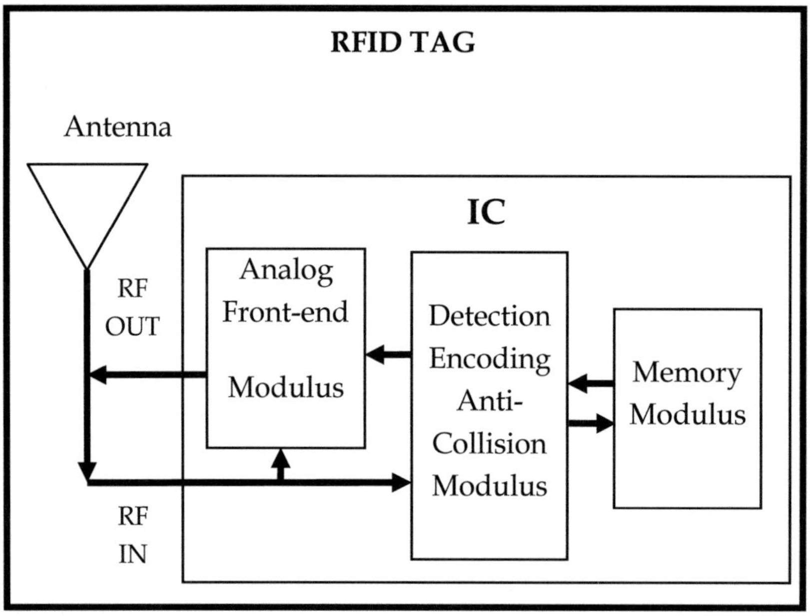

Fig. 3. Chip configuration

Tags can be classified according to: the power source, frequency, functionality and protocols that they belong to.
Depending on the source of the power, tags are classified as:

- Passive;
- Semi-active or semi-passive (also called battery assisted passive tags- BAP or battery assisted tag - BAT);
- Active.

The **passive RFID tag** has no internal power source. The passive tag's read range is limited by the amount of power that can be obtained from the RF waves from the reader. The benefits of passive RFID tags are that they are smaller, cheaper (<0.5$), unlimited in life span because power does not have to be supplied. The reading range, however, is shortened to around 10cm up to a few meters (<6m) (Clampitt, 2010). So, if tags are placed outside of the electromagnetic field, these devices do not work to detect. The disadvantages of passive tags are a lower read range and high power readers. The passive tags are mainly used to: item, box, or case level tracking; low-value assets; identification (passports, badges, etc.); managing DVDs, documents and library checkout, baggage tracking, point of sale, blood supply, drug packages, livestock, pets, etc. (RFID, 2010)

The **semi-active tag** has a battery on it but no radio transmitter. The battery powers its integrated circuit (IC), which helps it to modulate the reflected signal. The reflected signal is required because the tag does not have a radio transmitter. The advantage of this type of tag is that you do not need to power the tag from the reader. Therefore, one can use low-power readers and store more data on the tag. This type of tag is used to get longer read range (up

to 50m) or to couple the tag with environment sensors such as temperature, pressure, relative humidity, Global Positioning System (GPS), etc. Since the sensors require continuous power, the battery is required on the tag. The disadvantages of these tags are higher cost, larger and heavier tag and limited life due to battery. (HALAYCI, 2009)
The active RFID tag has its own power source and transmitters. This tag communicates at a longer distance because it is not dependent on a reflected signal. Its communication distance ranges from 100m to 225m. It has more memory, up to 128Kbytes. However, the cost is high (>20 $), the size is larger, and the weight is higher (Kinkenzeller, 2003). The active tag's life is between 2 and 5 years and it depends on the battery. The active tag stops working when the battery dies.
The active tags can be used in various applications like: box, pallet, or container level tracking; people tracking (such as patients); real-time location; long-range monitoring; area monitoring; security, sensor monitoring and others.
Tag characteristics are different according with the frequency bands in which the tag is designed to operate.
In tag design there are used four frequency bands:

- Low Frequency - LF 125 - 135 KHz;
- High Frequency - HF 13.56 MHz;
- Ultra High Frequency - UHF 860-960 MHz;
- Microwave Frequency - 2.45 and 5.8 GHz.

Low Frequency tags- LF 125 to 135 KHz have a very short read range –up to 40 cm with low-read speed. These tags are used in: access control; animal tagging; inventory control and car immobilizer
High Frequency tags - HF 13.553 to 13.567 MHz have a short to medium read range –30cm to 1m with medium-read speed. These tags are used in: Smart Cards and item tagging.
Ultra High Frequency tags - UHF 860 to 960 MHz have a medium read range 60 to 6m with high-read speed. These tags are used in pallet tagging. The tag costs are high.
Microwave frequency tags have a medium read range -60cm to 15m with high read speed. As UHF tags, Microwave tags are very expensive.
The library RFIDs mainly operate in the high-frequency (HF) 13.56 MHz band.
According to the tags read/ write capabilities, memory capacities, power sources and communication capabilities, these are classified in six functionality classes:

- Class 0 – including passive –read only tags (data are written once by manufacturing),
- Class 1 – including passive –read only after initial programming tags (field programmable only once),
- Class 2 – including passive tags with read and write functionality. These tags are rewritable by reprogramming,
- Class 3 – including semi-passive tags with read and write functionality,
- Class 4 –including active and reprogrammable tags,
- Class 5 – including readers and read/write functionality tags which can power class 0, 1 and 2 tags.

2.2 RFID interrogator or reader

The second important part of the RFID system is the Interrogator or Reader (fig.3). The RFID reader sends a pulse of radio energy to the tag and listens for the tag's response. The tag detects this energy and sends back a response that contains the tag's serial number and possibly other information as well (Garfinkel, 2005). The reader can be fixed in adequate place or hand -held according to ensure the best conditions to read the tags by passing them through the interrogation zone.

A hand-held reader is a small, lightweight device that is used to receive quickly and accurately information from the tag (fig.4).

A fixed reader is installed on a stationary point like a wall or a ceiling to read movement, location, or internal data of objects in the area (fig.5). The reader collects the information continuously. Depending on the reader size (especially its antenna), the range and accuracy is greater than hand-held readers. (KIM, 2007)

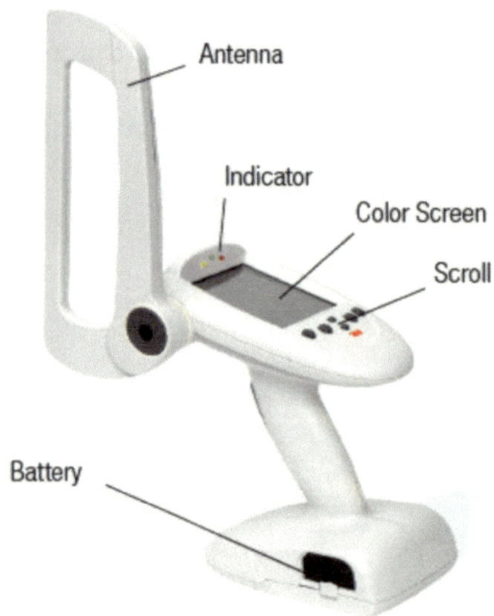

Fig. 4. Hand-held reader (www.3M.com/uk/library)

Fig. 5. Fixed Reader (www.us.ute.com)

There are two main classes of RFID readers: read-only, an example being those that operate with the purely passive Class 1 tags, and read/write, which can write new information back to a tag that has been equipped with a read/write memory (WARD, 2006). According with the main functionality, the readers must demodulate and decode the information received from the tags, and also these must assure the best conditions to communicate with the tags by supplying the necessary energy.

ISO Standard	Title	Status
ISO 11784	Radio frequency identification of animals -- Code structure	Published standard - 1996
ISO 11785	Radio frequency identification of animals -- Technical concept	Published Standard -1996
ISO/IEC 14443	Identification cards - Contactless integrated circuit(s) cards - Proximity cards	Published Standard 2000
ISO/IEC 15693	Identification cards - Contactless integrated circuit(s) cards - Vicinity cards	Published Standard 2000
ISO/IEC 18001	Information Technology - AIDC Techniques - RFID for Item Management - Application Requirement Profiles	Published Standard 2004
ISO/IEC 18000-1	Generic Parameters for Air Interface Ccmmunication for Globally Accepted Frequencies	Published Standard 2004
ISO/IEC 18000-2	Parameters for Air Interface Communications below 135KHz	Published Standard 2004
ISO/IEC 18000-3	Parameters for Air Interface Communications at 13.56 MHz	Published Standard 2004
ISO/IEC 18000-4	Parameters for Air Interface Communications at 2.45GHz	Final Draft International Standard
ISO/IEC 18000-6	Parameters for Air Interface Communications at 860-930 MHz	Published Standard 2004
ISO/IEC 15961	RFID for Item Management - Data protocol: Application interface	Published Standard 2004
ISO/IEC 15962	RFID for Item Management - Protocol: Data encoding rules and logical memory functions	Published Standard 2004
ISO/IEC 15963	RFID for Item Management - Unique Identification of RF Tag	Final Draft International Standard

Table 1. Common ISO Passive RFID Standards

2.3 RFID standards

The reader and the tag can communicate with each other through the protocols establish during the manufacturing. To assure communication between readers and tags from different manufacturers there are defined standardized protocols.

Two organizations are most involved in drafting standards for RFID technology: the International Organization for Standardization (ISO) and EPC global. ISO represents global interests and has been involved with different RFID technologies for many years (Table 1). Most of the work has been through various sub-groups of Joint Technical Committee One (JTC12), for drafting standards for information technology.

EPC global's mission started with the vision to identify every item with a unique electronic product code (EPC). The plan is to have a global network implemented making every item visible throughout the supply chain. A great amount of research and development resources

have been invested in creating specification and standardization of the EPC tags and the required infrastructure EPC global's efforts are primarily focused on UHF. (Team, 2010).
ISO developed a new series of standards—the ISO 18000 family—that addresses how tags and readers communicate in a number of item identification applications. One of these, ISO 18000 Part 3, identifies 13.56 MHz as the frequency for tag-reader communication in these applications. ISO 18000 Part 3 Mode 1 is the type of tag commonly used in many of these applications, including libraries (3M, 2011)
The ISO has formed an international working group to develop applications standards that will allow global interoperability. At this time ISO developed three different working drafts standards of standards, called ISO/WD 28560 part x.

- **Part 1** describes in general the data elements that can be used for libraries.
- **Part 2** describes the object based encoding drawn from ISO 15962. The only mandatory data element is the Primary Item Identifier (Barcode). If more optional elements are needed like e.g. owner of library, item set information, shelf location etc. an object index is required that the library system knows the particular elements that can be accessed on the tag. The advantage is the flexible memory size of tag due to the data elements that are stored.
- **Part 3** describes the fixed length encoding similar as already used e.g.in Denmark. Five data elements are mandatory (the Danish model includes 8 mandatory elements). (3M, 2011)

3. RFID technologies in libraries

RFID application in Library must be able to assure the maximum efficiency in operations, such as:

- loans and refund of materials (assisted by librarian or as self-serving);
- collection inventory;
- identify materials and rapidly finding of material that are wrongly placed on the shelves;
- collection security;
- automatic sorting for putting on the shelves.

In libraries, activities such as making an inventory of the book involves a lot of work and time spent by library staff, so the RFID system is suitable for identification, inventory and management books, magazines, CD's, tapes, videos and DVDs sites. Such a system significantly reduces repetitive operations; the books were quickly counted by scanning simultaneously and directly to the shelves.
Another application of RFID systems is the introduction of direct services offered to users, allowing the loan and return documents in faster and easier way, thanks to self-service workstations and the possibility of booking through the Internet. Based on the entry permit with a RFID chip inserted, readers can use these self-service stations.
Library readers have easier access to traditional library activities (booking, loan return) by using e-mail and by automatically generated SMS messages thus amplifying the degree of communication with the library. Using the library automatisation it is possible to send: automatic warning messages that are intended to remind to the reader the obligation to return the loan documents in time and messages for the extension of the loan (the user can extend his loan period, if there is no other request for the document and this extension is via the Internet and the information are automatically update in the system) Automat

documents booking system allows the automat detection of the reserved document by any RFID workstation and the system will deliver a message to protect the reservation.
The most important elements of RFID system for libraries are: door sensor, auto-loan unit and the librarian work-unit. These components are independent of each other, and to implemented computer circulation system. Since the components are "intelligent", there is no need for a server and adding components allow additional elements themselves, with the development system.
Librarian work-unit allows the following functions:

- operating loans / refunds;
- programming (write) labels;
- conversion barcode labels in RFID tags.

Gate sensor -RFID reader (fig.6)
It is designed to detect and read information from RFID tags passing through the area. The gate read the EPC or serial number (given by the library) and can tell if a book was escape or not. The reader consists of two antennas placed in parallel, plus an electronic reader. The distance between the two antennas can be 90 cm, while the three antennas can reach 1.8 m.
Auto-loan **unit** (fig.7)
After identifying the user, which may be based on an RFID identification card, it can put documents (books, CDs, video discs, etc.) on the reading surface to be recorded on his behalf and scheduled in "loan". The chip will be placed on the "quiet" mode and no alarm output will be active. It is possible that the return of books to be made also in auto-loan unit. The user can check more books to return an average reader can read 25 cm, so that depending on the thickness of the book, you can find out how many books can be returned in a single reading.

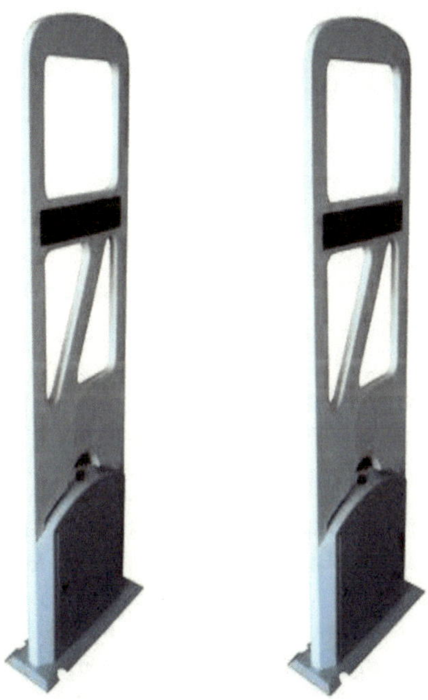

Fig. 6. RFID reader - Security gate (www.cfnewsads.thomasnet.com).

Fig. 7. Sef check unit (www.blog.library.villanova.edu/news)

RFID systems for **info documentary application assure a lot of advantages**

- Rapid check-out / check-in;
- Simplified self check-out / check-in;
- High-speed inventorying;
- High reliability;
- Better inter-library facilities, more efficient reservation facilities;
- More staff available for assistance;
- Long tag life;
- Automated materials handling.

RFID systems used in libraries provide the following features:

- High reliability;
- Compatibility with other RFID chips;
- Compatibility between different generations of RFID chips of different manufacturers(ISO 15693);
- No additional security elements (electro-magnetic tapes, etc..);

The system can be extended to access cards, access control in various areas of the library, copying or other payments to national library services, Internet access etc.

Components (gateway, readers, etc.) can be exchanged during operation without the need for library RFID system interference.

4. Application at Transilvania University of Brasov, Advanced Mechatronics Systems Research Department SIPTEH project

The project SIPTEH (Shared Integrated System for Processing of Technical Content) was approved and funded by the National Centre of Project Management in the summer of 2008, integrated in the National Plan for Research-Development and Innovation II 2007-2013, 4th Programme - Partnerships in major fields, 1st Research Direction - Information technology and communications. The project is to be carried out in the next three years, with July 2011 as deadline, and comprises a partnership between Dunarea de Jos University of Galati (coordinator), Transilvania University of Brasov, Politehnica University of Bucharest and Lucian Blaga University of Sibiu (partners).

The team recommended the use of a pre-indexing, indexing, storage and digital information retrieving system as in the figure 8

The four partners work independently but the results will be found on a common platform and will meet the users' needs, namely finding information sooner and more easily (Fig 8).

The project activities were:

- the workflows corresponding to harvesting, digitizing, indexing, storing and retrieving information in the institutional repository;
- modelling and simulation of the autonomous partners systems of digital information processing;
- modelling and simulation of the integrated system of digital information processing;
- specific software design of information processing in the integrated system.

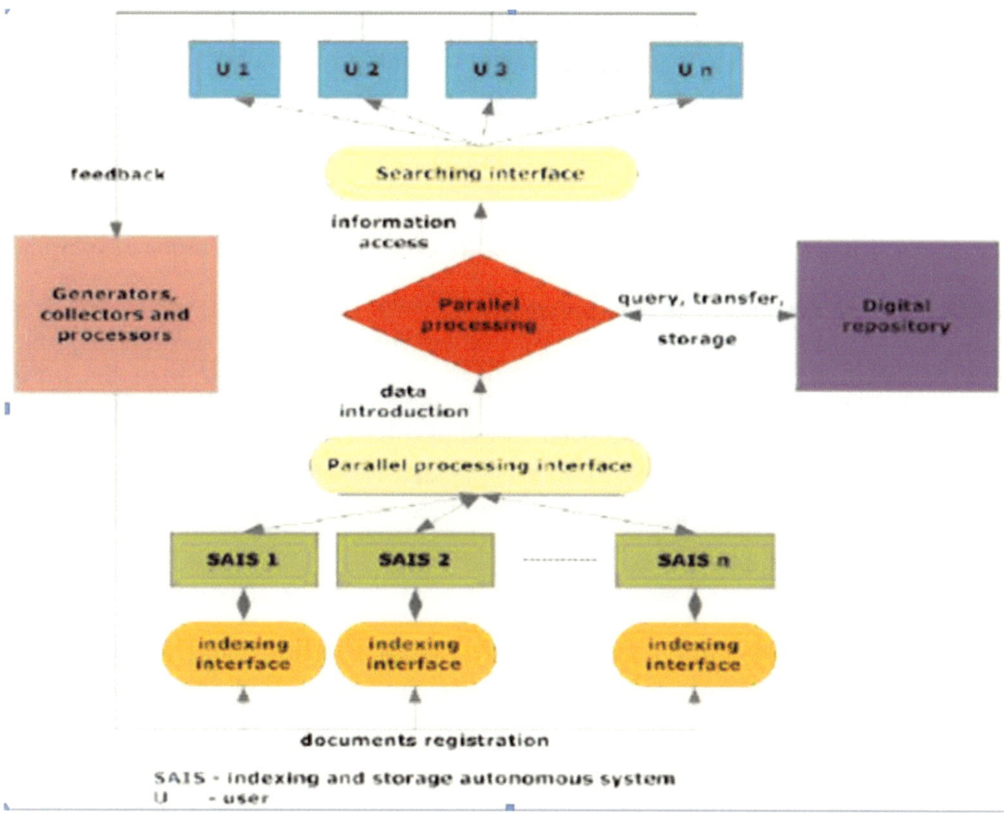

Fig. 8. On-line indexing and sharing integrated system of digital documents (SIPTEH, 2008)

Transilvania University of Brasov has already implemented Dspace as open source repository software.

The scientific research is based on the activity of information and documentation and is finalised by promoting the researches findings of everybody who can use them in order to develop knowledge and innovation. Using open sources and developing new models and systems that are based on them lead to scientific progress.

Transilvania University of Brasov proposed a model of harvesting, digitizing, indexing, storing and retrieving information using the principles of open access to information and the advantages of using open access sources. The model is projected so that it should allow integration within an integrated system by using RFID technology for documents accounts, circulation and periodical inventory, achieving, with a minimum cost price, the computerization of small libraries in universities research departments and visible improvement of the department scientific production by publishing the staff's papers in open access.

4.1 Open source initiative in digital preservation: the need for an open source digital repository and preservation system

The *open source* software means sets of programmes which are developed by a community, by a company or by an individual and are offered to be used under the General Public License ; they are characterized by their users' freedom to utilize, copy, redistribute, study, alter and improve them.

The *open source* software is free but the one that develops it can offer for a fee certain services, such as system instalment, staff training, technical support, data conversion.

Advantages of the Open source software:

- *Ability of being adapted and suited to local needs*: The software source code can be altered and improved so that it can meet the own needs of its user.
- *There is no restriction in using it*: There is no contractual restriction concerning the manner in which the software is used.
- *Low costs*: The software itself is free; there are costs only for an ulterior development, staff training, data conversion etc.

Disadvantages of the Open source software:

- *Big and unexpected efforts*: A library will find out that, in order to succeed in getting a good result, the necessary efforts will have to be much bigger than the initially supposed ones in adapting the software to its needs.
- *Lack of coordination*: The decentralized development of open source software is made rather chaotically and there are delays in solving programming errors.
- *Inappropriate technical support*: The documentation tends to be limited and destined to the developers especially. There are limitations concerning documentation for the users of this type of software.
- *Personalization*: The open source software is possible to not offer the desired level of personalization compared to a commercial soft.

The system proposed by Transilvania University is materialized into an integrated system of managing the documents of the research department library with a digital repository consisting of the university scientific production together with a RFID system of documents managing and accounting and a security system as well (Fig.9).

Fig. 9. Open source integrated system proposed at Transilvania University of Brasov

4.2 Open source library integrated systems

The library integrated systems that are used in the Romanian university libraries are software products that were acquired from different suppliers. The cheapest is Liberty which is used at Transilvania University Library; Liberty is useful in libraries that possess up to 1 million volumes.

To computerize the department library which is being developed but it is not to exceed 10,000 volumes in the near future; such an investment is useless now when we have at our disposal open sources library integrated systems that are used by libraries worldwide.

In order to choose some open source software dedicated to library I took into consideration the following:

- To exist current active developments of this product.
- To contain *modules* of cataloguing, circulation, users' access, acquisitions and serials control.
- To support MARC standards.
- The current source code and its documentation should be used under the *General Public License* (GPL)
- The product should already be used in libraries.
- *Scalability* – to support a big volume of data loading and to allow its extension.
- To be able to be *adapted* and in the same time to be a *friendly* system for the user.

KOHA

After having studied the market, we chose Koha. It is the first open-source Integrated Library System (ILS). In its use worldwide, its development is steered by a growing community of libraries collaborating to achieve their technology goals. The reasons for which we chose this system are presented below:

- *Koha is an Integrated Library System*. It is a complete system. Being an open source software product, there is no costs for license. It can be installed and used free of charge, it can be adapted to local needs; it has the same characteristics as a commercial software product has.
- It *was initially developed* in New Zeeland by Katipo Communications together with Horowhenua Library Trust.
- At present it is *maintained* by a dedicated team of software suppliers and by the technical staff of some libraries from all over the world.
- By adopting this software, the client becomes a kind of *joint owner* of this product. He can install the new versions if he likes it or not. Particularly, he can finance an ulterior development of his product or he can do it through his own efforts.
- Koha has been tried and tested and it has proved its stability and scalability, now it is used in hundreds of libraries worldwide.
- It is an economical alternative compared to the commercial software items which are expensive. The costs for commercial software include: the software acquisition with all its modules, ulterior acquisition of license and costs for training and technical assistance. There are no initial costs for Koha.
- Characteristics: Koha has all the characteristics of a complete commercial software product.
- It motivates and encourages the technical staff to be creative!

Requirements of KOHA system are not so expensive and can be easily achieved. There are:

- Koha Free 3.0 stable version (Available on Internet http://www.koha.org/)
- Server Web Apache, Free 2.0.58 (Available on Internet http://www.apache.org)
- MySQL. Relational Database Management System http://www.mysql.com Free
- Module Perl 5.8 http://www.cpan.org Free
- Zebra 2.0 http://www.indexdata.com/zebra Free
- Linux (RHEL 3.0, 4.0, 5.0) or any other variant of LINUX or WINDOWS Server

Requirements for the operator's skills are:

- Koha interfaces are logically projected and can be used extremely easily.
- *Staff and readers* need only basic competences in using computer; in order to use the system efficiently these competences are gained quickly.
- *Cataloguing module* needs *understanding cataloguing practice,* such as *knowing MARC standards,* and also using the instruments of finding with the help of Z39.50.
- *Administrators* must know the operating system (Linux, etc) for maintenance, have some cataloguing knowledge which is useful for the initial settings of preferences system (for setting branches, access rights, types of documents, types of borrowing, categories of readers etc).

Koha have all integrated systems modules. In our model we are focused only on few modules. One of them is OPAC Module with the next characteristics:

- Koha offers a public catalogue (OPAC) with complete functions.
- OPAC users can search on ten fields (topics, authors, titles, publishers, barcodes etc)
- The OPAC users who are authenticated as members can make online reservations for the library documents.
- Biblio basket: the members who are logged can select various registrations which they can send and find in the email afterwards; these can be saved in a usual form, text or in ISO 2709 format, and subsequently read with the help of *End Note*-type software.
- OPAC users can send suggestions concerning acquisitions; they will automatically be announced through email by any action taken in accordance with their suggestions.

  Another very interested module is KOHA – Cataloguing Module, with the next characteristics:
- Cataloguing module is one of Koha main strengths. There can be defined some forms in order to catalogue different types of publications: monographs, electronic resources, periodical magazines, etc.
- *Export / Import:* it allows the export and import of some registrations in MARC format; they can be found in the catalogues of some other libraries that use Koha and, with the help of Z39.50 protocol, they can be brought in the library own catalogue, thus achieving a quick cataloguing, subsequently only the local data will be filled in.
- *Searching:* it allows doing various searching, which can be done on any MARC field of the registration.
- *Barcode generation*- permits developing RFID solutions.

We implemented the Koha system at the research department of Transilvania University of Brasov.

We started to use cataloguing module and add records from our own library. We started with one of the author's book.

We started to grow Koha collection and we generated barcode for all the registered documents.

Fig. 10. Library integrated system at Transilvania University of Brasov. Koha modules

Fig. 11. Records in Koha system

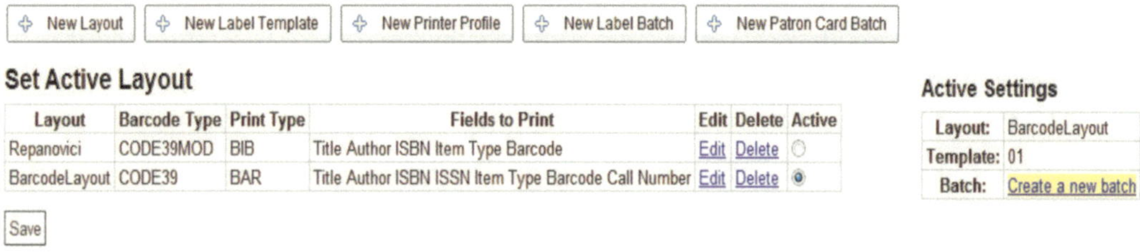

Fig. 12. Generating barcode for Koha records

4.3 Open source institutional repository technology

Dspace

The universities and the research centers from the whole world are very active in planning and implementing the digital repositories. A help guide has been published for the organizations which plan the implementation of IR by the presentation and selection of the software systems which best satisfy the needs of the institution.

All the presented systems satisfy three criteria:

- They are available free of charge by an Open Source license, namely they are available for free and can be modified, updated and redistributed.
- They are compatible with Open Archives Initiative-OAI and by any implementation participation to the global network of the institutional inter-operable repositories is possible
- They are recently made and available to the public
- The presented systems are Archimede, ARNO, CDSware, Dspace, Eprints, Fedora, i-Tor, MyCoRe and OPUS.

4.3.1 Content of the digital repository

The Dspace software, the open source used by Transilvania University offers the capacity of stocking and saving the following types of documents:

- Articles, pre-prints, e-prints;
- Technical reports;
- Research reports;
- Conference proceedings;
- Video, audio-video materials, images;
- Teaching materials;
- Digitalized materials;
- Bachelor degrees, master's degree, PhD theses.

The implementation team will be a partnership between:

- Faculty of Economic Sciences;
- Faculty of Mechanical Engineering;

The selection criteria of the implementation team are:

- Friendly departments to the mission of the PILOT DIGITAL REPOSITORY;
- Diversity in the area of disciplines;
- Diversity in content and formats;
- Management examples of the different forms of intellectual property;
- Archiving small-size collections for a start;
- Team with strong bonds and confidence.

The members of the team worked in other projects, too. Each of them has got expertise in their activities.

Technologies for choosing the platform and the software

As a consequence of a technical analysis, it is considered that Dspace is the solution for the open source where the PILOT REPOSITORY was implemented in the Transilvania University.

The basic services that are offered are:

- Archiving management for ensuring a long-term conservation,
- Persistent stocking, for back-ups and recuperation procedures included.
- Attribution of a unique identifier to each document for citing.

We have three research departments involved in development of repository.

Browsing "Books" at Advanced Mechatronics Research Department we choose the same book we generated barcode in Koha system.

In order to exemplify we chose the document for which we generated barcode in Koha library integrated system, fig.15.

Fig. 13. Aspect Dspace institutional repository at Transilvania University of Brasov

Fig. 14. Books at Mechatronic Advanced Department in Aspect Dspace repository

The implemented digital repository allows us to visualize the resources access. This way we can know the geographical area of interest or how many downloads were initiated. The example in the figure shows the total visits per month, countries and cities that were interested for the book *Promovarea productiei stiintifice prin depozite digitale* in english *Scientific production promotion by institutional repositories*. This thing can be noticed in the monthly activity.

DSpace at Transilvania University >
Advanced Mechatronic Systems >
Books (Advanced Mechatronics) >

Please use this identifier to cite or link to this item: `http://hdl.handle.net/123456789/166`

Title:	PROMOVAREA PRODUCTIEI ŞTIINTIFICE PRIN DEPOZITE DIGITALE
Authors:	REPANOVICI, Angela
Keywords:	marketing educaţional strategii de marketing cercetări de marketing acces deschis depozite digitale instituţionale evaluare academică analiza citărilor scientometrie cercetare ştiinţifică producţie ştiinţifică
Issue Date:	Sep-2010
Publisher:	Editura Academiei Române
Citation:	Angela Repanovici: Promovarea producţiei ştiinţifice prin depozite digitale, Editura Academiei Române, Bucureşti, 2010, 193 p., ISBN 978-973-27-1932-9

Fig. 15. Bibliographic description of the book in Aspect Dspace repository

	Views
PROMOVAREA PRODUCTIEI ŞTIINTIFICE PRIN DEPOZITE DIGITALE	18

Total Visits per Month

	septembrie 2010	octombrie 2010	noiembrie 2010	decembrie 2010	ianuarie 2011	februarie 2011	martie 2011
PROMOVAREA PRODUCTIEI ŞTIINTIFICE PRIN DEPOZITE DIGITALE	0	0	0	0	0	11	7

File Downloads

	Views
Promovarea productiei, Repanovici.PDF	5

Top Country Views

	Views
Romania	16
Netherlands	1

Top City Views

	Views
Brasov	12
Bacau	3

Fig. 16. Statistics of download and visibility of the book in Aspect Dspace repository

4.4 RFID system implementation

Implementing RFID system at the model achieved from combining the two open sources supposes the transformation of the barcode generated by the label that is integrated in RFID technology.

In Romania one of the university libraries which functions in an integrated system and uses RFID technology is the library of the university in Sibiu.

We implemented the system used in this library.

RFID (Radio Frequency ID) tags are the centerpiece of any RFID system. Applied to packages, pallets or individual items, RFID tags store the item data essential to any RFID-based tracking system.

Difference between Barcode and RFID (LIS Links : A Virtual Community of Indian LIS Professionals, 2011):

- Information can be read from RFID tags much faster than from barcodes
- Several items in a stack/counter can be read at the same time using RFID
- Items do not have to be handled one-by-one nor removed from the shelves
- Inventory-taking is no longer a tedious operation (LibBest, 2010)
- RFID can stand more than 10,000 read/write
- RFID can have theft bit which can be in two states "ON/OFF"
- Shelf verification/rectification can be done on daily basis
- More information can be written in the RFID tag on incremental basis
- Need not open/remove books to capture information
- Items are identified on upper and lower shelves more comfortably (Radio-frequency identification (RFID), 2010)

5. Transilvania university research department library and RFID system

The components of RFID are:

1. Tag or **intelligent label**, contains a microchip that is attached to an aerial, they are encapsulated, in accordance with the utility, into different materials, paper or plastic.

Characteristics:

- They are rewritable
- They permit ID material stocking (inventory number, barcode) for protecting the readers' information having a personal character.
- They permit stocking of the borrowing state directly on the tag
- dimensions: 48mm x 57mm
- price of RFID 1024 bits tags- (0.6 – 0.9 Euro / piece)

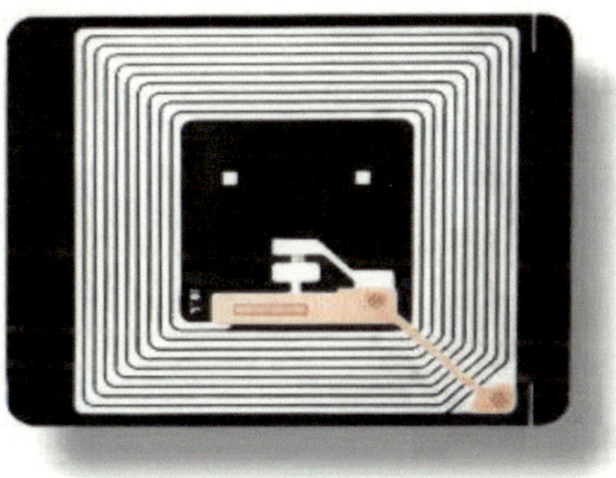

Fig. 17. Tag or intelligent label

2. RFID scanner or RFID mobile reader sends a radio impulse and waits for the tags answer. The tag captures the energy sent by the reader and uses it in sending itself its unique identification number and also other pieces of information that have been memorized previously. RFID readers can function in the stand-alone manner (independently) or can be connected to an IT network.

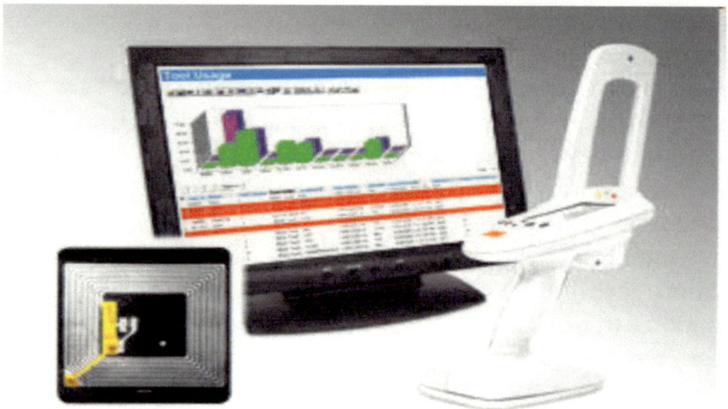

Fig. 18. Scanner RFID

3. Librarian - Koha system application work unit allows the following functionalities:

- operates book borrowings/book returns;
- programmes (writing) tags;
- converts barcode tags into RFID tags;
- allows identification of reader license by using both barcode technology and MIFARE-type cards;
- ergonomic (reading support placed at the down side).

4. RFID security gate or Gateway–Dual Technology (RF+EM) electronic system, which uses electromagnetic technology in combination with radiofrequency technology, thus allowing the combination of the advantages of both these technologies: flexibility in using and high grade of detecting. On the books in circulation fund there will be used activated/deactivated tags through contact or at distance, by panels set on circulation desk.

All the books with open access at the shelf have both barcodes and RFID tags; they are places as follows: the barcode inside the cover-front, down, and the RFID tag inside the cover-back, in 3 positions: down, medium and up, at a distance of maximum 2 cm from the back;

Fig. 19. RFID tag and barcode used on a document

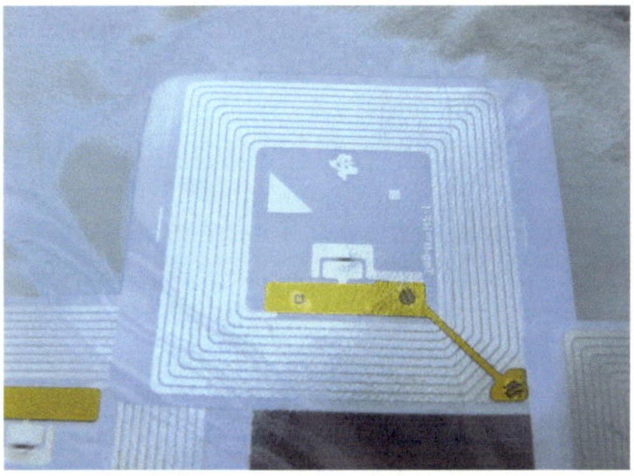

Fig. 20. RFID Tag

Initially we will work using barcode in parallel with RFID tag, namely:
- barcode - it is applied at Koha cataloguing module, when barcode is generated.
- then the tag is applied on the book and the barcode is written in the tag memory by the local RFID librarian station;

The tags that will be used are the ones used by the Library of The University of Sibiu, namely RFID 3 M tags with the following dimensions: 57,15 x 49,3 mm, 1000 pieces / box, capacity 1024 bits (Fig.20)

6. Conclusion

Compared to classical systems like Barcode where all the operations are made manually, the biggest benefit brought by RFID to library is rapidity with which information can be read/registered from/in a RFID Tag and possibility of operating with more objects in the same time as well.

Introducing the RFID system does not mean direct elimination, from the first day, of the barcode system. The two systems can exist simultaneously, the RFID readers having the capacity of reading barcodes, which allows library an easy and gradually transition towards the implementation of RFID system.

The system allows flexibility and modularity and thus the library can start introducing simple RFID, permitting ulterior adding of new products or activities according to requirements and necessities. RFID, its application, standardization, and innovation are constantly changing. Its adoption is still relatively new and hence there are many features of the technology that are not well understood by the general populace."RFID is increasing in popularity among libraries, as the early adopters of this technology have shown that it makes good economic sense, both for large and small libraries." (Narayanan A.)

The proposed solution offers the model with a minimum cost using open sources software and it offers a useful instrument for the university researchers opening new opportunities for master students and PhD students.

7. References

3M. (2011). Library systems. Retrieved January 29, 2011, from http://solutions.3m.co.uk/wps/portal/3M/en_GB/Library_Systems/Library_System/?WT.mc_id=www.3m.com/uk/library

Clampitt, H. (2010). The RFID HandBook 7th Edition . Retrieved March 3, 2011, from http://www.rfidhandbook.blogspot.com/
Course Hero. (2009) Retrieved March 3, 2011, from http://www.coursehero.com/file/2110010/basatsabris200612mast/
Create a new library. (2010). Retrieved March 2, 2011, from RFID 401 Tag quality and reliability:http://multimedia.3m.com/mws/mediawebserver?mwsId=66666UuZjc%20FSLXTtnXMVo8%20T2EVuQEcuZgVs6EVs6E666666--&fn=RFID%20401.pdf
Garfinkel, S. (2005). Understanding RFID Technology. Garfinkel.book.
Halayci, S. (2009). Design of a radio frequency identification (RFID) ANTENNA. Retrieved February 12, 2011, from http://etd.lib.metu.edu.tr/upload/12610554/index.pdf
Kim, G. K. (2007). Locating and tracking assets using RFID. Thesis.
Kinkenzler, K. (2003). RFID Handbook: Fundamentals and Application in Contactless Smart Cards and Identification. New York: John Wiley and Sons.
LibBest. (2010). Library RFID Management System. Retrieved January 15, 2011, from http://www.rfid-library.com/en/default_e.html
LIS Links (2011) : A Virtual Community of Indian LIS Professionals. (2011). Retrieved March 3, 2011, from http://lislinks.com/forum/topics/security-system-for-library
Narayanan A., S. S. (2009). Implementing RFID in Library: Methodologies, Advantage and Disadvantage. Retrieved March 3, 2011, from http://library.igcar.gov.in/readit-2005/conpro/lgw/s5-8.pdf
Radio-frequency identification (RFID). (2010). Retrieved January 11, 2011, from http://rfid-chip.org/
RFID (2010). Retrieved March 5, 2011, from http://www.scribd.com/doc/4098031/radio-frequency-identification-RFID
SIPTEH (2008). Integrated system for indexing and share online for digitized technical documents (SIPTEH). Galati: Dunarea de Jos University (in Romanian)
Team, G. R. (2010). RFID Asset Tracking. Retrieved December 21, 2010, from Understanding RFID:http://www.gaorfidassettracking.com/RFID_Asset_Tracking_Resources/rfid_understanding/
Ward, M. K. (2006). RFID: Frequency, standards, adoption and innovation. JISC Technology and Standard Watch.